Agrochemical Spectrum in India

A Complete Guide to Insecticides, Acaricides, Nematicides, Rodenticides, Molluscicides, and Biopesticides

Dr. Baswaraj V. Bhede
Assistant Professor (Entomology)
Vasantrao Naik Marathwada Krishi Vidyapeeth
Parbhani 431 402 (Maharashtra, India)

Made with ♥ on the Notion Press Platform
www.notionpress.com

DISCLAIMER:

The content of this book is compiled and offered strictly for informational purposes. Although reasonable care has been taken in its preparation, the author does not warrant or guarantee the accuracy, completeness, or usefulness of the information, or its reliability in achieving a user's objective.

Information presented is believed to be accurate at the time of publication. However, subsequent changes in conditions may affect its accuracy, and the author offers no assurance regarding the enduring accuracy of any advice or data contained herein.

Users bear the sole responsibility for independently assessing the relevance and precision of the book's content.

The author shall not be liable for any expenses, costs, losses, or damages incurred or arising from any individual's use of or reliance upon the information contained in this publication.

Product trade names are used in this book only as examples. Their inclusion does not imply a preference for, or endorsement of, that specific product or manufacturer over any other equivalent product available.

References and external links are provided for convenience; users are responsible for evaluating the accuracy and reliability of information on linked sites. The author does not accept responsibility for the content of external materials and does not necessarily endorse the organizations referenced.

Preface

The journey of Indian agriculture has always been inextricably linked with the challenge of feeding a massive and ever-growing population from finite and often resource-stressed land. A critical pillar supporting the success of the Green Revolution and subsequent agricultural intensification has been the use of agrochemicals—substances vital for protecting crops and ensuring sustained productivity against the relentless threat of pests, diseases, and weeds. Without effective crop protection, an estimated 20-40% of potential food production could be lost annually, jeopardizing food security and farm incomes.

In the diverse agro-climatic zones of India, farmers face a spectrum of biological challenges, necessitating a specialized arsenal of crop protection tools. This book, "Agrochemical Spectrum in India: A Complete Guide to Insecticides, Acaricides, Nematicides, Rodenticides, Molluscicides, and Biopesticides," is born from the necessity to provide a single, comprehensive, and up-to-date resource on the crucial chemicals and biopesticides employed in the country.

I move beyond a simple listing of products to offer a detailed exploration of the various pest control categories: Insecticides, Acaricides, Nematicides, and Molluscicides and Rodenticides. While conventional chemical agents remain indispensable for their efficacy and cost-effectiveness, their indiscriminate use has led to critical concerns: the rise of pesticide resistance in target pests, adverse impacts on non-target organisms (like beneficial pollinators), and the presence of toxic residues in food and the environment.

This reality has precipitated a global and national shift toward more sustainable practices. Therefore, a significant focus of this guide is on Biopesticides—formulations derived from natural sources such as microbes, plants, and animals. These eco-friendly alternatives are gaining momentum in India, driven by government support for sustainable agriculture and a growing consumer demand for residue-free food. We delve into their mechanisms, advantages, and the practical challenges related to their adoption, such as formulation stability and farmer awareness.

This book is crafted as an essential reference for a broad audience:

- Farmers and Agriculturists: To facilitate informed and judicious use of crop protection chemicals and bio-agents.
- Students and Researchers: To provide a foundational and detailed understanding of the chemistry, toxicology, and application of different agrochemicals.
- Industry Professionals and Policymakers: To offer insights into the current market dynamics, regulatory landscape, and the trajectory towards sustainable pest management solutions, including the adoption of Integrated Pest Management (IPM) strategies.

By synthesizing knowledge on both the chemical and biological spectrum of pest control agents, I hope to contribute to a future where Indian agriculture is both highly productive and ecologically sound. I invite the reader to explore this guide as a vital tool in mastering the art and science of crop protection in the Indian context.

February 2026

Baswaraj V. Bhede

INDEX

PART I
Basic Information of Pesticides

PART II
Sole Insecticides

PART III

Insecticide Combinations (Two or more Insecticides)

PART IV

Acaricides (Sole and Combinations)

PART V

Insecticide and Fungicide Combinations

PART VI

Nematicides

PART VII

Rodenticides

PART VIII

Molluscicide

PART IX

Biopesticides

PART X

PART I

Pesticide use scenario in India

Classification of Pesticide and Insecticide

Pesticide Formulations

Safe Use and Handling of Pesticides

Pesticide use scenario in India

India's highly diverse climate, while supporting an extensive array of crops, concurrently presents significant biological challenges from weeds, insect pests, and fungal diseases. These threats collectively account for substantial annual crop losses, estimated at approximately 45% due to weeds, 35% from insects, and 20% caused by fungal pathogens. To mitigate these devastating losses, farmers have increasingly relied on pesticides, leading to steady usage rates per hectare across the country.

The impact of weeds is particularly severe for staple crops like rice, wheat, and maize. As manual weeding becomes less feasible and more expensive due to labour shortages, herbicide usage has grown rapidly. However, the emerging issue of herbicide resistance threatens to drive further increases in chemical applications per hectare. In contrast, insecticide consumption has remained relatively stable, playing a crucial, consistent role in combating damaging pests that pose a key threat to productivity and yield. The push for increased cultivation intensity and efforts to boost overall food production are further contributing to this trend of higher per-hectare pesticide use.

Pesticide application in India remains a complex and evolving topic. Characterized by relatively low per-hectare usage compared to many developed nations, it is nevertheless marked by significant regional variations and persistent concerns regarding environmental safety, regulatory adherence, and pesticide residue levels in produce.

The key aspects of pesticide use in India is summarized as below.

1. Consumption Pattern and Intensity

- **Low Per-Hectare Use:** India's per-hectare pesticide consumption is among the lowest globally. Recent data suggests it is around 0.37 kg/ha to 0.5 kg/ha of Gross Cropped Area (GCA), which is significantly lower than countries like China (over 1.8 kg/ha) or the global average.
- **Total Consumption:** Total consumption has shown fluctuations but has generally trended downward in recent years. For instance, it declined from around 63,284 tonnes in 2021-2022 to about 52,466 tonnes in 2022-2023. The Indian pesticide market was valued at INR 260.0 Billion in 2024. Projections indicate significant growth, with the market expected to reach INR 440.1 Billion by 2033, representing a Compound Annual Growth Rate (CAGR) of 5.72% between 2025 and 2033.
- **Dominant Type of Pesticide:** Unlike global trends where herbicides dominate, insecticides traditionally hold the largest share of the pesticide market in India, followed by fungicides and bactericides. However, the share of herbicides and fungicides is growing due to rising labour costs for manual weeding and increased production of fruits and vegetables.

Global Pesticide Consumption

Pesticide Category	Approximate Share of Global Consumption (by Volume)	Key Global User
Herbicides	45% - 50%	North America, South America
Insecticides	30% - 35%	Asia-Pacific (especially China, India)
Fungicides	15% - 20%	Europe, Asia-Pacific (for high-value crops)

Source: FAOSTAT data for 2022; Market reports for 2023 projections.

Indian Pesticide Consumption

Pesticide Category	Approximate Share of Indian Consumption (by Volume)	Growth Drivers
Insecticides	~50% to 55%	High pest attacks on cotton, rice, and pulses.
Herbicides	~25% to 30%	Rapidly increasing due to high labour costs and shortages for manual weeding.
Fungicides	~15% to 20%	Stable use for disease control in rice, fruits, and vegetables.

Source: GOI, Ministry of Agriculture and Farmers Welfare (Rajya Sabha data for 2022-23 and 2023-24 estimates).

- **Highest Consuming States:** Pesticide consumption varies widely among states. Historically, states like Maharashtra, Uttar Pradesh, Punjab, Telangana, and Haryana have shown the highest total consumption. However, in terms of consumption intensity (kg/ha), states like Jammu & Kashmir, Punjab, Haryana, and Telangana often lead.

2. Major Crops

The highest consumption of pesticides in Indian agriculture is typically concentrated in a few major crops. Pesticide use intensity in India-measured as kilograms of active ingredient per hectare-differs considerably across states and crops. Among major crops, cotton and vegetables usually have the highest pesticide use rates, followed by paddy (rice). Below is a comparative summary:

Pesticide Use Intensity by Major Crop and State

State	Cotton (kg/ha)	Paddy (kg/ha)	Vegetables/ Fruits (kg/ha)	Notes
Punjab	1.5–2.0	1.0–1.6	1.5–2.5	Cotton and vegetables lead; intensive pesticide use region-wide
Maharashtra	2.0–2.6	0.9–1.3	1.2–2.3	Highest cotton intensity; paddy/veg moderate
Telangana	2.1–2.4	1.2–1.5	1.1–2.3	Cotton-intensive; chillies/veg high
Andhra Pradesh	1.8–2.1	1.1–1.4	1.9–2.1	Cotton and chilli highest in the south
Gujarat	1.5–1.9	0.8–1.0	0.9–1.8	Cotton major in Saurashtra region
West Bengal	-	1.1–1.8	2.0–2.6	Paddy/vegetables lead, especially in peri-urban belts
Jammu & Kashmir	-	-	3.6+ (apples/ orchards)	Among highest statewide (fruits); over 3 kg/ha in apples
Tamil Nadu	1.3–1.6	1.0–1.3	1.3–1.7	Cotton/veg/paddy comparable

Key Insights

- Cotton fields in central and southern India (Maharashtra, Telangana, Andhra Pradesh, Gujarat) consistently receive over 1.5–2.5 kg/ha of pesticides, with rates at the upper end when pest outbreaks occur.
- Paddy (rice) pesticide intensity is highest in Punjab, Haryana, West Bengal, and Tamil Nadu, ranging 1–1.8 kg/ha, driven by intensive cropping cycles.
- Vegetables and horticultural crops often exceed 2 kg/ha in states with large peri-urban production (West Bengal, Punjab, J&K).
- Jammu & Kashmir leads all states in overall intensity (over 3 kg/ha), especially for apples and other temperate fruits.
- States like Sikkim and some North Eastern states have very low chemical pesticide intensity due to organic agriculture adoption.
- National average pesticide use in India is about 0.5 kg/ha, but this average disguise the much higher use rates in concentrated crop belts.
- The use of pesticides is particularly high for crops with high pest pressure and economic value, explaining why cotton and vegetables lead in most states.
- Regulatory measures, government IPM programs, and climate factors also influence intensity by region.

In summary, pesticide use intensity is highest for cotton and fruits (such as apples), with the northern and central states leading for rice and vegetables, and significant regional differences reflecting crop, climate, and farming practices.

3. Regulatory Framework

- **Primary Legislation:** The use, manufacture, sale, transport, and import of pesticides are regulated by the Insecticides Act, 1968 and the Insecticides Rules, 1971.
- **Key Authority:** The Central Insecticides Board & Registration Committee (CIB & RC) is responsible for registering pesticides after evaluating their efficacy and safety to humans, animals, and the environment.
- **Pesticides Management Bill, 2020:** A proposed bill to replace the 1968 Act, which aims to provide a more modern and stringent regulatory framework, including provisions for penalizing the sale of spurious pesticides, regulating pesticide residues in food (in line with FSSAI standards), and compensating farmers for losses. The bill is still under review.
- **Banned/Restricted Pesticides:** The CIB & RC regularly reviews pesticides, leading to the banning or restriction of several chemicals, particularly those classified as Highly Hazardous Pesticides (HHPs).

4. Challenges and Trends

- **Residue Contamination:** Despite low per-hectare use, India faces significant issues with pesticide residue contamination in the food chain, which is a major public health and export concern.
- **Farmer Safety:** There are serious concerns about farmer health due to over-reliance on pesticides, poor knowledge of safe handling, and lack of use of Personal Protective Equipment (PPE). India records a high number of deaths annually from pesticide poisoning.

- **Promoting Alternatives:** The government promotes the Integrated Pest Management (IPM) program and encourages the use of Bio-pesticides (plant-based and microbial alternatives) to reduce reliance on chemical pesticides.
- **Biopesticide Growth:** The use of biopesticides is still low but is showing a positive growth trend and is viewed as a safer, sustainable alternative.

5. Key Drivers for Growth of Biopesticdes

The increasing adoption of biopesticides is driven by several factors:

- **Growing Awareness:** Increased consciousness among farmers and consumers about the harmful effects of chemical residues on food and the environment.
- **Government Initiatives:** Policies and subsidies from the government that promote organic and sustainable farming practices.
- **Export Requirements:** Stricter Maximum Residue Limit (MRL) compliance pressure from major international export markets (like the EU and USA) is forcing farmers to switch to residue-free alternatives.
- **Integrated Pest Management (IPM):** The growing practice of IPM encourages the use of biological control agents.

The trend indicates a clear shift towards more environmentally friendly and sustainable crop protection methods in Indian agriculture. The Indian crop protection chemicals market is witnessing significant growth and transformation, driven by increasing pest pressure, labour shortages, and changing agricultural practices.

*****

Classification of Pesticide and Insecticide

Pesticide

A pesticide is a substance or mixture of substances intended for preventing, destroying, repelling, or mitigating any pest.
In a very broad sense, a pesticide is anything designed to control a form of life that is deemed a nuisance or harmful to humans, their crops, or their animals.
Pesticides can be classified based on their target pests

Classification of pesticides based on the target organisms

Pesticide Type	Target Pests and Pesticide Function (Use)	Examples
Insecticides	Insects	Chlorpyriphos, Imidacloprid
Herbicides	Unwanted plants (weeds)	2,4-D, Glyphosate
Fungicides	Fungi (molds, mildews, rusts)	Mancozeb, Bordeaux mixture
Rodenticides	Rodents (rats, mice)	Warfarin, Zinc phosphide
Acaricides	Mites and Ticks	Fenpyroximate, Abamectin
Nematicides	Nematodes (microscopic worms)	Carbofuran, Fluopyrum
Bactericides	Bacteria	Streptomycin
Molluscicides	Snails and Slugs	Metaldehyde
Avicides	Birds	Fenthion, Strychnine

Insecticide

An insecticide is a type of pesticide specifically designed to control, repel, or destroy one or more species of insects.

Classification of insecticides

A. Based on mode of entry

This classification describes how the toxic chemical physically enters the insect's body to cause death.

1. Stomach Poison : A toxicant that kills the insect when it is ingested (eaten) by feeding on a treated surface (e.g., leaves, bait). The poison acts upon or is absorbed from the digestive system.
Target Pests: Primarily chewing insects (caterpillars, beetles).
Examples: Malathion, Spinosad.

2. Contact Poison : A toxicant that kills the insect by direct physical contact with its body surface (cuticle). The poison is absorbed through the exoskeleton.
Target Pests: Wide range of pests, especially those easily covered by spray (flies, aphids).
Examples: Fenvalerate (a pyrethroid).

3. Fumigant : A toxicant that enters the insect's body as a vapor or gas through the tracheal system (respiratory system), leading to death. These are primarily used in sealed spaces.

Target Pests: Stored-product pests (weevils, beetles).
Examples: Aluminium phosphide.
4. Systemic Poison : Chemicals that are absorbed by the plant (roots or leaves), penetrate the plant tissues, and are translocated (moved) throughout the vascular system (xylem and phloem). The insect is killed when it feeds on the poisoned plant sap or tissue.
Target Pests: Piercing-sucking pests and some leaf miners.
Examples: Dimethoate, Imidacloprid (a neonicotinoid).

B. Based on Mode of Action

This classification groups insecticides based on the specific physiological or biochemical process they disrupt within the insect's body.
1. Physical Poisons : These exert a physical effect rather than a chemical one. They often kill the insect by causing a loss of body moisture (desiccation) or by blocking spiracles (breathing tubes).
Mechanism: Physical effect; body moisture loss
Examples: Activated clay (inert dusts), Heavy oils
2. Protoplasmic Poisons : These toxicants kill by precipitating or coagulating proteins within the insect's cells, thereby destroying the cell structure and function (protoplasm).
Mechanism: Protein precipitation
Examples: Arsenicals
3. Respiratory Poisons : These insecticides interfere with the insect's ability to breathe by inactivating key respiratory enzymes (like cytochrome oxidase), thus blocking cellular respiration and energy production.
Mechanism: Inactivates respiratory enzymes.
Examples: Aluminium phosphide
4. Nerve Poisons : This is the largest and most common group. These compounds disrupt the insect's nervous system by interfering with the transmission of electrical signals or impulses, leading to uncontrolled excitation, paralysis, and death.
Mechanism: Inhibits impulse conduction (e.g., by targeting sodium channels or neurotransmitter enzymes).
Examples: Organophosphorus compounds, Pyrethroids
5. Chitin Inhibitors (Growth Regulators) : These are developmental poisons that disrupt the insect's molting process by inhibiting the synthesis of chitin, a primary component of the insect's exoskeleton. This prevents the formation of a new, healthy cuticle during molting.
Mechanism: Inhibits chitin synthesis
Examples: Diflubenzuron

C. Insecticide Resistance Action Committee (IRAC) Classification

Insecticide Resistance Action Committee (IRAC) works to facilitate communication and education on insecticide resistance and to promote the development and implementation of insecticide resistance management (IRM) strategies. IRAC develops Insecticide Mode of Action Classification System which classifies insecticides according to their target site or mode of action. This system is used to help growers select insecticides with different modes of action to rotate and thus reduce the risk of resistance development.

IRAC MoA Classification

Main Group / Primary Site of action	Sub Group, Class	Active Ingredient
1. Acetylcholinesterase (AChE) inhibitors	1A. Carbamate	Benfuracarb, Carbofuran, Carbosulfan, Fenobucarb, Methomyl, Thiodicarb
	1B. Organophosphate	Acephate, Chlorpyrifos, Dimethoate, Ethion, Malathion, Monocrotophos, Oxydemeton-methyl, Phenthoate, Phosalone, Phosmet, Profenofos, Quinalphos,
2. GABA-gated chloride channel blockers	2A. Cyclodiene organochlorines	Endosulfan
	2B. Phenylpyrazoles (Fiproles)	Fipronil
3. Sodium channel modulators	3A. Pyrethroids / Pyrethrins	Alphacypermethrin, Beta-cyfluthrin, Bifenthrin, Cypermethrin, Deltamethrin, Etofenoprox, Fenpropathrin, Fenvalerate, Fluvalinate, Lambda-cyhalothrin, Permethrin,
4. Nicotinic acetylcholine receptor (nAChR) competitive modulators	4A. Neonicotinoids	Acetamiprid, Clothi,anidin, Dinotefuran, Imidacloprid, Thiacloprid, Thiamethoxam
	4B. Nicotine	Nicotine
	4C. Sulfoximines	Sulfoxaflor
	4D. Butenolides	Flupyradifurone
	4E. Mesoionics	Triflumezopyrim
	4F. Pyridylidenes	Flupyrimin
5. Nicotinic acetylcholine receptor (nAChR) allosteric modulators - Site I	Spinosyns	Spinosad, Spinetoram
6. Glutamate-gated chloride channel (GluCl) allosteric modulators	Avermectins, Milbemycins	Abamectin, Emamectin benzoate, Milbemectin
7. Juvenile hormone mimics	7C. Pyriproxyfen	Pyriproxyfen
8. Miscellaneous non-specific (multi-site) inhibitors	8A. Alkyl halides	Methyl bromide
	8D. Borates	Boric acid
	8F. Methyl isothiocyanate generators	Dazomet
9. Chordotonal organ TRPV channel modulators	9B. Pyridine azomethine derivatives	Pymetrozine, Pyrifluquinazon
	9D. Pyropenes	Afidopyropen

10. Mite growth inhibitors affecting CHS1	10A. Hexythiazox	Hexythiazox
	10B. Etoxazole	Etoxazole
11. Microbial disruptors of insect midgut membranes	11A. *Bacillus thuringiensis* and the insecticidal proteins they produce	*Bacillus thuringiensis* Bt crop proteins: Cry1Ab, Cry1Ac, Cry1Fa, Cry1A.105, Cry2Ab, Vip3A, mCry3A, Cry3Ab, Cry3Bb, Cry34Ab1 / Cry35Ab1
	11B. *Bacillus sphaericus*	*Bacillus sphaericus*
12. Inhibitors of mitochondrial ATP synthase	12A. Diafenthiuron	Diafenthiuron
	12C. Propargite	Propargite
13 Uncouplers of oxidative phosphorylation via disruption of the proton gradient	Pyrroles	Chlorfenapyr
14. Nicotinic acetylcholine receptor (nAChR) channel blockers	Nereistoxin analogues	Cartap Hydrochloride, Thiocyclam Hydrogen Oxalate
15. Inhibitors of chitin biosynthesis affecting CHS1	Benzoylureas	Chlorfluazuron, Diflubenzuron, Flufenoxuron, Lufenuron, Novaluron
16. Inhibitors of chitin biosynthesis, type 1	Buprofezin	Buprofezin
18. Ecdysone receptor agonists	Diacylhydrazines	Chromafenozide, Methoxyfenozide
20. Mitochondrial complex III electron transport inhibitors – Qo site	20D. Bifenazate	Bifenazate
21 Mitochondrial complex I electron transport inhibitors	21A. METI acaricides and insecticides	Fenazaquin, Fenpyroximate, Pyridaben, Tolfenpyrad
	21B. Rotenone	Rotenone (*Derris*)
22. Voltage-dependent sodium channel blockers	22A. Oxadiazines	Indoxacarb
	22B. Semicarbazones	Metaflumizone
23. Inhibitors of acetyl CoA carboxylase	Tetronic and Tetramic acid derivatives	Spiromesifen, Spirotetramat
24. Mitochondrial complex IV electron transport inhibitors	24A. Phosphides	Aluminum phosphide, Zinc Phosphide
25. Mitochondrial complex II electron transport inhibitors	25A. beta-Ketonitrile derivatives	Cyenopyrafen, Cyflumetofen

28. Ryanodine receptor modulators	Diamides	Chlorantraniliprole, Cyantraniliprole, Flubendiamide, Cyclaniliprole, Tetraniliprole
29. Chordotonal organ nicotinamidase inhibitors	Flonicamid	Flonicamid
30. GABA-gated channel allosteric modulators	Meta-diamides Isoxazolines	Broflanilide, Fluxametamide, Isocycloseram
31. Baculoviruses Host-specific occluded pathogenic viruses	Nucleopolyhedroviruses (NPVs)	*Heliocoverpa armigera* NPV
UN Compounds	Benzpyrimoxan	Benzpyrimoxan
	Dicofol	Dicofol
		Pyridalyl
		Sulfur
UNF Fungal agents		*Beauveria bassiana, Metarhizium brunneum, Paecilomyces fumosoroseus*

D. Based on Hazard (Toxicity)

Classification based on toxicity levels as per the CIB and RC classification are as below.

Sr. No.	Classification of the Insecticides	Medium lethal dose LD_{50} mg/kg body weight of test animals		Colour of identification band on the label	Example
		Oral route Acute Toxicity	Dermal route Dermal Toxicity		
1	Extremely toxic	1-50	1-200	Bright Red	Monocrotophos
2	Highly toxic	51-500	201-2000	Bright Yellow	Imidacloprid, Fipronil
3	Moderately toxic	501-5000	2001-20000	Bright Blue	Acephate, Thiamethoxam
4	Slightly toxic	More than 5000	More than 20000	Bright Green	Chlorantraniliprole

The World Health Organization (WHO) classifies pesticides by acute toxicity to humans, which is crucial for handling and regulation. This uses the LD50 (Lethal Dose 50%) value.

Class	Hazard Level	LD50 for Rats (mg/Kg body weight)		Examples
		Oral	Dermal	
Ia	Extremely Hazardous	<5	<50	Parathion, Phorate, Aldicarb
Ib	Highly Hazardous	5-50	50-200	Carbofuran, Methomyl, Dichlorvos, Aldrin
II	Moderately Hazardous	50-2000	200-2000	Cypermethrin, DDT, Chlordane, Endosulfan
III	Slightly Hazardous	Over 2000	Over 2000	Malathion
U	Unlikely to Present Acute Hazard	5000 or Higher		Carbetamide, Cycloprothrin

E. Based on Chemical Nature

A) Inorganic insecticides :

These are chemicals that do not contain carbon-hydrogen bonds and were among the earliest insecticides used. Their use has significantly declined due to environmental and toxicity concerns.

- Arsenicals : Paris green, lead arsonate, calcium arsenate, zinc arsenate.
- Fluorine compounds : Sodium fluoride, calcium and magnesium fluosilicates, cryolite.
- Sulphur, lime sulphur, Zinc phosphide, Boric acid, Silica gel, Thalium sulphate

B) Organic insecticides :

These are chemicals that contain carbon-hydrogen bonds.

1) **Hydrocarbon oils :** Mineral or petroleum oils, coal tar oil.
2) **Animal originated insecticides :**
   - Nereistoxin – Originated from a marine annelid, Lumbrineris heteropoda.
   - Synthetic derivaties - Cartap hydrochloride, bensultap, thiocyclam.
3) **Plant originated insecticides :**
   - Nicotine - Tobacco leaves (*Nicotiana tabacum 2-5%, N. rustica -5-20%*).
   - Rotenone- Roots of plants, *Derris chinensis, Derris elliptica* (8-10% rotenone) *Lonchocarpus* spp., Inhibits respiratory metabolism, extremely toxic to fish.
   - Sabadilla – Seeds of lily sabadilla, Schoenocaulon officinale.
   - Ryanodine- Roots and stem of ryania plant, Ryania speciosa.
   - Pyrethrum- flowers of Chrysanthemum cinerariaefolium. Pyrethrins are toxic constituent; pyrethroids are synthetic analogues of pyrethrins.
   - Neem – Seed kernel of neem, Azadirachta indica. Azadirachtin ingredient insecticidal activity.
4) **Microbial origin insecticides** e.g. *Bacillus thuringiensis, Beauveria bassiana, Metarhizium anisopliae,* HaNPV, SlNPV
5) **Synthetic organic insecticides**
   - **Chlorinated hydrocarbon insecticide** e.g. DDT (Dichlorodiphenyl trichloroethane), Methoxychlor, Dicofol (Kelthane), HCH (Hexachlorocyclohexane), Aldrin, Dieldrin, Chlordane, Helptachlor, Endosulfan, Mirex.
   - **Organophosphates insecticides** : e.g. Dichlorvos, Phosphamidon, Monocrotophos, Dichrotophos, Chlorfenvinphos, TEPP, Naled, Trichlorophon, butonate, Parathion, Methyl parathion, Fenthion, Fenitrothion, Chlorpyriphos, Diazinon, Methyl demeton, Quinalphos, Profenofos, Demeton-s-methyl, oxydemeton methyl, Phorate, Ethion, Thiometon, Dimethoate, Malathion, Formothion, Phenthoate, Phosalone, Mephosfolan, Phenamiphos, Leptophos, EPN, Surecide, Acephate, Methamidophos, Fonofos, Dyfonate.
   - **Carbamate insecticides** : e.g. Dimetan, Isolan, Pyrolan, Primicarb, Carbaryl, Propoxur, Carbofuran, Methiocarb, Aminocarb, Bufencarb, Bendicarb, Aldicarb, Methomyl, Aldoxycarb, Carbosulfan, Thiodicarb, Benfuracarb
   - **Synthetic pyrethroids** : e.g. Allethrin, Phenothrin, Resmethrin, Bioallethrin, Fenpropathrin, Permethrin, Cypermethrin, Fenvelarate, Deltamethrin, Fluvalinate, Cyfluthrin, Beta-cyfluthrin.
   - **Neonicotinoids** : e.g. Imidacloprid, Thiamethoxam, Acetamiprid, Clothianidin, Thiacloprid

- **Diamide :** e.g. Chlorantraniliprole, Cyantraniliprole, Flubendiamide, Cyclaniliprole, Tetraniliprole
- **Meta-diamides :** e.g. Isoxazolines, Broflanilide, Fluxametamide, Isocycloseram
- **Spinosyns** e.g. Spinosad
- **Avermectin (Macrocyclic lactones)** e.g. Avermectin, Emamectin benzoate
- **Tertiary amines** e.g. Cartap hydrochloride
- **Oxadiazines** : e.g. Indoxacarb
- **Formamidines** e.g. Amitraz
- **Phenylpyrazoles** e.g. Fipronil
- **Pyrroles** e.g. Chlorfenapyr
- **Pyrazoles** e.g. Fenpyroximate, Tebufenpyrad
- **Pyridazinones** e.g. Pyridaben
- **Acylurea compounds** e.g. Diflubenzuron, Teflubenzuron
- **Tetronic and Tetramic acid derivatives** : e.g. Spiromesifen, Spirotetramat
- **Benzoylureas** : Chlorfluazuron, Diflubenzuron, Flufenoxuron, Lufenuron, Novaluron

*****

Pesticide Formulations

A pesticide formulation is the combination of the active ingredient and other materials (called inert ingredients or co-formulants) that makes the product suitable for storage, handling, application, and ultimately, effective pest control.

Key Components

Active Ingredient (AI): The chemical component in the product that actually controls the target pest (prevents, kills, repels, etc.).

Inert Ingredients: Materials added to the formulation that do not have direct pesticidal activity but serve essential functions, such as:

- **Carriers/Solvents:** Liquids (like water or oil) or solids (like clay or talc) used to dissolve or dilute the active ingredient.
- **Adjuvants:** Chemicals that enhance the pesticide's performance or safety. Examples include:
  - **Emulsifiers:** Allow oil-based active ingredients to mix evenly with water.
  - **Wetting Agents/Surfactants:** Help the spray spread out on the target surface.
  - **Stickers:** Help the pesticide adhere to the target surface.

Types of Formulations

A. Solid formulation : GR / G, DS, SG, SP, Tablet, WDG, WP, WS, D, B

GR / G (Granule) :

The active ingredient is coated or absorbed onto porous, coarse, ready to use particles (e.g., clay, plant material). Not mixed with water.

Advantages of GR/G:

- Ready to use.
- Low drift hazard.
- Simple application equipment
- Passes through foliage to soil.

Disadvantages of GR/G:

- Does not stick to foliage.
- Not for foliar treatment.
- Can be visible.

DS (Powder for dry seed treatment) :

This formulation is designed to be applied directly to seeds in its dry, powdered form, without the addition of water.

Advantages of DS:

- Ready to use (no mixing with water required).
- Simple application equipment needed (e.g., drum or specialized treater).

- Can be highly effective at protecting the seed and seedling immediately after planting.
- Very stable with a long shelf life when stored in a cool, dry place.

Disadvantage of DS:

- High inhalation hazard for the applicator when handling the powder.
- Can be difficult to achieve uniform coverage on the seed surface.
- Dusting off is a major concern; the powder can easily rub off the seed and lead to uneven control.
- The fine powder form creates a high drift potential if handled outside a controlled environment.

SG (Water soluble granule) :

It is a solid pesticide formulation consisting of granules that, when added to water, dissolve completely to form a true, clear solution. Unlike Wettable Powders (WP) or Water Dispersible Granules (WG), which form a suspension (particles floating in water), SG formulations leave no solid particles behind once mixed.

Advantages of SG

- High active ingredient loading in concentrated form.
- Low dust generation due to granule form, reducing inhalation hazards.
- No tank residues as they dissolve completely, avoiding nozzle blockages.
- Easy packaging, transport, and long storage stability.
- No flash point, enhancing safety in handling and storage.
- Can reduce environmental contamination due to uniform application.

Disadvantages of SG

- Manufacturing cost can be higher due to drying and granulation steps.
- Possible slow dissolution in cold water, requiring proper agitation.
- Dosing accuracy can be challenging due to density variation among granules.
- Not suitable for formulations requiring suspension or dispersible delivery.
- May require careful storage to prevent moisture absorption and caking.

SP (Water soluble powder) :

It is a dry, finely ground pesticide formulation that looks like dust but is designed to dissolve completely in water to form a true, clear solution. Unlike Wettable Powders (WP), which require constant agitation to keep the powder suspended in the water, an SP formulation disappears into the water (similar to salt or sugar) and does not settle out.

Advantages of SP

- Forms a true solution, so no agitation needed after mixing.
- Easy to store, handle, and transport.
- Less abrasive to equipment, reducing wear and nozzle clogging.
- Less dermal absorption hazard compared to some liquid formulations.
- Leaves little visible residue on treated surfaces.

Disadvantages of SP

- High inhalation hazard during handling and mixing due to fine powder form.
- Requires precise weighing for accurate dosing.
- Limited availability of pesticides in this form because few active ingredients dissolve readily in water.
- May require good quality water as hardness can affect solubility.
- Can be difficult to measure for small area applications

TB (Tablet) :

It is a pesticide formulation where the active ingredient and inert materials are compressed into a solid, pre-measured block or disk. While broadly labeled as TB, these tablets usually function in one of two specific ways:

For Spraying (Water Soluble/Dispersible): The tablet is dropped into a spray tank where it dissolves (ST) or disperses (WT) to form a mix.

For Direct Application ("Jumbo" Tabs) (DT): Used in rice farming; large tablets are thrown directly into paddy water, where they dissolve and spread on their own without spraying.

Advantages of TB

- Easy to handle, transport, and store with minimal spillage or dust exposure.
- Accurate dosing and minimal waste due to pre-measured tablet size.
- Reduced operator exposure and inhalation risk compared to powders.
- Longer shelf life and less susceptibility to moisture than powders.
- Convenient for precise applications, including small-scale treatments.

Disadvantages of TB

- May have slower dissolution in cold water or require complete stirring.
- Risk of premature disintegration if exposed to moisture during storage.
- Manufacturing cost is relatively higher due to compression and formulation steps.
- Limited flexibility in adjusting dose once tablets are pre-made.
- Not suitable for all pesticide active ingredients, especially those that require suspension formulations.

WG / WDG (Water dispersible granules)

It is a solid pesticide formulation consisting of small, dust-free granules that are designed to disintegrate quickly when added to water. Unlike SG (which dissolves completely), WG granules break apart to form a suspension (fine solid particles floating in the water). It is often also referred to as DF (Dry Flowable).

Advantages of WG / WDG

- Lower dust generation reduces inhalation hazards and contamination risks.
- Easy to measure, handle, transport, and store due to dry, free-flowing granules.
- More concentrated formulations with high active ingredient content.
- Rapid dispersion in water produces uniform spray suspensions.
- Reduced nozzle clogging compared to wettable powders.
- Safer and cleaner to use than powders.

Disadvantages of WG / WDG

- Requires constant agitation during spraying to maintain suspension.
- Can have slower dispersion in cold water.
- Granule size variation can cause segregation in containers.
- Not fully dissolved, so may leave some residue in spray equipment.
- Some risk of sedimentation in spray tank if not agitated properly.

WP (Wettable powder) :

It is a dry, finely ground pesticide formulation that looks like dust but contains a wetting agent that allows it to mix with water. Unlike SP (which dissolves completely) or EC (which forms a milky emulsion), WP particles do not dissolve. Instead, they hang suspended in the water, forming a cloudy mixture called a suspension.

Advantages of WP

- High active ingredient loading (usually 50% or more).
- Good residual control and efficacy on pests.
- Long shelf life as a dry product.
- Lower cost compared to some liquid formulations.
- Suitable for various spray equipment where agitation is possible.

Disadvantages of WP

- High dust content during handling and mixing, posing inhalation risks.
- Difficult to mix thoroughly and requires constant agitation during spraying.
- May clog spray nozzles if not properly mixed.
- Can leave visible residues on treated surfaces.
- Exposure risk to applicators during preparation.

WS (Water dispersible powder for slurry seed treatment) :

It is a specialized powder formulation designed specifically for treating seeds. Unlike standard Wettable Powders (WP) used for spraying crops, WS powders are mixed with a small amount of water to create a thick, soup-like mixture called a slurry. This slurry is then coated onto seeds before planting.

Advantages of WS

- Easy to produce and mix into water forming uniform slurry.
- Good storage stability as a dry powder.
- Water-dilutable, allowing easy plant and equipment cleaning.
- Effective seed coverage and pest protection.
- Can include safety dyes for treated seed identification.

Disadvantages of WS

- Slurry can be messy to prepare and handle.
- Requires constant stirring during slurry preparation and application.
- Potential settling of particles if slurry is left standing.
- Dustiness during handling, although less than wettable powders.
- Sensitive to moisture and temperature for storage.

DP (Dustable powder) :

It is a dry, free-flowing pesticide formulation designed to be applied directly as a dust, without mixing with water. Unlike Wettable Powders (WP) or Soluble Powders (SP), which are meant to be mixed into a spray tank, DP is ready-to-use straight from the package. It typically consists of a low concentration of active ingredient (often 1-10%) mixed with a fine, inert carrier like talc, clay, or chalk.

Advantages of DP

- Ready-to-use formulation without requiring dilution.
- Easy to apply to surfaces, cracks, soil, or seeds.
- Useful in dry conditions where spraying is not feasible.
- Good for spot treatments or hard-to-reach areas.
- Long shelf life and storage stability.

Disadvantages of DP

- High dust generation during handling and application causes inhalation hazards.
- Dust drift can lead to off-target contamination and environmental pollution.
- Low percentage of active ingredient requires large amounts for effectiveness.

- Less uniform coverage compared to liquid sprays.
- Not effective in windy or rainy conditions due to poor adhesion.

BR (Briquette) :

It is a solid, block-shaped pesticide formulation designed to release its active ingredient slowly over a long period when placed in water. Unlike tablets (TB) or granules (WG) which dissolve relatively quickly to treat a volume of water immediately, Briquettes are designed for controlled, slow release (often lasting 30 days or more). They are most famously used in public health for mosquito control.

Advantages of BR

- Provides controlled and prolonged release of active ingredients.
- Easy to handle, transport, and apply with minimal dust or spillage.
- Reduces operator exposure due to solid form.
- Suitable for targeted pest habitats or soil application.
- Can improve efficiency by reducing the frequency of applications.

Disadvantages of BR

- Manufacturing process can be complex and costly.
- Release rate depends on environmental factors (moisture, temperature).
- Not suitable for foliar applications.
- Limited flexibility in dose adjustment once briquette size is fixed.
- Potential persistence in soil might affect non-target organisms if not managed properly.

CB (Bait concentrate) :

Bait concentrate (CB) is a pesticide formulation in which the active ingredient is supplied in a concentrated form to be mixed with food or an attractant to prepare a toxic bait. It is not ready to use right out of the bottle or packet. It must be diluted or mixed with a specific food substance (the "carrier" or "attractant," such as wheat bran, sugar syrup, corn cobs, or protein hydrolysate) to create a finished bait that pests will eat.

Target Pests: Commonly used for fruit flies, rodents, cutworms, or ants where the pest needs to be lured to the poison.

Advantages of CB

- Flexible: the same CB can be mixed with different food bases to target different pests or situations.
- Economical in transport and storage because it is concentrated and has lower bulk than ready-to-use baits.
- Allows control over bait strength, palatability ingredients, and sometimes colour or markers.

Disadvantages of CB

- Requires correct on-farm or on-site mixing; mistakes in concentration can cause control failures or safety risks.
- More handling steps for the user, increasing operator exposure compared with ready-to-use baits.
- Dependent on the quality and attractiveness of locally mixed food carriers, so field performance may be more variable.

RB (Bait - ready for use) :
Ready-for-use Bait (RB) is a pesticide formulation where the active ingredient is already mixed with food or attractants and is packaged for direct application without requiring any on-site mixing.

Advantages of RB

- Convenient and easy to apply with no preparation or mixing needed.
- Reduces operator exposure and risk of handling concentrated toxicants.
- Consistent bait formulation ensures uniform efficacy.
- Ideal for use by non-technical personnel or in emergency control situations.
- Minimizes errors related to improper mixing or dilution.

Disadvantages of RB

- Bulkier and less economical to transport and store compared to concentrates.
- Limited flexibility as formulation is fixed and cannot be altered in the field.
- Potentially shorter shelf life due to pre-mixed food attractants.
- May not suit all pest species or sites if attractant type is not optimal.
- Higher initial cost compared to concentrate products due to convenience.

B. Liquid formulation

1. Simple solution (SL, UV)

SL (Soluble Concentrate) :
An SL is a clear liquid formulation where the active ingredient is dissolved in water or a water-miscible solvent.
It is designed to be diluted with water in a spray tank. When mixed with water, it forms a true solution. Unlike cloudy white mixtures (emulsions), an SL mix remains clear (like sugar dissolved in water).

Advantages of SL

- Easy to handle, store, and measure due to clear liquid form.
- Forms a true solution in water, so no agitation required after mixing.
- Less abrasive to spray equipment compared to suspensions and wettable powders.
- Leaves little to no visible residue on plants or surfaces.
- Relatively low phytotoxicity compared to some other liquid formulations.

Disadvantages of SL

- Some SL formulations may pose inhalation hazards during mixing.
- Limited availability as only a few active ingredients are water-soluble.
- Certain SLs can cause corrosion or deterioration of rubber and plastic parts.
- Potential phytotoxicity if misused or overdosed.
- Some SL formulations may be sensitive to hard or alkaline water.

UV (Ultra Low Volume [ULV]Liquids) :
Ultra Low Volume (ULV) liquids are pesticide formulations designed to be applied in very small volumes per hectare (usually less than 5 liters per hectare) using specialized equipment that creates very fine droplets for efficient pest control.

Advantages of ULV liquids

- Ready to use formulations requiring no dilution or mixing, reducing operator exposure and handling risks.

- Efficient coverage with small droplet sizes improves contact with targets and pesticide use efficiency.
- Saves water and time by applying very low volumes, making it ideal for large areas or remote sites.
- Reduced risk of pesticide runoff and environmental contamination due to low carrier volume.
- Suitable for rapid response and difficult terrain where conventional spraying is challenging.

Disadvantages of ULV liquids

- Require specialized and often expensive application equipment that may not be widely available.
- Requires operator skill for proper calibration and application to avoid drift and waste.
- Higher concentrations of active ingredients per volume increase environmental hazard risks.
- Potential for drift of fine droplets leading to off-target contamination.
- Some ULV formulations (often oil-based) can be harder to clean from equipment and may cause wear on rubber/plastic parts.

2. Solutions for Dispersion (DC, EC)

DC (Dispersible concentrate) :

A Dispersible Concentrate (DC) is a liquid pesticide formulation where the active ingredient (AI) is dissolved in a polar, water-miscible solvent in its concentrated form. When the concentrate is added to water in the spray tank, the active ingredient precipitates out of the solvent as fine, crystalline solid particles, which are then stabilized in the water as a suspension.

Advantages of DC

- Simple to manufacture and physically stable, with storage stability comparable to emulsifiable concentrates.
- Can contain active ingredients with low melting points that may be difficult to formulate otherwise.
- High biological and chemical activity due to good dispersion of active ingredient particles.
- Water-free formulation in concentrated form reduces degradation risks.
- Good long-term physical stability compared to other suspension or oil dispersion formulations.
- Easy to measure and mix before application.

Disadvantages of DC

- Limited number of active ingredients are soluble and suitable for DC formulations.
- Moderately low active ingredient content (typically less than 350 g/L).
- Requires water-free solvents and surfactants which can complicate formulation.
- Some solvent odours and toxicity concerns.
- Requires thorough mixing before and during application to maintain dispersion.
- Complex formulation requirements to keep surfactants miscible and control particle size.

EC (Emulsifiable concentrate) :

Emulsifiable Concentrate (EC) is a liquid pesticide formulation where the active ingredient is dissolved in an organic solvent along with emulsifiers, allowing it to mix with water to form an emulsion for spraying. In an EC, the active ingredient (AI) is dissolved in a petroleum-based or other organic, water-insoluble solvent, and then an emulsifying agent (surfactant) is added. When this concentrate is mixed with water in the spray tank, the emulsifier causes the oil/solvent droplets containing the AI to disperse uniformly throughout the water, forming a stable, milky-white emulsion.

Advantages of EC

- Easy to handle, store, and transport because of the concentrated liquid form.
- Forms a stable, milky emulsion upon dilution with water, requiring minimal agitation during spraying.
- Excellent coverage and penetration due to small droplet formation.
- High active ingredient concentration, making it cost-effective for large-scale use.
- Versatile, usable with various application equipment including ground sprayers and aircraft.
- Leaves little visible residue and dries quickly on plants.

Disadvantages of EC

- Contains flammable organic solvents posing fire and explosion hazards.
- Potentially toxic and corrosive to skin and eyes, requiring protective equipment.
- Can deteriorate rubber and plastic parts in spray equipment over time.
- Risk of phytotoxicity (plant damage) if overdosed or used incorrectly.
- Oil-based residue can stain certain surfaces like plastics or vinyl siding.
- Can be messy to mix and requires careful calibration to avoid over- or under-dosing.

3. Emulsions (ES, EW, ME)

ES (Emulsion for seed treatment) :

Emulsion for Seed Treatment (ES) is a liquid formulation where the active ingredient is dissolved in a solvent and then dispersed as fine droplets in water (an oil-in-water emulsion). It is a type of liquid seed treatment formulation.

Advantages of ES

- Reduced dust hazard compared with dry powders (DS, WS), improving operator safety and reducing off-target contamination.
- Good adhesion and smooth coating on the seed surface, giving uniform distribution of active ingredient and better handling/flow of treated seed.
- Water-based continuous phase (like EW) gives lower solvent odour, no flashpoint issue, and generally lower phytotoxicity and environmental impact than EC-type solvent systems.

Disadvantages of ES

- Formulation development is more complex; stability of the emulsion (creaming, phase separation) must be carefully controlled.
- Storage stability can be affected by temperature extremes, and viscosity must be balanced for pumpability and coating quality.
- Requires appropriate seed-treater equipment and correct calibration to avoid over- or under-coating and potential effects on germination or seed flow.

EW (Emulsion, oil in water) :

In an EW formulation, the oil-soluble active ingredient (AI) is first dissolved in a small amount of oil/solvent and then this oily solution is dispersed as very fine droplets in a continuous water phase, stabilized by surfactants (emulsifiers).

Advantages of EW

- Reduced dermal toxicity and lower environmental impact compared to solvent-based emulsifiable concentrates (EC) due to water continuous phase.
- Less likely to harm treated plants or surfaces because they lack organic solvents and have lower phytotoxicity.
- Good coverage and rain fastness; oil droplets evaporate more slowly, improving pesticide retention on plant surfaces.
- Safer and easier to handle, store, and transport as water-based emulsions have lower fire and explosion risk.
- Lower odour and less equipment corrosion compared to EC formulations.

Disadvantages of EW

- Can be more prone to sedimentation or phase separation, requiring formulation stabilizers and proper agitation before and during application.
- May have lower active ingredient concentration compared to ECs, potentially requiring higher volumes.
- Larger droplet size can reduce spray coverage on undersides of leaves or dense foliage.
- Sensitive to freezing or temperature extremes impacting stability and performance.
- Some formulations can cause nozzle clogging or require special sprayer cleaning.

ME (Micro-emulsion) :

ME formulations are characterized by having an active ingredient dissolved in an oil/solvent and then spontaneously dispersed in water to form incredibly small droplets (typically 10 to 100 nanometers). Because these droplets are smaller than the wavelength of visible light, a true Micro-Emulsion appears optically clear or translucent, unlike a regular emulsion (EW or EC), which appears milky.

Advantages of ME

- Thermodynamically stable, providing long shelf life and physical stability without phase separation.
- High bio-efficacy at lower active ingredient doses due to very fine droplet size and increased surface area.
- Good water dilution properties and ease of application.
- Lower solvent levels and higher flash points compared to traditional emulsifiable concentrates, improving safety.
- Improved solubility and compatibility of poorly soluble active ingredients, enhancing pesticide effectiveness.
- Reduced environmental impact owing to water-based nature and better dispersion.

Disadvantages of ME

- Higher formulation complexity requiring precise selection of surfactants and solvents.
- Generally higher production costs compared to conventional formulations.
- Requirement for specialized raw materials which can increase formulation expense.
- Potential toxicity or side effects from surfactants or co-surfactants used.
- Stability of some microemulsions may be affected by water hardness or temperature changes.

4. Suspensions (CS, FS, OD, SC)

CS (Capsule suspension) :
Capsule Suspension (CS) is one of the most advanced liquid pesticide formulations. It consists of an active ingredient (AI) encapsulated within tiny, protective polymeric shells (microcapsules), which are then suspended in water. The microcapsules are designed to release the active ingredient over time, making it a "smart" or controlled-release formulation.

Advantages of CS

- Controlled or delayed release of the active ingredient, improving bio-efficacy and reducing required doses.
- Improved safety during handling and application as the pesticide is enclosed within capsules.
- Protection of the active ingredient from environmental degradation such as UV light.
- Reduced environmental contamination and leaching risk.
- Uniform and stable suspension upon dilution, aiding in even application.
- Low or no organic solvent content, making CS formulations more environmentally friendly.

Disadvantages of CS

- Complex and costly formulation process compared to conventional pesticide formulations.
- Possible incompatibility issues between the active ingredient and encapsulating polymers.
- Lower concentration of active ingredients compared to some other formulations.
- Requires careful process control to maintain capsule integrity and release characteristics.
- Limited flexibility for modifications once formulated.

FS (Flowable concentrate for seed treatment) :
FS is a modification of a Suspension Concentrate (SC) formulation, specifically designed for application directly onto seeds. It consists of very finely ground solid active ingredients (AI) suspended in a water-based liquid, with specialized additives such as dispersing agents, binders or adhesion agents and colorants or pigment.

Advantages of FS

- Water-based and dispersible in water, making them safer and more convenient than solvent-based formulations.
- Low potential for seed germination issues.
- Good retention and uniform coating on seeds with reduced dust problems.
- Easy cleanup of seed treatment machinery after use.
- Improved flowability and handling of treated seeds, aiding in planting efficiency.
- Supports use of multiple active ingredients or biological agents in one formulation.

Disadvantages of FS

- Storage stability can be affected by temperature extremes, potentially causing agitation or viscosity issues.
- High loading of actives and binders may cause seed stickiness, reducing seed flowability if not properly formulated.
- Formulation development is complex, requiring careful balance of dispersants, binders, and pigments.

- May require sophisticated seed coating equipment and process control for uniform application.

OD (Oil dispersion) :

Oil Dispersion (OD) is an advanced, non-aqueous (water-free) liquid formulation where a solid active ingredient (AI) is finely ground and uniformly suspended in an oil medium (e.g., mineral oil, vegetable oil, or methylated seed oils). The OD formulation is designed to be diluted with water in the spray tank, where it spontaneously forms a stable oil-in-water emulsion. It is often a modern replacement for older, solvent-heavy Emulsifiable Concentrates (ECs) or water-based Suspended Concentrates (SCs).

Advantages of OD

- Suitable for hydrolytically unstable active ingredients that degrade in water, extending shelf life.
- The oil phase acts as a built-in adjuvant enhancing penetration and spray retention on plant surfaces.
- OD reduces need for external adjuvants, simplifying application and improving efficacy.
- Oil-based formulation improves rain fastness and pesticide adherence, reducing runoff.
- Less environmental contamination due to reduced pesticide wash-off.
- Non-flammable and safer compared to solvent-based emulsifiable concentrates.
- Enhances performance in water-scarce areas due to low water requirement in application.
- Can combine multiple actives or biologicals with synergistic effects.

Disadvantages of OD

- May experience sedimentation or phase separation if not properly formulated.
- Requires careful ingredient balance and high-shear homogenization during manufacture.
- Viscosity can vary, affecting handling and pumpability.
- Generally more complex and costly formulation process.
- Potential issues in cleaning and equipment maintenance due to oil residues.
- Large droplets in spray might reduce coverage uniformity on dense foliage or undersides of leaves.

SC (Suspension concentrate / flowable concentrate) :

Suspension Concentrate (SC), often referred to as a Flowable Concentrate (F), is a liquid pesticide formulation where the solid active ingredient is finely ground into tiny particles and suspended in a liquid medium, usually water. SC formulations are widely adopted as a modern, safer alternative to older powder formulations (like Wettable Powders - WPs) and solvent-heavy liquid formulations (like Emulsifiable Concentrates - ECs).

Advantages of SC/F

- High concentration of active ingredient in a liquid form, allowing accurate and easy measurement.
- Non-flammable and generally safer to handle compared to solvent-based formulations.
- Reduced dust hazard compared to dry powders.
- Good suspension stability with proper formulation, minimizing sedimentation.
- Easy to mix with water to form uniform sprays without clogging nozzles.
- Environmentally friendlier due to less use of organic solvents.
- Can incorporate multiple active ingredients or adjuvants for synergistic effects.

Disadvantages of SC/F

- Potential sedimentation if formulation or storage conditions are not optimal, requiring agitation.
- Higher viscosity than some liquid formulations, potentially affecting sprayability.
- Not suitable for all active ingredients; some are challenging to formulate as stable suspensions.
- Requires specialized manufacturing to achieve fine particle size and stability.
- More complex cleanup of spray equipment than soluble concentrates due to solid residue.

5. Dual character liquid formulation (SE)

SE (Suspo-emulsion) :

Suspo-Emulsion (SE) is a complex and highly specialized liquid pesticide formulation that is essentially a blend of two different formulation types:

a. **Suspension Concentrate (SC) Phase:** Contains one or more solid, water-insoluble active ingredients finely ground and suspended in water.
b. **Emulsion (EW) Phase:** Contains one or more liquid or oil-soluble solid active ingredients dissolved in an oil/solvent and emulsified as fine droplets in the same continuous water phase.

The resulting product is a concentrated, water-based formulation containing both suspended solid particles and dispersed oil droplets.

Advantages of SE

- Allows formulation of multiple active ingredients with different physical properties, broadening spectrum of pest control.
- Eliminates tank-mix incompatibility issues, providing convenience and reliability.
- Water-based system with lower solvent content, reducing dermal toxicity and environmental impact.
- Enables combination of water-soluble suspended particles and oil-based emulsified actives.
- Improved efficacy by simultaneous delivery of diverse actives.

Disadvantages of SE

- Complex and challenging formulation requiring careful surfactant selection for stability of both suspension and emulsion phases.
- Potential physical instability issues like sedimentation, creaming, or phase separation over time.
- Higher development and manufacturing costs compared to simpler formulations.
- Can have higher viscosity, affecting ease of handling and spraying.
- Requires shaking before use to maintain homogeneity.

6. Multicharacter liquid formulation (ZC, ZW, ZE)

ZC (A mixed formulation of CS and SC) :

The ZC formulation, standing for Mixed formulation of Capsule Suspension (CS) and Suspension Concentrate (SC), is one of the most advanced and complex liquid formulations in the agrochemical industry.

It combines two different physical forms of active ingredients (AIs) into a single, water-based product:

a. **CS Component (Microcapsules)**: One or more AIs are encapsulated within tiny polymer shells for controlled release, UV protection, or reduced toxicity.
b. **SC Component (Suspended Particles)**: One or more different AIs are present as fine, solid particles suspended freely in the same aqueous liquid.

This blend is designed to be diluted in water prior to application.

Advantages of ZC

- Combines controlled release properties of CS with high bioavailability and dispersibility of SC.
- Improved efficacy from complementary release profiles of actives.
- Enhanced safety due to encapsulation reducing exposure to highly toxic ingredients.
- Versatile formulation allowing multiple active ingredients with different solubility or release requirements.
- Good physical stability and ease of application with proper formulation.

Disadvantages of ZC

- More complex and costly formulation and manufacturing processes compared to single-type formulations.
- Potential formulation stability challenges requiring careful optimization of capsule properties and suspension stability.
- Higher development time and quality control needs.
- Possible higher viscosity affecting pumpability and sprayability.
- Limited flexibility for post-formulation adjustments.

ZE (A mixed formulation of CS and SE) :

The ZE formulation (Mixed formulation of CS and SE) is one of the most sophisticated and technically challenging liquid pesticide formulations. It is essentially a three-in-one system suspended in a single water phase.

It combines three forms of active ingredients (AIs):

a. **CS (Capsule Suspension)**: Microcapsules containing one or more AIs (usually liquid or oil-soluble) for controlled/slow release.
b. **SC (Suspension Concentrate)**: Fine, solid particles of one or more AIs suspended freely for immediate effect.
c. **EW (Emulsion, Oil in Water)**: Fine oil droplets containing one or more AIs (usually liquid or oil-soluble) dispersed for immediate or rapid effect.

This creates a stable aqueous dispersion of microcapsules, solid particles, and oil droplets.

Advantages of ZE

- Provides controlled release from capsule suspension combined with broad-spectrum multi-active delivery in suspo-emulsion.
- Enhanced efficacy by combining slow-release and immediate-release active ingredients.
- Improved safety and reduced operator exposure due to encapsulation of toxic actives.
- Water-based, environmentally friendlier system with versatile application options.
- Reduces tank mixing incompatibilities by combining different actives in one stable formulation.

Disadvantages of ZE

- Highly complex formulation requiring expertise to ensure physical and chemical stability of both capsule and emulsion/suspension phases.
- Higher manufacturing costs and longer development times.
- Potential challenges in maintaining homogeneity and preventing phase separation.
- May have higher viscosity affecting ease of handling and spraying.
- Limited flexibility for adjustments after formulation.

ZW (A mixed formulation of CS and EW) :

The ZW formulation is a modern, complex liquid pesticide that is a Mixed Formulation of Capsule Suspension (CS) and Emulsion, Oil in Water (EW).

It is an aqueous (water-based) system that combines two types of active ingredients (AIs) in one product:

a. **CS Component**: One AI is microencapsulated for controlled/slow release.
b. **EW Component**: A second AI is dissolved in oil droplets and dispersed in water for immediate release.

Advantages of ZW

- Offers controlled release of active ingredients from capsules with the good coverage and low toxicity of EW emulsions.
- Water-based continuous phase reduces flammability and environmental hazards compared to solvent-based EC products.
- Improved safety due to encapsulation reducing operator exposure to hazardous actives.
- Good rain fastness and adherence from oil droplets in EW phase combined with prolonged release from capsules.
- Facilitates combination of multiple active ingredients with different physicochemical properties.

Disadvantages of ZW

- Complex formulation requiring advanced expertise to maintain suspension stability of capsules and emulsion stability.
- Potential for higher viscosity affecting sprayability and handling.
- Risk of phase separation or instability over time without appropriate stabilizers and quality control.
- Higher manufacturing costs due to combined technologies.
- Limited options for post-formulation modification or dose adjustment.

7. Formulations prepared as devices (MC, MV, LV)

MC (Mosquito coils) :

Solid formulations made from pyrethroid or other insecticides combined with binders, shaped into coils that burn slowly to release insecticidal smoke.

Advantages: Effective for indoor and outdoor mosquito control, easy to use and transport, relatively low cost.

Disadvantages: Smoke can be irritating or allergenic, not suitable in confined unventilated spaces, potential health concerns from prolonged inhalation.

MV (Vaporising mats) :
Thin mats impregnated with insecticides (often pyrethroids) that release insecticide vapours when heated by an electric vaporiser device.
Advantages: Odourless or low odour, low indoor air pollution, convenient for continuous use, safe with proper ventilation.
Disadvantages: Requires electricity and vaporiser device, limited effectiveness outdoors, replacement frequency needed for mats.

LV (Liquid vaporisers) :
Liquid insecticides in refill bottles or cartridge form for use with electric vaporiser heaters; release insecticide vapours continuously to keep mosquitoes away.
Advantages: Continuous and controlled insecticide release, easy dose adjustment by changing liquid concentration or heating duration, suitable for indoor use.
Disadvantages: Requires electricity and vaporiser device, potential for chemical exposure, recurring cost of liquid refills.

8. Other formulations (AE, PA, CB)

AE (Aerosol dispenser) :
Aerosol Dispensers (AE) are pesticide formulations contained in pressurized canisters that release active ingredients as fine droplets or mist for insect control in enclosed spaces.

Advantages of AE

- Ready-to-use with no need for mixing, offering convenience and ease of application.
- Produces fine droplets for effective coverage and rapid knockdown of flying and crawling insects.
- Portable and easy to handle for household or commercial use.
- Allows controlled, repeatable dosing through spray bursts.
- Minimal residue and odour with properly formulated products.

Disadvantages of AE

- Contains volatile organic compounds and propellants that contribute to air pollution and greenhouse gas emissions.
- Potential health hazards from inhalation or skin contact, requiring proper ventilation and careful use.
- Non-refillable, adding to environmental waste and disposal challenges.
- Flammable propellants create fire and explosion risks during storage and use.
- May have limited use outdoors due to spray drift and weather effects.

PA (Paste) :
Paste (PA) pesticide formulations are viscous, semi-solid preparations containing active ingredients mixed with solvents, binders, and fillers to form a thick, spreadable consistency suitable for localized application.

Advantages of PA

- Easy and precise application on specific sites such as cracks, crevices, or tree trunks.
- Reduces pesticide wastage by targeted use.
- Minimizes environmental contamination and non-target exposure.
- Long-lasting due to slow release of active ingredients.
- Good adhesion to treated surfaces even in harsh weather conditions.
- Useful for termite control, baits, and localized pest management.

Disadvantages of PA

- Limited to spot or localized treatments, not suitable for wide-area application.
- Can be messy or difficult to handle if not formulated well.
- Drying time may vary, affecting residual activity.
- May require specialized application tools for some uses.
- Potential stability issues depending on formulation and storage conditions.

CB (Bait concentrate) :

Bait Concentrate (CB) formulations are concentrated pesticide products designed to be mixed with food or attractants on-site to prepare ready-to-use toxic baits for pest control.

Advantages of CB

- Flexible formulation allowing mixing with various food bases tailored to target different pests and environments.
- Economical to transport and store due to concentrated form with lower bulk.
- User control over bait strength, palatability, and tracking markers.
- Reduced risk during storage and transport compared to ready baits.

Disadvantages of CB

- Requires accurate on-site mixing, with potential for errors leading to under- or overdosing.
- Additional handling steps increase operator exposure risk.
- Performance depends on the quality and attractiveness of the locally mixed carrier.
- On-site mixing may not be convenient or feasible for all users or situations.

*****

Safe Use and Handling of Pesticides

I. Before You Apply (Preparation)

- **Read the Entire Label**: Check the signal word (e.g., DANGER, POISON, WARNING, CAUTION) and read all instructions for use, mixing, application, and safety precautions.
- **Choose the Right Product**: Ensure the pesticide is registered for the specific pest you are trying to control and the area you are treating (e.g., indoor, outdoor, garden, lawn).
- **Use Personal Protective Equipment (PPE)**: The label specifies the minimum PPE required, which often includes:
    - Gloves: Chemical-resistant gloves (unlined, usually made of nitrile or neoprene).
    - Eye Protection: Goggles or a face shield, especially when mixing or spraying.
    - Body Protection: Long-sleeved shirt, long pants, and closed-toe shoes/boots.
    - Respiratory Protection: A respirator or mask, if required by the label, especially for fine sprays or dusty materials.
- **Remove or Protect Non-Targets**: Before application, remove children, pets, toys, food, dishes, and pet bowls from the treatment area. Cover or put away sensitive items.

II. During Application

Mix Safely:

- Mix pesticides outdoors or in a well-ventilated area.
- Mix only the exact amount needed to avoid storing or disposing of excess.
- Never use food or drink containers for mixing or storage.
- Pour liquids slowly to avoid splashing, keeping the container below eye level.

Apply Correctly:

- Do not use more than recommended. Applying twice the amount does not mean twice the effect, but it does increase the risk.
- Apply only to the target area specified on the label.
- Ensure good ventilation when applying indoors.
- If applying outdoors, avoid windy or rainy conditions to prevent drift or runoff into water sources (wells, drains, ponds).
- Do not eat, drink, or smoke while handling or applying pesticides.

III. After Application (Cleanup & Storage)

Decontaminate Yourself:

- Wash hands and face thoroughly with soap and water immediately after use and before eating, smoking, or using the restroom.
- Remove contaminated clothing immediately. Wash work clothes separately from other laundry.
- Shower or bathe at the end of the day.

Clean Equipment:

- Rinse application equipment (e.g., sprayer tank) and apply the rinse water to a labelled treatment site.

- Wash boots and other non-absorbent PPE. Discard absorbent items that have been contaminated (like leather gloves or shoes).

IV. Safe Storage and Disposal

Storage:

- **Keep in the Original Container**: Never transfer a pesticide to an unlabelled container, especially one that could be mistaken for food or drink.
- **Secure and Lock**: Store all pesticides in a locked cabinet or room out of reach of children and pets.
- **Location**: Store in a cool, dry, well-ventilated area, away from food, animal feed, heat sources, or flammable materials.
- **Position**: Keep containers upright and check them periodically for leaks.

Disposal:

- **Empty Containers**: Follow the label instructions. Many empty liquid containers should be triple-rinsed (water added, shaken, and the rinse poured into the sprayer for use) before disposal or recycling.
- **Do NOT Reuse Containers**: Never reuse empty pesticide containers.
- **Leftover Pesticides**: Never pour pesticides down the sink, toilet, sewer, or storm drain, as this can pollute water sources. Check with your local waste management or environmental agency for hazardous waste collection programs.

*****

PART II

Sole Insecticides

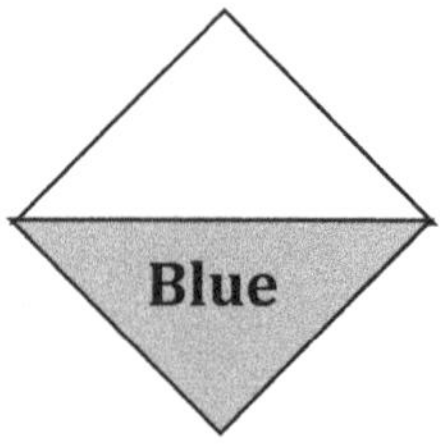

ACEPHATE

Available formulations in India :
1. **Acephate 75.00 % SP**
2. **Acephate 95.00 % SG**
3. **Acephate 97.00 % DF**

Classification on the basis of

Target organism : Insecticide
Mode of entry : Stomach and Contact poison, Systemic
Mode of action : Nerve poison
Chemical group : Organophosphate

IRAC classification

Main Group and Primary Site of Action : 1 Acetylcholinesterase (AChE) inhibitors
Sub-group or exemplifying Active Ingredient : 1B Organophosphates

Category of toxicity : Moderately toxic
Toxicity label colour : Blue

- Acephate is a systemic organophosphate insecticide widely used in agriculture and other settings to control a broad range of chewing and sucking insects.
- **Mode of Action:** Acephate works by inhibiting acetylcholinesterase (AChE) in the nervous system of insects, leading to an accumulation of acetylcholine, causing overstimulation, paralysis, and death. It has both contact and systemic activity, meaning it can kill pests upon direct contact and is absorbed by plants, making them toxic to insects that feed on them.
- **Metabolism:** In plants and insects, acephate can be metabolized into methamidophos, another organophosphate insecticide which is a more potent AChE inhibitor. In mammals, acephate is primarily excreted unchanged, with a smaller amount being metabolized, limiting its toxicity compared to insects.
- Acephate is particularly susceptible to alkaline hydrolysis. Its degradation rate increases significantly as the pH of the water becomes more alkaline (above pH 7).

Label claim (DPPQ & S, Faridabad)

Acephate 75.00 % SP

Crop	Pests	Dose / ha		Waiting period (Days)
		Quantity of formulation (g)	Volume of water (Lit.)	
Cotton	Jassids	390	500-1000	15
	Bollworms	780	500-1000	15
Safflower	Aphids	780	500-1000	15
Rice	Stem borer, Leaf folder, Plant hoppers, Green leaf hopper	666-1000	300-500	15

Trade Names

Sr. No.	Trade Name	Company
1	Sritaf	Crystal Crop Protection Ltd.
2	Acemain	ADAMA India Private Limited
3	Asataf	Rallis India Limited
4	Hilphate	HIL (India) Limited
5	Tomodachi	IFFCO-MC Crop Science Private Ltd.
6	Starthene	Swal Corporation Ltd.
7	Tremor	Biostadt India Ltd.
8	Ortain	Coromandel International Ltd.
9	X-Pride	Auxin Crop Care
10	Miltaf	Insecticides (India) Ltd.
11	Assault	Modern Insecticides Ltd.
12	i-Phate	Sydney Formulation & Chemicals
13	Acegold	Crop Chemicals India Ltd.
14	Hemcid	Hemani Industries Ltd., Ankleshwar
15	Chettak	GSP Crop Science Pvt. Ltd., Ahemadabad
16	Rock Star	GFG Crop Science Pvt. Ltd.
17	Acetop	Tusova Agro Pvt. Ltd.
18	Goldstar	Herenba Industries Ltd.
19	Sikthen	Sikko Industries Ltd.
20	Striker	Nichino India Pvt. Ltd.
21	Sinethane	Silver Sine Bio Tech
22	Acetaf	Natural Crop Science
23	Tagace	Tropical Agrosystem India (P) Ltd.
24	Atank	Essential Biosciences
25	Battle 75	Synwood Agro Ltd.
26	Martin	Global Crop Care
27	Luma	Green India Care
28	Ace Gold	Linuxcrop Chemicals Pvt. Ltd.
29	Acetate	RACCOLTO Agritech India Pvt. Ltd.

Acephate 95.00 % SG

Crop	Pests	Dose / ha		Waiting period (Days)
		Quantity of formulation (g)	Volume of water (Lit.)	
Rice	Yellow stem borer, Leaf folder, Brown plant hopper	592	500	30
Cotton	Jassid	790	500	18
Chilli	Thrips, Aphid, Fruit borer	790	500	07
Cardamom	Thrips, Shoot, Panicle and Capsule borer	1000	1000	14

Trade Names

Sr. No.	Trade Name	Company
1	Hunk	Rallis India Limited
2	Hump	Sydney Formulation & Chemicals
3	Macho	Crop Chemicals India Ltd.
4	Takatwar	Paras Crops Ltd.
5	Sikphate	Sikko Industries Ltd.
6	Shine Power	Silver Sine Bio Tech
7	Linda	Green India Care
8	Raccophate	RACCOLTO Agritech India Pvt. Ltd.

Acephate 97.00 % DF

Crop	Pests	Dose / ha		Waiting period (Days)
		Quantity of formulation (g)	Volume of water (Lit.)	
Cotton	Jassid, Bollworm complex	450 - 600	500	48
Rice	Yellow stem borer, Leaf folder, Plant hoppers, Green leaf hopper	750	500	21

Trade Names

Sr. No.	Trade Name	Company
1	Perito Ultimos	Swal Corporation Ltd.

Yellow

ACETAMIPRID

Available formulations in India : **Acetamiprid 20 % SP**

Classification on the basis of

Target organism : Insecticide
Mode of entry : Systemic and Contact poison
Mode of action : Nerve poison
Chemical group : Neonicotinoid

IRAC classification

Main Group and Primary Site of Action : 4 Nicotinic acetylcholine receptor (nAchR) competitive modulators
Sub-group or exemplifying Active Ingredient : 4A Neonicotinoids

Category of toxicity : Highly toxic
Toxicity label colour : Yellow

- Acetamiprid is a systemic insecticide belonging to the neonicotinoid family, widely employed to control various sucking insects.
- **Mode of action :** Its primary mode of action involves binding to and activating the nicotinic acetylcholine receptor (nAChR) in the insect's central nervous system. This agonistic activity disrupts normal nerve signal transmission, ultimately causing paralysis and death.
- Notably, acetamiprid demonstrates systemic and translaminar properties, enabling it to be absorbed by the plant and translocated throughout its tissues, providing protection to new growth and even the lower surfaces of leaves.
- Additionally, it exhibits both contact and stomach poisoning effects.
- Acetamiprid is hydrolytically stable at pH 4, 5, and 7. This means it doesn't break down significantly in these conditions.

Label claim (DPPQ & S, Faridabad)

Acetamiprid 20 % SP

Crop	Pests	Dose / ha		Waiting period (Days)
		Quantity of formulation (g)	Volume of water (Lit.)	
Cotton	Aphid, Jassid	50	500-600	15
	Whitefly	100	500-600	15
Cabbage	Aphid	75	500-600	07
Okra	Aphid	75	500-600	03
Chilli	Thrips	50-100	500-600	03
Rice	Brown plant hopper	50-100	500-600	07

Trade Names

Sr. No.	Trade Name	Company
1	Hilprid	HIL (India) Limited
2	Manik	Rallis India Limited
3	Rapid	Crystal Crop Protection Ltd.
4	Harrier	ADAMA India Private Limited
5	Arystaprid	UPL Ltd.
6	Kaitaku	IFFCO-MC Crop Science Private Ltd.
7	Rider, Wapkil	Biostadt India Ltd.
8	Polar	Gharda Chemicals Ltd.
9	Dhanpreet	Dhanuka Agritech Ltd.
10	Scuba	Coromandel International Ltd.
11	Prime	Hyderabad Chemical Products Ltd., Crop Chemicals India Ltd.
12	Lift	Indofil Industries Ltd.
13	Prime Gold	Alfa Crop Science
14	Aux-Pride	Auxin Crop Care
15	Magik	Anu Products Ltd.
16	Sharp	Insecticides (India) Ltd.
17	Portar	Modern Insecticides Ltd.
18	Nagarjuna Ennova	NACL Industries Ltd.
19	Active	Arysta Lifescience
20	Excel Acetacel	Excel Crop Care Ltd.
21	Qvidick	Crop Life Science Ltd.
22	Ankush	Sydney Formulation & Chemicals
23	Sarvotam	Parijat Industries India Pvt. Ltd.
24	Passion	Sumitomo Chemical India Ltd.
25	Permit	Isagro Asia Agrochemicals Pvt. Ltd.
26	Active	GFG Crop Science Pvt. Ltd.
27	Spirit	Tusova Agro Pvt. Ltd.
28	Dyken	Herenba Industries Ltd.
29	Sikprid	Sikko Industries Ltd.
30	Pioneer	Silver Sine Bio Tech
31	Natural Pride	Natural Crop Science
32	Ride	Tropical Agrosystem India (P) Ltd.
33	Proton	Essential Biosciences
34	Arron	Synwood Agro Ltd.
35	Control	Shivalik Crop Sciences (P) Ltd.
36	Ekka	Krishi Rasayan
37	Master	Global Crop Care
38	Cicada	Green India Care
39	Venus	JU Agri Sciences Pvt. Ltd.
40	Linux Pride	Linuxcrop Chemicals Pvt. Ltd.
41	Shark	Synovita Crop science Pvt. Ltd.
42	Exylon Exypride	Torrent Crop Science
43	Prado	Preventis Chem India (P) Ltd.
44	Prizeman	RACCOLTO Agritech India Pvt. Ltd.

Blue

AFIDOPYROPEN

Available formulations in India : **Afidopyropen 50 G/L DC**

Classification on the basis of

Target organism : Insecticide
Mode of entry : Systemic, Contact poison
Mode of action : Nerve poison
Chemical group : Pyropenes

IRAC classification

Main Group and Primary Site of Action : 9 Chordotonal organ TRPV channel modulators
Sub-group or exemplifying Active Ingredient : 9D Pyropenes

Category of toxicity : Moderately toxic
Toxicity label colour : Blue

- Afidopyropen, a novel insecticide in the pyropene chemical class discovered from microbial metabolite research, effectively controls various sucking insects.
- **Mode of action :** Its unique mode of action involves modulating the transient receptor potential vanilloid (TRPV) channel in insect chordotonal organs – stretch sensors crucial for balance and coordination. This disruption leads to disorientation, rapid cessation of feeding, and ultimately death by dehydration and starvation.
- This novel mode of action is valuable for resistance management, demonstrating efficacy against pests resistant to neonicotinoids and organophosphates.
- Additionally, afidopyropen exhibits a generally favourable safety profile for non-target organisms like honeybees, natural enemies, and mammals when used as directed.
- While primarily a contact insecticide, some evidence suggests limited systemic movement in plants.
- Afidopyropen is considered hydrolytically stable at pH 4 and 7. It is known to be susceptible to hydrolysis under alkaline conditions.

Label claim (DPPQ & S, Faridabad)

Afidopyropen 50 G/L DC

Crop	Pests	Dose / ha		Waiting period (Days)
		Quantity of formulation (ml)	Volume of water (Lit.)	
Cotton	Whitefly, Jassid	1000	500-750	25
Brinjal	Whitefly, Jassid	1000	500-750	01
Cucumber	Whitefly	700-1000	500	05

Trade Names

Sr. No.	Trade Name	Company
1	Sefina	BASF Crop Protection

Yellow

ALPHACYPERMETHRIN

Available formulations in India :
1. **Alphacypermethrin 10 % EC**
2. **Alphacypermethrin 10 % SC**

Classification on the basis of

Target organism : Insecticide

Mode of entry : Stomach and Contact poison

Mode of action : Nerve poison

Chemical group : Pyrethroid

IRAC classification

Main Group and Primary Site of Action : 3 Sodium channel modulators

Sub-group or exemplifying Active Ingredient : 3A Pyrethroid

Category of toxicity : Highly toxic

Toxicity label colour : Yellow

- Alpha-cypermethrin stands out as a potent broad-spectrum type II pyrethroid insecticide. This formulation specifically utilizes a racemic mixture of the two most active isomers (1R Cis S and 1S Cis R) found within cypermethrin, resulting in a significantly enhanced field activity, being two to three times more effective than standard cypermethrin.
- **Mode of action :** Its dual action as a contact and stomach poison allows it to effectively target insects by disrupting their nervous system. Like other pyrethroids, it acts as an axonic poison, preventing the closure of sodium channels in neuronal membranes, leading to rapid paralysis (knockdown effect) and subsequent death.
- This makes it a valuable tool for controlling a wide array of insect pests across numerous crops.
- Nevertheless, its use requires careful consideration due to its extreme toxicity to fish and aquatic invertebrates and high toxicity to bees, particularly during direct application or when residues are present on flowering plants.
- Alpha-cypermethrin stays stable in neutral and acidic water, typically at a pH of 7 or 4. However, it breaks down quickly in alkaline conditions, with a pH of 8 or higher.

Label claim (DPPQ & S, Faridabad)

Alphacypermethrin 10 % EC

Crop	Pests	Dose / ha		Waiting period (Days)
		Quantity of formulation (ml)	Volume of water (Lit.)	
Cotton	Bollworms	165-280	600-1000	07

Trade Names

Sr. No.	Trade Name	Company
1	Alpha Cypermethrin	Rallis India Limited
2	Stop 10 EC	Biostadt India Ltd.
3	Alphaguard 10 EC	Gharda Chemicals Ltd.
4	Gem	Indofil Industries Ltd.
5	Alpha Plus	Anu Products Ltd.
6	Rubal	Modern Insecticides Ltd.
7	Amazin	Sydney Formulation & Chemicals
8	Alfagold	Crop Chemicals India Ltd.
9	Alpha Shakti	Herenba Industries Ltd., Mumbai
10	Alfasik	Sikko Industries Ltd., Ahemadabad
11	Sursa	AC Agro care India Pvt. Ltd., Bangalore
12	Anithrin	Hemani Industries Ltd., Ankleshwar
13	Safari	Super Crop Safe Ltd., Ahemadabad
14	Dolphin	GSP Crop Science Pvt. Ltd., Ahemadabad
15	Forward	Isagro Asia Agrochemicals Pvt. Ltd.
16	Pather	Tusova Agro Pvt. Ltd.
17	Alfasik	Sikko Industries Ltd.
18	Alpha Cyprox	Natural Crop Science
19	Thril	Tropical Agrosystem India (P) Ltd.
20	Alpha Star	Synwood Agro Ltd.
21	Kri-alpha	Krishi Rasayan
22	Alpha	Global Crop Care
23	Alpha	JU Agri Sciences Pvt. Ltd.
24	AlphaOne	RACCOLTO Agritech India Pvt. Ltd.

Alphacypermethrin 10 % SC

Crop	Pests	Dose / ha		Waiting period (Days)
		Quantity of formulation (ml)	Volume of water (Lit.)	
Cotton	Bollworms	250-300	500-1000	10

Trade Names

Sr. No.	Trade Name	Company
1	Tata Alpha	Rallis India Limited
2	Pendulum	Gharda Chemicals Ltd.
3	Aster	Sydney Formulation & Chemicals
4	Spectra	Herenba Industries Ltd., Mumbai
5	Alpha-10	Sikko Industries Ltd.
6	Thriller Plus	Tropical Agrosystem India (P) Ltd.
7	Alsyn	Synwood Agro Ltd.
8	HouseCare Max	Shivalik Crop Sciences (P) Ltd.
9	Kri-alpha++	Krishi Rasayan
10	AlphaTen	RACCOLTO Agritech India Pvt. Ltd.

Red

ALUMINIUM PHOSPHIDE

Available formulations in India	:	**1. Aluminium phosphide 56.00% (3g Tablet, 10g Pouch)** **2. Aluminium phosphide 15.00% (12g Tablet)** **3. Aluminium phosphide 77.50% GR** **4. Aluminium phosphide 06.00% Tablet**
Classification on the basis of		
Target organism	:	Insecticide, Rodenticide and Acaricide
Mode of entry	:	Respiratory poison
Mode of action	:	Nerve poison
Chemical group	:	Phosphides
IRAC classification		
Main Group and Primary Site of Action	:	24 Mitochondrial complex IV electron transport inhibitors
Sub-group or exemplifying Active Ingredient	:	24A Phosphides
Category of toxicity	:	Extremely toxic
Toxicity label colour	:	Red

- Aluminium phosphide (AlP) is a highly toxic inorganic compound with the molecular formula AlP. This fumigant is utilized as a rodenticide, insecticide, and acaricide for pest management in stored grains and other agricultural commodities.
- **Mechanism of Action:** The toxicity of AlP is attributed to the hydrolysis and subsequent release of phosphine gas (PH_3). Phosphine acts as a potent metabolic inhibitor, primarily targeting the mitochondrial cytochrome c oxidase enzyme (complex IV of the electron transport chain), thereby disrupting oxidative phosphorylation and leading to cellular hypoxia and systemic organ failure. Additionally, phosphine induces free radical formation and exhibits direct corrosive effects on biological tissues.
- Its primary application is as a fumigant in enclosed environments, including stored cereal grains, animal feeds, processed food products, non-food commodities (e.g., tobacco, timber), empty warehouses, silos, and various storage structures, as well as within the confines of ships and aircraft.
- AlP is also employed for the control of rodent populations within burrows in non-domestic, non-cropland, and agricultural settings.
- Furthermore, it has limited applications in the control of moles and other small verminous mammals.
- Due to its inherent reactivity and flammability, aluminium phosphide is classified as a restricted-use chemical and presents a significant fire and explosion hazard.

Label claim (DPPQ & S, Faridabad)

Aluminium phosphide 56.00% (3 g Tablet, 10 g Pouch)

Commodity	Target pests	Dose	Exposure	Waiting period & others
Stored Whole Cereals and Seed, Grains Millet, Pulses, Dry Fruits, Nuts, Spices & Oil Seeds	Rice Weevil, Lesser Grain Borer, Khapra Beetle, Rust Red Flour Beetle, Saw Toothed Grain Beetle, Caddle Beetle, Drug Store Beetle, Cigarette Beetle, Pulse Beetle	03 tablets (03 g) per ton or 150 g per 100 m^3 or 10 g Pouch per ton of commodity or 150 g per 100 m^3	Minimum 05 Days (Rice Weevil) or 07 Days (Khapra Beetle)	One hour of partial aeration in case non- polyethylene packed commodities allowed by 6-8 hrs of full aeration. For polyethylene packed commodities minimum aeration period is 48 hrs. The waiting period for the release of stock is 48 hrs in both the cases. Recommendation for bag stock 15 days.
Mild Products: De-oiled Cakes, Rice Bran Flour, Grain Animal & Poultry Food Split Pulses (Dal) & other Processed Food	Long Headed Floor Beetle, Coffee Borer, Dried Fruit Beetle, Flat Grain Beetle, Carpet Beetle	03 tablets/1 pouch per ton or 225 g/100 m^3	5 days	Aeration is waiting Period 07 days to be checked PH3 detector strips.
Empty Godowns & Sheds	Rice Moth, Almond Moth, Mites, Fruit Fly, Granary Weevil, Caddle or Flour worm, Red Flour Beetle, Indian Meal Moth, Larger cabinet Moth, Wheat Kernel Damage in the field Cockroach	14 tablets/1000 m^3 or 150 g/100 m^3 or 4 pouch 10 g each/1000 m^3	72 hrs.	Aeration Period 24 hrs Detector strips or phosphine detect tubes should be used in the premises to signal safety of atmosphere.
Rodents Burrows	Rodents	01 Tablet/ Burrow	--	--

Trade Names

Sr. No.	Trade Name	Company
1	Celphos	Sumitomo Chemical India Ltd.
2	Quickphos	UPL Ltd.
3	Sanphos	Sandhya Organic Chemical Pvt. Ltd.
4	Quickill	Global Crop Care
5	Grianphos	Royal EDB Tube Mfg. Co.
6	Pestphos	Intech Organics Ltd.

Aluminium phosphide 15.00% (12 g Tablet)

Commodity	Target pests	Dose	Exposure	Waiting period & others
Stored Whole Cereals and Seed Grains	Rice Weevil, Rust Red Flour Beetle	1 tablet (12 g) per ton or 600 /100 m^3	Non polythene Packed commodities: ▪ Partial-1 hour. ▪ Full-(6-8) hour. Polythene Packed commodities: ▪ Minimum 48 hrs.	07-14 days
Millets, pulses, dry fruits, nuts, spices & oilseeds (Air tight cover or godowns)	Lesser Grain Borer, Khapra Beetle, Saw Toothed Grain Beetle, Rice Moth, Almond Moth	900 / 100 m^3	--	05
Milled products: De-oiled cakes, Rice bran	Rust red flower beetle	3 tablets / ton	48 hrs.	05
Flour Suji meals and Crushed grain (Animal & poultry feed), Split Pulses Dals)	Saw Toothed Grain, Beetle, Rice Moth, Almond Moth, long headed flour beetle & Mites	900 /100 m^3	48 hrs.	03
Other processed food and Empty Godowns & Sheds (under air tight condition)	All insect pests	14 tablets / 1000 tons or	48 hrs.	03
		600 g / 1000 m^3	24 hrs.	03

Trade Names

Sr. No.	Trade Name	Company
1	Celphos 15% Tablet	Sumitomo Chemical India Ltd.
2	Quickphos 15%	UPL Ltd.
3	Sanphos	Sandhya Organic Chemical Pvt. Ltd.

Aluminium phosphide 77.50% GR

Commodity	Target pests	Dose	Exposure	Waiting period & others
Stored Grain	Red Rust Flour Beetle, Lesser Grain Borer, Rice Weevil, Khapra Beetle	3.35 g	07 days	24 hrs.

Trade Names

Sr. No.	Trade Name	Company
1		

Aluminium phosphide 06.00% Tablet

Commodity	Target pests	Dose
Crop & Non-Crop area	Field rodents	One tablet of 12 g / burrow

Trade Names

Sr. No.	Trade Name	Company
1		

Yellow

BETA-CYFLUTHRIN

Available formulations in India : **Beta-cyfluthrin 02.45 % SC**

Classification on the basis of

Target organism : Insecticide

Mode of entry : Stomach and Contact poison

Mode of action : Nerve poison

Chemical group : Pyrethroid

IRAC classification

Main Group and Primary Site of Action : 3 Sodium channel modulators

Sub-group or exemplifying Active Ingredient : 3A Pyrethroid

Category of toxicity : Highly toxic

Toxicity label colour : Yellow

- Beta-cyfluthrin is a synthetic pyrethroid insecticide employed in agriculture, horticulture, and public health for broad-spectrum pest control.
- **Mode of action :** This non-systemic insecticide exhibits both contact and stomach action. As a sodium channel modulator, it disrupts the insect's nervous system, resulting in paralysis and death.
- Beta-cyfluthrin provides a rapid knockdown effect coupled with long-lasting efficacy.
- Notably, it is highly toxic to mammals and may act as a neurotoxin. Furthermore, it poses a high toxicity risk to fish, aquatic invertebrates, aquatic plants, and honeybees, while exhibiting slightly lower toxicity to birds, algae, and earthworms.
- It is most stable under neutral and slightly acidic conditions, while it breaks down quickly in alkaline environments.

Label claim (DPPQ & S, Faridabad)

Beta-cyfluthrin 02.45 % SC

Crop	Pests	Dose / ha		Waiting period (Days)
		Quantity of formulation (ml)	Volume of water (Lit.)	
Cotton	Bollworm	500-750	500-1000	20

Trade Names

Sr. No.	Trade Name	Company
1	Responsar	Bayer Crop Science Ltd.

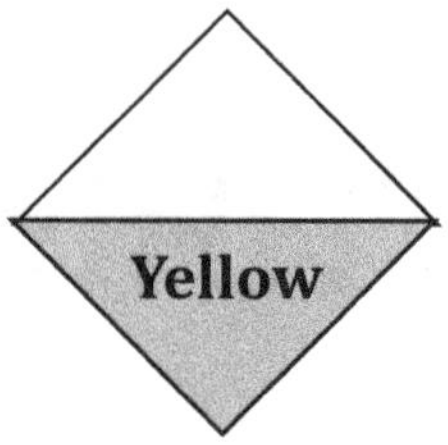

BENFURACARB

Available formulations in India :
1. **Benfuracarb 03.00 % GR**
2. **Benfuracarb 40.00 % EC**

Classification on the basis of

Target organism : Insecticide, Nematicide
Mode of entry : Systemic, Contact and Stomach poison
Mode of action : Nerve poison
Chemical group : Carbamate

IRAC classification

Main Group and Primary Site of Action : 1 Acetylcholinesterase (AChE) inhibitors
Sub-group or exemplifying Active Ingredient : 1A Carbamate

Category of toxicity : Highly toxic
Toxicity label colour : Yellow

- Benfuracarb is a broad-spectrum carbamate insecticide and nematicide used in agriculture to control various pests in crops.
- **Mode of action :** It exhibits systemic and translocating action within the plant, providing long-lasting efficacy against a wide range of pests, including certain resistant species. Benfuracarb acts by inhibiting acetylcholinesterase in insects, disrupting their nervous system.

Label claim (DPPQ & S, Faridabad)
Benfuracarb 03.00 % GR

Crop	Pests	Dose / ha		Waiting period (Days)
		Quantity of formulation (g)	Volume of water (Lit.)	
Rice	Stem borer, Leaf folder, Brown plant hopper	3300	--	20

Trade Names

Sr. No.	Trade Name	Company
1	Tadaaki	Insecticides (India) Ltd.

Benfuracarb 40.00 % EC

Crop	Pests	Dose / ha		Waiting period (Days)
		Quantity of formulation (ml)	Volume of water (Lit.)	
Red gram (Tur / Pigeonpea)	Pod borer	2500	500	20

Trade Names

Sr. No.	Trade Name	Company
1	Oncol	Coromandel International Ltd.

Blue

BENZPYRIMOXAN

Available formulations in India : **Benzpyrimoxan 10.00 % SC**

Classification on the basis of

Target organism : Insecticide
Mode of entry : Contact and Stomach poison
Mode of action : Growth regulator
Chemical group : Pyrimidine derivative

IRAC classification

Main Group and Primary Site of Action : Unknown
Sub-group or exemplifying Active Ingredient : Unknown

Category of toxicity : Moderately toxic
Toxicity label colour : Blue

- Benzpyrimoxan is a novel pyrimidine derivative insecticide, first reported around 2007 and registered in 2019 by Nihon Nohyaku Co., Ltd. It offers a unique mode of action valuable for resistance management.
- **Mode of action :** Unlike other insect growth regulators (like buprofezin, which inhibits chitin biosynthesis), Benzpyrimoxan interferes with the normal metabolism of ecdysone (molting hormone) in insects. It causes ecdysteroid levels to remain higher than normal during the molting process, disrupting the molting mechanism.
- It primarily acts on nymphal stages of target pests, causing abnormal molting and death. It shows low or no adulticidal activity.

Label claim (DPPQ & S, Faridabad)

Benzpyrimoxan 10.00 % SC

Crop	Pests	Dose / ha		Waiting period (Days)
		Quantity of formulation (ml)	Volume of water (Lit.)	
Rice	Brown plant hopper, White backed plant hopper	750-1000	500	31

Trade Names

Sr. No.	Trade Name	Company
1	Orchestra	Nichino India Private Ltd.

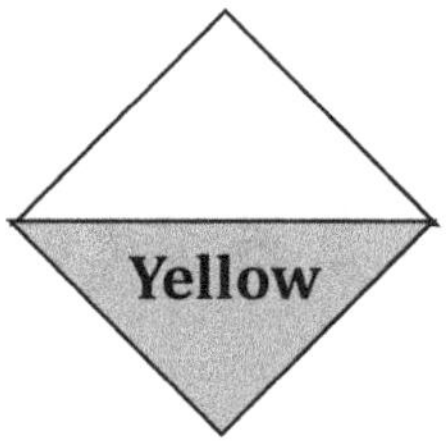

BIFENTHRIN

Available formulations in India :
1. **Bifenthrin 08.00 % SC**
2. **Bifenthrin 08.80 % CS**
3. **Bifenthrin 10.00 % EC**
4. **Bifenthrin 02.50 % EC**

Classification on the basis of

Target organism : Insecticide, Acaricide
Mode of entry : Stomach and Contact poison
Mode of action : Nerve poison
Chemical group : Pyrethroid

IRAC classification

Main Group and Primary Site of Action : 3 Sodium channel modulators
Sub-group or exemplifying Active Ingredient : 3A Pyrethroid

Category of toxicity : Highly toxic
Toxicity label colour : Yellow

- Bifenthrin, a synthetic pyrethroid insecticide and acaricide, is a man-made analog of natural insecticides found in chrysanthemum flowers. Widely employed in agriculture, horticulture, and for pest control in and around residential and commercial buildings, bifenthrin effectively targets a broad spectrum of pests.
- **Mode of action :** Its mode of action is through contact and ingestion. Bifenthrin disrupts nerve signal transmission in insects by acting on voltage-gated sodium channels, which are essential for nerve cell function. By binding to these channels and interfering with their normal opening and closing, it causes paralysis and ultimately death. Classified as a Type I pyrethroid, bifenthrin induces a shorter opening of the sodium channels compared to Type II pyrethroids.
- This insecticide provides a relatively rapid knockdown effect along with residual control.

Label claim (DPPQ & S, Faridabad)

Bifenthrin 08.00 % SC

Crop	Pests	Dose / ha		Waiting period (Days)
		Quantity of formulation (ml)	Volume of water (Lit.)	
Rose	Red spider mite, Tea mosquito bug	500	400	11
Apple	Mite	7.5 ml / tree	10 lit/ tree	21

Trade Names

Sr. No.	Trade Name	Company
1	Leothrin Gold	Crop Life Science Ltd.

Bifenthrin 08.80 % CS

Crop	Pests	Dose / ha		Waiting period (Days)
		Quantity of formulation (ml)	Volume of water (Lit.)	
Rice	Stem borer, Leaf folder	500	500	21

Trade Names

Sr. No.	Trade Name	Company
1	Rockstar	Insecticides (India) Ltd.

Bifenthrin 10.00 % EC

Crop	Pests	Dose / ha		Waiting period (Days)
		Quantity of formulation (ml)	Volume of water (Lit.)	
Cotton	Bollworms, Whitefly	800	500	15
Rice	Stem borer, Leaf folder, Green leaf hopper	500	500	21
Sugarcane	Termite	1000	500	300

Trade Names

Sr. No.	Trade Name	Company
1	Imperial	ADAMA India Private Limited
2	Disect	UPL Ltd.
3	Hamada	IFFCO-MC Crop Science Private Ltd.
4	Hectastar	Swal Corporation Ltd.
5	Wanted	Gharda Chemicals Ltd.
6	Marker	Dhanuka Agritech Ltd.
7	Canister	Coromandel International Ltd.
8	Talstar	FMC Corporation
9	Bannerr	Tropical Agrosystem India (P) Ltd.
10	Hit Star	Auxin Crop Care
11	A-Star	Anu Products Ltd.
12	Super Star	Insecticides (India) Ltd.
13	Marshal 10 EC	Modern Insecticides Ltd.
14	Leothrin	Crop Life Science Ltd.
15	Bicep 10 EC	Crop Chemicals India Ltd.
16	Pharlap	Sumitomo Chemical India Ltd.
17	Anicid	Hemani Industries Ltd., Ankleshwar
18	Highlight	Isagro Asia Agrochemicals Pvt. Ltd.
19	Crystal	GFG Crop Science Pvt. Ltd.
20	Enthrone	Tusova Agro Pvt. Ltd.
21	Finstar	Herenba Industries Ltd.
22	Bifensik	Sikko Industries Ltd.
23	Coma	Nichino India
24	Bison 10	KP Agro Industries

25	Excellent	Greenwell Biotech
26	Defenthrin	Somnath Demeter Agritron Ltd.
27	Bifen 10	Natural Crop Science
28	Bifenzo	Katyayani Organics
29	Bannerr	Tropical Agrosystem India (P) Ltd.
30	Synthrin	Synwood Agro Ltd.
31	Glowflex	Today's Global Crop Protection
32	Star+	Vajra Crop Science
33	Rock-10	Krishi Rasayan
34	Taffy	Tata Rallis India Ltd.
35	Lockdown	Green India Care
36	Century	JU Agri Sciences Pvt. Ltd.
37	B-10	Linuxcrop Chemicals Pvt. Ltd.
38	Startis	Synovita Crop Science Pvt. Ltd.
39	Bifenforce	Mankind Agritech
40	Big Star	Preventis Chem India (P) Ltd.
41	Jarmar	RACCOLTO Agritech India Pvt. Ltd.

Bifenthrin 02.50 % EC

1. **Pre and post construction**: Bifenthrin 2.5% EC shall be applied at 0.05% a.i. conc. i.e. 20.0 ml formulated product diluted in 1 liter of water for the control of termites in building during pre and post construction. Treatment should be as per IS 6313 (Part 2) :2001 for pre-construction chemical treatment and IS 6313 (Part-3) : 2001 for post construction treatment of the existing building.
2. Recommendation for use of control of wood borer (Powder post beetle) in plywood, veneer and wood.

Use	Method of application	Dose a.i.	Dilution
Plywood	Glue Line Poisoning	10 g/m^3 of wood	400 ml formulation per m^3 of wood
	Dipping	0.025% Solution	Mix 01 lit of formulation in 99 lit of water to make 0.025% solution
Veneer	Dipping	0.025% Solution	Mix 01 lit of formulation in 99 lit of water to make 0.025% solution
Wood	Dipping / Brushing	0.025% Solution	Mix 01 lit of formulation in 99 lit of water to make 0.025% solution

Trade Names

Sr. No.	Trade Name	Company
1	Marshal 2.5 EC	Modern Insecticides Ltd.
2	Bicep 2.5 EC	Crop Chemicals India Ltd.
3	Termifinn	Herenba Industries Ltd.
4	Bifeno	Sikko Industries Ltd.
5	Bison 2.5	KP Agro Industries
6	Bannerr	Tropical Agrosystem India (P) Ltd.
7	Method	RACCOLTO Agritech India Pvt. Ltd.

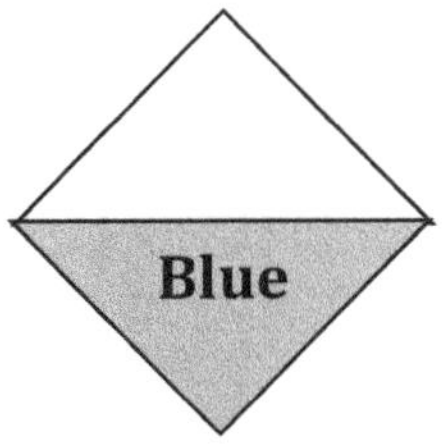

BROFLANILIDE

Available formulations in India :
1. **Broflanilide 300 g/l SC**
2. **Broflanilide 20.00 % SC**

Classification on the basis of

Target organism	:	Insecticide
Mode of entry	:	Stomach and contact poison
Mode of action	:	Nerve poison
Chemical group	:	Benzamide
IRAC classification		
Main Group and Primary Site of Action	:	30 GABA-gated chloride channel allosteric modulators
Sub-group or exemplifying Active Ingredient	:	Meta-diamides
Category of toxicity	:	Moderately toxic
Toxicity label colour	:	Blue

- Broflanilide is a relatively new insecticide that has gained attention for its novel mode of action and effectiveness against a wide range of chewing insect pests, including those that have developed resistance to other insecticides.
- **Mode of action :** It functions as a noncompetitive antagonist of the insect resistant-to-dieldrin (RDL) gamma-aminobutyric acid (GABA) receptor.
- In simpler terms, GABA is a neurotransmitter in the insect's nervous system that normally transmits inhibitory signals. Broflanilide (or its metabolite, desmethyl-broflanilide) binds to a unique site on the GABA receptor, preventing GABA from transmitting these inhibitory signals. This leads to overexcitation of the insect's nervous system, causing paralysis, convulsions, and ultimately death.
- Its novel mode of action is particularly significant because it means there is no known cross-resistance with existing insecticide classes, making it a valuable tool for insecticide resistance management.
- Its stability is high at neutral and acidic pH. The increased degradation in alkaline conditions suggests that in environments with higher pH, its breakdown may be somewhat faster.

Label claim (DPPQ & S, Faridabad)

Broflanilide 300 g/l SC

Crop	Pests	Dose / ha		Waiting period (Days)
		Quantity of formulation (ml)	Volume of water (Lit.)	
Chilli	Fruit borer	42-62	500	01
	Thrips	62-84	500	01
Brinjal	Shoot and fruit borer	42-62	500	01

	Thrips	62-84	500	01
Tomato	Fruit borer, Leaf miner	62-84	500	01
Soybean	Gram pod borer, Tobacco leaf eating caterpillar, Semilooper	42-62	500	37
Pigeonpea	Pod borer, Spotted pod borer	42-62	500	25

Trade Names

Sr. No.	Trade Name	Company
1	Exponus	BASF India Ltd.

Broflanilide 20.00 % SC

Crop	Pests	Dose / ha		Waiting period (Days)
		Quantity of formulation (ml)	Volume of water (Lit.)	
Brinjal	Shoot and fruit borer, Thrips, Jassid	125	500	01
Cabbage	Diamond back moth, Tobacco leaf eating caterpillar	125	500	01
Chilli	Fruit borer, Thrips, Jassid	125	500	01
Okra	Fruit borer, Thrips, Jassid	125	500	01
Maize	Fall armyworm	125	500	29
Cotton	Thrips, Jassid	125	500	10
Pomegranate	Thrips, Fruit borer	125	500	01
Soybean	Pod borer, Semilooper, Tobacco caterpillar	125	500	23
Tomato	Thrips, Leaf miner, Fruit borer	125	500	01

Trade Names

Sr. No.	Trade Name	Company
1	Brofreya	PI Industries Ltd.

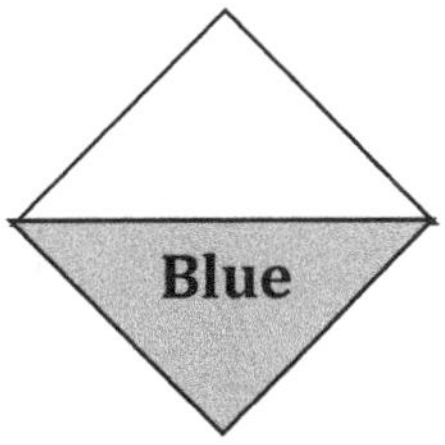

BUPROFEZIN

Available formulations in India	:	**1. Buprofezin 25.00 % SC** **2. Buprofezin 70.00 % DF**
Classification on the basis of		
Target organism	:	Insecticide
Mode of entry	:	Stomach and contact poison
Mode of action	:	Growth regulator (Inhibition of chitin synthesis)
Chemical group	:	Thiadiazine
IRAC classification		
Main Group and Primary Site of Action	:	16 Inhibitors of chitin biosynthesis, type 1
Sub-group or exemplifying Active Ingredient	:	Buprofezin
Category of toxicity	:	Moderately toxic
Toxicity label colour	:	Blue

- Buprofezin provides effective control against pests such as mealybugs, leafhoppers, and whiteflies by primarily targeting their immature nymph stages through contact.
- **Mode of action :** It disrupts their molting process by inhibiting the formation of chitin, a vital component of their exoskeleton. This interference prevents successful shedding of the old skin, leading to abnormal development and the eventual death of the nymphs. Beyond its direct impact on nymphs, buprofezin also disrupts the pest lifecycle by reducing egg viability and suppressing the egg-laying capacity of adults. This multi-faceted approach contributes to a reduction in future insect populations and offers sustained protection for plants.
- A notable advantage of buprofezin is its relative safety for beneficial insects like ladybird beetles and lacewings, making it a suitable option for integrated pest management (IPM) programs that prioritize a balanced ecosystem.
- While buprofezin may exhibit moderate toxicity to some wildlife and plants, it presents a low risk to honeybees and mammals via oral exposure.

Label claim (DPPQ & S, Faridabad)

Buprofezin 25 % SC

Crop	Pests	Dose / ha		Waiting period (Days)
		Quantity of formulation (ml)	Volume of water (Lit.)	
Cotton	Whitefly, Aphid, Jassid, Thrips	1000	500-750	20
Chilli	Yellow mite	300-600	500-750	05
Mango	Hoppers	1-2 ml / lit.	5 –15 lit./ tree	20
Grape	Mealy bug	1000-1500	500-1000	07

Rice	Brown plant hopper, Green leaf hopper, White backed plant hopper	800	400-500	20

Trade Names

Sr. No.	Trade Name	Company
1	Tribune	Crystal Crop Protection Ltd.
2	Bipimain	ADAMA India Private Limited
3	Applaud	Rallis India Limited
4	Yukata	IFFCO-MC Crop Science Private Ltd.
5	Banzo, Sprize	Biostadt India Ltd.
6	Machan	Gharda Chemicals Ltd.
7	Apple	Dhanuka Agritech Ltd.
8	Flotis	Bayer Crop Science Ltd.
9	Ninja	Coromandel International Ltd.
10	Super Star	Anu Products Ltd.
11	Phentom, Hakko	Insecticides (India) Ltd.
12	Kick-Out	Modern Insecticides Ltd.
13	Balloon	Sydney Formulation & Chemicals
14	Bufen	Crop Chemicals India Ltd.
15	Bravo	Tusova Agro Pvt. Ltd.
16	Heralaud	Herenba Industries Ltd.
17	Bufisik	Sikko Industries Ltd.
18	Allfall	Nichino India
19	Kuprozin	KP Agro Industries
20	Allowed	Silver Sine Bio Tech
21	Bupro 25	Natural Crop Science
22	Voltage	Tropical Agrosystem India (P) Ltd.
23	B-Pro	Essential Biosciences
24	Lumix	Synwood Agro Ltd.
25	Apollo	HPM Chemicals and Fertilizers Ltd.
26	Bencher	RACCOLTO Agritech India Pvt. Ltd.

Buprofezin 70 % DF

Crop	Pests	Dose / ha		Waiting period (Days)
		Quantity of formulation (ml)	Volume of water (Lit.)	
Cotton	Jassid, Whitefly	357-429	500	20
Rice	Brown plant hopper	250	500	24
Okra	Jassid	286	500	20

Trade Names

Sr. No.	Trade Name	Company
1	Stalin	Anu Products Ltd.
2	Zeeny	Rallis India Ltd.
3	Bencher Plus	RACCOLTO Agritech India Pvt. Ltd.

CARBOFURAN

Available formulations in India	:	**Carbofuran 03.00 % CG**
Classification on the basis of		
Target organism	:	Insecticide, Nematicide, Acaricide
Mode of entry	:	Stomach and Contact poison, Systemic
Mode of action	:	Nerve poison
Chemical group	:	Carbamate
IRAC classification		
Main Group and Primary Site of Action	:	1 Acetylcholinesterase (AChE) inhibitors
Sub-group or exemplifying Active Ingredient	:	1A Carbamate
Category of toxicity	:	Highly toxic
Toxicity label colour	:	Yellow

- Carbofuran is a highly toxic carbamate insecticide, nematicide, and acaricide extensively employed in agriculture for controlling both soil and foliage-dwelling pests. As a systemic insecticide, it is absorbed by plant roots and translocated throughout the plant tissues, offering protection against feeding pests. It also exhibits contact activity.
- **Mode of action :** Its mode of action involves systemic, contact, and stomach poisoning. Carbofuran functions by inhibiting acetylcholinesterase (AChE) in the nervous systems of insects and other animals. AChE is a vital enzyme responsible for the breakdown of acetylcholine, a neurotransmitter. By blocking AChE, carbofuran causes an accumulation of acetylcholine at nerve synapses and neuromuscular junctions, resulting in overstimulation, paralysis, and ultimately death. This mechanism of action is comparable to that of highly toxic nerve agents.
- Notably, carbofuran is highly toxic to birds, fish, and bees.

Label claim (DPPQ & S, Faridabad)

Carbofuran 03.00 % CG

Crop	Pests	Dose / ha		Waiting period (Days)
		Quantity of formulation (Kg)	Volume of water (Lit.)	
Sorghum	Shoot fly	33.30	--	--
	Stem borer	8.30	--	--
Bajara	Shoot fly	50.00	--	--
Maize	Stem borer, Shoot fly, Thrips	33.30	--	--
Wheat	Ear cockle nematode	10.00	--	--
	Cereal cyst nematode	66.60	--	--
Barley	Aphid, Cyst nematode	33.30	--	--
	Jassid	41.60	--	--

Rice	Brown plant hopper, Green leaf hopper, Gall midge, Stem borer, Hispa	25.00	--	--
	Nematode	50.00	--	--
Groundnut	Pod borer	50.00	--	--
	White grub	33.30	--	--
Soybean	Root knot nematode	50.00	--	--
Mustard	Leaf miner	66.60	--	--
	Whitefly	33.30	--	--
Sugarcane	Top shoot borer	66.60	--	--
Jute	Nematodes	33.30	--	--
Okra	Jassid	33.30	--	--
Brinjal	Root knot nematode, Reniform nematode	66.60	--	--
Chilli	Aphid, Thrips	33.30	--	--
Cabbage	Nematode	50.00	--	--
French bean	White grub	23.30	--	--
	Grey & Stem weevil	33.30	--	--
Potato	Aphid	16.60	--	--
	Jassid	33.30	--	--
Tomato	Whitefly	40.00	--	--
Pea	Shoot fly, Aphid	33.30	--	--
Citrus	Nematode	12.00	--	--
	Leaf miner	50.00	--	--
Mandarin	Soft green scale	13.30 g / plant	--	--
Banana	Rhizome weevil	33 g / sucker	--	--
	Aphid	166g / sucker	--	--
	Nematode	50 g / Sucker	--	--
Apple	Wooly aphid	166 g / tree	--	--
Peach	Leaf curl aphid	33.30	--	--
Tea	Cock chafer grub	33.10 g / plant	--	--

Trade Names

Sr. No.	Trade Name	Company
1	Furadan, Furadan Ultra	Crystal Crop Protection Ltd.
2	Carbomain	ADAMA India Private Limited
3	Furadan 3 G	FMC Corporation
4	Carbogran	Hyderabad Chemical Products Ltd.
5	Anufuran 3G	Anu Products Ltd.
6	Fury	NACL Industries Ltd.
7	Carbogran 3G	Nichino India
8	Kurbo 3	KP Agro Industries
9	Carbman	Global Crop Care
10	Bond	Green India Care
11	Furious 3G	Synovita Crop science Pvt. Ltd.

Yellow

CARBOSULFAN

Available formulations in India	:	**1. Carbosulfan 06.00 % G** **2. Carbosulfan 25.00 % EC** **3. Carbosulfan 25.00 % DS**

Classification on the basis of

Target organism	:	Insecticide, Acaricide, Nematicide
Mode of entry	:	Stomach and Contact poison
Mode of action	:	Nerve poison
Chemical group	:	Carbamate
IRAC classification		
Main Group and Primary Site of Action	:	1 Acetylcholinesterase (AChE) inhibitors
Sub-group or exemplifying Active Ingredient	:	1A Carbamate
Category of toxicity	:	Highly toxic
Toxicity label colour	:	Yellow

- Carbosulfan, a carbamate insecticide, nematicide, and acaricide, is utilized in agriculture to manage a broad spectrum of pests across various crops.
- Functioning as a pro-insecticide, it undergoes metabolism within the target pest, converting into carbofuran.
- **Mode of action :** Its mode of action is systemic, with both contact and stomach activity. Following its conversion to carbofuran, it inhibits acetylcholinesterase (AChE) in the nervous systems of insects and other animals. This inhibition results in the accumulation of acetylcholine, leading to overstimulation, paralysis, and ultimately death.
- Notably, carbosulfan exhibits high ecotoxicity to a wide range of species.
- It is less stable in acidic and alkaline conditions and more stable in neutral media.

Label claim (DPPQ & S, Faridabad)

Carbosulfan 06.00% Granules

Crop	Pests	Dose / ha		Waiting period (Days)
		Quantity of formulation (Kg)	Volume of water (Lit.)	
Rice	Stem borer, Gall midge, Green leaf hopper, Leaf folder	16.70	--	37

Trade Names

Sr. No.	Trade Name	Company
1	Hindustan 6 G	Hindustan Agro Life

Carbosulfan 25.00% EC

Crop	Pests	Dose / ha		Waiting period (Days)
		Quantity of formulation (ml)	Volume of water (Lit.)	
Rice	Green leaf hopper, White backed plant hopper, Brown plant hopper, Stem borer, Gall midge, Leaf folder	800-1000	500-1000	14
Cotton	Aphid, Thrips	1250	500	70
Chilli	White aphid	800-1000	500-1000	08
Brinjal	Shoot and fruit borer	1250	500	05
Cumin	Aphid, Thrips	1250	500	17

Trade Names

Sr. No.	Trade Name	Company
1	Aaatank	Dhanuka Agritech Ltd.
2	Marshal	FMC Corporation
3	Master	Gujarat Crop Care
4	Carbosik	Sikko Industries Ltd.

Carbosulfan 25.00% DS

Crop	Pests	Dose / ha		Waiting period (Days)
		Quantity of formulation (g)	Volume of water (Lit.)	
Cotton	Aphid, Jassid, Thrips	60 g / Kg seed	--	--

Trade Names

Sr. No.	Trade Name	Company
1	Aayudh	Coromandel International Ltd.
2	Advantage	FMC Corporation

Yellow

CARTAP HYDROCHLORIDE

Available formulations in India :
1. **Cartap hydrochloride 04.00 % G**
2. **Cartap hydrochloride 50.00 % SP**
3. **Cartap hydrochloride 75.00 % SG**

Classification on the basis of

Target organism : Insecticide
Mode of entry : Stomach and Contact poison, Systemic
Mode of action : Nerve poison
Chemical group : Thiocarbamate

IRAC classification

Main Group and Primary Site of Action : 14 Nicotinic acetylcholine receptor (nAChR) channel blockers
Sub-group or exemplifying Active Ingredient : Nereistoxin analogues

Category of toxicity : Highly toxic
Toxicity label colour : Yellow

- Cartap hydrochloride, a systemic insecticide in the nereistoxin analogue group, effectively controls a broad spectrum of chewing and sucking insect pests across various crops.
- **Mode of action :** Its mode of action is systemic, with both contact and stomach activity. It disrupts nerve transmission in insects by binding to receptors within their nervous system. This ganglionic blocking action in the central nervous system ultimately leads to insect paralysis and death, impacting all developmental stages: egg, larva, and adult.
- While providing persistent, long-term pest control, cartap hydrochloride's high water solubility and low volatility suggest a reduced tendency for soil accumulation, presenting a potential environmental advantage.
- Cartap hydrochloride is stable in acidic conditions but prone to hydrolysis in neutral and alkaline solutions.

Label claim (DPPQ & S, Faridabad)

Cartap Hydrochloride 04.00% Granules

Crop	Pests	Dose / ha		Waiting period (Days)
		Quantity of formulation (Kg)	Volume of water (Lit.)	
Rice	Stem borer	18.75	--	--
	Leaf folder, Whorl maggot	18.75 – 25.00	--	--

Trade Names

Sr. No.	Trade Name	Company
1	Nidan	Crystal Crop Protection Ltd.
2	Boregan G	ADAMA India Private Limited

3	Kaardon 4g	UPL Ltd.
4	Cartox G	Rallies India Limited
5	Hilcartap	HIL (India) Limited
6	Katsu	IFFCO-MC Crop Science Private Ltd.
7	Startop	Swal Corporation Ltd.
8	Dartriz 4 G	Biostadt India Ltd.
9	Kraft	Gharda Chemicals Ltd.
10	Caldan 4 G	Dhanuka Agritech Ltd.
11	Parry Ratna	Coromandel International Ltd.
12	Beacon GR	Indofil Industries Ltd.
13	Anudan 4G	Anu Products Ltd.
14	Indan 4G	Insecticides (India) Ltd.
15	Counter 4G	Modern Insecticides Ltd.
16	Nagarjuna 4G	NACL Industries Ltd.
17	Carvan	Arysta Lifescience
18	Car-Hydro	Sydney Formulation & Chemicals
19	Kanjo	Parijat Industries India Pvt. Ltd.
20	Hemtox	Hemani Industries Ltd., Ankleshwar
21	Dhaal	Isagro Asia Agrochemicals Pvt. Ltd.
22	Cargo 4G	Herenba Industries Ltd.
23	Cardon 4G	Natural Crop Science
24	Fast	Tropical Agrosystem India (P) Ltd.
25	Prime	Apex Agro Industries
26	Kritap-4G	Krishi Rasayan
27	Istoxin GR	Global Crop Care
28	Cartap GR	JU Agri Sciences Pvt. Ltd.
29	Note-4G	Linuxcrop Chemicals Pvt. Ltd.
30	Sartap	Saraswati Agro Life Science (India) Pvt. Ltd.
31	Stellar 4G	Synovita Crop science Pvt. Ltd.
32	Cartakind GR	Mankind Agritech
33	Prodan Gr	Preventis Chem India (P) Ltd.
34	Dhaan 4G	Chemley Agritech Pvt. Ltd.
35	Dilogue-4G	RACCOLTO Agritech India Pvt. Ltd.

Cartap Hydrochloride 50.00% SP

Crop	Pests	Dose / ha		Waiting period (Days)
		Quantity of formulation (g)	Volume of water (Lit.)	
Rice	Stem borer, Leaf folder	1000	500-1000	--

Trade Names

Sr. No.	Trade Name	Company
1	Boregan SP	ADAMA India Private Limited
2	Cartop, Dartriz SP	Biostadt India Ltd.
3	Caldan 50 SP	Dhanuka Agritech Ltd.

4	Josh SP	Coromandel International Ltd.
5	Beacon SP	Indofil Industries Ltd.
6	Anudan 50 SP	Anu Products Ltd.
7	Indan SP	Insecticides (India) Ltd.
8	Quick 50 SP	NACL Industries Ltd.
9	Carvan Plus 50	Arysta Lifescience
10	Kanjo SP	Parijat Industries India Pvt. Ltd.
11	CAP-X 50 SP	Sumitomo Chemical India Ltd.
12	Cargo	Herenba Industries Ltd.
13	Cartsik	Sikko Industries Ltd.
14	Wartap 50 SP	Nichino India
15	Captionin	Silver Sine Bio Tech
16	Cardon 50	Natural Crop Science
17	Cartashield 50	Katyayani Organics
18	Fast	Tropical Agrosystem India (P) Ltd.
19	Tapride	Essential Biosciences
20	Kritap-50SP	Krishi Rasayan
21	Cinos	Global Crop Care
22	Killer	Green India Care
23	Elect SP	JU Agri Sciences Pvt. Ltd.
24	Note-50	Linuxcrop Chemicals Pvt. Ltd.
25	Stellar 50 SP	Synovita Crop science Pvt. Ltd.
26	Kraft SP	Gharda Chemicals Ltd.
27	Cartakind	Mankind Agritech
28	Prodan	Preventis Chem India (P) Ltd.
29	Dilogue-50 SP	RACCOLTO Agritech India Pvt. Ltd.

Cartap Hydrochloride 75.00% SG

Crop	Pests	Dose / ha		Waiting period (Days)
		Quantity of formulation (g)	Volume of water (Lit.)	
Rice	Yellow stem borer, Leaf folder	425-500	250-500	35-89

Trade Names

Sr. No.	Trade Name	Company
1	Mortor	Dhanuka Agritech Ltd.
2	Cartap-75	Sikko Industries Ltd.
3	Istoxin SG	Global Crop Care
4	Cygnus	Synovita Crop science Pvt. Ltd.
5	Prodan 75 SG	Preventis Chem India (P) Ltd.
6	Hydro-75	RACCOLTO Agritech India Pvt. Ltd.

Green

CHLORANTRANILIPROLE

Available formulations in India :

1. **Chlorantraniliprole 18.50 % SC**
2. **Chlorantraniliprole 00.40 % GR**
3. **Chlorantraniliprole 00.53 % GR**
4. **Chlorantraniliprole 35.00 % WG**
5. **Chlorantraniliprole 05.00 % DT**
6. **Chlorantraniliprole 47.85 % SC**
7. **Chlorantraniliprole 50.00 % WS**

Classification on the basis of

Target organism : Insecticide

Mode of entry : Stomach and Contact poison

Mode of action : Nerve and muscle poison

Chemical group : Diamides

IRAC classification

Main Group and Primary Site of Action : 28 Ryanodine receptor modulators

Sub-group or exemplifying Active Ingredient : Diamides

Category of toxicity : Slightly toxic

Toxicity label colour : Green

- Chlorantraniliprole, a synthetic insecticide in the anthranilic diamide class, is extensively used in agriculture and horticulture for broad-spectrum control of chewing and certain sucking pests.
- **Mode of action :** Its primary mode of action is through ingestion, supplemented by some contact activity. By binding to ryanodine receptors in insect muscles, it triggers an uncontrolled release of calcium, leading to paralysis, cessation of feeding, lethargy, and ultimately death. This insecticide is effective against the egg, larval, and adult stages of susceptible insects.
- Notably, the structural differences in ryanodine receptors between insects and mammals result in chlorantraniliprole binding more tightly to the insect version, thereby minimizing the risk to mammals.
- Furthermore, its primary route of toxicity through ingestion helps to reduce exposure risks for applicators and beneficial insects.

Label claim (DPPQ & S, Faridabad)

Chlorantraniliprole 18.50% SC

Crop	Pests	Dose / ha		Waiting period (Days)
		Quantity of formulation (ml)	Volume of water (Lit.)	
Rice	Stem borer, Leaf folder	150	500	47

Cotton	American & Spotted bollworm, Tobacco leaf eating caterpillar	150	500	09
Sugarcane	Termite	500-625	1000	208
	Early shoot borer, Top shoot borer	375	1000	208
Pigeonpea	Pod borer, Pod fly	150	500	22
Bengal gram (Chickpea)	Pod borer	125	500	11
Black gram (Udid)	Pod borers	100	500	20
Soybean	Green semilooper, Stem fly, Girdle beetle	150	500-750	22
Tomato	Fruit borer	150	500	03
Chilli	Fruit borer	150	500	03
Brinjal	Shoot and fruit borer	200	500-750	22
Cabbage	Diamondback moth	50	500	03
Bittergaurd	Fruit borer, Caterpillars	100-125	500	05
Okra	Fruit borer	125	500	05
Maize	Spotted stem borer	200	500	10
Groundnut	Tobacco caterpillar	150	500	28
Green gram	Pod borers (Pod borer, Tobacco caterpillar, Spotted pod borer)	125	400-600	14

Trade Names

Sr. No.	Trade Name	Company
1	Coragen	DuPont Crop Protection FMC Coprporation
2	Cosko SC	PI Industries
3	Cover Liq	Dhanuka Agritech Ltd.
4	Clorentis	ATPL
5	Raise	Multiplex Bio-Tech Pvt. Ltd.
6	Shenzi	UPL Ltd.
7	Harcos	Sahib Pesticides and Seeds Ltd.
8	Helipto	GSP Crop Science Pvt. Ltd.
9	Tathastu	HPM Chemicals and Fertilizers Ltd.
10	Citigen	Best Agrolife Ltd.
11	Jumbo	United Chemical Industries
12	Mission	Insecticides India Ltd.
13	Draggon Super Power	Perfect Cropscience Pvt. Ltd.
14	Sudarshan	Sudarshan Farm Chemicals India Pvt. Ltd.
15	Coffeeza	Hok Agrichem Private Limited

16	Pepora	Titan Agritech Limited
17	Vamos	Sulphur Mills Limited
18	Klogen	Shivalik Crop Sciences Pvt. Ltd.
19	Tarzan	Krishi Rasayan
20	Exyprole	Torrent Crop Science
21	Gateway	Tata Rallis
22	Vortex	Swastik Agri Science
23	OMG	Gharda Chemicals Limited
24	Chakraveer	Katyayani Organics
25	Yunico Flow	Sumitomo Chemical India Ltd.
26	Kinsta SC	Swal Corporation Ltd.
27	Catgen	Chemley Agritech Private Limited
28	Cargar	Essential Biosciences
29	Arashi	Nichino India Pvt. Ltd.
30	Power Ctrl	Berrysun Agro Science Pvt. Ltd.
31	Carzone	R K Chemicals
32	Cosayr	ADAMA India Private Limited
33	Pevota	Chambal Fertilizers and Chemicals Limited
34	Tagzen	Tropical Agro
35	Aconite	Today's Global Crop Protection
36	Cygnus	FORBES Impex & Agricare
37	Natgen	NATCO Pharma Limited
38	Conical	Shree Ram Agro

Chlorantraniliprole 00.40% GR

Crop	Pests	Dose / ha		Waiting period (Days)
		Quantity of formulation (Kg)	Application	
Rice	Yellow stem borer, Leaf folder	10.00	Broadcast	53
Sugarcane	Early shoot borer, Top shoot borer	18.75	Broadcast	147
Maize	Fall armyworm, stem borer, Pink stem borer	12.50	Whorl application	65

Trade Names

Sr. No.	Trade Name	Company
1	Ferterra	FMC Corporation
2	Cosko GR	PI Industries
3	Cover Gr	Dhanuka Agritech Ltd.
4	Enfuse	ADAMA India Private Limited
5	Vistara	Best Agrolife Ltd.
6	Cellterra	Cellgene Bio Science
7	Solid	Anmol Biotech Pvt. Ltd.
8	Fatat	Spiral Crop Inductry
9	Forex	Angel Crop Science

10	Faster	Agri Science Ltd.
11	Fantastic	Katyayani Organics
12	Fastek	Tatva Crop Science
13	Fortune	Perfect Crop Science
14	Smash GR	Global Crop care
15	Cover king	Agriwell Crop Science
16	Fallcon	Green Agritech
17	Silterra	Silver Sine Bio Tech
18	Fattera+	Natural Crop Science
19	Zen GR	Tropical Agrosystem India (P) Ltd.
20	Fertap-Gold	Essential Biosciences
21	Syntera	Synwood Agro Ltd.
22	Krinox	Krishi Rasayan
23	Gateway GR	Tata Rallis India Ltd.
24	JU Farrera	JU Agri Sciences Pvt. Ltd.
25	Catgen GR	Chemley Agritech Pvt. Ltd.

Chlorantraniliprole 00.53% GR

Crop	Pests	Dose / ha		Waiting period (Days)
		Quantity of formulation (Kg)	Application	
Rice	Yellow stem borer, Leaf folder	7.50	Broadcast	63

Trade Names

Sr. No.	Trade Name	Company
1		

Chlorantraniliprole 05.00% DT

Crop	Pests	Dose / ha		Waiting period (Days)
		Quantity of formulation (No.)	Volume of water (Lit.)	
Rice	Yellow stem borer, Leaf folder	1000-1250 Tablets		47

Trade Names

Sr. No.	Trade Name	Company
1		

Chlorantraniliprole 35.00% WG

Crop	Pests	Dose / ha		Waiting period (Days)
		Quantity of formulation (g)	Volume of water (Lit.)	
Okra	Fruit borer, Shoot and fruit borer	71	500	05
Tomato	Fruit borer	86	500	03

Trade Names

Sr. No.	Trade Name	Company
1	Corprima	FMC Corporation

Chlorantraniliprole 50.00% FS

Crop	Pests	Dose / Kg seed		Waiting period (Days)
		Quantity of formulation (ml)	Volume of water (Lit.)	
Rice	Yellow stem borer, Leaf folder	1.2	--	--
Maize	Spotted stem borer	0.64	--	--
	Fall armyworm	1.12		

Trade Names

Sr. No.	Trade Name	Company
1	Lumivia	Corteva Agriscience

Chlorantraniliprole 47.85 % SC

Crop	Pests	Dose / ha		Waiting period (Days)
		Quantity of formulation (ml)	Volume of water (Lit.)	
Red gram	Pod borer, Spotted pod borer, Pod fly	50	500-750	31
Chickpea	Pod borer Tobacco caterpillar	41.66	500	26
Sugarcane	Early shoot borer, Top borer, Stalk borer	125	1000	214
Soybean	Tobacco caterpillar, Green semilooper, Girdle beetle, Stem fly	50	500	44
Maize	Fall armyworm, Stem borer, Pink stem borer	67-83	500	87
Rice	Yellow stem borer, Rice leaf folder	50	500	47
Cotton	American bollworm, Tobacco caterpillar	50	500	67
Chilli	Fruit borer, Tobacco caterpillar, Beet armyworm	50	500	03
Groundnut	Tobacco caterpillar, Leaf miner	50	400-500	34
Tomato	Tobacco caterpillar, Fruit borer	50	400-500	03
Green gram	Pod borer, Tobacco caterpillar, Green semilooper, Spotted pod borer	41.67	400-600	26

Trade Names

Sr. No.	Trade Name	Company
1	Tuventa	FMC Corporation

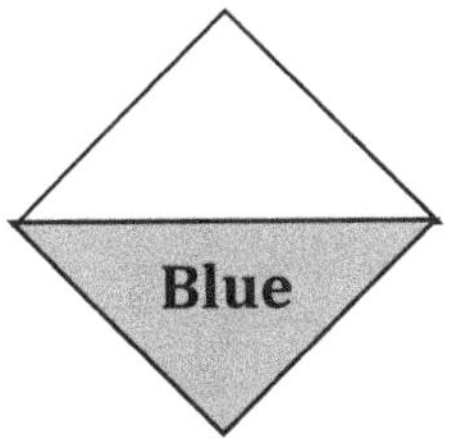

CHLORFENAPYR

Available formulations in India : 1. **Chlorfenapyr 10.00 % SC**
2. **Chlorfenapyr 240 g/L SC**

Classification on the basis of

Target organism	:	Insecticide, Acaricide
Mode of entry	:	Stomach and Contact poison
Mode of action	:	Energy metabolism
Chemical group	:	Pyrroles

IRAC classification

Main Group and Primary Site of Action	:	13 Uncouplers of oxidative phosphorylation via disruption of the proton gradient
Sub-group or exemplifying Active Ingredient	:	Pyrroles

Category of toxicity : Moderately toxic
Toxicity label colour : Blue

- Chlorfenapyr, a pro-insecticide in the pyrrole class, requires metabolic activation within the insect to exert its effects.
- **Mode of Action:** Chlorfenapyr disrupts the production of adenosine triphosphate (ATP) in insect cells. Following ingestion or contact, it is metabolized into an active compound (CL 303268). This active metabolite uncouples oxidative phosphorylation in the mitochondria, leading to energy depletion, cellular death, and ultimately, insect mortality.
- This novel mode of action makes chlorfenapyr effective against pests resistant to common insecticide classes like pyrethroids, carbamates, and organophosphates, rendering it a valuable tool for resistance management.
- As a broad-spectrum insecticide and acaricide, chlorfenapyr effectively controls various pests. The combination of its translaminar movement (ability to penetrate plant leaves) and unique mode of action typically results in pest mortality within 72 hours.
- Chlorfenapyr's development originates from dioxapyrrolomycin, a natural product discovered in the bacterium *Streptomyces fumanus*.
- Chlorfenapyr is reported to be stable to hydrolysis at pH 4, 7, and 9 even at elevated temperatures (e.g., 50°C).

Label claim (DPPQ & S, Faridabad)

Chlorfenapyr 10.00 % SC

Crop	Pests	Dose / ha		Waiting period (Days)
		Quantity of formulation (ml)	Volume of water (Lit.)	
Cabbage	Diamondback moth	750-1000	500	07
Chilli	Yellow mite	750-1000	500	05

Trade Names

Sr. No.	Trade Name	Company
1	Intrepid	BASF Crop Protection
2	Microbs	Global Crop Care
3	Clodihuge	Mankind Agritech

Chlorfenapyr 240 g/L SC

Crop	Pests	Dose / ha		Waiting period (Days)
		Quantity of formulation (ml)	Volume of water (Lit.)	
Red gram	Pod borer, Spotted pod borer	800-1000	500	41
Cotton	Tobacco caterpillar, Thrips	312.5-425	500	48
Chickpea	Pod borer	800-1000	500	49
Brinjal	Shoot and fruit borer, Mite	800-1000	500	10
Chilli	Mite, Thrips, Tobacco caterpillar	312.5-425	500	07
	Fruit borer	800-1000	500	07
Soybean	Tobacco caterpillar, Green semilooper	312.5-425	500	21

Trade Names

Sr. No.	Trade Name	Company
1	Implicite	Parijat Industries (India) Ltd.
2	Pirate	BASF India Limited
3	Krid-24	Krishi Rasayan

CHLORFLUAZURON

Available formulations in India : **Chlorfluazuron 05.40 % EC**

Classification on the basis of

Target organism : Insecticide
Mode of entry : Stomach and Contact poison
Mode of action : Insect Growth Regulator (IGR)
Chemical group : Benzoylureas

IRAC classification

Main Group and Primary Site of Action : 15 Inhibitors of chitin biosynthesis affecting CHS1
Sub-group or exemplifying Active Ingredient : Benzoylureas

Category of toxicity : Slightly toxic
Toxicity label colour : Green

- Chlorfluazuron, an insect growth regulator (IGR) in the benzoylurea class discovered and developed by Ishihara Sangyo Kaisha (ISK) in the late 1980s, acts as an anti-molting agent.
- **Mode of action :** Its primary mode of action involves inhibiting chitin biosynthesis, a vital component of the insect cuticle (exoskeleton). This interference disrupts the molting process, causing a loss of cuticle elasticity and firmness, ultimately leading to abortive molting and the death of insect larvae and pupae.
- Primarily a stomach poison effective through ingestion, Chlorfluazuron also exhibits some contact activity at higher rates.
- While non-systemic, it demonstrates translaminar movement, allowing it to penetrate leaves and affect pests on the underside.
- Characterized by a slower action compared to neurotoxic insecticides due to its reliance on the insect's molting cycle, Chlorfluazuron offers the benefit of long residual activity in controlling various chewing pests, particularly Lepidopteran larvae (caterpillars), across a wide range of crops.

Label claim (DPPQ & S, Faridabad)

Chlorfluazuron 05.40 % EC

Crop	Pests	Dose / ha		Waiting period (Days)
		Quantity of formulation (ml)	Volume of water (Lit.)	
Cotton	American bollworm, Tobacco caterpillar	1500-2000	500	10
Cabbage	Diamondback moth, Tobacco caterpillar	1500	500	07

Soybean	Gram pod borer, Tobacco caterpillar	1500	500	34
Chilli	Fruit borer, Tobacco caterpillar	2000	500	03
Tomato	Fruit borer, Tobacco caterpillar	2000	500	03

Trade Names

Sr. No.	Trade Name	Company
1	Atabron	UPL Ltd.
2	DSP-99	Swal Corporation Ltd.

Yellow

CHLORPYRIFOS

Available formulations in India :

1. **Chlorpyrifos 10.00 % G**
2. **Chlorpyrifos 75.00 % WG**
3. **Chlorpyrifos 19.00 % ME**
4. **Chlorpyrifos 20.00 % EC**
5. **Chlorpyrifos 50.00 % EC**
6. **Chlorpyrifos 01.50 % DP**

Classification on the basis of

Target organism : Insecticide
Mode of entry : Stomach, Contact and Respiratory poison
Mode of action : Nerve poison
Chemical group : Organophosphates

IRAC classification

Main Group and Primary Site of Action : 1 Acetylcholinesterase (AChE) inhibitors
Sub-group or exemplifying Active Ingredient : 1B Organophosphates

Category of toxicity : Highly toxic
Toxicity label colour : Yellow

- Chlorpyrifos is an organophosphorus compound exhibiting insecticidal and acaricidal properties. It finds widespread application in agricultural and select non-agricultural pest management strategies.
- **Mode of action :** The primary mode of action involves the irreversible inhibition of acetylcholinesterase (AChE), a key enzyme mediating cholinergic neurotransmission in insects (and also in mammals). This inhibition results in the synaptic accumulation of acetylcholine, leading to cholinergic crisis characterized by nervous system overstimulation, culminating in paralysis and mortality in target insect populations.
- Chlorpyrifos demonstrates contact, oral, and respiratory toxicity against susceptible pests.

Label claim (DPPQ & S, Faridabad)

Chlorpyrifos 10.00% GR

Crop	Pests	Dose / ha		Waiting period (Days)
		Quantity of formulation (Kg)	Volume of water (Lit.)	
Rice	Stem borer, Leaf folder, Gall midge	10.00	--	30

Trade Names

Sr. No.	Trade Name	Company
1	Starban	Swal Corporation Ltd.

2	Pratap	Sydney Formulation & Chemicals
3	Goldban GR	Crop Chemicals India Ltd.
4	Bansik-GR	Sikko Industries Ltd.
5	Force-10	Silver Sine Bio Tech
6	Chlormex-10G	Natural Crop Science
7	Chloro GR	Katyayani Organics
8	Tagban Gold	Tropical Agrosystem India (P) Ltd.
9	Seylon-GR	Synwood Agro Ltd.
10	Chloridon	Preventis Chem India (P) Ltd.
11	Termiphos 10 GR	RACCOLTO Agritech India Pvt. Ltd.

Chlorpyrifos 75.00% WG

Crop	Pests	Dose / ha		Waiting period (Days)
		Quantity of formulation (g)	Volume of water (Lit.)	
Rice	Yellow stem borer	500-533	500-1000	15

Trade Names

Sr. No.	Trade Name	Company
1	Made India	Om Agro Organics

Chlorpyrifos 19.00% ME

Crop	Pests	Dose / ha		Waiting period (Days)
		Quantity of formulation (ml)	Volume of water (Lit.)	
Rice	Stem borer	1425	500	30

Trade Names

Sr. No.	Trade Name	Company
1	Chlorpyrifos 19.00% ME	Hindustan Crop Science

Chlorpyrifos 20.00% EC

Crop	Pests	Dose / ha		Waiting period (Days)
		Quantity of formulation (ml)	Volume of water (Lit.)	
Rice	Stem borer, Hispa, Gall midge, Whorl maggot	1250	500-1000	--
	Leaf folder	1875	500-1000	--
Sugarcane	Early shoot borer, Stalk borer	1250-1500	500-1000	--
	Pyrilla	1500	500-1000	--
	Black bug	750	500-1000	--
Cotton	Aphid, Whitefly, Bollworm	1250	500-1000	--
	Cutworm	3750	500-1000	--
Groundnut	Aphid	1000	500-1000	--
	Root grub	1125	500-1000	--

Mustard	Aphid	500	500-1000	--
Gram	Cutworm	2500	500-1000	--
Beans	Pod borer, Black bug	3000	500-1000	--
Brinjal	Shoot and fruit borer	1000	500-1000	--
Cabbage	Diamondback moth	2000	500-1000	--
Onion	Root grub	5000	500-1000	--
Apple	Aphid	3750-5000	1500-2000	--
Ber	Leaf hopper	2250-3000	1500-2000	--
Citrus	Black citrus aphid	1500-2000	1500-2000	--
Tobacco	Ground beetle	1750	500-1000	--
Termite control	▪ **Non cropped area** : Building (Pre & Post construction treatment) and Forestry @1.0% a.i. (500 ml in 10 lit. water) ▪ **Cropped area** : Wheat (3-4 ml/Kg seed), Barley (4-6 ml/Kg seed), Gram (15-30 ml/Kg seed) ▪ **Soil treatment** : Wheat (2-3 lit/ha.), Sugarcane (6.25 lit/ha.)			

Trade Names

Sr. No.	Trade Name	Company
1	Dursban	Crystal Crop Protection Ltd.
2	Premain	ADAMA India Private Limited
3	Tafaban	Rallis India Limited
4	Strike 20 EC	Biostadt India Ltd.
5	Chlorguard 20 EC, Navigator	Gharda Chemicals Ltd.
6	Trishul	Coromandel International Ltd.
7	Tricel	Excel Crop Care Ltd.
8	CP Gold	Alfa Crop Science
9	Clorex-20	Auxin Crop Care
10	Clearout	Anu Products Ltd.
11	Lethal TC	Insecticides (India) Ltd.
12	Linker 20	Modern Insecticides Ltd.
13	Force	NACL Industries Ltd.
14	Deviban	Devidayal Agro Chemicals Ltd.
15	Chlorasis	Sydney Formulation & Chemicals
16	Goldban-20	Crop Chemicals India Ltd.
17	Kemtrek EC	Sumitomo Chemical India Ltd.
18	Suldrin	Super Crop Safe Ltd., Ahemadabad
19	Radar	Isagro Asia Agrochemicals Pvt. Ltd.
20	Chlorotop 20 TC	Gujarat Pesti-Chem Industries
21	Clorex	GFG Crop Science Pvt. Ltd.
22	Contra	Tusova Agro Pvt. Ltd.
23	Heraban	Herenba Industries Ltd.
24	Prahar	Gujarat Crop Care
25	Sikkoban	Sikko Industries Ltd.
26	Fotis	Nichino India
27	Darban 20	Silver Sine Bio Tech

28	Clormex	Natural Crop Science
20	Chloro 20	Katyayani Organics
21	Tagban	Tropical Agrosystem India (P) Ltd.
22	Chlormites	Essential Biosciences
22	Syn-20	Synwood Agro Ltd.
23	Akhanda 200	Koronation Bio-Science
24	Terminator	Agrosis
25	Kri-Shan	Krishi Rasayan
26	Dawco	Global Crop Care
27	Chloro-20	JU Agri Sciences Pvt. Ltd.
28	Clone-20	Linuxcrop Chemicals Pvt. Ltd.
29	SA-Safaban	Saraswati Agro Life Science (India) Pvt. Ltd.
30	Catoban	Chemley Agritech Pvt. Ltd.
31	Termiphos-TC	RACCOLTO Agritech India Pvt. Ltd.

Chlorpyrifos 50.00% EC

Crop	Pests	Dose / ha		Waiting period (Days)
		Quantity of formulation (ml)	Volume of water (Lit.)	
Rice	Stem borer, Leaf folder	750-800	500-1000	15
Cotton	Bollworms	1000-1200	500-1000	30

- **For non- agricultural use :** For protecting building from termite attack at pre and post construction stages, apply Chlorpyriphos 50% EC @ 0.5% and 1.0% concentration (100 ml and 200 ml / 10 lit. water).

Trade Names

Sr. No.	Trade Name	Company
1	Predator	Crystal Crop Protection Ltd. Dow AgroSciences India Pvt. Ltd.
2	Hibiki	IFFCO-MC Crop Science Private Ltd.
3	Ecoguard	Gharda Chemicals Ltd.
4	Integer	Coromandel International Ltd.
5	Durmet Force	FMC Corporation
6	Action 500	Tropical Agrosystem India (P) Ltd.
7	Clorex-505	Auxin Crop Care
8	Clearout 50	Anu Products Ltd.
9	Tricel Power	Excel Crop Care Ltd.
10	Chlofos	Sydney Formulation & Chemicals
11	Killer 5000	Crop Chemicals India Ltd.
12	Kemtrek Force EC	Sumitomo Chemical India Ltd.
13	Hem 5	Hemani Industries Ltd., Ankleshwar
14	Terror RL	Isagro Asia Agrochemicals Pvt. Ltd.
15	Troban	Isagro Asia Agrochemicals Pvt. Ltd.
16	Vardhan	Herenba Industries Ltd.
17	Prahar Plus	Gujarat Crop Care
18	Chlorsik	Sikko Industries Ltd.

19	Xact	Nichino India
20	Darban	Silver Sine Bio Tech
21	Clormex Silver	Natural Crop Science
22	Chloro 50	Katyayani Organics
23	Action-500	Tropical Agrosystem India (P) Ltd.
24	Chlormites pro	Essential Biosciences
25	Syn-50	Synwood Agro Ltd.
26	Harrier	Indian Agrochemicals
27	Impact	Global Crop Care
28	Chloro-50	JU Agri Sciences Pvt. Ltd.
29	Clone-50	Linuxcrop Chemicals Pvt. Ltd.
30	Exylon Cyclone 50	Torrent Crop Science
31	Strong 50 EC	Preventis Chem India (P) Ltd.
32	Humble-50	RACCOLTO Agritech India Pvt. Ltd.

Chlorpyrifos 01.50% DP

Crop	Pests	Dose / ha		Waiting period (Days)
		Quantity of formulation (Kg)	Volume of water (Lit.)	
Rice	Stem borer, Green leaf hopper, Brown plant hopper, Leaf folder, Gall midge, Grasshopper	25.00	--	07
Bengal gram / Chickpea	Pod borer	25.00	--	07

Trade Names

Sr. No.	Trade Name	Company
1	Chlorodol	Anu Products Ltd.
2	Deviban 1.5 DP	Arysta Lifescience
3	Chloro-DP	Sikko Industries Ltd.
4	Hindol	HPM Chemicals & Fertilizers Ltd.
5	Chloro Dust	Katyayani Organics
6	Impact DP	Global Crop Care
7	Termi-Dust	RACCOLTO Agritech India Pvt. Ltd.

Blue

CHROMAFENOZIDE

Available formulations in India : **Chromafenozide 80.00 % WP**

Classification on the basis of

Target organism : Insecticide
Mode of entry : Stomach and Contact poison
Mode of action : Growth regulator
Chemical group : Bisacylhydrazines

IRAC classification

Main Group and Primary Site of Action : 18 Ecdysone receptor agonists
Sub-group or exemplifying Active Ingredient : Diacylhydrazines

Category of toxicity : Moderately toxic
Toxicity label colour : Blue

- Chromafenozide, a synthetic insecticide of the bisacylhydrazine class, targets a wide range of lepidopteran pests in agriculture.
- **Mode of action :** Its mechanism of action centers on its role as an ecdysone receptor agonist. By mimicking the insect molting hormone, ecdysone, chromafenozide binds to the ecdysone receptors in susceptible larvae. This binding initiates a premature and incomplete molt, a disruptive process that halts development, stops feeding, causes paralysis, and ultimately leads to death.
- The insecticide primarily works through ingestion and also has contact activity, often exhibiting selectivity towards lepidopteran larvae with a low impact on many beneficial insects, mammals, and birds.

Label claim (DPPQ & S, Faridabad)
Chromafenozide 80.00 % WP

Crop	Pests	Dose / ha		Waiting period (Days)
		Quantity of formulation (g)	Volume of water (Lit.)	
Rice	Stem borer, Leaf folder	94-125	500	32

Trade Names

Sr. No.	Trade Name	Company
1	Dodger	PI Industries Ltd.

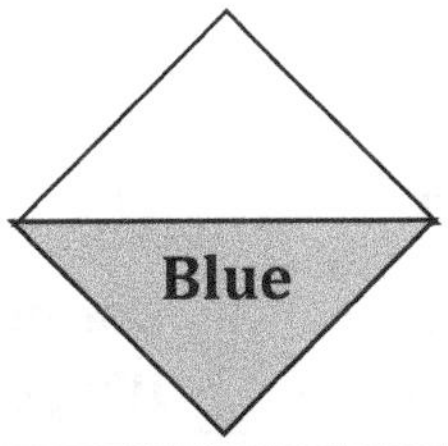

CLOTHIANIDIN

Available formulations in India :
1. **Clothianidin 00.50 % GR**
2. **Clothianidin 50.00 % WDG**

Classification on the basis of

Target organism : Insecticide
Mode of entry : Systemic, Stomach and Contact poison
Mode of action : Nerve poison
Chemical group : Neonicotinoid

IRAC classification

Main Group and Primary Site of Action : 4 Nicotinic acetylcholine receptor (nAchR) competitive modulators
Sub-group or exemplifying Active Ingredient : 4A Neonicotinoids

Category of toxicity : Moderately toxic
Toxicity label colour : Blue

- Clothianidin, a neonicotinoid insecticide from Takeda and Bayer, shares chemical similarities with nicotine, thiamethoxam, and imidacloprid.
- **Mode of action :** It functions by acting as an agonist at the nicotinic acetylcholine receptor (nAChR) in the insect central nervous system. This binding disrupts normal nerve signal transmission, causing overstimulation, paralysis, and death, as clothianidin is not susceptible to breakdown by acetylcholinesterase.
- Notably, clothianidin exhibits translaminar and root-systemic activity, enabling it to move throughout the plant after application, offering comprehensive protection against a broad spectrum of sucking and chewing pests in both agricultural and non-agricultural uses through ingestion or contact.

Label claim (DPPQ & S, Faridabad)

Clothianidin 00.50 % GR

Crop	Pests	Dose / ha		Waiting period (Days)
		Quantity of formulation (Kg)	Volume of water (Lit.)	
Okra	Jassid, Whitefly	8-12	--	01

Trade Names

Sr. No.	Trade Name	Company
1	Santana	Sumitomo Chemical India Pvt. Ltd.

Clothianidin 50.00 % WDG

Crop	Pests	Dose / ha		Waiting period (Days)
		Quantity of formulation (g)	Volume of water (Lit.)	
Rice	Brown plant hopper	20-24	500	12
Cotton	Jassid	30-40	500	20
	Whitefly	40-50	500	20
Cotton (Soil drench)	Aphid, Jassid, Thrips, Whitefly	200-250	1000	76
Sugarcane (Soil drench)	Termite	250	1000	310
Tea	Mosquito bug	120	500	05
Grapes (Soil drench)	Thrips, Jassid, Mealybug	250	1000	15
Groundnut (Soil drench)	White grub	250	1000	62

Trade Names

Sr. No.	Trade Name	Company
1	Dantotsu	Sumitomo Chemical India Pvt. Ltd.
2	Ghotu	Best Agrolife Ltd.
3	Den	Tropical Agrosystem India (P) Ltd.
4	Clothiright	Mankind Agritech
5	Dotsu 50	Chemley Agritech Pvt. Ltd.

Green

CYANTRANILIPROLE

Available formulations in India :
1. **Cyantraniliprole 10.26 % OD**
2. **Cyantraniliprole 48.00 % FS**
3. **Cyantraniliprole 18.41 % SC**

Classification on the basis of

Target organism : Insecticide
Mode of entry : Systemic, Stomach and Contact poison
Mode of action : Nerve and muscle poison
Chemical group : Diamides

IRAC classification

Main Group and Primary Site of Action : 28 Ryanodine receptor modulators
Sub-group or exemplifying Active Ingredient : Diamides

Category of toxicity : Slightly toxic
Toxicity label colour : Green

- Cyantraniliprole, a modern insecticide in the anthranilic diamide class, provides effective control against a broad spectrum of chewing and sucking pests across numerous crops.
- **Mode of action :** Its primary mode of action involves disrupting the insect's nervous system by acting as a ryanodine receptor modulator. Specifically, it binds to and activates these receptors in insect muscle cells, triggering an uncontrolled release of calcium from internal stores. This leads to sustained muscle contraction, resulting in paralysis and subsequent mortality of the pest.
- While exhibiting both systemic and contact activity, cyantraniliprole demonstrates greater potency when ingested by the target insect.
- Its unique mode of action makes it a valuable tool in insecticide resistance management strategies, particularly when used in rotation with insecticides possessing different mechanisms of action.

Label claim (DPPQ & S, Faridabad)

Cyantraniliprole 10.26 % OD

Crop	Pests	Dose / ha		Waiting period (Days)
		Quantity of formulation (ml)	Volume of water (Lit.)	
Grape	Thrips, Flea beetle	700	1000	05
Pomegranate	Thrips, Pomegranate butterfly	750	1000	05
	Whitefly, Aphid	900	1000	05

Cabbage	Cabbage Aphid, Mustard Aphid, Diamond back moth, Tobacco caterpillar	600	500	05
Chilli	Thrips, Fruit borer, Tobacco caterpillar	600	500	03
Tomato	Leaf miner, Aphid, Thrips, Whitefly, Fruit borer	900	500	03
Gherkins	Leaf miner, Red pumpkin beetle, Aphid, Thrips, Whitefly, Pumpkin caterpillar, Fruit fly	900	500	05
Okra	Whitefly, Aphid, Shoot and fruit borer, Tobacco caterpillar, Fruit borer	900	500	03
Brinjal	Whitefly, Aphid, Thrips, Shoot and fruit borer	900	500	03
Cotton	Whitefly, Aphid, Thrips, Tobacco caterpillar, American bollworm, Spotted bollworm	900	500	07
Bittergourd, Ridgegourd	Leaf miner, Aphid, Thrips, Whitefly, Pumpkin caterpillar	900	500	05
Watermelon	Leaf miner, Aphid, Thrips, Whitefly	900	500	05
Potato	Thrips, Aphid	600	500	12
Capsicum	Thrips, Tobacco caterpillar, Fruit borer	600	500	03
Citrus	Citrus psylla, Thrips	600	500	05

Trade Names

Sr. No.	Trade Name	Company
1	Benevia	FMC Corporation
2	Diamide	Global Crop Care

Cyantraniliprole 48.00 % FS

Crop	Pests	Dose / Kg seed		Waiting period (Days)
		Quantity of formulation (ml)	Volume of water (Lit.)	
Maize	Fall armyworm	3	--	--
	Stem borer, Cutworm	2	--	--

Trade Names

Sr. No.	Trade Name	Company
1	Fortenza	Syngenta India Pvt. Ltd.

Cyantraniliprole 18.41 % SC

Crop	Pests	Dose / 1000 seedlings		Waiting period (Days)
		Quantity of formulation (ml)	Volume of water (Lit.)	
Chilli	Thrips, Whitefly, Tobacco caterpillar	43.75	10	80
Capsicum	Thrips, Aphid, Whitefly, Leaf miner, Tobacco caterpillar	43.75	10	61
Tomato	Thrips, Aphid, Whitefly, Leaf miner	43.75	10	75
Cucumber	Thrips, Aphid, Whitefly, Leaf miner, Cucumber moth	62.50	50	42
Watermelon	Thrips, Aphid, Whitefly, Leaf miner	75	50	47

Trade Names

Sr. No.	Trade Name	Company
1		

Green

CYCLANILIPROLE

Available formulations in India :
1. **Cyclaniliprole 09.30 % DC (10.00 % DC)**
2. **Cyclaniliprole 08.00 % SL**

Classification on the basis of

Target organism : Insecticide
Mode of entry : Stomach and Contact poison
Mode of action : Nerve and muscle poison
Chemical group : Diamides

IRAC classification

Main Group and Primary Site of Action : 28 Ryanodine receptor modulators
Sub-group or exemplifying Active Ingredient : Diamides

Category of toxicity : Slightly toxic
Toxicity label colour : Green

- Cyclaniliprole is a synthetic insecticide belonging to the diamide and pyridylpyrazole chemical classes. It's used to control a wide range of insect pests on various crops, including fruits, vegetables, and ornamental plants.
- **Mode of action :** This insecticide works by targeting the ryanodine receptor (RyR) in an insect's muscle cells. The RyR is a crucial calcium channel that regulates muscle contraction. When cyclaniliprole binds to this receptor, it forces the channel to open, causing an uncontrolled release of calcium ions. This leads to immediate feeding cessation, muscle paralysis, and ultimately, the death of the insect. Because this mode of action is highly specific to insects, cyclaniliprole has very little effect on the RyR of mammals.
- Cyclaniliprole is a robust and stable compound that is not significantly affected by the pH of the water (4-9 pH) used for spraying, which helps ensure its consistent performance in the field.

Label claim (DPPQ & S, Faridabad)

Cyclaniliprole 09.30 % DC (10.00 % DC)

Crop	Pests	Dose / ha		Waiting period (Days)
		Quantity of formulation (ml)	Volume of water (Lit.)	
Rice	Stem borer, Leaf folder	350-400	500	40

Trade Names

Sr. No.	Trade Name	Company
1	Prachand	Koromandal International
2	Pressido	Pil Industries Ltd.

Cyclaniliprole 08.00 % SC

Crop	Pests	Dose / ha		Waiting period (Days)
		Quantity of formulation (ml)	Volume of water (Lit.)	
Rice	Stem borer, Leaf folder	437.5-500	500	30

Trade Names

Sr. No.	Trade Name	Company
1		

Yellow

CYPERMETHRIN

Available formulations in India :
1. **Cypermethrin 00.25 % DP**
2. **Cypermethrin 10.00 % EC**
3. **Cypermethrin 25.00 % EC**

Classification on the basis of

Target organism : Insecticide, Acaricide
Mode of entry : Stomach and Contact poison
Mode of action : Nerve poison
Chemical group : Pyrethroid

IRAC classification

Main Group and Primary Site of Action : 3 Sodium channel modulators
Sub-group or exemplifying Active Ingredient : 3A Pyrethroid

Category of toxicity : Highly toxic
Toxicity label colour : Yellow

- Cypermethrin is a widely used synthetic pyrethroid insecticide and acaricide. It's effective against a broad range of insect pests in agriculture, households, and animal health.
- **Mode of action :** Cypermethrin is a non-systemic insecticide with contact and stomach action. It acts as a fast-acting neurotoxin in insects by disrupting the normal functioning of the sodium channels in their nervous systems. This prolonged opening of sodium channels leads to hyperexcitability, causing muscle spasms, paralysis, and ultimately, the death of the insect. At higher concentrations, it may also affect GABA-gated chloride channels.

Label claim (DPPQ & S, Faridabad)

Cypermethrin 00.25% DP

Crop	Pests	Dose / ha		Waiting period (Days)
		Quantity of formulation (Kg)	Volume of water (Lit.)	
Brinjal	Shoot and fruit borer	20.00 - 24.00	--	03

Trade Names

Sr. No.	Trade Name	Company
1	Cis-Fire	Sydney Formulation & Chemicals
2	Cypra Dust	Herenba Industries Ltd.
3	Sikkorin-DP	Sikko Industries Ltd.
4	Cyper Dust	Katyayani Organics
5	Pralaya	Essential Biosciences
6	Cyper Dust	RACCOLTO Agritech India Pvt. Ltd.

Cypermethrin 10.00% EC

Crop	Pests	Dose / ha		Waiting period (Days)
		Quantity of formulation (ml)	Volume of water (Lit.)	
Cotton	American, Spotted and Pink bollworms	550-760	150-1000	07
Wheat	Shoot fly	550	500-800	14
Sunflower	Bihar hairy caterpillar	550-760	500-700	14
Cabbage	Diamondback moth	650-760	100-400	07
Okra	Fruit borer	550-760	100-400	03
Brinjal	Shoot and fruit borer	550-760	100-400	03

Trade Names

Sr. No.	Trade Name	Company
1	Ustaad	UPL Ltd.
2	Cyperguard 10 EC	Gharda Chemicals Ltd.
3	Superkiller 10%EC	Dhanuka Agritech Ltd.
4	Challenger 10	Tropical Agrosystem India (P) Ltd.
5	Cypero-10	Auxin Crop Care
6	Anumite	Anu Products Ltd.
7	Ripple 10 EC	Modern Insecticides Ltd.
8	Devicyper 10	Arysta Lifescience
9	Goldcyp-10	Crop Chemicals India Ltd.
10	Hemthrin	Hemani Industries Ltd., Ankleshwar
11	Lacer	Isagro Asia Agrochemicals Pvt. Ltd.
12	Viper	Tusova Agro Pvt. Ltd.
13	Cypra	Herenba Industries Ltd.
14	Sikkorin-10	Sikko Industries Ltd.
15	Kyper 10	KP Agro Industries
16	Cyprox	Natural Crop Science
17	Ninja Cyper	Essential Biosciences
18	Cyper-10	JU Agri Sciences Pvt. Ltd.
19	Parthrin	RACCOLTO Agritech India Pvt. Ltd.

Cypermethrin 25.00% EC

Crop	Pests	Dose / ha		Waiting period (Days)
		Quantity of formulation (ml)	Volume of water (Lit.)	
Cotton	Bollworms	160-280	400-800	--
	Jassid, Thrips	80-120	200-300	--
Okra	Shoot and fruit borer, Jassid	150-200	500	03
Brinjal	Shoot & fruit borer, Jassids, *Epilachna* grub (Hadda beetle)	150-200	500	01

Trade Names

Sr. No.	Trade Name	Company
1	Colt	PI Industries
2	Auzar 25 EC	Biostadt India Ltd.
3	Cyperguard 25 EC	Gharda Chemicals Ltd.
4	Superkiller 25% EC	Dhanuka Agritech Ltd.
5	Cyperkill	Coromandel International Ltd.
6	Super Fighter, Cymbush	Syngenta India Ltd.
7	Challenger 25	Tropical Agrosystem India (P) Ltd.
8	Auxi Cyper-25	Auxin Crop Care
9	Anukill	Anu Products Ltd.
10	Ripple 25 EC	Modern Insecticides Ltd.
11	Hexakill	Crop Life Science Ltd.
12	Cyclone	Sydney Formulation & Chemicals
13	Action-25	Crop Chemicals India Ltd.
14	Hemthrin Plus	Hemani Industries Ltd., Ankleshwar
15	Trofy	Isagro Asia Agrochemicals Pvt. Ltd.
16	Corolla 25	GFG Crop Science Pvt. Ltd.
17	Viper 25	Tusova Agro Pvt. Ltd.
18	Cypra Plus	Herenba Industries Ltd.
19	Sikkorin	Sikko Industries Ltd.
20	Angel	Nichino India
21	Kyper 25	KP Agro Industries
22	Sine Bush	Silver Sine Bio Tech
23	Cyprox Super	Natural Crop Science
24	Challenger 25	Tropical Agrosystem India (P) Ltd.
25	Metcyper	Essential Biosciences
26	Bighit	Synwood Agro Ltd.
27	Cyano	Global Crop Care
28	Cyper-25	JU Agri Sciences Pvt. Ltd.
29	Exylon Cypher	Torrent Crop Science
30	Cymax 25% EC	Preventis Chem India (P) Ltd.
31	Rakothrin	RACCOLTO Agritech India Pvt. Ltd.

Yellow

DELTAMETHRIN

Available formulations in India :

1. **Deltamethrin 11.00 % EC**
2. **Deltamethrin 25.00 % Tablet**
3. **Deltamethrin 01.80 % EC**
4. **Deltamethrin 02.50 % WP**
5. **Deltamethrin 02.80 % EC**

Classification on the basis of

Target organism : Insecticide
Mode of entry : Stomach and Contact poison
Mode of action : Nerve poison
Chemical group : Pyrethroid

IRAC classification

Main Group and Primary Site of Action : 3 Sodium channel modulators
Sub-group or exemplifying Active Ingredient : 3A Pyrethroid

Category of toxicity : Highly toxic
Toxicity label colour : Yellow

- Deltamethrin is a highly effective synthetic pyrethroid insecticide, belonging to a class of human-made compounds that mimic natural pyrethrins found in chrysanthemum flowers.
- Widely utilized in agriculture, public health, and households, deltamethrin offers a broad spectrum of activity and acts relatively quickly against target pests.
- **Mode of action :** Its mode of action involves disrupting the normal function of the insect nervous system by acting as a neurotoxin. Specifically, deltamethrin interferes with the sodium channels in nerve cells, causing them to remain open for an extended duration. This prolonged opening leads to hyperexcitation, resulting in rapid paralysis (the "knockdown effect") and subsequent death of the insect.

Label claim (DPPQ & S, Faridabad)

Deltamethrin 11.00% EC

Crop	Pests	Dose / ha		Waiting period (Days)
		Quantity of formulation (ml)	Volume of water (Lit.)	
Cotton	Bollworms	125	400-600	30
Rice	Stem borer, Leaf folder, Green leaf hopper, Whorl maggot	150	500	13
	Leaf folder	150-187.5	500	13
Tea	Thrips	100	400	15
Tomato	Fruit borers	100-125	375-500	03

Okra	Fruit borers	100-125	375-500	03
Chilli	Fruit borers	175	500	05
Onion	Thrips	150	500	05

Trade Names

Sr. No.	Trade Name	Company
1	Delta	Alfa Crop Science
2	Deuce Power	Isagro Asia Agrochemicals Pvt. Ltd.
3	Delta Super	Herenba Industries Ltd.
4	Deltacrop	Plantix Crop Care, Rajkot
5	Deltasik	Sikko Industries Ltd.
6	Delta Sine	Silver Sine Bio Tech
7	Deltarin	Natural Crop Science
8	Tagcis 100	Tropical Agrosystem India (P) Ltd.
9	Del-11	Essential Biosciences
10	Tridon Plus	Synwood Agro Ltd.
11	Prime	Global Crop Care
12	Descert	Green India Care
13	Fire	Linuxcrop Chemicals Pvt. Ltd.
14	Boxerr	Gharda Chemicals Ltd.
15	Agrikind Delta	Mankind Agritech
16	Exylon Defender	Torrent Crop Science
17	Tugon	RACCOLTO Agritech India Pvt. Ltd.

Deltamethrin 25.00% Tablet

Crop	Pests	Dose / ha		Waiting period (Days)
		Quantity of formulation (g)	Volume of water (Lit.)	
Cotton	Bollworms	50	400-600	30

Trade Names

Sr. No.	Trade Name	Company
1	Deltamethrin 25% T	Tagros Chemicals India Pvt. Ltd.
2	Hogotab	RACCOLTO Agritech India Pvt. Ltd.

Deltamethrin 01.80% EC

Crop	Pests	Dose / ha		Waiting period (Days)
		Quantity of formulation (ml)	Volume of water (Lit.)	
Cotton	Bollworms	781	400-600	30
	Jassid	625	400-600	30
Rice	Stem borer, Leaf folder	625-780	500	07

Trade Names

Sr. No.	Trade Name	Company
1	Delrox	Chemet Wets & Flows Pvt. Ltd.
2	Hogo	RACCOLTO Agritech India Pvt. Ltd.

Deltamethrin 02.50% WP

Crop	Pests	Dose / ha		Waiting period (Days)
		Quantity of formulation (g)	Volume of water (Lit.)	
Wheat & Rice (Grain & seed in stacks)	Rice weevil, Leaser grain borer, Khapra beetle, Red flour beetle, Saw toothed grain beetle, Rice moth, Almond moth	1200	1 lit. / 30 m^2	--
Walls, ceilings, floors of Godowns	Rice weevil, Leaser grain borer, Khapra beetle, Red flour beetle, Saw toothed grain beetle, Rice moth, Almond moth	1200	1.5-2.5 lit. / 50 m^2	--
Public health	Mosquito	25000-50000	--	--

Trade Names

Sr. No.	Trade Name	Company
1	Deltaflow	Herenba Industries Ltd.
2	Chemothrine	Chemet Wets & Flows Pvt. Ltd.
3	Savier	Kalyani Industries Limited
4	Del-25	Katyayani Organics
5	T-Delta	Tropical Agrosystem India (P) Ltd.
6	Power-25	Krishi Rasayan
7	Liner	Green India Care
8	Setai	RACCOLTO Agritech India Pvt. Ltd.

Deltamethrin 02.80% EC

Crop	Pests	Dose / ha		Waiting period (Days)
		Quantity of formulation (ml)	Volume of water (Lit.)	
Cotton	Bollworm	500	400-600	--
	Sucking insects	400	400-600	--
Groundnut	Leaf miner	500	400-600	03
Pigeonpea (Tur / Red gram)	Pod borer, Pod fly	500	500	10
Okra	Shoot and fruit borer	400-600	400-600	01
	Jassid	400	400-600	01
Chilli	Fruit borer	400-500	400-600	05
Brinjal	Shoot and fruit borer	400-500	400-600	03
Tea	Thrips, caterpillar	120-150	400-600	03
	Leaf folder	400	400-600	03
	Looper	100-150	400-600	03
Mango	Hopper	0.33-0.5 ml / lit.	As per requirement	01

Trade Names

Sr. No.	Trade Name	Company
1	Decis	Bayer Crop Science Ltd.
2	Decaguard	Gharda Chemicals Ltd.
3	Chemistar 2.8 EC	Modern Insecticides Ltd.
4	Dose	Crop Chemicals India Ltd.
5	Deuce	Isagro Asia Agrochemicals Pvt. Ltd.
6	Deltajet	Herenba Industries Ltd.
7	Delta-208	Sgree Chem India
8	Deltahit	Ganesh Chemical Industries
9	Deltar	Arbuda Agrochemicals Pvt. Ltd.
10	San Delta Super	Sandhya Organic Chemicals Ltd.
11	Reltam	Redox Industries Ltd.
12	DelMate	Fengel Crop Sciences
13	Deltachem	Chemet Wets & Flows Pvt. Ltd.
14	Raven	Kalyani Industries Limited
15	Del-28	Katyayani Organics
16	Tagcis Gold	Tropical Agrosystem India (P) Ltd.
17	EBS Del-28	Essential Biosciences
18	Power-2.8	Krishi Rasayan
19	Darthrin	RACCOLTO Agritech India Pvt. Ltd.

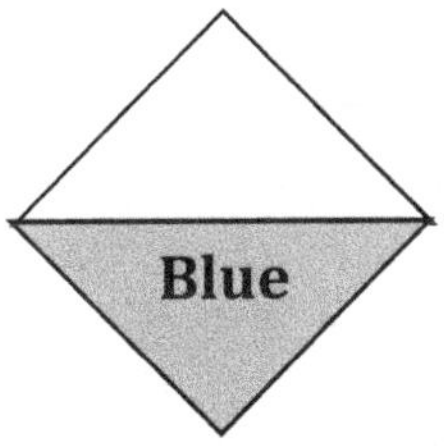

DIAFENTHIURON

Available formulations in India :
1. **Diafenthiuron 50.00 % WP**
2. **Diafenthiuron 47.80 % SC**

Classification on the basis of

Target organism : Insecticide, Acaricide

Mode of entry : Stomach and Contact poison

Mode of action : Energy metabolism

Chemical group : Thiourea

IRAC classification

Main Group and Primary Site of Action : 12 Inhibitors of mitochondrial ATP synthase

Sub-group or exemplifying Active Ingredient : 12A Diafenthiuron

Category of toxicity : Moderately toxic

Toxicity label colour : Blue

- Diafenthiuron, a thiourea insecticide and acaricide, provides effective control against a wide spectrum of sap-feeding pests and mites across diverse crops. Its notable characteristic lies in its distinct mode of action and efficacy against pests exhibiting resistance to other insecticide groups.
- **Mode of Action :** Diafenthiuron itself is not the biologically active molecule. It undergoes conversion into its active form, diafenthiuron carbodiimide, through either photochemical reactions or metabolic processes within the target pest or on the plant surface, particularly when exposed to sunlight. This active metabolite functions by inhibiting mitochondrial ATP synthase in insects and mites. This enzyme plays a vital role in cellular energy production (ATP). By blocking ATP synthesis, diafenthiuron disrupts the pest's energy homeostasis, ultimately leading to paralysis and mortality.
- Diafenthiuron demonstrates both contact and stomach poison activity. Certain formulations also exhibit translaminar movement (penetration and movement within leaf tissues) and vapor action, enhancing its ability to reach concealed pests.
- In contrast to some neurotoxic insecticides that induce rapid knockdown, diafenthiuron typically exhibits a slower effect. While pests may become immobile and cease feeding relatively quickly, mortality can take 3-4 days to occur, sometimes described as a "hidden knockdown effect."

Label claim (DPPQ & S, Faridabad)

Diafenthiuron 50.00% WP

Crop	Pests	Dose / ha		Waiting period (Days)
		Quantity of formulation (g)	Volume of water (Lit.)	
Cotton	Whitefly, Aphid, Jassid, Thrips	600	500-1000	21

Cabbage	Diamondback moth	600	500-750	07
Okra	Whitefly, Red spider mite	600	500	05
Tomato	Whitefly, Red spider mite	600	500	05
Chilli	Mites	600	500-750	03
Brinjal	Whitefly	600	500-750	03
Cardamom	Thrips, Capsule borer	800	1000	07
Citrus	Mite	2 g / lit.	2-3 lit. / tree	30
Watermelon	Whitefly, Red spider mite	600	500	05

Trade Names

Sr. No.	Trade Name	Company
1	Agas, Mainstar	ADAMA India Private Limited
2	Shoku	IFFCO-MC Crop Science Private Ltd.
3	Derby, Engage	Biostadt India Ltd.
4	Gainer, Stick	Gharda Chemicals Ltd.
5	Pager	Dhanuka Agritech Ltd.
6	Billion	Bharat Insecticides Ltd.
7	Ferotia	Coromandel International Ltd.
8	Polo, Pegasus	Syngenta India Ltd.
9	Panda-505	Auxin Crop Care
10	Jimmy, Jimmy Delux	Anu Products Ltd.
11	Logo	Insecticides (India) Ltd.
12	Spice, Nagarjuna Fibre	NACL Industries Ltd.
13	Yuddham	Crop Life Science Ltd.
14	Diatron	Parijat Industries India Pvt. Ltd.
15	Ludo	Crop Chemicals India Ltd.
16	Agzact, Award	Sumitomo Chemical India Ltd.
17	Chiliron (Chilli), Fenthuron (Cotton)	Herenba Industries Ltd.
18	Olosik	Sikko Industries Ltd.
19	Wishh, Duel	Nichino India
20	Kifenth	KP Agro Industries
21	Perasut	Natural Crop Science
22	Jio-Veg	Tropical Agrosystem India (P) Ltd.
23	Fifty Horses	Essential Biosciences
24	Weera	Synwood Agro Ltd.
25	Krijet	Krishi rasyan
26	Longere	Global Crop Care
27	Quest	Tata Rallis India Ltd.
28	Dice	Green India Care
29	Xylo Green	JU Agri Sciences Pvt. Ltd.
30	SC-Bodyguard	Katyayani Organics
31	Diffence	Preventis Chem India (P) Ltd.
32	Code	Preventis Chem India (P) Ltd.
33	Dirofen	Chemley Agritech Pvt. Ltd.
34	Polomino	RACCOLTO Agritech India Pvt. Ltd.

Diafenthiuron 47.80% SC

Crop	Pests	Dose / ha		Waiting period (Days)
		Quantity of formulation (ml)	Volume of water (Lit.)	
Cotton	Whitefly, Aphid, Jassid, Thrips	500	500	41
Potato	Whitefly, Leafhopper, Mite	500	500	44
Soybean	Whitefly, Mite	500	500	44

Trade Names

Sr. No.	Trade Name	Company
1	Olo-SC	Sikko Industries Ltd.
2	Bodyguard	Katyayani Organics
3	Dhoom	Greenwell Biotech
4	Aakrosh	Astha Chemicals
5	Rise Plus	HPM Chemicals and Fertilizers Ltd.
6	D-Fenon	D G Enterprise
7	White Out	Eagle Plant Protect Pvt. Ltd.
8	Zeep	Synergy
9	Deside	Silver Sine Bio Tech
10	Natural Difen	Natural Crop Science
11	Moleth	Global Crop Care
12	Diffence Ultra	Preventis Chem India (P) Ltd.
13	Code Ultra	Preventis Chem India (P) Ltd.
14	Polon Pro	RACCOLTO Agritech India Pvt. Ltd.

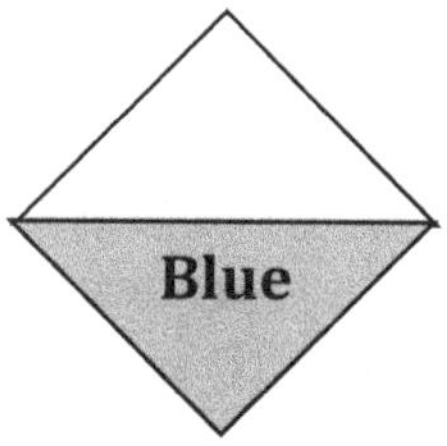

DIFLUBENZURON

Available formulations in India : **Diflubenzuron 25.00 % WP**

Classification on the basis of

Target organism : Insecticide
Mode of entry : Stomach and Contact poison
Mode of action : Growth regulator
Chemical group : Benzoylureas

IRAC classification

Main Group and Primary Site of Action : 15 Inhibitors of chitin biosynthesis affecting CHS1
Sub-group or exemplifying Active Ingredient : Benzoylurea

Category of toxicity : Moderately toxic
Toxicity label colour : Blue

- Diflubenzuron is a benzoylurea insecticide that acts as an Insect Growth Regulator (IGR).
- **Mode of action :** Diflubenzuron disrupts the formation of chitin in insects. Chitin is a vital component that insects need to build their exoskeleton and grow. When insects come into contact with diflubenzuron, they cannot properly form their cuticle, leading to death during molting or when hatching from eggs. It is particularly effective against larval stages of insects, but it also exhibits some ovicidal activity. It typically does not have a direct effect on adult insects.

Label claim (DPPQ & S, Faridabad)

Diflubenzuron 25.00 % WP

Crop	Pests	Dose / ha		Waiting period (Days)
		Quantity of formulation (g)	Volume of water (Lit.)	
Cotton	Bollworms	300	500-1000	--
	Tobacco caterpillar	300-350	500-1000	--

Trade Names

Sr. No.	Trade Name	Company
1	Track-N-Kill	Modern Insecticides Ltd.
2	B-Larv	Bayer Crop Science Ltd.
3	Naiger	Shri Usha Enterprises, Krishna
4	Adept	UPL India Ltd.

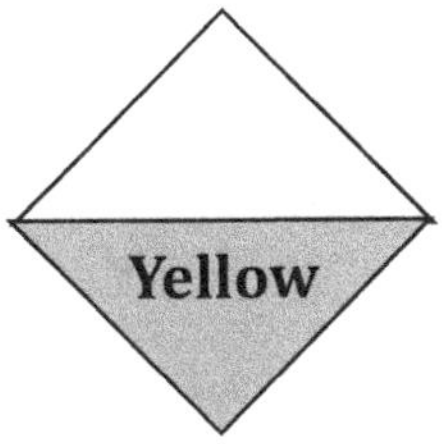

DIMETHOATE

Available formulations in India	:	**Dimethoate 30.00 % EC**
Classification on the basis of		
Target organism	:	Insecticide
Mode of entry	:	Stomach and Contact poison
Mode of action	:	Nerve poison
Chemical group	:	Organophosphates
IRAC classification		
Main Group and Primary Site of Action	:	1 Acetylcholinesterase (AChE) inhibitors
Sub-group or exemplifying Active Ingredient	:	1B Organophosphates
Category of toxicity	:	Highly toxic
Toxicity label colour	:	Yellow

- Dimethoate is a systemic organophosphate insecticide and acaricide. This means it is absorbed by the plant and transported throughout its tissues, providing protection against sucking and some chewing pests that feed on the treated plant.
- It also has contact activity, killing pests that come into direct contact with the spray.
- **Mode of action :** Dimethoate acts as a cholinesterase inhibitor. It interferes with the enzyme acetylcholinesterase (AChE) in the insect's nervous system. AChE is crucial for breaking down the neurotransmitter acetylcholine, which transmits nerve impulses. By inhibiting AChE, dimethoate causes an accumulation of acetylcholine at nerve synapses, leading to overstimulation of the nervous system, muscle twitching, paralysis, and ultimately death of the insect or mite.

Label claim (DPPQ & S, Faridabad)

Dimethoate 30.00 % EC

Crop	Pest	Dose / ha		Waiting period (Days)
		Quantity of formulation (ml)	Volume of water (Lit.)	
Bajra	Milky weed bug	594-660	500-1000	--
Maize	Stem borer	660	500-1000	--
	Shoot fly	1155	500-1000	--
Sorghum	Midge fly	1650	500-1000	--
Castor	Jassid, Mite	825	500-1000	--
	Semilooper	1155	500-1000	--
Mustard	Leaf miner, Aphid, Sawfly	660	500-1000	--
Safflower	Aphid	660	500-1000	--
Cabbage	Aphid	660	500-1000	--

Cauliflower	Painted bug, Mustard aphid	660	500-1000	--
Okra	Aphid	2310	500-1000	--
	Leaf hopper (Jassid)	1980	500-1000	--
Tomato	Aphid, Whitefly	990	500-1000	--
Brinjal	Jassid	1980	500-1000	--
	Shoot borer	660	500-1000	--
Chilli	Thrips	660	500-1000	--
	Mites	990	500-1000	--
Onion	Thrips	660	500-1000	--
Potato	Thrips, Aphid	660	500-1000	--
Apple	Stem borer	1485-1980	1500-2000	In March, April & June
Apricot	Aphid	1485-1980	1500-2000	Prebloom
Banana	Aphid	1485-1980	1500-2000	At 3 months old
	Lace wing bug	1485-1980	1500-2000	--
Citrus	Black aphid	1485-1980	1500-2000	--
Fig	Jassid	1485-1980	1500-2000	--
Mango	Mealy bug, Hopper	2475-3300	1500-2000	--
Rose	Scale	2475	500-1000	--
	Thrips	1320	500-1000	--

Trade Names

Sr. No.	Trade Name	Company
1	Tafgor	Rallis India Limited
2	Anugor	Anu Products Ltd.
3	Rokker	Sydney Formulation & Chemicals
4	Chemgor	Crop Chemicals India Ltd.
5	Isagor	Isagro Asia Agrochemicals Pvt. Ltd.
6	Heragor	Herenba Industries Ltd.
7	Rogor	Cheminova
8	Dimat	KP Agro Industries
9	Rigor Super	Natural Crop Science
10	Demat	Katyayani Organics
11	Tagor	Tropical Agrosystem India (P) Ltd.
12	Defender-30	Essential Biosciences
13	Fiper	Global Crop Care
14	I-Rogor	Green India Care
15	Agrikind Demon	Mankind Agritech
16	Exylon Dronex	Torrent Crop Science
17	Bardot	RACCOLTO Agritech India Pvt. Ltd.

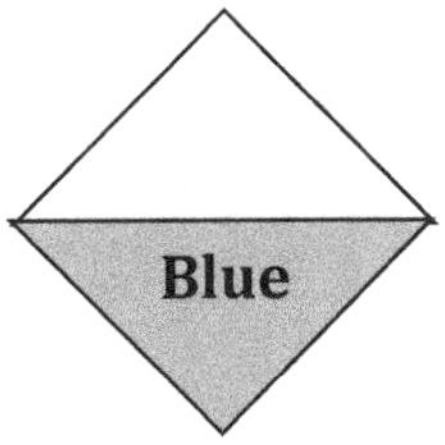

DIMPROPYRIDAZ

Available formulations in India : **Dimpropyridaz 120 g/l SL**

Classification on the basis of

Target organism : Insecticide

Mode of entry : Stomach and Contact poison, Translaminar

Mode of action : Nerve action, Disruption of sensory chordotonal organs

Chemical group : Pyridazine pyrazolecarboxamides

IRAC classification

Main Group and Primary Site of Action : 36 Chordotonal organ modulators – undefined target site

Sub-group or exemplifying Active Ingredient : Pyridazine pyrazolecarboxamide

Category of toxicity : Moderately toxic

Toxicity label colour : Blue

- Dimpropyridaz, a novel pro-insecticide in the pyridazine pyrazolecarboxamide class, boasts a unique mode of action: it disrupts the function of sensory chordotonal organs in insects.
- This interference effectively controls various sucking pests.
- **Mode of Action :** Upon entering the insect, dimpropyridaz is metabolized into an active secondary amide metabolite. This metabolite targets and disrupts the chordotonal organs, crucial sensory structures responsible for hearing, balance, gravity perception, acceleration, and kinesthesia. The disruption leads to rapid feeding cessation, which may also reduce transmission of viruses by sucking pests.

Label claim (DPPQ & S, Faridabad)

Dimpropyridaz 120 g/l SL

Crop	Pests	Dose / ha		Waiting period (Days)
		Quantity of formulation (ml)	Volume of water (Lit.)	
Birnjal	Whitefly	900-1000	500	03
	Jassid, Aphid	700	500	03
Cotton	Whitefly	900-1000	500	31
	Jassid, Aphid	700	500	31
Cucumber	Whitefly	900-1000	500	03
	Jassid, Aphid	700	500	03
Tomato	Whitefly, Jassid, Aphid	900-1000	500	03
Chilli	Whitefly, Aphid	900-1000	500	03

Trade Names

Sr. No.	Trade Name	Company
1	Efficon	BASF India Limited

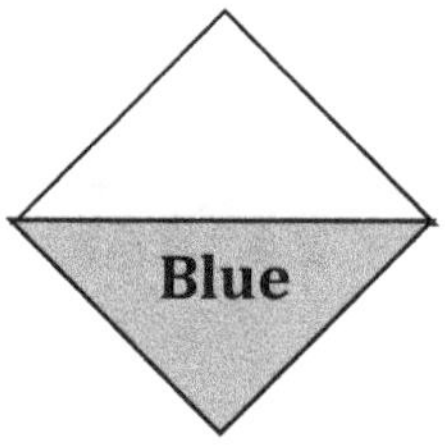

DINOTEFURON

Available formulations in India :
1. **Dinotefuran 20.00 % SG**
2. **Dinotefuran 70.00 % WG**

Classification on the basis of

Target organism : Insecticide
Mode of entry : Systemic, Stomach and Contact poison
Mode of action : Nerve poison
Chemical group : Neonicotinoid

IRAC classification

Main Group and Primary Site of Action : 4 Nicotinic acetylcholine receptor (nAchR) competitive modulators
Sub-group or exemplifying Active Ingredient : 4A Neonicotinoids

Category of toxicity : Moderately toxic
Toxicity label colour : Blue

- Dinotefuron, a relatively recent third-generation neonicotinoid belonging to the furanicotinyl subclass, stands out as a broad-spectrum insecticide.
- Its efficacy extends across diverse applications, including agriculture, residential pest control, and veterinary medicine.
- **Mode of action :** Dinotefuron functions as an agonist of insect nicotinic acetylcholine receptors (nAChRs) within the insect's central nervous system. By binding to these receptors, it triggers an overstimulation of nerve cells, ultimately leading to paralysis and the insect's demise.
- While sharing a similar mode of action with other neonicotinoids, dinotefuron is thought to possess a slightly different binding mechanism, contributing to its effectiveness against certain pests exhibiting resistance to other neonicotinoids.
- Furthermore, it exhibits both contact and stomach toxicity and demonstrates systemic action, allowing plant absorption and translocation for protection of new growth. Its translaminar activity also enables movement within the leaf, providing control on both sprayed and unsprayed surfaces.
- Notably, dinotefuron distinguishes itself from other insecticide classes by not inhibiting cholinesterase or interfering with sodium channels.

Label claim (DPPQ & S, Faridabad)

Dinotefuran 20.00 % SG

Crop	Pests	Dose / ha		Waiting period (Days)
		Quantity of formulation (g)	Volume of water (Lit.)	
Rice	Brown plant hopper	150-200	500	21
Cotton	Whitefly, Aphid, Jassid, Thrips	125-150	500	15

Trade Names

Sr. No.	Trade Name	Company
1	Osheen	PI Industries
2	Senpai	IFFCO-MC Crop Science Private Ltd.
3	Ossum	Biostadt India Ltd.
4	Domar	Dhanuka Agritech Ltd.
5	Token	Indofil Industries Ltd.
6	Dominant	Insecticides (India) Ltd.
7	Diron	Best Agrolife Ltd.
8	Damdar-20	Silver Sine Bio Tech
9	Ossein	Natural Crop Science
10	Apocalypse	Katyayani Organics
11	Tagocean	Tropical Agrosystem India (P) Ltd.
12	Aura-20	Essential Biosciences
13	Todino 20	Synwood Agro Ltd.
14	Batman	Koronation Bio-Science
15	Mawsim	Today's Global Crop Protection
16	Oasis	Indian Agrochemicals
17	Krown	Krishi Rasayan
18	Don	Global Crop Care
19	Trilok	JU Agri Sciences Pvt. Ltd.
20	Dinoforce	Mankind Agritech
21	Dixiron	Chemley Agritech Pvt. Ltd.

Dinotefuran 70.00% WG

Crop	Pests	Dose / ha		Waiting period (Days)
		Quantity of formulation (g)	Volume of water (Lit.)	
Cotton	Aphid, Whitefly, Leaf hopper	87	500	17
Brinjal	Jassid, Aphid, Whitefly	87	500	07
Rice	Brown plant hopper, White backed plant hopper, Green leaf hopper	88	500	32

Trade names

Sr. No.	Trade Name	Company
1	Osheen Ultra	PI Industries
2	Hogger	Indofil Industries Ltd.

Yellow

EMAMECTIN BENZOATE

Available formulations in India : **1. Emamectin benzoate 05.00 % SG**
2. Emamectin benzoate 01.90 % EC

Classification on the basis of

Target organism : Insecticide
Mode of entry : Stomach and Contact poison
Mode of action : Nerve and muscle poison
Chemical group : Avermectins (Macrocyclic lactones)

IRAC classification

Main Group and Primary Site of Action : 6 Glutamate-gated chloride channel (GluCl) allosteric modulators
Sub-group or exemplifying Active Ingredient : Avermectins

Category of toxicity : Highly toxic
Toxicity label colour : Yellow

- Emamectin benzoate, a widely employed non-systemic insecticide from the avermectin family, is a semi-synthetic derivative of abamectin, a natural compound produced by the soil bacterium *Streptomyces avermitilis*.
- It demonstrates particular effectiveness against a variety of lepidopteran pests (caterpillars, moths) and certain other insects and mites in agricultural settings.
- **Mode of action:** Emamectin benzoate primarily acts through ingestion (stomach action) with some contact activity and is non-systemic. Its mode of action involves disrupting the insect nervous system. Specifically, it functions as an allosteric modulator of glutamate-gated chloride (GluCl) channels found in insect nerve and muscle cells. The binding to these channels triggers an influx of chloride ions, leading to hyperpolarization of nerve and muscle cells. Consequently, the insect experiences paralysis, ceases feeding shortly after exposure, and typically dies within 2 to 4 days. Furthermore, emamectin benzoate enhances the binding of the neurotransmitter gamma-aminobutyric acid (GABA) to its receptors, further disrupting nerve signal transmission.
- Notably, it exhibits translaminar activity, allowing it to penetrate leaf tissues and create a reservoir within the leaf, thus providing lasting control against pests feeding on the underside of leaves.

Label claim (DPPQ & S, Faridabad)

Emamectin benzoate 05.00% SG

Crop	Pests	Dose / ha		Waiting period (Days)
		Quantity of formulation (g)	Volume of water (Lit.)	
Cotton	Bollworms	190-220	500	10
Pigeonpea	Pod borer	220	500-750	14

Chickpea	Pod borer	220	500	14
Okra	Fruit and shoot borer	135-170	500	05
Cabbage	Diamondback moth	150-200	500	03
Chilli	Fruit borer, Thrips, Mite	200	500	03
Brinjal	Shoot and fruit borer	200	500	03
Grape	Thrips	220	500-1000	05
Tea	Tea looper	200	500	01

Trade Names

Sr. No.	Trade Name	Company
1	Missile, Proclaim	Crystal Crop Protection Ltd.
2	Amnon	ADAMA India Private Limited
3	Spolit	UPL Ltd.
4	Egao	IFFCO-MC Crop Science Private Ltd.
5	Starclaim	Swal Corporation Ltd.
6	Bioclaim, Inclaim	Biostadt India Ltd.
7	EM-1	Dhanuka Agritech Ltd.
8	Benzer	Coromandel International Ltd.
9	Hectar	Cropnosys India Pvt. Ltd.
10	Elpida	Godrej Agrovet Ltd.
11	Volax	Indofil Industries Ltd.
12	Agent Plus	Indofil Industries Ltd.
13	Ema Gold	Alfa Crop Science
14	Tag Emboz	Tropical Agrosystem India (P) Ltd.
15	Pro Monk	Auxin Crop Care
16	Rudra	Anu Products Ltd.
17	Xplode	Insecticides (India) Ltd.
18	Punch	Modern Insecticides Ltd.
19	Nagarjuna Trust	NACL Industries Ltd.
20	Robot	Excel Crop Care Ltd.
21	Larbotin	Crop Life Science Ltd.
22	Rilon	Rallis India Ltd.
23	Perform	Parijat Industries India Pvt. Ltd.
24	Eagle	Crop Chemicals India Ltd.
25	On A Top	Sumitomo Chemical India Ltd.
26	Pocket	Isagro Asia Agrochemicals Pvt. Ltd.
27	Procura	Gujarat Pesti-Chem Industries
28	Benzo	Herenba Industries Ltd.
29	O-Claim	Sikko Industries Ltd.
30	Irma	Best Agrolife Ltd.
31	Track	Nichino India
32	Provin	Silver Sine Bio Tech
33	Pro Clean	Natural Crop Science
34	Ema5	Katyayani Organics
35	Emaan	Essential Biosciences
36	Lazo	Synwood Agro Ltd.
37	K-Pro	Koronation Bio-Science
38	U-Sure	Today's Global Crop Protection
39	Macho	Krishi Crop Science Pvt. Ltd.
40	Kri-Star	Krishi Rasayan

41	Claim	Global Crop Care
42	EBA Green	Green India Care
43	E-Mate	JU Agri Sciences Pvt. Ltd.
44	E-Mactron	Linuxcrop Chemicals Pvt. Ltd.
45	SA-Inmectin	Saraswati Agro Life Science (India) Pvt. Ltd.
46	Amect	Synovita Crop science Pvt. Ltd.
47	Emaneed	Mankind Agritech
48	Exylon Mectimax 5	Torrent Crop Science
49	Fusion	Advance Pesticides
50	Attack	Preventis Chem India (P) Ltd.
51	Emachem	Chemley Agritech Pvt. Ltd.

Emamectin benzoate 01.90% EC

Crop	Pests	Dose / ha		Waiting period (Days)
		Quantity of formulation (ml)	Volume of water (Lit.)	
Cotton	Bollworms	580	500	15
Chickpea (Gram)	Pod borer	375	500	14
Rice	Leaf folder, Hispa	425	500	48
Soybean	Green semilooper, Pod borer, Tobacco leaf eating caterpillar, Girdle beetle	425	500	20
Chilli	Fruit borer, Thrips	375	500	03

Trade Names

Sr. No.	Trade Name	Company
1	Billo	Crystal Crop Protection Ltd.
2	Thor-99	Auxin Crop Care
3	Libra	Crop Life Science Ltd.
4	Khanjar	Sikko Industries Ltd.
5	Badhiya	Somnath Demeter Agritron Ltd.
6	Larvasine	
7	Larvi Clean	Natural Crop Science
8	Ema19	Katyayani Organics
9	Turbo	Tropical Agrosystem India (P) Ltd.
10	Emaan-19	Essential Biosciences
11	Lazo Pro	Synwood Agro Ltd.
12	Garuda	Koronation Bio-Science
13	Emma	Shivalik Crop Science (P) Ltd.
14	Boombar	Skyn Crop Care Pvt. Ltd.
15	Lipido	Global Crop Care
16	Vinashak	Green India Care
17	Larva	Linuxcrop Chemicals Pvt. Ltd.
18	Exylon Electra	Torrent Crop Science
19	Shareman	RACCOLTO Agritech India Pvt. Ltd.

ETHION

Available formulations in India : **Ethion 50.00 % EC**

Classification on the basis of

Target organism : Insecticide, Acaricide
Mode of entry : Stomach and Contact poison
Mode of action : Nerve poison
Chemical group : Organophosphates

IRAC classification

Main Group and Primary Site of Action : 1 Acetylcholinesterase (AChE) inhibitors
Sub-group or exemplifying Active Ingredient : 1B Organophosphates

Category of toxicity : Highly toxic
Toxicity label colour : Yellow

- Ethion is a broad-spectrum organophosphate insecticide and acaricide, meaning it controls both insects and mites.
- **Mode of action** : Its mode of action is primarily through contact, and it works by inhibiting acetylcholinesterase (AChE), an essential enzyme in the nervous system of insects and mites. This disruption leads to a buildup of acetylcholine, causing overstimulation, paralysis, and ultimately death.
- Ethion is resistant to hydrolysis in neutral and acidic conditions but degrades significantly faster in alkaline solutions.

Label claim (DPPQ & S, Faridabad)

Ethion 50.00 % EC

Crop	Pests	Dose / ha		Waiting period (Days)
		Quantity of formulation (ml)	Volume of water (Lit.)	
Cotton	Whitefly	1500-2000	500-1000	--
	Bollworms	2000	500-1000	25
Chickpea	Pod borer	1000-1500	500-1000	21
Pigeonpea	Pod borer	1000-1500	500-1000	21
Soybean	Girdle beetle, Stem fly	1500	500-1000	30
Chilli	Mites, Thrips	1500-2000	500-1000	05
Tea	Red spider mites, Purple mites, Yellow mite, Thrips, Scale	500	500-1000	03

Trade Names

Sr. No.	Trade Name	Company
1	Fosmite	PI Industries
2	Tafethion	Rallis India Limited

3	Hilmite	HIL (India) Limited
4	Mit-505	Swal Corporation Ltd.
5	Webber	Modern Insecticides Ltd.
6	Deviastra 50	Arysta Lifescience
7	Goldmit-50	Crop Chemicals India Ltd.
8	G-Might	Tusova Agro Pvt. Ltd.
9	Heron	Herenba Industries Ltd.
10	Samrat	Gujarat Crop Care
11	Sikmite	Sikko Industries Ltd.
12	Ethion Super	Natural Crop Science
13	Novathion	Tropical Agrosystem India (P) Ltd.
14	Ethon Hunt	Essential Biosciences
15	Myti-Nash	Synwood Agro Ltd.
16	Mig 50	Koronation Bio-Science
17	SA-Safthion	Saraswati Agro Life Science (India) Pvt. Ltd.
18	Pedasos	RACCOLTO Agritech India Pvt. Ltd.

ETHOFENPROX

Available formulations in India : **Ethofenprox 10.00 % EC**

Classification on the basis of

Target organism : Insecticide
Mode of entry : Stomach and Contact poison
Mode of action : Nerve poison
Chemical group : Pyrethroid

IRAC classification

Main Group and Primary Site of Action : 3 Sodium channel modulators
Sub-group or exemplifying Active Ingredient : 3A Pyrethroid

Category of toxicity : Highly toxic
Toxicity label colour : Yellow

- Ethofenprox, a non-ester pyrethroid insecticide, distinguishes itself from many ester-based pyrethroids with its ether linkage. This structural difference contributes to its unique characteristics, notably a relatively lower toxicity to fish compared to some ester pyrethroids.
- Exhibiting both contact and stomach action, ethofenprox effectively controls a broad spectrum of insect pests.
- **Mode of action :** Ethofenprox kills insects through direct contact and by ingestion. Similar to other pyrethroids, it disrupts the insect nervous system by interfering with the function of voltage-gated sodium channels in nerve axons. This interference causes repetitive nerve impulses, ultimately leading to paralysis and the insect's death.
- It is also known for its relatively rapid knockdown effect.

Label claim (DPPQ & S, Faridabad)

Ethofenprox 10.00 % EC

Crop	Pests	Dose / ha		Waiting period (Days)
		Quantity of formulation (ml)	Volume of water (Lit.)	
Rice	Brown plant hopper, Green leaf hopper, Stem borer, Leaf folder, Gall midge, Whorl maggot, White backed plant hopper	500-750	500-1000	15

Trade Names

Sr. No.	Trade Name	Company
1	Trebon Excel 10 EC	Biostadt India Limited
2	Time Out	Modern Insecticides Ltd.
3	Primo	Isagro Asia Agrochemicals Pvt. Ltd.

Blue

ETHYLENE DICHLORIDE + CARBON TETRACHLORIDE

Available formulations in India : **Ethylene dichloride + Carbon tetrachloride (3:1)**

Classification on the basis of

Target organism : Insecticide
Mode of entry : Fumigant
Mode of action : Respiratory poison
Chemical group : Chlorinated hydrocarbon

IRAC classification

Main Group and Primary Site of Action :
Sub-group or exemplifying Active Ingredient :

Category of toxicity : Moderately toxic
Toxicity label colour : Blue

- The EDCT mixture is a liquid fumigant typically composed of 3 parts Ethylene Dichloride (EDC) to 1 part Carbon Tetrachloride (CCl_4) by volume.
- **Primary use :** Fumigating bulk-stored grain and sealed containers to control stored product pests.
- **Safety feature:** CCl_4 is added to the flammable EDC to reduce the fire hazard and ensure safer application.
- **Mode of action :** The mixture rapidly volatilizes, and the vapor enters the insect via the spiracles and tracheal system. Both components are Chlorinated Hydrocarbons (CHCs) and are highly lipophilic. They act as membrane disruptors and narcotics, easily penetrating the nerve cell membranes. They interfere with nerve impulse transmission in the Central Nervous System (CNS), leading to narcosis (paralysis) and ultimately death from system collapse.
- **Residue/Safety :** EDC residues can persist in treated grain for months, necessitating caution due to its potential toxicity and carcinogenicity to minimize human exposure.

Label claim (DPPQ & S, Faridabad)

Ethylene dichloride + Carbon tetrachloride (3:1)

Crop	Pests	Condition	Weight of volume	Exposure period	Conc. in air (ppm)	Aeration / Waiting
Stored whole cereals, Millets, Pulses	Rice weevil, Lesser grain Borer, Khapra Beetle, Rust red flour beetle, Pulse	Air tight cover	300 - 400 gm/m^3 (230 - 307 ml)	48 - 72 Hr. for cover fumigation	10 ppm	Partial aeration For at least 1 hr. followed by 24 hr. complete Aeration

	beetle, Dried fruit Beetle					waiting period of 24 hr.
Godown fumigation	Rice weevil, Lesser grain Borer, Khapra Beetle, Rust red flour beetle, Pulse beetle, Dried fruit Beetle	Air tight cover	150 gm/m3	07 days	10 ppm	Partial aeration For at least 1 hr. followed by 24 hr. complete Aeration waiting period of 24 hr.

Trade Names

Sr. No.	Trade Name	Company
1	Granol	Ultima Search
2	Grainguard	Royal EDB Tube Mfg. Co.

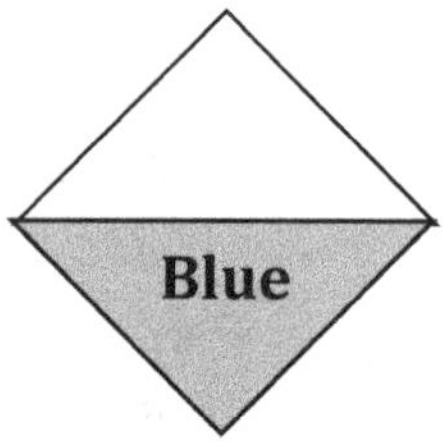

FENMEZODITIAZ

Available formulations in India : **Fenmezoditiaz 200 g/l SC**

Classification on the basis of

Target organism	:	Insecticide
Mode of entry	:	Systemic poison
Mode of action	:	Nerve poison
Chemical group	:	Mesoionics

IRAC classification

Main Group and Primary Site of Action	:	4 Nicotinic acetylcholine receptor (nAChR) competitive modulators
Sub-group or exemplifying Active Ingredient	:	4E Mesoionics

Category of toxicity : Moderately toxic
Toxicity label colour : Blue

- Fenmezoditiaz is a cutting-edge, systemic insecticide developed by BASF. It belongs to a new class of chemistry known as mesoionics and is specifically designed to tackle difficult-to-control piercing and sucking pests, primarily in rice cultivation.
- **Mode of action**: It acts as a competitive modulator of the nicotinic acetylcholine receptor (nAChR). It disrupts nerve signal transmission, leading to rapid feeding cessation, paralysis, and death of the pest.
- Pests stop feeding almost immediately after exposure, protecting the crop from instant damage and viral transmission.
- It has systemic and translaminar action. The chemical moves throughout the plant tissue, ensuring that even hoppers hidden at the base of the plant or under leaves are controlled.

Label claim (DPPQ & S, Faridabad)

Fenmezoditiaz 200 g/l SC

Crop	Pests	Dose / ha		Waiting period (Days)
		Quantity of formulation (ml)	Volume of water (Lit.)	
Rice	Brown plant hopper, White backed plant hopper	125-150	500	37

Trade Names

Sr. No.	Trade Name	Company
1	Valexio	BASF India Ltd.

Yellow

FENOBUCARB (BPMC)

Available formulations in India : **Fenobucarb 50.00 % EC**

Classification on the basis of

Target organism : Insecticide
Mode of entry : Stomach and Contact poison
Mode of action : Nerve poison
Chemical group : Carbamates

IRAC classification

Main Group and Primary Site of Action : 1 Acetylcholinesterase (AChE) inhibitors
Sub-group or exemplifying Active Ingredient : 1A Carbamates

Category of toxicity : Highly toxic
Toxicity label colour : Yellow

- Fenobucarb, also widely known as BPMC (2-sec-butylphenyl methylcarbamate), is a carbamate insecticide.
- This broad-spectrum insecticide is employed in agriculture and pest control, particularly for managing Hemipteran pests (sucking insects).
- Acting as both a contact and stomach poison, fenobucarb exhibits a relatively fast killing action and moderately long residual activity.
- **Mode of action :** Its mode of action involves the inhibition of acetylcholinesterase (AChE), a non-systemic mechanism that disrupts the normal functioning of the insect nervous system. By inhibiting the enzyme that breaks down the neurotransmitter acetylcholine, fenobucarb causes a buildup at nerve synapses, leading to overstimulation, paralysis, and ultimately, the insect's death.

Label claim (DPPQ & S, Faridabad)

Fenobucarb 50.00 % EC

Crop	Pests	Dose / ha		Waiting period (Days)
		Quantity of formulation (ml)	Volume of water (Lit.)	
Rice	Brown plant hopper, Green leaf hopper	500-1500	500	30

Trade Names

Sr. No.	Trade Name	Company
1	Feno-50	Essential Biosciences
2	Feno	India Agro Seeds
3	Warden	Maxah Bio Tech

4	Fen Care	Neolect Crop Science
5	Feno Rex	Raxima Bio Tech Co.
6	Fenhop-50	Nauxima Bio Tech
7	Fenox	Agri Science Co.
8	Tofan	SP Agrochemicals Industries
9	Fenob	Goldenox Crop care
10	Blast	Sudershan Bio-Crop Science
11	Josh	Anu Products Ltd.
12	Bipkil	Nichino India
13	Killer Fen	Natural Crop Science
14	BPMC	Synwood Agro Ltd.
15	Laher	Global Crop Care
16	Pinaka	Green India Care
17	Fenochem	Chemley Agritech Pvt. Ltd.
18	Parwin	RACCOLTO Agritech India Pvt. Ltd.

Yellow

FENPROPATHRIN

Available formulations in India :
1. **Fenpropathrin 10.00 % EC**
2. **Fenpropathrin 30.00% EC**
3. **Fenpropathrin 10.00 % EW**

Classification on the basis of

Target organism : Insecticide
Mode of entry : Stomach and Contact poison
Mode of action : Nerve poison
Chemical group : Pyrethroid

IRAC classification

Main Group and Primary Site of Action : 3 Sodium channel modulators
Sub-group or exemplifying Active Ingredient : 3A Pyrethroid

Category of toxicity : Highly toxic
Toxicity label colour : Yellow

- Fenpropathrin, a synthetic pyrethroid insecticide and acaricide, is commonly used in agriculture and homes to manage a variety of pests.
- It is part of the pyrethroid class, synthetic analogues of natural pyrethrins from pyrethrum flowers.
- **Mode of action :** Classified as a Type II pyrethroid due to its alpha-cyano group, fenpropathrin generally exhibits greater potency in disrupting insect nerve function compared to Type I pyrethroids. Functioning as both a contact and stomach poison, fenpropathrin, like other pyrethroids, disrupts sodium channels in insect nerve cells, leading to excitation, convulsions, paralysis, and ultimately, the death of the pest by interfering with nerve impulse transmission.

Label claim (DPPQ & S, Faridabad)

Fenpropathrin 10.00% EC

Crop	Pests	Dose / ha		Waiting period (Days)
		Quantity of formulation (ml)	Volume of water (Lit.)	
Cotton	Pink bollworm, Spotted bollworm, American bollworm	750-1000	750-1000	14

Trade Names

Sr. No.	Trade Name	Company
1	Danitol	Sumitomo Chemical India Ltd.
2	Manitol	Makam Agrochem Pvt. Ltd.
3	Teraxin	Synovita Crop science Pvt. Ltd.

Fenpropathrin 30.00% EC

Crop	Pests	Dose / ha		Waiting period (Days)
		Quantity of formulation (ml)	Volume of water (Lit.)	
Cotton	Pink bollworm, Spotted bollworm, American bollworm, Whitefly	250-340	750-1000	14
Chilli	Thrips, Whitefly, Mite	250-340	750-1000	07
Brinjal	Whitefly, Shoot and fruit borer, Mite	250-340	750-1000	10
Okra	Whitefly, Shoot and fruit borer, Mite	250-340	750-1000	07
Tea	Mite	165-200	400-500	07
Paddy (Rice)	Yellow stem borer, Leaf folder	333	500	30

Trade Names

Sr. No.	Trade Name	Company
1	Meothrin	Sumitomo Chemical India Ltd.
2	Macothrin	Makam Agrochem Pvt. Ltd.
3	Triothrin	Global Crop Care

Fenpropathrin 10.00% EW

Crop	Pests	Dose / ha		Waiting period (Days)
		Quantity of formulation (ml)	Volume of water (Lit.)	
Rice	Stem borer, Leaf folder	1000	500	58

Trade Names

Sr. No.	Trade Name	Company
1	Danitol Nxt	Sumitomo Chemical India Ltd.

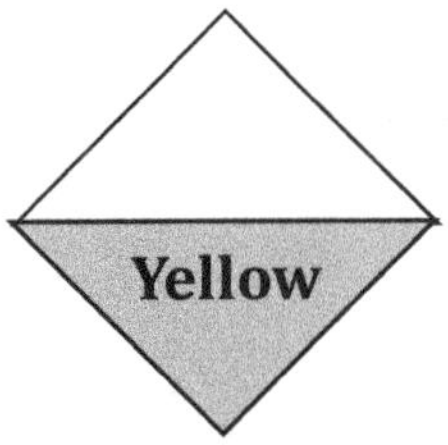

FENPYROXIMATE

Available formulations in India : 1. **Fenpyroximate 05.00% EC**
2. **Fenpyroximate 05.00 % SC**

Classification on the basis of

Target organism : Insecticide, Acaricide

Mode of entry : Stomach and Contact poison

Mode of action : Energy metabolism

Chemical group : Pyrazole

IRAC classification

Main Group and Primary Site of Action : 21 Mitochondrial complex I electron transport inhibitors

Sub-group or exemplifying Active Ingredient : 21A METI acaricides and insecticides

Category of toxicity : Highly toxic

Toxicity label colour : Yellow

- Fenpyroximate is a synthetic acaricide (miticide) used to control various types of mites in agriculture. While it primarily targets mites, it can also have some insecticidal activity against certain pests.
- **Mode of action :** It belongs to the METI acaricides and insecticides group (Mitochondrial Electron Transport Inhibitors) and specifically targets Mitochondrial complex I electron transport. Fenpyroximate disrupts the energy generation process in the mitochondria of the target pest's cells. It inhibits mitochondrial complex I electron transport, specifically blocking ubiquinone reduction, which is similar to the action of rotenone (another insecticide/acaricide). This disruption leads to a rapid cessation of all biological activities, including feeding and reproduction, ultimately causing paralysis and death of the mites.
- It has a quick knockdown effect on larvae, nymphs, and adults.
- It also exhibits molting and oviposition inhibitory action on nymphs.
- Fenpyroximate is a stable acaricide, particularly at neutral to slightly alkaline pH levels. While it is less stable in acidic environments.

Label claim (DPPQ & S, Faridabad)

Fenpyroximate 05.00% EC

Crop	Pests	Dose / ha		Waiting period (Days)
		Quantity of formulation (ml)	Volume of water (Lit.)	
Tea	Red spider mite, Pink mite, Purple mite	300-600	400-500	07
Chilli	Yellow mite	300-600	300-500	07
Cotton	Jassid, Mite	750	500	15

Coconut	Eriopyid mite (Root feeding)	10 ml / Lit.	As required	-
	Eriopyid mite (Spraying)	0.75-1.00 ml / Lit.	As required	-

Trade Names

Sr. No.	Trade Name	Company
1	Neon	Tata Rallis India Limited
2	Pyromite	Excel Crop Care Ltd.
3	Formite	Parijat Industries India Pvt. Ltd.
4	Mitigate	Isagro Asia Agrochemicals Pvt. Ltd.
5	Akio	Nichino India
6	Fenprox	Natural Crop Science
7	Akhanda	Koronation Bio-Science
8	K-Mit	Krishi Rasayan
9	Maximite	Global Crop Care
10	Signet	Preventis Chem India (P) Ltd.

Fenpyroximate 05.00% SC

Crop	Pests	Dose / ha		Waiting period (Days)
		Quantity of formulation (ml)	Volume of water (Lit.)	
Tea	Red spider mite, Pink mite, Purple mite	600-1200	400	07
Chilli	Yellow mite	300-600	500-750	03

Trade Names

Sr. No.	Trade Name	Company
1	Sedna	Tata Rallis India Limited

Yellow

FENVALERATE

Available formulations in India :
1. **Fenvalerate 20.00 % EC**
2. **Fenvalerate 02.00% EC**
3. **Fenvalerate 00.40% DP**

Classification on the basis of

Target organism : Insecticide, Acaricide
Mode of entry : Stomach and Contact poison
Mode of action : Nerve poison
Chemical group : Pyrethroid

IRAC classification

Main Group and Primary Site of Action : 3 Sodium channel modulators
Sub-group or exemplifying Active Ingredient : 3A Pyrethroid

Category of toxicity : Highly toxic
Toxicity label colour : Yellow

- Fenvalerate is a synthetic pyrethroid insecticide and acaricide. It's a man-made chemical designed to mimic the insecticidal properties of natural pyrethrins found in chrysanthemum flowers.
- This insecticide works through both contact and stomach action, effectively controlling a wide range of pests.
- Fenvalerate is actually a mixture of four optical isomers. Among these, esfenvalerate is the most potent, making up about 23% of commercial fenvalerate products.
- **Mode of action** : Like other pyrethroids, fenvalerate targets the insect's nervous system. It disrupts the normal function of sodium channels in nerve cells, leading to continuous and excessive nerve impulses. This overstimulation results in excitation, convulsions, paralysis, and ultimately, the death of the insect.
- Fenvalerate is classified as a Type II pyrethroid. At toxic doses in mammals, Type II pyrethroids are known to cause a "CS syndrome," characterized by choreoathetosis (involuntary movements) and salivation.

Label claim (DPPQ & S, Faridabad)

Fenvalerate 20.00 % EC

Crop	Pests	Dose / ha		Waiting period (Days)
		Quantity of formulation (ml)	Volume of water (Lit.)	
Cauliflower	Diamond back moth	300-375	600-750	07
Cotton	Bollworm	375-500	700-900	07
	Aphid, Jassid, Thrips	125-200	250-400	07
Brinjal	Shoot and fruit borer, Aphid	375-500	600-800	05
Okra	Shoot and fruit borer, Jassid	300-375	600-750	07

Trade Names

Sr. No.	Trade Name	Company
1	Tatafen	Rallis India Limited
2	Fencor-20	Modern Insecticides Ltd.
3	Goldfen-20	Crop Chemicals India Ltd.
4	Sumicidin	Sumitomo Chemical India Ltd.
5	Fenval	Isagro Asia Agrochemicals Pvt. Ltd.
6	Fazer	GFG Crop Science Pvt. Ltd.
7	Superfen	Tusova Agro Pvt. Ltd.
8	Fenta	Herenba Industries Ltd.
9	Sikkofen	Sikko Industries Ltd.
10	Krifen-20 EC	Krishi Rasayan

Fenvalerate 02.00 % Conc.

Crop	Pests	Dose / ha		Waiting period (Days)
		Quantity of formulation (ml)	Volume of water (Lit.)	
Cotton	Spotted, Spiny, Pink, American, Egyptian bollworm	4000-5000	--	--

Trade Names

Sr. No.	Trade Name	Company
1		

Fenvalerate 00.40 % DP

Crop	Pests	Dose / ha		Waiting period (Days)
		Quantity of formulation (ml)	Volume of water (Lit.)	
Cotton	Spotted bollworm, Pink bollworm	20000-25000	--	07

Trade Names

Sr. No.	Trade Name	Company
1	Anufen	Anu Products Ltd.
2	Fencor-D	Modern Insecticides Ltd.
3	Fenval Dust	Isagro Asia Agrochemicals Pvt. Ltd.
4	Fenwell	Excel Crop Care ltd.
5	Sikkofen-DP	Sikko Industries Ltd.
6	Flow Fen	Silver Sine Bio Tech
7	Krifen-0.4	Krishi Rasayan
8	Ghatak	Global Crop Care
9	Feni Dust	Green India Care

FIPRONIL

Available formulations in India :
1. **Fipronil 05.00 % SC**
2. **Fipronil 18.87 % SC**
3. **Fipronil 02.92 % EC**
4. **Fipronil 00.30 % GR**
5. **Fipronil 00.60 % GR**
6. **Fipronil 80.00 % WG**
7. **Fipronil 00.60 % WG**

Classification on the basis of

Target organism	:	Insecticide
Mode of entry	:	Contact and Stomach poison
Mode of action	:	Nerve poison
Chemical group	:	Phenylpyrazoles (Fiproles)
IRAC classification		
Main Group and Primary Site of Action	:	2 GABA-gated chloride channel blockers
Sub-group or exemplifying Active Ingredient	:	2B Phenylpyrazoles (Fiproles)
Category of toxicity	:	Highly toxic
Toxicity label colour	:	Yellow

- Fipronil is a broad-spectrum, synthetic insecticide from the phenylpyrazole chemical family.
- It's extensively used in agriculture and pest control to manage a wide array of insect pests.
- Primarily a non-systemic insecticide, fipronil works when pests directly contact or ingest it. However, some formulations can offer limited systemic activity.
- **Mode of action** : Fipronil disrupts the central nervous system of insects by blocking chloride ion passage through two crucial pathways: gamma-aminobutyric acid (GABA) receptors and glutamate-gated chloride (GluCl) channels. This blockage overexcites the insect's nerves and muscles, leading to paralysis and eventual death.
- What makes fipronil effective is its selective toxicity to insects. It binds much more strongly to insect GABA receptors than to those in mammals. Plus, fipronil acts on GluCl channels, which aren't even present in mammals, further enhancing its specificity.
- There's some debate in scientific literature over how systemic fipronil is, especially when compared to highly systemic products like neonicotinoids. Nonetheless, it's frequently used in ways that rely on its movement within plants.

Label claim (DPPQ & S, Faridabad)

Fipronil 05.00 % SC

Crop	Pests	Dose / ha		Waiting period (Days)
		Quantity of formulation (ml)	Volume of water (Lit.)	
Rice	Stem borer, Brown plant hopper, Green leaf hopper, White backed hopper, Rice gall midge, Whorl maggot	1000-1500	500	32
Cabbage	Diamond back moth	800-1000	500	07
Chilli	Thrips, Aphid, Fruit borer	800-1000	500	07
Sugarcane	Early shoot borer, Root borer	1500-2000	500	270
Cotton	Aphid, Thrips, Jassid, Whitefly	1500-2000	500	06
	Bollworms	2000	500	07

Trade Names

Sr. No.	Trade Name	Company
1	Agadi SC	ADAMA India Private Limited
2	Unipro SC	UPL Ltd.
3	Sonic Flo	Rallis India Limited
4	Shinzen Plus	IFFCO-MC Crop Science Private Ltd.
5	Star Gazette	Swal Corporation Ltd.
6	Mahaveer SC	Gharda Chemicals Ltd.
7	Fax SC	Dhanuka Agritech Ltd.
8	Regent SC	Bayer Crop Science Ltd.
9	Salvo SC	Coromandel International Ltd.
10	Aleenta	Alfa Crop Science
11	Fip Max-5	Auxin Crop Care
12	Dial	Anu Products Ltd.
13	Sargent	Insecticides (India) Ltd.
14	Devigent Plus	Arysta Lifescience
15	Rabid	Crop Life Science Ltd.
16	Mortel SC	Parijat Industries India Pvt. Ltd.
17	Rippen Super	Crop Chemicals India Ltd.
18	Efil	Sumitomo Chemical India Ltd.
19	Hemgent Plus	Hemani Industries Ltd., Ankleshwar
20	Chariot	Isagro Asia Agrochemicals Pvt. Ltd.
21	Heranil Plus	Herenba Industries Ltd.
22	Siknil	Sikko Industries Ltd.
23	Fipro Ultra Max	Somnath Demeter Agritron Ltd.
24	A-Fighter	Silver Sine Bio Tech
25	Fipromex	Natural Crop Science

26	Agent	Tropical Agrosystem India (P) Ltd.
27	Kavach	Essential Biosciences
28	Mebaz-SC	Synwood Agro Ltd.
29	Exylon Fiproguard	Torrent Crop Science
30	Revolt	Preventis Chem India (P) Ltd.
31	Moolan	RACCOLTO Agritech India Pvt. Ltd.

Fipronil 18.87 % SC

Crop	Pests	Dose / ha		Waiting period (Days)
		Quantity of formulation (ml)	Volume of water (Lit.)	
Cotton	Thrips	375	375-500	21
Chilli	Thrips, Aphid, Fruit borer	250	500	05
Rice	Stem borer, Leaf folder, Brown plant hopper	250	500	46

Trade Names

Sr. No.	Trade Name	Company
1	Regent Gold	Bayer Crop Science Ltd.
2	Reagent Super	Natural Crop Science
3	Kiwi	Koronation Bio-Science
4	Praz	Global Crop Care
5	Johnny	Gharda Chemicals Ltd.
6	Revolt Max	Preventis Chem India (P) Ltd.
7	Eerlik	RACCOLTO Agritech India Pvt. Ltd.

Fipronil 02.92 % EC

Crop	Pests	Dose		Waiting period (Days)
		Quantity of formulation (ml)	Volume of water (Lit.)	
Pre and Post construction (Building)	Termite	100	1	--

Trade Names

Sr. No.	Trade Name	Company
1	Agenda 25 EC	Bayer Crop Science Ltd.
2	Fepro-2.92	Sikko Industries Ltd.
3	Chili Ultra	Linuxcrop Chemicals Pvt. Ltd.
4	Pronil	RACCOLTO Agritech India Pvt. Ltd.

Fipronil 00.30 % GR

Crop	Pests	Dose / ha		Waiting period (Days)
		Quantity of formulation (Kg)	Volume of water (Lit.)	
Rice	Stem borer, Brown plant hopper, Green leaf	16.67-25.00	--	32

	hopper, White backed hopper, Rice gall midge, Whorl maggot			
Sugarcane	Early shoot borer, Root borer	25.00-33.30	--	09
Wheat	Termite	20.00	--	91

Trade Names

Sr. No.	Trade Name	Company
1	Agadi G	ADAMA India Private Limited
2	Sonic	Rallis India Limited
3	Shinzen	IFFCO-MC Crop Science Private Ltd.
4	Janbaaz GR	Biostadt India Ltd.
5	Mahaveer GR	Gharda Chemicals Ltd.
6	Regent GR	Bayer Crop Science Ltd.
7	Salvo Gr	Coromandel International Ltd.
8	Fip Max-G	Auxin Crop Care
9	Dial	Anu Products Ltd.
10	Quencher	Modern Insecticides Ltd.
11	Mortel GR	Parijat Industries India Pvt. Ltd.
12	Rippen	Parijat Industries India Pvt. Ltd.
13	Hemgent	Hemani Industries Ltd., Ankleshwar
14	Heranil Gr	Herenba Industries Ltd.
15	Fiprosik	Sikko Industries Ltd.
16	Feast	KP Agro Industries
17	Fenil-3	Silver Sine Bio Tech
18	ReAgent	Natural Crop Science
19	Agent GR	Tropical Agrosystem India (P) Ltd.
20	EBS Gold	Essential Biosciences
21	Ruler-Gr	Krishi Rasayan
22	Fipronil GR	JU Agri Sciences Pvt. Ltd.
23	SA-Grover GR	Saraswati Agro Life Science (India) Pvt. Ltd.
24	Revolt Gr	Preventis Chem India (P) Ltd.
25	Moolan GR	RACCOLTO Agritech India Pvt. Ltd.

Fipronil 00.60 % GR

Crop	Pests	Dose / ha		Waiting period (Days)
		Quantity of formulation (Kg)	Volume of water (Lit.)	
Rice	Stem borer, Leaf folder	10	--	65
Sugarcane	Early shoot borer, Termite	12.5	--	229
Wheat	Pink stem borer, Armyworm	10	--	106
	Termite	10	--	117

Trade Names

Sr. No.	Trade Name	Company
1	Dial Ultra	Anu Products Ltd.
2	Mortel Plus	Parijat Industries India Pvt. Ltd.
3	Diesal	Gharda Chemicals Ltd.
4	Regent Ultra	Bayer Crop Science Ltd.
5	Fenil-6	Silver Sine Bio Tech
6	ReAgent Plus	Natural Crop Science
7	Mebaz-G	Synwood Agro Ltd.
8	Ruler Power	Krishi Rasayan
9	JU Ultra	JU Agri Sciences Pvt. Ltd.
10	SA Grohit	Saraswati Agro Life Science (India) Pvt. Ltd.
11	Agrikind Fipro GR	Mankind Agritech
12	Portal Gr	RACCOLTO Agritech India Pvt. Ltd.

Fipronil 80.00 % WG

Crop	Pests	Dose / ha		Waiting period (Days)
		Quantity of formulation (g)	Volume of water (Lit.)	
Rice	Stem borer, Leaf folder	50-62.50	375-500	19
Grapes	Thrips	50-62.50	750-1000	10
Onion	Thrips	75	500	15
Cabbage	Diamond back moth	93.75	500	15
Chilli	Thrips	50-62.50	500	05
Cotton	Thrips	75	375-500	14

Trade Names

Sr. No.	Trade Name	Company
1	Agadi Super	ADAMA India Private Limited
2	Quartz	Gharda Chemicals Ltd.
3	Jump WG	Bayer Crop Science Ltd.
4	Ahead	Parijat Industries India Pvt. Ltd.
5	Fipro-WG	Sikko Industries Ltd.
6	A-Fightetr 80	Silver Sine Bio Tech
7	Flip Jump	Natural Crop Science
8	Trump	Tropical Agrosystem India (P) Ltd.
9	Jet	Koronation Bio-Science
10	Ruler-80	Krishi Rasayan
11	Gaba-N	Global Crop Care
12	JU Hemp	JU Agri Sciences Pvt. Ltd.
13	Invix	Synovita Crop science Pvt. Ltd.
14	Mankind Fipro	Mankind Agritech
15	Exylon Falcon	Torrent Crop Science
16	Fida	Preventis Chem India (P) Ltd.
17	Fipnil	RACCOLTO Agritech India Pvt. Ltd.

Fipronil 00.60 % WG

Crop	Pests	Dose / ha (Direct broadcasting)		Waiting period (Days)
		Quantity of formulation (Kg)	Volume of water (Lit.)	
Rice	Yellow stem borer	8.33-12.50	--	49
Sugarcane	Early shoot borer, Termite	13.33	--	238

Trade Names

Sr. No.	Trade Name	Company
1		

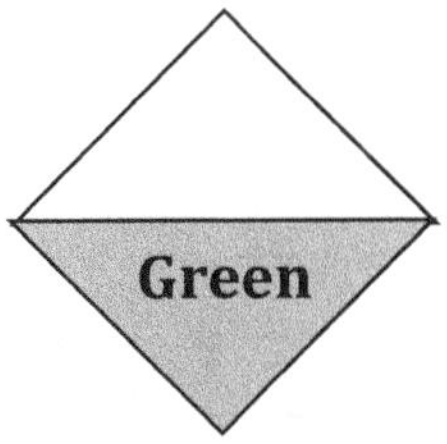

FLONICAMID

Available formulations in India : **Flonicamid 50.00 % WG**

Classification on the basis of

Target organism : Insecticide

Mode of entry : Systemic, Stomach and Contact poison

Mode of action : Nerve action

Chemical group : Pyridinecarboxamide

IRAC classification

Main Group and Primary Site of Action : 29 Chordotonal organ nicotinamidase inhibitors

Sub-group or exemplifying Active Ingredient : Flonicamid

Category of toxicity : Slightly toxic

Toxicity label colour : Green

- Flonicamid is a relatively new, selective systemic insecticide from the pyridinecarboxamide chemical class.
- It's primarily used to control piercing-sucking insects like aphids, whiteflies, psyllids, and some thrips.
- As a systemic insecticide, flonicamid is absorbed by the plant and transported throughout its tissues, protecting various parts of the plant. It also exhibits translaminar activity, moving from one leaf surface to the other.
- **Mode of action** : Flonicamid is classified under IRAC Group 29 (selective feeding blockers) due to its unique and still-elucidated mode of action. It disrupts insect chordotonal organs. These organs are crucial for sensing vibrations, hearing, balance, and movement. By affecting these, flonicamid affects the feeding behavior of susceptible insects, causing them to rapidly stop feeding upon exposure. This cessation of feeding ultimately leads to starvation and the eventual death of the pest.
- While the exact biochemical target isn't fully understood, it's thought to interfere with the stylet penetration and sap ingestion process in these insects.
- Additionally, flonicamid shows some ovicidal (egg-killing) and nymphicidal (young insect-killing) activity.

Label claim (DPPQ & S, Faridabad)

Flonicamid 50.00 % WG

Crop	Pests	Dose / ha		Waiting period (Days)
		Quantity of formulation (g)	Volume of water (Lit.)	
Rice	Brown plant hopper, White backed plant hopper, Green leaf hopper	150	500	36

Cotton	Aphid, Jassid, Thrips, Whitefly	150	500	25
Okra	Aphid, Jassid, Whitefly	200	500	10
Brinjal	Aphid, Jassid, Whitefly	200	500	15
Mango	Mango hopper	400	1000	15
Tomato	Aphid, Jassid, Whitefly	200	500	05
Chilli	Aphid, Jassid, Whitefly	200	500	05
Soybean	Aphid, Jassid, Whitefly	200	500	35
Potato	Aphid, Jassid, Whitefly	200	500	20

Trade Names

Sr. No.	Trade Name	Company
1	Ulala	UPL Ltd.
2	Panama	Swal Corporation Ltd.
3	Urrja	Green Bharat Chemical Pvt. Ltd.
4	Hideki	Bharat Certis Agriscience Ltd.
5	Renkli	Bayer Crop Science Ltd.

FLOMETOQUIN

Available formulations in India	:	**Flometoquin 10.00 % SC**
Classification on the basis of		
Target organism	:	Insecticide
Mode of entry	:	Contact and Stomach poison, Translaminar
Mode of action	:	Energy metabolism
Chemical group	:	Phenoxy-quinoline
IRAC classification		
Main Group and Primary Site of Action	:	Mitochondrial complex III electron transport inhibitors - Qi site
Sub-group or exemplifying Active Ingredient	:	Flometoquin
Category of toxicity	:	
Toxicity label colour	:	

- Flometoquin is a novel, high-performance insecticide developed by Nippon Kayaku and Mitsui Chemicals.
- **Mode of action** : Flometoquin works by inhibiting the mitochondrial electron transport chain (specifically at the Qi site of Complex III). It prevents the insect's cells from producing ATP (energy). Once energy production stops, the insect experiences a rapid knockdown effect. It stops moving, feeding, and dies shortly after exposure.
- It works through both contact and ingestion. It possesses translaminar activity (moves from the top of the leaf to the bottom), which is crucial for thrips that hide in the folds of leaves or flowers.
- It is highly effective against pests that have developed resistance to neonicotinoids or pyrethroids.

Label claim (DPPQ & S, Faridabad)

Flometoquin 10.00 % SC

Crop	Pests	Dose / ha		Waiting period (Days)
		Quantity of formulation (ml)	Volume of water (Lit.)	
Cabbage	Diamond back moth	750-1000	500	10
Chilli	Thrips	750-1000	500	10

Trade Names

Sr. No.	Trade Name	Company
1	Kajura	Nippon Kayaku and Mitsui Chemicals

Green

FLUBENDIAMIDE

Available formulations in India :

1. **Flubendiamide 20.00 % WG**
2. **Flubendiamide 39.35 % SC**
3. **Flubendiamide 00.70 % GR**

Classification on the basis of

Target organism : Insecticide
Mode of entry : Stomach and Contact poison
Mode of action : Muscle poison
Chemical group : Diamides

IRAC classification

Main Group and Primary Site of Action : 28 Ryanodine receptor modulators
Sub-group or exemplifying Active Ingredient : Diamides

Category of toxicity : Slightly toxic
Toxicity label colour : Green

- Flubendiamide is a synthetic insecticide in the diamide class, highly effective against a broad range of lepidopteran pests (moths and butterflies) and some coleopteran pests (beetles). It generally has a favourable safety profile for non-target organisms.
- This insecticide is non-systemic, primarily working through ingestion, though it also offers some contact activity. It provides ovicidal (egg-killing), larvicidal (larvae-killing), and some adulticidal (adult-killing) activity against susceptible pests.
- **Mode of action :** Flubendiamide's specific mode of action targets the ryanodine receptors in insect muscles, which are crucial for muscle contraction. It acts as a ryanodine receptor agonist, targeting crucial calcium release channels in insect muscle cells. By binding to these receptors, it triggers their uncontrolled activation, leading to an excessive release of calcium ions. This disrupts normal muscle function, causing uncontrolled muscle contraction, paralysis, and rapid feeding cessation, ultimately resulting in the insect's death.
- While flubendiamide itself is very stable to pH changes, a slightly acidic to neutral is generally recommended for spray solutions.

Label claim (DPPQ & S, Faridabad)

Flubendiamide 20.00 % WG

Crop	Pests	Dose / ha		Waiting period (Days)
		Quantity of formulation (g)	Volume of water (Lit.)	
Rice	Stem borer, Leaf folder	125	500	30
Cotton	American bollworm	250	500	30
Pigeonpea	Pod borer	250	500	30

Cabbage	Diamond back moth	62.5	500	07
Tomato	Fruit borer	250	500	05
Tea	Semilooper	150	400	07
Chilli	Fruit borer	250-300	500	05
Soybean	Tobacco caterpillar, Semilooper	250-300	500	29
Groundnut	Tobacco caterpillar	300	500	31
Black gram	Tobacco caterpillar, Spotted pod borer (*Maruca* sp.)	300	500	23
Bengal gram	Pod borer	250	500	15
Sugarcane	Early shoot borer	375	500-750	204
Maize	Fall armyworm	250	500	55

Trade Names

Sr. No.	Trade Name	Company
1	Takumi	Rallis India Limited
2	Invade	UPL Ltd.
3	Fluton	PI Industries
4	Takibi	IFFCO-MC Crop Science Private Ltd.
5	Kabukey	Swal Corporation Ltd.
6	Katana	Nichino India
7	Yakumi	Global Crop Care

Flubendiamide 39.35 % SC

Crop	Pests	Dose / ha		Waiting period (Days)
		Quantity of formulation (ml)	Volume of water (Lit.)	
Rice	Stem borer, Leaf folder	50	375-500	40
Cotton	American bollworm, Spotted bollworm	100-125	375-500	25
Pigeonpea	Pod borer	100	500	10
Cabbage	Diamond back moth	37.5-50	375-500	07
Tomato	Fruit borer	100	375-500	05
Brinjal	Shoot and fruit borer	150-187.5	500	05
Chilli	Fruit borer	100-125	500	07
Soybean	Defoliaters (Gram pod borer, Tobacco caterpillar, Semilooper	150	500	17
Black gram	Pod borer	100	500	11
Bengal gram	Pod borers (Gram pod borer, Tobacco caterpillar)	150	500	05
Gherkin	Fruit borer	75-100	500	05
Cardamom	Capsule borer	15 ml / 100 lit water	500	15

Trade Names

Sr. No.	Trade Name	Company
1	Fame	Bayer Crop Science Ltd.
2	Voter	Biostadt India Ltd.
3	Bandhu	Anu Products Ltd.
4	Flubensik	Sikko Industries Ltd.
5	Fluben	Katyayani Organics
6	Glory	Global Crop Care

Flubendiamide 00.70 % GR

Crop	Pests	Dose / ha		Waiting period (Days)
		Quantity of formulation (Kg)	Volume of water (Lit.)	
Rice	Stem borer	12.14-14.28	--	25

Trade Names

Sr. No.	Trade Name	Company
1	Shori	Nichino India
2	Zygant	Tata Rallis India Ltd.

Blue

FLUPYRADIFURONE

Available formulations in India	:	**Flupyradifurone 17.09 % SL**
Classification on the basis of		
Target organism	:	Insecticide
Mode of entry	:	Systemic
Mode of action	:	Nerve poison
Chemical group	:	Butenolides
IRAC classification		
Main Group and Primary Site of Action	:	4 Nicotinic acetylcholine receptor (nAChR) competitive modulators
Sub-group or exemplifying Active Ingredient	:	4D Butenolides
Category of toxicity	:	Moderately toxic
Toxicity label colour	:	Blue

- Flupyradifurone is a modern insecticide. It's a key tool in pest management due to its specific mode of action and efficacy against various sucking insect pests.
- **Mode of Action :** It acts as an agonist of the nicotinic acetylcholine receptor (nAChR) in the insect's central nervous system. This means it binds to and activates these receptors, similar to natural neurotransmitters, leading to an uncontrolled influx of calcium ions. This continuous overstimulation of the insect's nervous system ultimately results in paralysis and death.
- Flupyradifurone is a systemic insecticide, meaning it is absorbed by the plant and transported throughout its tissues. This allows it to reach pests that feed on different parts of the plant, including those on the underside of leaves or within plant tissues.
- Flupyradifurone is not significantly affected by changes in pH in typical environmental conditions (4-9 pH), making it a stable compound in water. Its primary degradation pathway in the environment is through photolysis (degradation by light), not hydrolysis.

Label claim (DPPQ & S, Faridabad)

Flupyradifurone 17.09 % SL

Crop	Pest	Dose / ha		Waiting period (Days)
		Quantity of formulation (ml)	Volume of water (Lit.)	
Okra	Jassid, Whitefly	1250	500	20
Tea	Mosquito bug	750	400-500	07

Trade Names

Sr. No.	Trade Name	Company
1	Sivanto Prime	Bayer Crop Science Ltd.

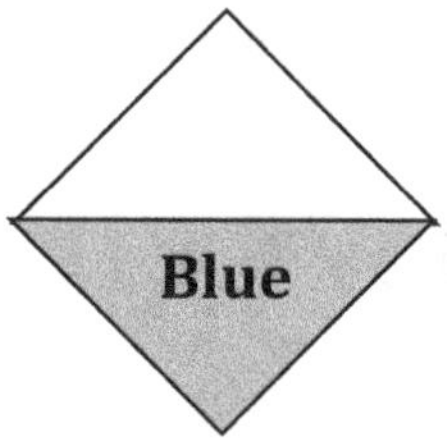

FLUPYRIMIN

Available formulations in India : 1. **Flupyrimin 02.00 % GR**
2. **Flupyrimin 10.00 % SC**

Classification on the basis of

Target organism : Insecticide
Mode of entry : Systemic, Contact poison
Mode of action : Nerve poison
Chemical group : Pyridinylidene

IRAC classification

Main Group and Primary Site of Action : 4 Nicotinic acetylcholine receptor (nAChR) competitive modulators
Sub-group or exemplifying Active Ingredient : 4F Pyridylidenes

Category of toxicity : Moderately toxic
Toxicity label colour : Blue

- Flupyrimin is a modern insecticide with key applications in controlling major insect pests in rice.
- Flupyrimin acts as a nicotinic acetylcholine receptor (nAChR) competitive modulator. It works as a nicotinic antagonist, binding to the insect's nAChRs and disrupting the transmission of nerve signals. This leads to the rapid paralysis and death of the insect.
- It is highly effective against major rice pests, particularly brown planthopper (BPH) and yellow stem borer (YSB), including populations that have developed resistance to neonicotinoids like imidacloprid.
- Flupyrimin exhibits good stability to hydrolysis.

Label claim (DPPQ & S, Faridabad)

Flupyrimin 02.00 % GR

Crop	Pests	Dose / ha		Waiting period (Days)
		Quantity of formulation (Kg)	Volume of water (Lit.)	
Rice	Stem borer, Brown plant hopper	5.0-7.5 Kg	--077	

Trade Names

Sr. No.	Trade Name	Company
1	Karmax	Swal Corporation Ltd.
2	Kevuka	UPL Ltd.

Flupyrimin 10.00 % SC

Crop	Pests	Dose / ha		Waiting period (Days)
		Quantity of formulation (ml)	Volume of water (Lit.)	
Rice	Brown plant hopper	750-1000	500	07

Trade Names

Sr. No.	Trade Name	Company
1	Viola	Swal Corporation Ltd.
2	Imagine	UPL Ltd.

Yellow

FLUVALINATE

Available formulations in India : **Fluvalinate 25.00 % EC**

Classification on the basis of

Target organism : Insecticide, Acaricide
Mode of entry : Stomach and Contact poison
Mode of action : Nerve poison
Chemical group : Pyrethroid

IRAC classification

Main Group and Primary Site of Action : 3 Sodium channel modulators
Sub-group or exemplifying Active Ingredient : 3A Pyrethroid

Category of toxicity : Highly toxic
Toxicity label colour : Yellow

- Fluvalinate is a synthetic pyrethroid insecticide and acaricide widely used to control a broad range of insect and mite pests in both agricultural and non-agricultural environments.
- **Mode of action :** It acts as a contact and stomach poison, effectively disrupting the central nervous system of pests. Fluvalinate achieves this by modulating sodium channels, leading to rapid paralysis and ultimately death.
- Technical fluvalinate is a mixture of four stereoisomers. However, commercial products typically utilize only one active isomer, tau-fluvalinate, which is the most potent form.
- One of its most significant applications is in the control of Varroa mites in honey bee colonies, where these mites pose a major threat to bee health and survival.

Label claim (DPPQ & S, Faridabad)

Fluvalinate 25.00 % EC

Crop	Pests	Dose / ha		Waiting period (Days)
		Quantity of formulation (ml)	Volume of water (Lit.)	
Cotton	Aphid, Jassid, Red cotton bug, Bollworm	200-400	500-1000	07

Trade Names

Sr. No.	Trade Name	Company
1	IPL-Contanol EC	IPL

Blue

FLUXAMETAMIDE

Available formulations in India : **Fluxametamide 10.00 % EC**

Classification on the basis of

Target organism : Insecticide

Mode of entry : Stomach and Contact poison, Translaminar

Mode of action : Nerve poison

Chemical group : Isoxazoline

IRAC classification

Main Group and Primary Site of Action : 30 GABA-gated chloride channel allosteric modulators

Sub-group or exemplifying Active Ingredient : Meta-diamides

Category of toxicity : Moderately toxic

Toxicity label colour : Blue

- Fluxametamide is a powerful, broad-spectrum insecticide from the isoxazoline chemical family. It's used to control a wide range of insect pests on various crops.
- **Mode of action :** Fluxametamide's unique mode of action makes it highly effective. It acts as an antagonist on the insect's nervous system, specifically targeting ligand-gated chloride channels (LGCCs). By disrupting the channels that are controlled by gamma-aminobutyric acid (GABA) and glutamate, it causes the insect's nervous system to become overstimulated. This leads to paralysis and, ultimately, death.

  A key advantage of fluxametamide is that it binds to a different site on the GABA-gated chloride channels than older insecticides. This makes it a valuable tool for managing pests that have developed resistance to other products. Additionally, it is highly selective, targeting insects while being safer for mammals.
- Fluxametamide works through both contact and stomach action. It also has translaminar activity, which allows it to penetrate the leaf surface and kill hidden pests on the underside of leaves.

Label claim (DPPQ & S, Faridabad)

Fluxametamide 10.00 % EC

Crop	Pests	Dose / ha		Waiting period (Days)
		Quantity of formulation (ml)	Volume of water (Lit.)	
Brinjal	Leaf hopper, Thrips, Shoot and fruit borer	400	500	05
Cabbage	Diamond back moth, Tobacco caterpillar, Semilooper	400	500	05

Chilli	Thrips, Fruit borer, Tobacco caterpillar	400	500	05
Okra	Leaf hopper, Thrips, Fruit borer	400	500	05
Red gram	Spotted pod borer, Pod borer	400	500	05
Tomato	Thrips, Fruit borer	400	500	05

Trade Names

Sr. No.	Trade Name	Company
1	Gracia	Godrej Agrovet
2	Shinwa	Insecticides (India) Ltd.

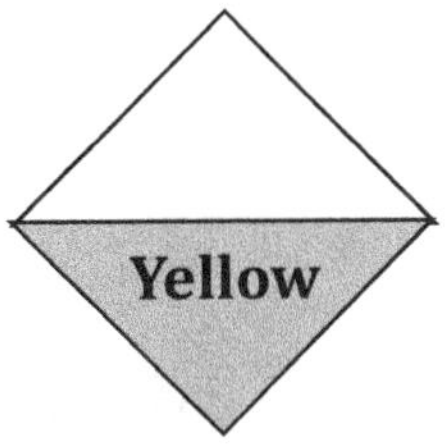

IMIDACLOPRID

Available formulations in India :

1. **Imidacloprid 70.00% WG**
2. **Imidacloprid 48.00% FS**
3. **Imidacloprid 70.00% WS**
4. **Imidacloprid 30.50% SC**
5. **Imidacloprid 17,80% SL**
6. **Imidacloprid 17.10% SL**
7. **Imidacloprid 00.30% GR**

Classification on the basis of

Target organism	:	Insecticide
Mode of entry	:	Systemic, Stomach and Contact poison
Mode of action	:	Nerve poison
Chemical group	:	Neonicotinoid
IRAC classification		
Main Group and Primary Site of Action	:	4 Nicotinic acetylcholine receptor (nAchR) competitive modulators
Sub-group or exemplifying Active Ingredient	:	4A Neonicotinoids
Category of toxicity	:	Highly toxic
Toxicity label colour	:	Yellow

- Imidacloprid is a widely used systemic insecticide belonging to the neonicotinoid class. It mimics nicotine, a natural insecticide found in tobacco. Being systemic, when applied to soil or leaves, it's absorbed by the plant and distributed throughout its tissues—including stems, leaves, fruit, and flowers. This effectively makes the entire plant toxic to any insects that feed on it.
- **Mode of action** : Imidacloprid acts as a neurotoxin, disrupting the insect's nervous system. It's effective through both contact and ingestion. Specifically, imidacloprid works by binding to nicotinic acetylcholine receptors (nAChRs) in the insect's central nervous system. This binding is significantly stronger in insects compared to mammals and birds, which makes it selectively more toxic to insects. The continuous activation of these receptors disrupts normal nerve signal transmission, leading to initial overstimulation and spontaneous nerve firing, followed by a blockage of nerve signals, ultimately causing paralysis and death of the insect.
- A major concern with imidacloprid is its high toxicity to honey bees and other beneficial insects.

Label claim (DPPQ & S, Faridabad)

Imidacloprid 70.00% WG

Crop	Pests	Dose / ha		Waiting period (Days)
		Quantity of formulation (g)	Volume of water (Lit.)	
Cotton	Jassid, Aphid, Thrips	30-35	375-500	07
Paddy	Brown plant hopper, White backed plant hopper	30-35	300-375	07
Okra	Jassid, Aphid, Thrips	30-35	300-375	03
Cucumber	Aphid, Jassid	35	500	05
Tomato	Thrips, Whitefly	50	500	30
Potato	Aphid, Whitefly	90	500	30

Trade Names

Sr. No.	Trade Name	Company
1	Looper	Crystal Crop Protection Ltd.
2	Cohigan WG	ADAMA India Private Limited
3	Intice	Biostadt India Ltd.
4	Ad-Fyre	Dhanuka Agritech Ltd.
5	Admire	Bayer Crop Science Ltd.
6	Anumida Gold	Anu Products Ltd.
7	Ado-70	Crop Life Science Ltd.
8	Noble	Crop Chemicals India Ltd.
9	M-Con WG	Sumitomo Chemical India Ltd.
10	Hemprid	Hemani Industries Ltd., Ankleshwar
11	Admit Super	Isagro Asia Agrochemicals Pvt. Ltd.
12	Brazil 70	GFG Crop Science Pvt. Ltd.
13	Mantra G	Herenba Industries Ltd.
14	Midaton	Sikko Industries Ltd.
15	Bolt 70% WG	Nichino India
16	Boom	KP Agro Industries
17	Emidex	Natural Crop Science
18	Tagprid	Tropical Agrosystem India (P) Ltd.
19	Imidapro	Essential Biosciences
20	Adfas	Frontline Agri Science
21	Advice 70	Green India Care
22	Mankind Imida 70	Mankind Agritech
23	Exylon Impactor	Torrent Crop Science
24	Merit	Preventis Chem India (P) Ltd.
25	Accusto Gold	RACCOLTO Agritech India Pvt. Ltd.

Imidacloprid 48.00% FS

Crop	Pests	Dose / Kg seed		Waiting period (Days)
		Quantity of formulation (ml)	Volume of water (Lit.)	
Cotton	Aphid, Jassid, Thrips, Whitefly	5.0-9.0	--	--

Okra	Jassid, Aphid	5.0-9.0	--	--
Sunflower	Jassid, Whitefly	5.0-9.0	--	--
Sorghum	Shoot fly	12.0	--	--
Pearl millet	Shoot fly, Termite	12.0	--	--
Soybean	Jassid	1.25	--	--
Maize	Shoot fly	1.0	--	--
Potato	Aphid. Jassid	0.0175	--	--
Rice	Thrips	0.25	--	--
Wheat	Aphid. Termite	0.35	--	--
Green Pea	Aphid, Jassid	0.2-0.3	--	--
Bengal Gram	Termite	0.25	--	--
Black gram	Aphid, Jassid, Whitefly	0.7	--	--
Cumin	Aphid, Jassid, Thrips	0.6-0.7	--	--
Groundnut	Termite	0.2	--	--
Chilli	Aphid, Jassid, Thrips	1.2-1.5	--	--

Trade Names

Sr. No.	Trade Name	Company
1	Gaucho 600 FS	Bayer Crop Science Ltd.
2	Imicon Preimum	Hyderabad Chemical Products Ltd.
3	Hi Mida FS	HPM Chemicals & Fertilizers Ltd.
4	Amida	Crystal Crop Protection Ltd.
5	Imigrow	Crop Life Sciences Ltd.
6	Tatamida 600 FS	Rallis India Limited
7	Seedflow FS	Chemet Chemicals Pvt. Ltd.
8	Mida FS	Sikko Industries Ltd.
9	Emidex	Natural Crop Science
10	Premix	Global Crop Care
11	Imidon FS	Preventis Chem India (P) Ltd.
12	Adingo	RACCOLTO Agritech India Pvt. Ltd.

Imidacloprid 70.00% WS

Crop	Pests	Dose / Kg seed		Waiting period (Days)
		Quantity of formulation (ml)	Volume of water (Lit.)	
Cotton	Aphid, Jassid, Thrips, Whitefly	5-10	--	--
Okra	Jassid, Aphid	5-10	--	--
Chilli	Jassid, Aphid, Thrips	10-15	--	--
Sunflower	Jassid, Whitefly	7	--	--
Sugarcane	Termite	10-15	--	--
Sorghum	Shoot fly	10	--	--
Pearl millet	Termite, Shoot fly	10	--	--
Mustard	Mustard sawfly, Painted bug	7	--	--

Trade Names

Sr. No.	Trade Name	Company
1	Amida 70 WS	Crystal Crop Protection Ltd.
2	Intice	Biostadt India Ltd.
3	Seed Touch WS	Chemet Chemicals Pvt. Ltd.
4	Imida WS	Modern Insecticides Ltd.
5	Midapower	Sikko Industries Ltd.
6	Magik 70 WS	Tropical Agrosystem India (P) Ltd.
7	Complir	RACCOLTO Agritech India Pvt. Ltd.

Imidacloprid 30.50% SC

Crop	Pests	Dose / ha		Waiting period (Days)
		Quantity of formulation (ml)	Volume of water (Lit.)	
Cotton	Aphid, Jassid, Thrips	60-75	500-750	26
Rice (Paddy)	Brown plant hopper, White backed plant hopper	60-75	500-750	37
Chilli	Aphid, Thrips	125-150	500	5
For non- agricultural use: For protecting building from termite attack at pre and post construction stages, apply Imidacloprid 30.5% SC @ 0.075% a.i. concentration.				

Trade Names

Sr. No.	Trade Name	Company
1	Ultimo Super	Biostadt India Ltd.
2	Media Super	Dhanuka Agritech Ltd.
3	Confidor Super	Bayer Crop Science Ltd.
4	A-Mida Super Plus	Auxin Crop Care
5	Anumida Super	Anu Products Ltd.
6	Imida SC	Modern Insecticides Ltd.
7	Imidacel Super	Excel Crop Care Ltd.
8	Interex	Crop Life Science Ltd.
9	Nagara Super	Parijat Industries India Pvt. Ltd.
10	Imidore Super	Crop Chemicals India Ltd.
11	M-Con Super	Sumitomo Chemical India Ltd.
12	Abatis	Isagro Asia Agrochemicals Pvt. Ltd.
13	Brazil Super	GFG Crop Science Pvt. Ltd.
14	Mantra Super	Herenba Industries Ltd.
15	Midasik	Sikko Industries Ltd.
16	Imicon Super	Nichino India
17	Demida	Somnath Demeter Agritron Ltd.
18	Super Sinemida	Silver Sine Bio Tech
19	Emida Gold	Natural Crop Science
20	Magik Super	Tropical Agrosystem India (P) Ltd.
21	Immediate	Essential Biosciences
22	Mida Plus	Synwood Agro Ltd.

23	Elmida Super	Elgon Crop Science
24	Fasmida	Frontline Agri Science
25	Samrat 30.5	JU Agri Sciences Pvt. Ltd.
26	Nisan Super	Linuxcrop Chemicals Pvt. Ltd.
27	Imida-Force	Mankind Agritech
28	Exylon Exymax	Torrent Crop Science
29	Amaze Super	Preventis Chem India (P) Ltd.
30	Accusto Super	RACCOLTO Agritech India Pvt. Ltd.

Imidacloprid 17.80% SL

Crop	Pests	Dose / ha		Waiting period (Days)
		Quantity of formulation (ml)	Volume of water (Lit.)	
Cotton	Aphid, Jassid, Thrips, Whitefly	100-125	500-700	40
Paddy (Rice)	Brown plant hopper, White backed plant hopper, Green leaf hopper	100-125	500-700	40
Chilli	Aphid, Jassid, Thrips	125-150	500-700	40
Sugarcane	Termite	350	1875	45
Mango	Hopper	2-4 ml/tree	10 lit/tree	45
Sunflower	Jassid, Thrips, Whitefly	100	500	30
Okra	Aphid, Jassid, Thrips	100	500	03
Citrus	Leaf miner, Psylla	50	Depending upon size of tree and protection equipment used	15
Groundnut	Aphid, Jassid	100-125	500	40
Tomato	Whitefly	150-175	500	03
Grapes	Flea beetle	300-400	1000	32

Trade Names

Sr. No.	Trade Name	Company
1	Confidor	Bayer Crop Science Ltd.
2	Tatamida SL	Rallis India Limited
3	Jumbo	PI Industries
4	Imidagold	UPL Ltd.
5	Cohigan	ADAMA India Private Limited
6	Hilmida	HIL (India) Limited
7	Isogashi	IFFCO-MC Crop Science Private Ltd.
8	Imidastar	Swal Corporation Ltd.
9	Ultimo 200 SL	Biostadt India Ltd.
10	Maharaja	Gharda Chemicals Ltd.
11	Media	Dhanuka Agritech Ltd.
12	Confidence 555	Crystal Crop Protection Ltd.

13	Parrymida	Coromandel International Ltd.
14	Atom	Indofil Industries Ltd.
15	Pure Mida	Alfa Crop Science
16	A-Mida Super	Auxin Crop Care
17	Anumida	Anu Products Ltd.
18	Counter	Modern Insecticides Ltd.
19	Nagarjuna Mida	NACL Industries Ltd.
20	Imidacel	Excel Crop Care Ltd.
21	Imigrow	Crop Life Science Ltd.
22	Nagara	Parijat Industries India Pvt. Ltd.
23	Imidore	Crop Chemicals India Ltd.
24	M-Con	Sumitomo Chemical India Ltd.
25	Sumida	Super Crop Safe Ltd., Ahemadabad
26	Admit	Isagro Asia Agrochemicals Pvt. Ltd.
27	Brazil	GFG Crop Science Pvt. Ltd.
28	G Sensor	Tusova Agro Pvt. Ltd.
29	Mantra	Herenba Industries Ltd.
30	Sikkomida	Sikko Industries Ltd.
31	Imicon	Nichino India
32	Sine Mida	Silver Sine Bio Tech
33	Emida Silver	Natural Crop Science
34	Magik	Tropical Agrosystem India (P) Ltd.
35	Imidasecure	Essential Biosciences
36	Synmida	Synwood Agro Ltd.
37	Revange	Agrosis Crop Science
38	Samrat	JU Agri Sciences Pvt. Ltd.
39	Nisan	Linuxcrop Chemicals Pvt. Ltd.
40	Imidhuge	Mankind Agritech
41	Exylon Exymida	Torrent Crop Science
42	Imidon	Preventis Chem India (P) Ltd.
43	Accusto	RACCOLTO Agritech India Pvt. Ltd.

Imidacloprid 00.30% GR

Crop	Pests	Dose / ha		Waiting period (Days)
		Quantity of formulation (Kg)	Volume of water (Lit.)	
Paddy (Rice)	Stem borer	15 Kg	--	26

Trade Names

Sr. No.	Trade Name	Company
1	Anumet	Anu Products Ltd.
2	Ultimate	Excel Crop Care Ltd.
3	Imida Super	Proxima Bio Tech Private Ltd.
4	Magic Gold	Tropical Agrosystem India (P) Ltd.
5	Kumi	RACCOLTO Agritech India Pvt. Ltd.

Imidacloprid 17.10% SL

Crop	Pests	Dose / ha		Waiting period (Days)
		Quantity of formulation (g)	Volume of water (Lit.)	
Cotton	Aphid, Jassid, Thrips, Whitefly	250	500	50
Rice	Brown plant hopper, White backed plant hopper, Green leaf hopper	300	500	39

Trade Names

Sr. No.	Trade Name	Company
1	Confidor	Bayer Crop Science Ltd.

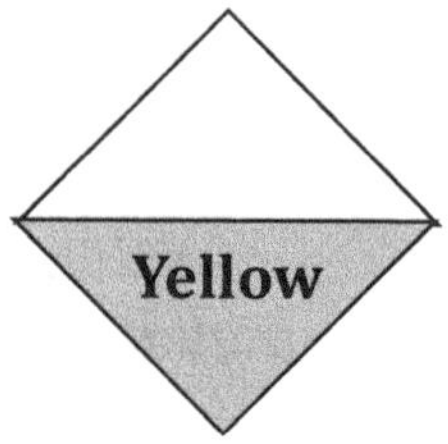

INDOXACARB

Available formulations in India :
1. **Indoxacarb 14.50 % SC**
2. **Indoxacarb 15.80 % EC**

Classification on the basis of

Target organism : Insecticide

Mode of entry : Stomach and Contact poison

Mode of action : Nerve poison

Chemical group : Oxadiazines

IRAC classification

Main Group and Primary Site of Action : 22 Voltage-dependent sodium channel blockers

Sub-group or exemplifying Active Ingredient : 22A Oxadiazines

Category of toxicity : Highly toxic

Toxicity label colour : Yellow

- Indoxacarb is an oxadiazine insecticide. It's a non-systemic insecticide primarily used to control lepidopteran larvae (like caterpillars) in various crops.
- **Mode of action** : Indoxacarb works through both contact and ingestion. It's a pro-insecticide, meaning the insect needs to metabolize it to unlock its full activity. Once an insect ingests or absorbs indoxacarb, enzymes, mainly in its gut, convert it into a more active form. This active metabolite then binds to voltage-dependent sodium channels in the insect's nerve cells. This binding disrupts the normal flow of sodium ions, which are crucial for nerve signal transmission. This disruption leads to tremors, the insect stops feeding, becomes paralyzed, and eventually dies.
- Indoxacarb's action on sodium channels is different from that of pyrethroid insecticides. This means that insects resistant to pyrethroids may still be susceptible to indoxacarb, offering an alternative for pest management.

Label claim (DPPQ & S, Faridabad)

Indoxacarb 14.50 % SC

Crop	Pests	Dose / ha		Waiting period (Days)
		Quantity of formulation (ml)	Volume of water (Lit.)	
Cotton	Bollworm	500	600-1000	16
Cabbage	Diamond back moth	200-266	400-750	07
Chilli	Fruit borer	333-400	300-600	05
Tomato	Fruit borer	400-500	300-600	05
Pigeonpea	Pod borer complex	333-400	500-1000	15
Rice	Yellow stem borer, Leaf folder	200-333	500	37

Soybean	Tobacco caterpillar	333	500	36
Chickpea	Pod borer	400-500	500	19

Trade Names

Sr. No.	Trade Name	Company
1	Avaunt	DuPont Crop Protection FMC Coprporation
2	King Dixa	Gharda Chemicals Ltd.
3	Queen Doxa	Auxin Crop Care
4	Winner	Modern Insecticides Ltd.
5	King Carb	Parijat Industries India Pvt. Ltd.
6	Isacarb	Isagro Asia Agrochemicals Pvt. Ltd.
7	Adora	Tusova Agro Pvt. Ltd.
8	Heroxa	Herenba Industries Ltd.
9	Sikkovant	Sikko Industries Ltd.
10	Easikill	Somnath Demeter Agritron Ltd.
11	Indoxa Prime	Silver Sine Bio Tech
12	Indoxa Gold	Natural Crop Science
13	Power	Tropical Agrosystem India (P) Ltd.
14	Vinashak	Essential Biosciences
15	Banner	Preventis Chem India (P) Ltd.
16	Makkam	RACCOLTO Agritech India Pvt. Ltd.

Indoxacarb 15.80 % SC

Crop	Pests	Dose / ha		Waiting period (Days)
		Quantity of formulation (ml)	Volume of water (Lit.)	
Cotton	American bollworm	500	500-1000	14
Cabbage	Diamond back moth	266	500-1000	05
Pigeonpea	Pod borer complex	333	500-700	12
Rice	Leaf folder	200	500	14
Soybean	Tobacco caterpillar, Gram pod borer, Green semilooper, Stem fly	333	500	31
Chickpea	Pod borer	333	500	18

Trade Names

Sr. No.	Trade Name	Company
1	Avaunt	DuPont Crop Protection FMC Coprporation
2	Dhava Gold	Dhanuka Agritech Ltd.
3	Indo gold	Star Chemicals
4	Ascend	Zuari Industries Ltd.
5	Zulu	RACCOLTO Agritech India Pvt. Ltd.

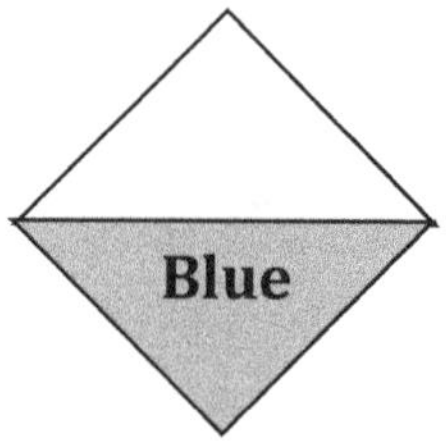

ISOCYCLOSERAM

Available formulations in India :
1. **Isocycloseram 09.20% (10.00%) DC**
2. **Isocycloseram 18.10 % (20.00 %) SC**

Classification on the basis of

Target organism : Insecticide, Acaricide
Mode of entry : Stomach and Contact poison
Mode of action : Nerve poison
Chemical group : Isoxazoline

IRAC classification

Main Group and Primary Site of Action : Group 30. GABA-gated channel allosteric modulators
Sub-group or exemplifying Active Ingredient : Isoxazolines

Category of toxicity : Slightly toxic
Toxicity label colour : Blue

- Isocycloseram is a relatively new and highly effective isoxazoline insecticide and acaricide.
- It offers broad-spectrum control against a wide range of agricultural and urban pests.
- **Mode of action :** This insecticide works by acting as a GABA-gated chloride channel allosteric modulator. This means it targets an insect's nervous system by binding to a unique site on the GABA (gamma-aminobutyric acid) receptor. This binding site is different from those targeted by older insecticides like fiproles and cyclodienes. By interfering with chloride ion influx, isocycloseram causes hyperexcitation of the insect's nervous system, leading to paralysis and eventual death.
- It's effective through both contact and ingestion. Some sources indicate it might have some systemic activity in plants, being absorbed and translocated to different parts.
- Its unique binding site on the GABA receptor is different from fiproles and organochlorines, making it potentially effective against pests resistant to these older insecticide classes.

Label claim (DPPQ & S, Faridabad)

Isocycloseram 09.20% (10.00%) DC

Crop	Pests	Dose / ha		Waiting period (Days)
		Quantity of formulation (ml)	Volume of water (Lit.)	
Brinjal	Jassid, Red spider mite	200	500	05
	Shoot and fruit borer	600	500	05
Cabbage	Leaf feeder, Diamond back moth	200-300	500	10

Chilli	Yellow mite	200	500	07
	Thrips, Fruit borer	600	500	07
Cotton	Jassid	200	500	37
	Thrips, Bollworm	600	500	37
Pigeonpea	Gram pod borer, Spotted pod borer	500-600	500	58
Groundnut	Leaf miner, Leaf feeder, Thrips, Jassid	500-600	500	48
Soybean	Leaf worm, Semilooper, Girdle beetle, Stem fly	600	500	35

Trade Names

Sr. No.	Trade Name	Company
1	Simodis	Syngenta India Ltd.

Isocycloseram 18.10% (20.00%) SC

Crop	Pests	Dose / ha		Waiting period (Days)
		Quantity of formulation (ml)	Volume of water (Lit.)	
Rice	Leaf folder	100	500	29
	Stem borer	300	500	29
Maize	Fall armyworm, Stem borer	300	500	48

Trade Names

Sr. No.	Trade Name	Company
1	Incipio	Syngenta India Ltd.

Yellow

LAMBDA-CYHALOTHRIN

Available formulations in India :
1. **Lambda-cyhalothrin 04.90 % CS**
2. **Lambda-cyhalothrin 02.50 % EC**
3. **Lambda-cyhalothrin 05.00 % EC**

Classification on the basis of

Target organism : Insecticide

Mode of entry : Stomach and Contact poison

Mode of action : Nerve poison

Chemical group : Pyrethroid

IRAC classification

Main Group and Primary Site of Action : 3 Sodium channel modulators

Sub-group or exemplifying Active Ingredient : 3A Pyrethroid

Category of toxicity : Highly toxic

Toxicity label colour : Yellow

- Lambda-cyhalothrin is a synthetic pyrethroid insecticide widely used in agriculture and public health for its broad-spectrum activity and rapid knockdown effect on various pests. It mimics the action of natural pyrethrins found in chrysanthemum flowers but is more stable and effective.
- **Mode of action :** Lambda-cyhalothrin is a non-systemic insecticide with contact, stomach, and repellent properties. It works by targeting the nervous system of insects. It disrupts the gating mechanism of voltage-gated sodium channels on the surface of insect nerve cells. This disruption keeps the sodium channels in an open position, leading to a prolonged influx of sodium ions into the nerve cells. This causes continuous nerve impulses, resulting in rapid knockdown, paralysis, and ultimately death in the insects.
- It is highly toxic to many fish and aquatic invertebrate species, and highly toxic to bees.

Label claim (DPPQ & S, Faridabad)

Lambda-cyhalothrion 04.90 % CS

Crop	Pest	Dose / ha		Waiting period (Days)
		Quantity of formulation (ml)	Volume of water (Lit.)	
Cotton	Bollworms	500	500	21
Paddy (Rice)	Stem borer, Leaf folder	250	500	15
Brinjal	Shoot and fruit borer	300	500	05
Okra	Fruit borer	300	500	05
Tomato	Fruit borer	300	500	05
Grapes	Thrips, Flea beetle	250	500-1000	07

Chilli	Thrips, Pod borer	500	500	05
Soybean	Stem fly, Semilooper	300	500	31
Pomegranate	Thrips, Fruit borer	400-800	500-1000	05
Cardamom	Shoot and capsule borer, Thrips	400	1000	34

Trade Names

Sr. No.	Trade Name	Company
1	Baaz	Crystal Crop Protection Ltd.
2	Yuri	IFFCO-MC Crop Science Private Ltd.
3	Jackal	Dhanuka Agritech Ltd.
4	Matador	Syngenta India Ltd.
5	Matchless	Anu Products Ltd.
6	Metro	Insecticides (India) Ltd.
7	Ladayak	Crop Life Science Ltd.
8	Mantra	Crop Chemicals India Ltd.
9	Jayam Plus	Herenba Industries Ltd.
10	Lacycap	Sikko Industries Ltd.
11	Lakra	Best Agrolife Ltd.
12	Lambda5X	Somnath Demeter Agritron Ltd.
13	Lamtone	Silver Sine Bio Tech
14	Jackshot	Natural Crop Science
15	Command Super	Tropical Agrosystem India (P) Ltd.
16	Cyhalomaster	Essential Biosciences
17	Lekso Pro	Synwood Agro Ltd.
18	Arjun	Indian Agrochemicals
19	Loc++	Krishi Rasayan
20	Keeper	Green India Care
21	Lambda CS	JU Agri Sciences Pvt. Ltd.
22	Bazz	Linuxcrop Chemicals Pvt. Ltd.
23	SA-Sisu	Saraswati Agro Life Science (India) Pvt. Ltd.
24	Exylon Luxor	Torrent Crop Science
25	Dhulka CS	RACCOLTO Agritech India Pvt. Ltd.

Lambda-cyhalothrion 02.50 % EC

Crop	Pests	Dose / ha		Waiting period (Days)
		Quantity of formulation (ml)	Volume of water (Lit.)	
Cotton	Bollworms, Jassid, Thrips	600-1000	400-600	21
Paddy (Rice)	Stem borer, Leaf folder, Green leaf hopper, Gall midge, Hispa, Thrips	500	400-600	15

Trade Names

Sr. No.	Trade Name	Company
1	Lamdex	ADAMA India Private Limited
2	Reeva 2.5	Rallis India Limited
3	Kozuka	Biostadt India Ltd.
4	Demand	Tropical Agrosystem India (P) Ltd.
5	Kabaddi	Anu Products Ltd.
6	Bravo 5000	Insecticides (India) Ltd.
7	Leokem 2.5	Sumitomo Chemical India Ltd.
8	Veera	Herenba Industries Ltd.
9	Cheetah	Nichino India Pvt. Ltd.
10	Lucky	KP Agro Industries
11	Lamsexy	Natural Crop Science
12	Owl	Linuxcrop Chemicals Pvt. Ltd.
13	Boss	Preventis Chem India (P) Ltd.
14	Sinfonia	RACCOLTO Agritech India Pvt. Ltd.

Lambda-cyhalothrion 05.00 % EC

Crop	Pests	Dose / ha		Waiting period (Days)
		Quantity of formulation (ml)	Volume of water (Lit.)	
Cotton	Bollworms, Jassid, Thrips	300-500	400-600	21
Paddy (Rice)	Stem borer, Leaf folder, Green leaf hopper, Gall midge, Hispa, Thrips	250	400-600	15
Brinjal	Shoot and fruit borer	300	400-600	04
Tomato	Fruit borer	300	400-600	04
Chilli	Thrips, Mite, Pod borer	300	400-600	05
Pigeonpea	Pod borer, Pod fly	400-500	400-600	15
Onion	Thrips	300	300-400	05
Okra	Jassid, Shoot and fruit borer	300	300-400	04
Chickpea	Pod borer	500	300-400	06
Groundnut	Thrips, Leafhopper, Leaf miner	200-300	400-500	10
Mango	Hoppers	0.5-1.0 ml / lit. water	-	07

Trade Names

Sr. No.	Trade Name	Company
1	Karate	Syngenta India Ltd.
2	Judo Plus	Crystal Crop Protection Ltd.
3	Lamdex Super	ADAMA India Private Limited

4	Reeva 5	Rallis India Limited
5	Lambda Star	Swal Corporation Ltd.
6	Rudra 5 EC	Gharda Chemicals Ltd.
7	Deva Shakti	Dhanuka Agritech Ltd.
8	Hillambda	HIL (India) Limited
9	Wrestler	Anu Products Ltd.
10	Excel Super Glow	Excel Crop Care Ltd.
11	Tanja-50	Crop Life Science Ltd.
12	Locslay 5	Parijat Industries India Pvt. Ltd.
13	Kataar	Crop Chemicals India Ltd.
14	Leokem Plus	Sumitomo Chemical India Ltd.
15	Hemrin 5	Hemani Industries Ltd., Ankleshwar
16	Judgement	Isagro Asia Agrochemicals Pvt. Ltd.
17	Storm	GFG Crop Science Pvt. Ltd.
18	Karateka	Tusova Agro Pvt. Ltd.
19	Jayam	Herenba Industries Ltd.
20	Lemsik	Sikko Industries Ltd.
21	Simha	Nichino India
22	Lamato	Volrich Crop Science
23	Katar-5	Silver Sine Bio Tech
24	Lamdexy Gold	Natural Crop Science
25	Command	Tropical Agrosystem India (P) Ltd.
26	Lambda Knight	Essential Biosciences
27	Lekso-5	Synwood Agro Ltd.
28	Fifty 50	Indian Agrochemicals
29	Mission 2.0	
30	Loc-5	Krishi Rasayan
31	Baton	Global Crop Care
32	Shooter-50	JU Agri Sciences Pvt. Ltd.
33	Eagle	Linuxcrop Chemicals Pvt. Ltd.
34	Exylon LamdaMax	Torrent Crop Science
35	Big Boss	Preventis Chem India (P) Ltd.
36	Holaro	Chemley Agritech Pvt. Ltd.
37	Banister	RACCOLTO Agritech India Pvt. Ltd.

Green

LUFENURON

Available formulations in India : **Lufenuron 05.40 % EC**

Classification on the basis of

Target organism : Insecticide

Mode of entry : Stomach and Contact poison

Mode of action : Growth regulator

Chemical group : Benzoylureas

IRAC classification

Main Group and Primary Site of Action : 15 Inhibitors of chitin biosynthesis affecting CHS1

Sub-group or exemplifying Active Ingredient : Benzoylurea

Category of toxicity : Slightly toxic

Toxicity label colour : Green

- Lufenuron is a benzoylurea insecticide that primarily acts as an insect growth regulator (IGR) by inhibiting chitin synthesis. Chitin is a vital component of the exoskeleton of insects and the cell walls of fungi.
- **Mode of action :** Its main mode of action is inhibiting chitin synthesis in insects. Chitin is a vital component of the insect exoskeleton (outer shell). Without the ability to produce chitin, insects cannot successfully molt (shed their old skin to grow) or develop a hard exoskeleton. Lufenuron disrupts the enzyme processes that insects use to create chitin. This inhibition is most effective against larval stages of insects, preventing them from successfully molting to the next developmental stage.
- Lufenuron can also affect insect eggs, preventing the larvae from hatching properly. This can occur when adult females are exposed to lufenuron, and it is passed into their eggs (transovarial effect).
- While primarily acting through ingestion (when insects feed on treated plants or or are ingested by fleas on a treated host), it can also have some contact activity.
- In fungi and yeast, the inhibition of chitin synthesis weakens their cell walls, potentially leading to their death.

Label claim (DPPQ & S, Faridabad)

Lufenuron 05.40 % EC

Crop	Pests	Dose / ha		Waiting period (Days)
		Quantity of formulation (ml)	Volume of water (Lit.)	
Cabbage	Diamond back moth	600	500	14
Cauliflower	Diamond back moth	600	500	05
Pigeonpea	Pod borer, Pod fly	600	500-1000	65
Cotton	American bollworm	600	500-750	48

Black gram	Pod borer	600	500	10
Chilli	Fruit borer	600	500	05

Trade Names

Sr. No.	Trade Name	Company
1	Lunox	Crystal Crop Protection Ltd.
2	Cigna	Syngenta India Ltd.
3	Modern Guard	Modern Insecticides Ltd.
4	K-One	Koronation Bio-Science
5	Lunor	Global Crop Care

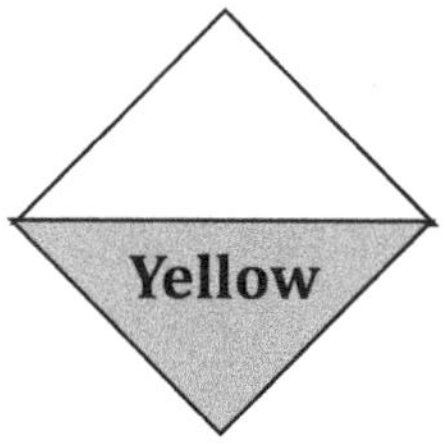

MALATHION

Available formulations in India :
1. **Malathion 05.00 % DP**
2. **Malathion 50.00 % EC**

Classification on the basis of

Target organism : Insecticide
Mode of entry : Stomach and Contact poison
Mode of action : Nerve poison
Chemical group : Organophosphates

IRAC classification

Main Group and Primary Site of Action : 1 Acetylcholinesterase (AChE) inhibitors
Sub-group or exemplifying Active Ingredient : 1B Organophosphates

Category of toxicity : Highly toxic
Toxicity label colour : Yellow

- Malathion is extensively used in agriculture to control a broad spectrum of insect pests on fruits, vegetables, grains, and other crops.
- Malathion belongs to the organophosphate class of insecticides.
- **Mode of action :** Its primary mechanism of action involves disrupting the nervous system of insects (and other organisms, including mammals, if exposed to sufficient amounts). Malathion itself is a "pro-insecticide." Once it enters the body of an insect (or mammal), it is metabolized by enzymes into a more potent form called malaoxon. Malaoxon then irreversibly binds to and inhibits the enzyme acetylcholinesterase (AChE). AChE is crucial for breaking down the neurotransmitter acetylcholine (ACh) in the synaptic cleft (the space between nerve cells, or between nerve cells and muscle cells). By inhibiting AChE, malaoxon causes acetylcholine to accumulate at these synapses. This continuous accumulation of acetylcholine leads to constant overstimulation of the nerve and muscle cells. In insects, this manifests as uncontrolled muscle spasms, tremors, paralysis, and ultimately death, often due to respiratory failure.
- Malathion is most stable in a slightly acidic to neutral pH range, specifically between pH 5 and 7. When the pH is greater than 7 (alkaline conditions), malathion degrades rapidly.

Label claim (DPPQ & S, Faridabad)

Malathion 05.00 % DP

Crop	Pests	Dose / ha		Waiting period (Days)
		Quantity of formulation (Kg)	Volume of water (Lit.)	
Paddy (Rice)	Rice hispa	25	--	--

Trade Names

Sr. No.	Trade Name	Company
1	Shield 5 DP	Modern Insecticides Ltd.
2	Caplon	Global Crop Care

Malathion 50.00 % EC

Crop	Pests	Dose / ha		Waiting period (Days)
		Quantity of formulation (ml)	Volume of water (Lit.)	
Paddy (Rice)	Rice hispa	1150	500-1000	--
Cabbage	Mustard aphid	1500	500-1000	--

Trade Names

Sr. No.	Trade Name	Company
1	Hilmala	HIL (India) Limited
2	Cythion	Coromandel International Ltd.
3	Malathion 50% EC	Hyderabad Chemical Products Ltd.
4	Anu Malathion	Anu Products Ltd.
5	Shield 50 EC	Modern Insecticides Ltd.
6	Devimalt	Devidayal Agro Chemicals Ltd.
7	Malathion 50 % EC	Gujarat Pesti-Chem Industries
8	Eagle	Herenba Industries Ltd.
9	Malasik	Sikko Industries Ltd.
10	Melon	KP Agro Industries
11	Melon-50	Natural Crop Science
12	Molothion	Essential Biosciences
13	Mallux	Global Crop Care

Yellow

METAFLUMIZONE

Available formulations in India	:	**Metaflumizone 22.00 % SC**
Classification on the basis of		
Target organism	:	Insecticide
Mode of entry	:	Stomach and Contact poison
Mode of action	:	Nerve poison
Chemical group	:	Semicarbazones
IRAC classification		
Main Group and Primary Site of Action	:	22 Voltage-dependent sodium channel blockers
Sub-group or exemplifying Active Ingredient	:	22B Semicarbazones
Category of toxicity	:	Highly toxic
Toxicity label colour	:	Yellow

- Metaflumizone is a modern insecticide belonging to the semicarbazone chemical class. It's known for its unique mode of action, making it a valuable tool in pest management.
- **Mode of action :** Metaflumizone works by acting on the insect nervous system. Specifically, it is a voltage-dependent sodium channel blocker. Unlike some older sodium channel modulators (like pyrethroids), metaflumizone blocks sodium channels by binding selectively to the slow-inactivated state of these channels. Voltage-gated sodium channels are crucial for the generation and transmission of nerve impulses in insects. By binding to and blocking these channels, metaflumizone prevents sodium ions from passing through the nerve cell membrane. This disruption leads to flaccid paralysis in the insect. The insect becomes unable to move or feed, and ultimately dies.
- An important distinction is that metaflumizone does not require bioactivation within the insect's body to become active, unlike some other sodium channel inhibitors (e.g., indoxacarb).
- Metaflumizone is most stable in neutral to alkaline aqueous solutions (7-9 pH) and degrades more rapidly in acidic conditions.

Label claim (DPPQ & S, Faridabad)

Metaflumizone 22.00 % SC

Crop	Pests	Dose / ha		Waiting period (Days)
		Quantity of formulation (ml)	Volume of water (Lit.)	
Cabbage	Diamond back moth	750-1000	500	03

Trade Names

Sr. No.	Trade Name	Company
1	Varismo	BASF India Ltd.

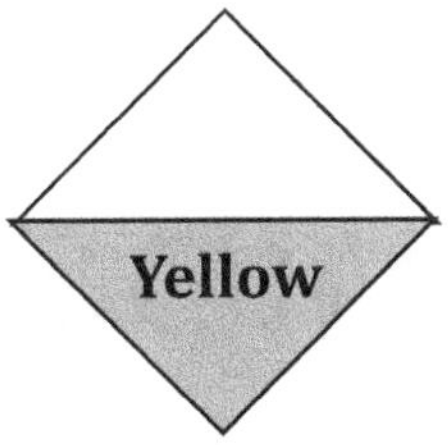

METHOXYFENOZIDE

Available formulations in India : **Methoxyfenozide 21.80 % SC**

Classification on the basis of

Target organism : Insecticide

Mode of entry : Stomach and Contact poison

Mode of action : Insect growth regulator

Chemical group : Diacylhydrazines

IRAC classification

Main Group and Primary Site of Action : 18 Ecdysone receptor agonists

Sub-group or exemplifying Active Ingredient : Diacylhydrazines

Category of toxicity : Highly toxic

Toxicity label colour : Yellow

- Methoxyfenozide is a synthetic insecticide belonging to the diacylhydrazine class. It is primarily used to control lepidopteran larvae, which are caterpillars of moths and butterflies.
- **Mode of action :** Methoxyfenozide is an insect growth regulator (IGR) that acts as an ecdysone receptor agonist. This means it mimics the action of 20-hydroxyecdysone, the natural insect molting hormone. When a caterpillar ingests or comes into contact with Methoxyfenozide, it binds to the ecdysone receptors, tricking the insect into a premature and incomplete molting process. This disrupts the insect's development, leading to it stopping feeding, becoming dehydrated, and eventually dying. This unique and targeted mode of action makes it effective against pests that have developed resistance to other types of insecticides.
- Methoxyfenozide exhibits strong pH stability, making it a reliable insecticide for spray applications using a variety of water sources.

Label claim (DPPQ & S, Faridabad)

Methoxyfenozide 21.80 % SC

Crop	Pests	Dose / ha		Waiting period (Days)
		Quantity of formulation (ml)	Volume of water (Lit.)	
Groundnut	Leaf eating (Tobacco) caterpillar, Leaf miner, Pod borer	875	500	26
Sugarcane	Early shoot borer	500-625	500	16

Trade Names

Sr. No.	Trade Name	Company
1	Intrepid	Corteva Agriscience

Red

METHYL BROMIDE

Available formulations in India : **Methyl bromide 98.00 %**

Classification on the basis of

Target organism : Insecticide
Mode of entry : Stomach and Contact poison
Mode of action : Multi-site action
Chemical group : 8A Alkyl halides

IRAC classification

Main Group and Primary Site of Action : 8 Miscellaneous non-specific (multi-site) inhibitors
Sub-group or exemplifying Active Ingredient : Alkyl halides

Category of toxicity : Extremely toxic
Toxicity label colour : Red

- Methyl bromide (also known as bromomethane) is a broad-spectrum fumigant that has been widely used to control a variety of pests, including fungi, weeds, insects, nematodes, and rodents. It's an odourless, colourless gas at room temperature, though at very high concentrations it may have a musty or fruity odour.
- **Mode of action :** Methyl bromide works by spreading throughout the treated area and penetrating the organisms. It disrupts their cellular processes through a process called "methylation," where it donates methyl groups that bind covalently to DNA bases and other cellular components, leading to cellular dysfunction and death. This broad-spectrum activity makes it effective against a wide range of pests.
- Methyl bromide is a highly toxic substance and poses significant health risks to humans, especially at fumigation sites. It is also classified as a Class I ozone-depleting substance (ODS).

Label claim (DPPQ & S, Faridabad)

Methyl bromide 98.00 %

Crop	Pests	Dose / Quantity of formulation (g)	Waiting period (Hrs.)
Stored whol cereals and seed, Millets, Pulses	Rice weevil, Lesser grain borer, Khapra beetle, Rust red flour beetle, Drug store beetle	Air tight cover 24 g/m3 (6-8 hrs)	24 hrs As when residues not to exceed 25 ppm
Milled products : Flour	Khapra beetle, Rust red flour beetle, Lesser grain borer	Air tight cover 24-32 g/m3 (12-24 hrs)	72 hrs As when residues not to exceed 25 ppm

Dry fruits, Nuts, Spices and Oil seeds	Rust red flour beetle	Air tight cover 24-32 g/m3 (24 hrs)	72 hrs As when residues not to exceed 25 ppm

Trade Names

Sr. No.	Trade Name	Company
1	Methyl bromide	Intech Organics Ltd.

Red

MONOCROTOPHOS

Available formulations in India : **Monocrotophos 15.00 % SG**

Classification on the basis of

Target organism : Insecticide
Mode of entry : Stomach and Contact poison
Mode of action : Nerve poison
Chemical group : Organophosphates

IRAC classification

Main Group and Primary Site of Action : 1 Acetylcholinesterase (AChE) inhibitors
Sub-group or exemplifying Active Ingredient : 1B Organophosphates

Category of toxicity : Extremely toxic
Toxicity label colour : Red

- Monocrotophos is a synthetic organophosphate insecticide and acaricide known for its broad-spectrum activity against various agricultural pests. It acts as a systemic insecticide, meaning it is absorbed by the plant and translocated throughout its tissues, as well as having contact and stomach action.
- Monocrotophos inhibits acetylcholinesterase, an enzyme essential for normal nerve impulse transmission in insects. This leads to an accumulation of acetylcholine at nerve synapses, causing overstimulation, paralysis, and ultimately the death of the insect.
- It is unstable in both acidic and alkaline conditions but particularly sensitive to alkaline pH.

Label claim (DPPQ & S, Faridabad)

Monocrotophos 15.00 % SG

Crop	Pests	Dose / ha		Waiting period (Days)
		Quantity of formulation (g)	Volume of water (Lit.)	
Cotton	Aphid, Jassid, Thrips, Whitefly	1333	500-1000	58

Trade Names

Sr. No.	Trade Name	Company
1	Kroto-15	KP Agro Industries
2	Tonal GR	RACCOLTO Agritech India Pvt. Ltd.

NITENPYRAM

Available formulations in India : **Nitenpyram 25.00 % SP**

Classification on the basis of

Target organism : Insecticide

Mode of entry : Systemic, Stomach and Contact poison

Mode of action : Nerve poison

Chemical group : Neonicotinoid

IRAC classification

Main Group and Primary Site of Action : 4 Nicotinic acetylcholine receptor (nAChR) competitive modulators

Sub-group or exemplifying Active Ingredient : 4A Neonicotinoids

Category of toxicity : Highly toxic

Toxicity label colour : Yellow

- Nitenpyram is a neonicotinoid insecticide that acts as a neurotoxin, primarily targeting the nervous systems of insects.
- Mode of action : It functions as an agonist of nicotinic acetylcholine receptors (nAChRs) in insects, binding irreversibly to these receptors in the central nervous system. This binding disrupts ion flow across postsynaptic neuron membranes, causing overstimulation, paralysis, and rapid death, with high selectivity for insect receptors over mammalian ones.
- Nitenpyram offers quick knockdown effects against piercing-sucking pests via contact, stomach, and systemic action.

Label claim (DPPQ & S, Faridabad)

Nitenpyram 25.00 % SP

Crop	Pests	Dose / ha		Waiting period (Days)
		Quantity of formulation (g)	Volume of water (Lit.)	
Cotton	Jassid, Whitefly	320	500	92
Paddy	Brown plant hopper, White backed plant hopper	600	500	37

Trade Names

Sr. No.	Trade Name	Company
1	Invicto	Coromandel

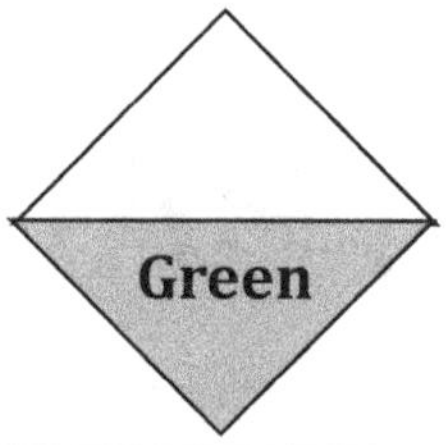

NOVALURON

Available formulations in India : **1. Novaluron 10.00% EC**
2. Novaluron 08.80% SC

Classification on the basis of

Target organism : Insecticide
Mode of entry : Stomach and Contact poison
Mode of action : Growth regulator
Chemical group : Benzoylureas

IRAC classification

Main Group and Primary Site of Action : 15 Inhibitors of chitin biosynthesis affecting CHS1
Sub-group or exemplifying Active Ingredient : Benzoylureas

Category of toxicity : Slightly toxic
Toxicity label colour : Green

- Novaluron is an insecticide belonging to the benzoylphenyl urea class and acts as an insect growth regulator (IGR).
- Novaluron primarily works by inhibiting chitin synthesis in insects. Chitin is a crucial component of the exoskeleton (cuticle) of insects, especially during the molting process when they shed their old skin and grow.
- By interfering with chitin production, Novaluron disrupts the normal growth and development of immature insects (larvae and nymphs), leading to abortive molting and ultimately death.
- It mainly acts through ingestion but also has some contact activity.

Label claim (DPPQ & S, Faridabad)

Novaluron 10.00% EC

Crop	Pests	Dose / ha		Waiting period (Days)
		Quantity of formulation (ml)	Volume of water (Lit.)	
Cotton	American bollworm	1000	500-1000	40
Cabbage	Diamond back moth	750	500-1000	05
Tomato	Fruit borer	750	500-1000	01-03
Chilli	Fruit borer, Tobacco caterpillar	375	500	03
Bengal gram	Pod borer	750	500	07

Trade Names

Sr. No.	Trade Name	Company
1	Pedestal	ADAMA India Private Limited
2	Remostar	Swal Corporation Ltd.
3	Rimon	Indofil Industries Ltd.
4	Uniqueron	Shree Ram Agro India
5	Uniron	UPL Ltd.
6	Novalon	Natural Crop Science
7	Nimon	Essential Biosciences
8	Rimo+	Global Crop Care
9	Rebel	Preventis Chem India (P) Ltd.

Novaluron 08.80% SC

Crop	Pests	Dose / ha		Waiting period (Days)
		Quantity of formulation (ml)	Volume of water (Lit.)	
Cotton	American bollworm, Tobacco caterpillar	1000	500-1000	20

Trade Names

Sr. No.	Trade Name	Company
1	Pedestal Supra	ADAMA India Private Limited

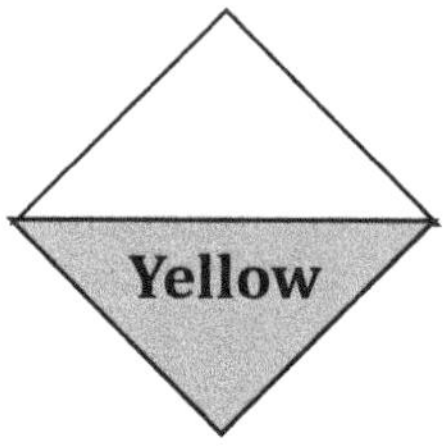

OXYDEMETON-METHYL

Available formulations in India	:	**Oxydemeton-methyl 02.45 % SC**
Classification on the basis of		
Target organism	:	Insecticide, Acaricide
Mode of entry	:	Systemic, Stomach and Contact poison
Mode of action	:	Nerve poison
Chemical group	:	Organophosphates
IRAC classification		
Main Group and Primary Site of Action	:	1 Acetylcholinesterase (AChE) inhibitors
Sub-group or exemplifying Active Ingredient	:	1B Organophosphates
Category of toxicity	:	Highly toxic
Toxicity label colour	:	Yellow

- Oxydemeton-methyl is an organothiophosphate insecticide and acaricide.
- It is a systemic insecticide and acaricide, meaning it is absorbed by the plant and translocated throughout its tissues, providing protection against pests that feed on various parts of the plant.
- It also has contact and stomach action.
- Oxydemeton-methyl is a cholinesterase inhibitor. It interferes with the enzyme acetylcholinesterase in the nervous system of insects. This overstimulates the nervous system, leading to symptoms like tremors, convulsions, paralysis, and ultimately death.

Label claim (DPPQ & S, Faridabad)

Oxydemeton-methyl 02.45 % SC

Crop	Pests	Dose / ha		Waiting period (Days)
		Quantity of formulation (ml)	Volume of water (Lit.)	
Rice	Green leaf hopper	500	500-1000	--
	White leaf hopper	1000	500-1000	--
Maize	Shoot fly	1000	500-1000	--
Sorghum	Shoot fly	1000	500-1000	--
Cotton	Aphid, Jassid	1200	500-1000	--
Groundnut	Aphid, Leaf miner	1000	500-1000	--
Mustard	Aphid	1000	500-1000	--
Sesamum	Leafhopper	1200	500-1000	--
Okra	Whitefly	1000	500-1000	--
	Jassid, Leaf beetle	1600	500-1000	--
Chilli	Aphid	1600	500-1000	--
	Mites	2000	500-1000	--

	Thrips	1000	500-1000	--
Onion	Thrips	1200	500-1000	--
Tomato	Whitefly	1000	500-1000	--
Potato	Aphid	1000	500-1000	--
Apple	Sanjose scale	4200-5600	1500-2000	--
	Wooly aphid	1500-2000	1500-2000	--
Banana	Tingid bug	1500-2000	1500-2000	--
	Aphid	3000-4000	1500-2000	--
Mango	Hoppers	1500-2000	1500-2000	--
Peach	Leaf curl aphid	1500-2000	1500-2000	--
Coffee	Green bug	2500	500-1000	--
	Leaf miner	4000	500-1000	--
Tobacco	Whitefly, Aphid	1000	500-1000	--

Trade Names

Sr. No.	Trade Name	Company
1	Metasystox	UPL Ltd.

Yellow

PERMETHRIN

Available formulations in India : **Permethrin 25.00% EC**

Classification on the basis of

Target organism : Insecticide
Mode of entry : Stomach and Contact poison
Mode of action : Nerve poison
Chemical group : Pyrethroid

IRAC classification

Main Group and Primary Site of Action : 3 Sodium channel modulators
Sub-group or exemplifying Active Ingredient : 3A Pyrethroid

Category of toxicity : Highly toxic
Toxicity label colour : Yellow

- Permethrin is a widely used synthetic pyrethroid insecticide and acaricide. It mimics the effects of natural pyrethrins, which are derived from chrysanthemum flowers.
- Permethrin is a neurotoxin. It disrupts the normal functioning of the nervous system in insects and mites. Specifically, it affects the sodium channels in nerve cell membranes, causing prolonged depolarization. This leads to overstimulation, paralysis, and ultimately death of the pest.
- It primarily acts through contact but also has some stomach action.

Label claim (DPPQ & S, Faridabad)

Permethrin 25.00% EC

Crop	Pests	Dose / ha		Waiting period (Days)
		Quantity of formulation (ml)	Volume of water (Lit.)	
Cotton	Bollworms	400-500	500-1000	--

Trade Names

Sr. No.	Trade Name	Company
1	Agniban	UPL Ltd., Arysta Lifescience
2	Permasect	Coromandel International Ltd.
3	Action Super	Tropical Agrosystem India (P) Ltd.
4	Parmeet	Anu Products Ltd.
5	Proud	Crop Chemicals India Ltd.
6	Signor	Herenba Industries Ltd., Mumbai
7	Rotus	Hemani Industries Ltd., Ankleshwar
8	Porsha	Chaman Lal Raj Kumar, Bhatinda
9	Benesik	Sikko Industries Ltd.
10	Permatis	Preventis Chem India (P) Ltd.
11	Stella	RACCOLTO Agritech India Pvt. Ltd.

Yellow

PHENTHOATE

Available formulations in India : **1. Phenthoate 02.00% DP**
2. Phenthoate 50.00% EC

Classification on the basis of

Target organism : Insecticide
Mode of entry : Stomach and Contact poison
Mode of action : Nerve poison
Chemical group : Organophosphates

IRAC classification

Main Group and Primary Site of Action : 1 Acetylcholinesterase (AChE) inhibitors
Sub-group or exemplifying Active Ingredient : 1B Organophosphates

Category of toxicity : Highly toxic
Toxicity label colour : Yellow

- Phenthoate is a broad-spectrum organophosphate insecticide and acaricide used to control a variety of insect pests in agriculture.
- Phenthoate is a non-systemic insecticide that works through contact and stomach action. Its primary mode of action is the inhibition of the enzyme acetylcholinesterase (AChE) in insects. This inhibition leads to the accumulation of acetylcholine in the nervous system, causing continuous stimulation, paralysis, and eventually death of the insect.
- It is known for its rapid "knockdown" action and also has ovicidal (egg-killing) and repellent properties, which helps prevent adult moths from laying eggs.
- Phenthoate is relatively stable in acidic to neutral solutions, specifically in a pH range from 3.9 to 7.8. However, it undergoes rapid alkaline hydrolysis when in contact with water at a pH above 7. The degradation rate significantly increases as the solution becomes more alkaline.

Label claim (DPPQ & S, Faridabad)

Phenthoate 02.00 % DP

Crop	Pests	Dose / ha		Waiting period (Days)
		Quantity of formulation (Kg)	Volume of water (Lit.)	
Sorghum	Red spider mite, Pink mite, Purple mite, Scarlet mite	20	--	After 90% Emergence of earhead
Safflower	Aphid	20	--	--

Trade Names

Sr. No.	Trade Name	Company
1	Pendo dp	Global Crop Care

Phenthoate 50.00 % EC

Crop	Pests	Dose / ha		Waiting period (Days)
		Quantity of formulation (ml)	Volume of water (Lit.)	
Paddy	Rice caseworm	1000	500-1000	--
Groundnut	Leaf webber	1000	500-1000	--

Trade Names

Sr. No.	Trade Name	Company
1	Phendal	Coromandel International Ltd.
2	Anusan	Anu Products Ltd.
3	Cidial	Isagro Asia Agrochemicals Pvt. Ltd.
4	Pendo	Global Crop Care
5	Phensan	Shivalik Crop Sciences Pvt. Ltd.
6	Phonex	Preventis Chem India (P) Ltd.

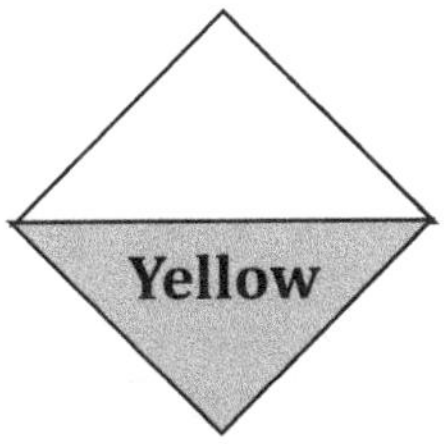

PHOSALONE

Available formulations in India :
1. **Phosalone 35.00 % EC**
2. **Phosalone 05.00 % DP**

Classification on the basis of

Target organism : Insecticide, Acaricide
Mode of entry : Stomach and Contact poison
Mode of action : Nerve poison
Chemical group : Organophosphates

IRAC classification

Main Group and Primary Site of Action : 1 Acetylcholinesterase (AChE) inhibitors
Sub-group or exemplifying Active Ingredient : 1B Organophosphates

Category of toxicity : Highly toxic
Toxicity label colour : Yellow

- Phosalone is an organophosphate insecticide and acaricide commonly used to control a wide range of pests on various crops such as fruits, vegetables, cotton, and ornamentals. It is non-systemic and works mainly through contact and ingestion.
- Like many organothiophosphates, phosalone is considered a pro-insecticide, meaning it is metabolized in the insect's body (and in animals) into a more potent oxygen analogue (phosphate), which is the active inhibitor.
- Mode of action : Its mode of action involves inhibition of the cholinesterase enzyme in the nervous system of insects, which disrupts nerve signal transmission leading to paralysis and death. This neurotoxic effect provides rapid knockdown of pests and is effective on both adults and larvae.

Label claim (DPPQ & S, Faridabad)

Phosalone 35.00 % EC

Crop	Pests	Dose / ha		Waiting period (Days)
		Quantity of formulation (ml)	Volume of water (Lit.)	
Barley	Aphid	1428	500-1000	--
Sorghum	Earhead midge	1143	500-100	--
Jute	Red spider mite	1000	500-1000	--
Brinjal	Feuit borer	1428	500-1000	--
Cabbage	Aphid	1428	500-1000	--
Tomato	Fruit borer	1285	500-1000	--
Tea	Aphid, Pink mite, Purple mite	1028	500-1000	--

Trade Names

Sr. No.	Trade Name	Company
1	Phase	
2	Phosalone 35% EC	Umbrella Life Science

Phosalone 05.00 % DP

Crop	Pests	Dose / ha		Waiting period (Days)
		Quantity of formulation (Kg)	Volume of water (Lit.)	
Sorghum	Earhead midge	25	--	--

Trade Names

Sr. No.	Trade Name	Company
1		

Yellow

PHOSMET

Available formulations in India : **Phosmet 50.00% EC**

Classification on the basis of

Target organism : Insecticide, Acaricide

Mode of entry : Stomach, Contact and Respiratory poison

Mode of action : Nerve poison

Chemical group : Organophosphates

IRAC classification

Main Group and Primary Site of Action : 1 Acetylcholinesterase (AChE) inhibitors

Sub-group or exemplifying Active Ingredient : 1B Organophosphates

Category of toxicity : Highly toxic

Toxicity label colour : Yellow

- Phosmet is a non-systemic organophosphate insecticide and acaricide derived from phthalimide.
- It is a broad-spectrum insecticide, meaning it can control a wide variety of pests.
- Phosmet acts through contact, ingestion, and inhalation.
- It works by inhibiting acetylcholinesterase (AChE), an enzyme crucial for nerve function in insects and mites, leading to overstimulation, paralysis, and death.

Label claim (DPPQ & S, Faridabad)

Phosmet 50.00% EC

Crop	Pests	Dose / ha		Waiting period (Days)
		Quantity of formulation (ml)	Volume of water (Lit.)	
Chilli	Aphid, Thrips, Fruit borer	1000	500	10
Rice	Yellow stem borer, Leaf folder	1200	500	45
Cotton	Jassid, Aphid, Whitefly, Bollworms	1200	500	47
Apple	Wooly apple aphid, Green aphid, Tortricid moth	01 ml / lit.	01 lit.	31

Trade Names

Sr. No.	Trade Name	Company
1	SA-Imidan	Saraswati Agro Life Science

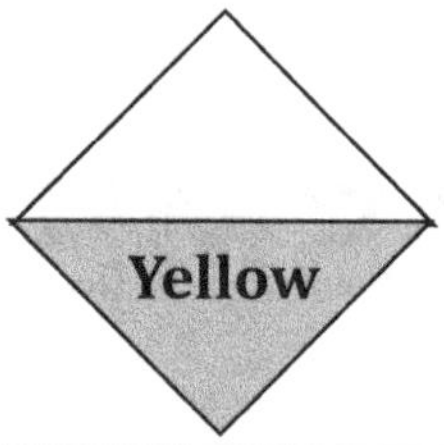

PROFENOFOS

Available formulations in India : **Profenofos 50.00% EC**

Classification on the basis of

Target organism : Insecticide
Mode of entry : Stomach and Contact poison
Mode of action : Nerve poison
Chemical group : Organophosphates

IRAC classification

Main Group and Primary Site of Action : 1 Acetylcholinesterase (AChE) inhibitors
Sub-group or exemplifying Active Ingredient : 1B Organophosphates

Category of toxicity : Highly toxic
Toxicity label colour : Yellow

- Profenofos is an organophosphate insecticide that is used to control a broad range of pests in agriculture.
- It is a non-systemic insecticide with contact and stomach action and it also exhibits a translaminar effect. It has ovicidal action.
- Its mode of action, like other organophosphates, is the inhibition of the acetylcholinesterase enzyme in the nervous system of insects. This leads to an accumulation of acetylcholine, causing overstimulation, paralysis, and death.

Label claim (DPPQ & S, Faridabad)

Profenofos 50.00% EC

Crop	Pests	Dose / ha		Waiting period (Days)
		Quantity of formulation (ml)	Volume of water (Lit.)	
Cotton	Bollworms	1500-2000	500-1000	15
	Jassid, Thrips, Aphid, Whitefly	1000	500-1000	15
Soybean	Semilooper, Girdle beetle	1000	500	40
Chilli	Fruit borer, Tobacco caterpillar, Thrips, Aphid	1000	500	03
Onion	Thrips	1000	500	80

Trade Names

Sr. No.	Trade Name	Company
1	Curacron	Syngenta India Ltd.
2	Profigan	ADAMA India Private Limited
3	Carina	PI Industries
4	Jashn	Rallis India Limited

5	Hilfos	HIL (India) Limited
6	Kusaking	IFFCO-MC Crop Science Private Ltd.
7	Prahar	Biostadt India Ltd.
8	Ajanta	Coromandel International Ltd.
9	Celcron	Excel Crop Care
10	Pro Gold	Alfa Crop Science
11	Curacron	Syngenta India Ltd.
12	Pro-Killer	Auxin Crop Care
13	Prefer	Anu Products Ltd.
14	Poast	Modern Insecticides Ltd.
15	Profex	NACL Industries Ltd.
16	Rana	Parijat Industries India Pvt. Ltd.
17	Nucron	Crop Chemicals India Ltd.
18	Kemcron	Sumitomo Chemical India Ltd.
19	Simcron	GSP Crop Science Pvt. Ltd., Ahemadabad
20	Profenophos 50 % EC	Gujarat Pesti-Chem Industries
21	Profit	GFG Crop Science Pvt. Ltd.
22	Ceracron	Tusova Agro Pvt. Ltd.
23	Progress	Herenba Industries Ltd.
24	Prosik	Sikko Industries Ltd.
25	Getrid	Nichino India Pvt. Ltd.
26	Silvercron 50	Silver Sine Bio Tech
27	Profmex	Natural Crop Science
28	Tagpro	Tropical Agrosystem India (P) Ltd.
29	Pramaan	Essential Biosciences
30	Professor	Synwood Agro Ltd.
31	Krifos	Krishi Rasayan
32	Hunter	Global Crop Care
33	Egg Kill 50	Green India Care
34	Proper	Linuxcrop Chemicals Pvt. Ltd.
35	Profos	Preventis Chem India (P) Ltd.
36	Kombat	RACCOLTO Agritech India Pvt. Ltd.

Yellow

PYMETROZINE

Available formulations in India : **Pymetrozine 50.00% WG**

Classification on the basis of

Target organism : Insecticide
Mode of entry : Systemic, Stomach poison
Mode of action : Chordotonal organs
Chemical group : Pyridine azomethine derivatives

IRAC classification

Main Group and Primary Site of Action : 9 Chordotonal organ TRPV channel modulators
Sub-group or exemplifying Active Ingredient : 9B Pyridine azomethine derivatives

Category of toxicity : Highly toxic
Toxicity label colour : Yellow

- Pymetrozine is a selective insecticide that controls piercing-sucking pests like planthoppers. It's part of the pyridine-azomethine chemical class and has a unique, non-neurotoxic mode of action.
- **Mode of action :** Unlike traditional insecticides that cause a rapid knockdown, pymetrozine works by inhibiting the insect's ability to feed. It specifically targets the chordotonal organs in the legs of sap-feeding insects. By disrupting the nerve impulses to these organs, pymetrozine causes the pest to lose coordination and prevents it from feeding. Within minutes of exposure, the insect stops feeding, and it eventually dies from starvation.
- As a systemic insecticide, it can be absorbed by a plant and transported through its vascular system, protecting new growth.
- It is most effective when applied as a foliar spray or soil treatment and is very stable in solutions with a pH between 7 and 9.

Label claim (DPPQ & S, Faridabad)

Pymetrozine 50.00% WG

Crop	Pests	Dose / ha		Waiting period (Days)
		Quantity of formulation (g)	Volume of water (Lit.)	
Rice	Brown plant hopper	300	500	19
Mango	Hopper	300	1000	36

Trade Names

Sr. No.	Trade Name	Company
1	Chess	Syngenta India Ltd.
2	Zapper	Swal Corporation Ltd.

3	Apply	Dhanuka Agritech Ltd.
4	Eltra	FMC Corporation
5	Full Charge	Anu Products Ltd.
6	Mahir	Insecticides (India) Ltd.
7	Hop-Maar	Parijat Industries India Pvt. Ltd.
8	Pymax	Best Agrolife Ltd.
9	Gohan	Nichino India
10	Impeach Pro	Mahindra Summit Agriscience
11	Chees 50	Natural Crop Science
12	Champ	Tropical Agro
13	Insectastop	Essential Biosciences
14	Charmi	Synwood Agro Ltd.
15	Checkerz	Today's Global Crop Protection
16	Konnect	Krishi Rasayan
17	Pyra BHP	Global Crop Care
18	Clue	Tata Rallis India Ltd.
19	Cutter	Green India Care
20	JU-Claim	JU Agri Sciences Pvt. Ltd.
21	Pymepure	Mankind Agritech
22	Viking	Preventis Chem India (P) Ltd.
23	Pymozo	Chemley Agritech Pvt. Ltd.

Blue

PYRIFLUQUINAZON

Available formulations in India : **Pyrifluquinazon 20% WG**

Classification on the basis of

Target organism : Insecticide
Mode of entry : Stomach and Contact poison
Mode of action : Effect on chordotonal organs
Chemical group : Pyridine azomethine

IRAC classification

Main Group and Primary Site of Action : Group 9. Chordotonal organ TRPV channel modulators
Sub-group or exemplifying Active Ingredient : 9B Pyridine azomethine

Category of toxicity : Moderately toxic
Toxicity label colour : Blue

- Pyrifluquinazon is a relatively new insecticide belonging to the pyridine azomethine chemical group (IRAC Group 9B). Its mode of action is unique: it's a chordotonal organ Transient Receptor Potential Vanilloid (TRPV) channel modulator. This means it disrupts the sensory system of insects, specifically affecting their stretch receptors, which are crucial for coordination, movement, feeding, and flight.
- It modifies insect behavior by rapidly stopping their feeding and eventually die from starvation and dehydration.
- It has contact and ingestion activity.
- It exhibits translaminar activity, meaning it can move within the leaf, providing control of pests on the underside.
- Pyrifluquinazon is unstable and rapidly degrades under alkaline (high pH) conditions. For maximum stability and efficacy, the pH of the spray solution should be kept between neutral to acidic (pH 5-7).

Label claim (DPPQ & S, Faridabad)

Pyrifluquinazon 20% WG

Crop	Pests	Dose / ha		Waiting period (Days)
		Quantity of formulation (g)	Volume of water (Lit.)	
Cotton	Whitefly	375-500	500	30

Trade Names

Sr. No.	Trade Name	Company
1	Clasto	Tata Rallies India Ltd.
2	Zingala	Nihon Nohyaku Co., Ltd.

Green

PYRIPROXYFEN

Available formulations in India : 1. **Pyriproxyfen 10.00% EC**
2. **Pyriproxyfen 10.00% EW**

Classification on the basis of

Target organism : Insecticide
Mode of entry : Stomach and Contact poison
Mode of action : Growth regulator
Chemical group : Pyridine ether

IRAC classification

Main Group and Primary Site of Action : 7 Juvenile hormone mimics
Sub-group or exemplifying Active Ingredient : 7C Pyridine ether

Category of toxicity : Slightly toxic
Toxicity label colour : Green

- Pyriproxyfen is an insect growth regulator (IGR) that mimics juvenile hormone in insects. It belongs to the pyridine ether chemical group.
- Pyriproxyfen acts as a mimic of juvenile hormone (JH), a naturally occurring hormone in insects that regulates development and molting. By mimicking JH, pyriproxyfen prevents immature insects (larvae and nymphs) from completing metamorphosis into fertile adults.
- It can also interfere with egg development and hatching (ovicidal action). Eggs laid by treated females often fail to hatch.
- In some cases, exposure to pyriproxyfen can lead to the sterilization of adult female insects, reducing their reproductive capacity.
- Pyriproxyfen primarily targets the immature stages of insects and has little direct killing effect on adult insects. However, it can reduce adult populations by preventing the emergence of new adults and by sterilizing females.
- It can be effective through both contact and ingestion by the insect.
- Some formulations exhibit translaminar activity, meaning the insecticide can move from one leaf surface to the other, offering better protection.

Label claim (DPPQ & S, Faridabad)

Pyriproxyfen 10.00 % EC

Crop	Pests	Dose / ha		Waiting period (Days)
		Quantity of formulation (ml)	Volume of water (Lit.)	
Brinjal	Whitefly, Jassid	500	300	07
Cotton	Whitefly	1000	500	31
Chilli	Whitefly, Aphid	500	300	07
Okra	Whitefly, Jassid	500	300	07

Trade Names

Sr. No.	Trade Name	Company
1	Komugi	IFFCO-MC Crop Science Private LTd.
2	Amadis	FMC Corporation
3	Lano	Sumitomo Chemical India Ltd.
4	Stoxx	Anu Products Ltd.
5	Daita	Parijat Industries India Pvt. Ltd.
6	Pyrisuperb	AgFarm India Pvt Ltd
7	Poros	Best Agrolife Ltd.
8	Pyripon-E	Multiplex Bio-Tech Pvt. Ltd.
9	Pyrimoon	Essential Biosciences
10	Leveno	Sahib Pesticides and Seeds Ltd.
11	Pyrigold	Star Chemicals
12	Dhadak	HPM Chemicals and Fertilizers Ltd.
13	Proxy-10	Nauxima Bio Tech
14	Lanoov	Natural Crop Science
15	Nano	Tropical Agrosystem India (P) Ltd.
16	Nanofas	Frontline Agri Science
17	JU-Plano	JU Agri Sciences Pvt. Ltd.
18	SA Florfen	Saraswati Agro Life Science (India) Pvt. Ltd.

Pyriproxyfen 10.00 % EW

Crop	Pests	Dose / ha		Waiting period (Days)
		Quantity of formulation (ml)	Volume of water (Lit.)	
Cotton	Whitefly, Jassid, Thrips	1000-2500	500	38

Trade Names

Sr. No.	Trade Name	Company
1		

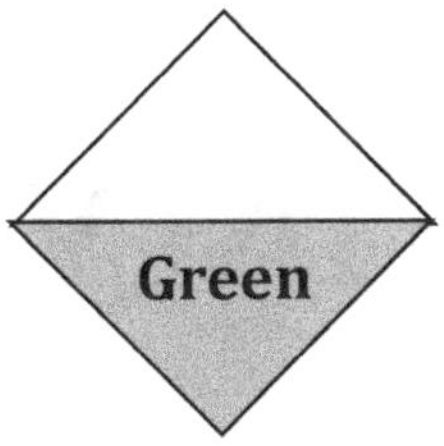

PYRIDABEN

Available formulations in India	:	**Pyridaben 20.00% WP**
Classification on the basis of		
Target organism	:	Insecticide, Acaricide
Mode of entry	:	Stomach and Contact poison
Mode of action	:	Energy metabolism
Chemical group	:	Pyridazinone
IRAC classification		
Main Group and Primary Site of Action	:	21 Mitochondrial complex I electron transport inhibitors
Sub-group or exemplifying Active Ingredient	:	21A METI acaricides and insecticides
Category of toxicity	:	Slightly toxic
Toxicity label colour	:	Green

- Pyridaben is a non-systemic contact acaricide and insecticide belonging to the pyridazinone chemical group. It is effective against various phytophagous mites and some insects.
- Pyridaben works by inhibiting mitochondrial electron transport at the NADH-coenzyme Q reductase (Complex I) site in the target pests. This disrupts their energy production, leading to rapid paralysis and death.

Label claim (DPPQ & S, Faridabad)

Pyridaben 20.00% WP

Crop	Pests	Dose / ha		Waiting period (Days)
		Quantity of formulation (g)	Volume of water (Lit.)	
Tea	Red spider mite	500	500	07
Cotton	Whitefly	500	500	28
Chilli	Yellow mite	375-500	500	05
Brinjal	Red spider mite	500	500	03
Apple	European red mite	5 ml	10 lit / tree	05

Trade Names

Sr. No.	Trade Name	Company
1	Hanabi	Godrej Agrovet

Green

PYRIDALYL

Available formulations in India : **Pyridalyl 10.00% EC**

Classification on the basis of

Target organism : Insecticide
Mode of entry : Stomach and Contact poison
Mode of action : Inhibition of Cellular Protein Synthesis
Chemical group : Pyridalyl

IRAC classification

Main Group and Primary Site of Action : Unknown - Compounds of unknown or uncertain MoA
Sub-group or exemplifying Active Ingredient : Pyridalyl

Category of toxicity : Slightly toxic
Toxicity label colour : Green

- The mode of action of Pyridalyl is not yet fully understood, classifying it under the IRAC UN (Unclassified) group of insecticides. However, research suggests several potential mechanisms:
- Inhibition of Cellular Protein Synthesis: Studies indicate that Pyridalyl can rapidly and significantly inhibit protein synthesis in insect cell lines but not in mammalian cell lines.
- Disruption of Insect Vigor: Treated insects exhibit a loss of vigor and typically die within a few hours.
- Effects on Molting: Some observations suggest that lower concentrations of Pyridalyl can cause burn-like scars on larvae, interfering with their ability to molt properly.

Label claim (DPPQ & S, Faridabad)

Pyridalyl 10.00% EC

Crop	Pests	Dose / ha		Waiting period (Days)
		Quantity of formulation (ml)	Volume of water (Lit.)	
Cotton	Bollworms	750-1000	500-750	07
Okra	Shoot and fruit borer	500-750	500-750	03
Cabbage	Diamond back moth	500-750	500-750	03
Maize	Fall armyworm	1000	750	27

Trade Names

Sr. No.	Trade Name	Company
1	Sumiplio	Sumitomo Chemical India Ltd.

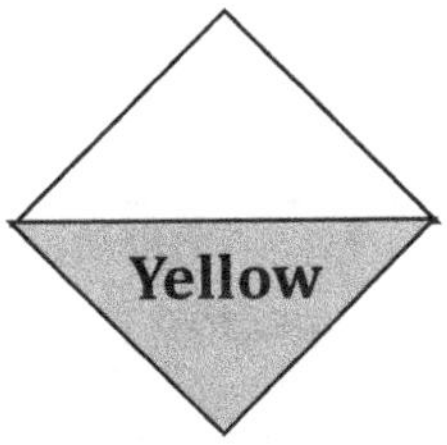

QUINALPHOS

Available formulations in India	:	**1. Quinalphos 25.00% Gel** **2. Quinalphos 05.00% Granules** **3. Quinalphos 20.00% AF** **4. Quinalphos 25.00% EC** **5. Quinalphos 01.50% DP**
Classification on the basis of		
Target organism	:	Insecticide
Mode of entry	:	Stomach and Contact poison
Mode of action	:	Nerve poison
Chemical group	:	Organophosphates
IRAC classification		
Main Group and Primary Site of Action	:	1 Acetylcholinesterase (AChE) inhibitors
Sub-group or exemplifying Active Ingredient	:	1B Organophosphates
Category of toxicity	:	Highly toxic
Toxicity label colour	:	Yellow

- Quinalphos is an organophosphate insecticide and acaricide with contact and stomach action.
- Quinalphos works by inhibiting the acetylcholinesterase (AChE) enzyme in the nervous system of insects and mites. This enzyme is crucial for breaking down the neurotransmitter acetylcholine (ACh). When AChE is inhibited, acetylcholine accumulates at the nerve synapses, leading to overstimulation of the nervous system, followed by paralysis and death of the pest.
- It also exhibits translaminar action, meaning it can penetrate plant tissues and move within the leaf to some extent.

Label claim (DPPQ & S, Faridabad)

Quinalphos 25.00 % Gel

Crop	Pests	Dose / ha		Waiting period (Days)
		Quantity of formulation (ml)	Volume of water (Lit.)	
Chilli	Aphid	1000	500-1000	--
Rice	Brown plant hopper, Leaf folder, Stem borer, Hispa	1000	500-1000	--

Trade Names

Sr. No.	Trade Name	Company
1		

Quinalphos 05.00 % Granules

Crop	Pests	Dose / ha		Waiting period (Days)
		Quantity of formulation (Kg)	Volume of water (Lit.)	
Rice	Gall midge, Stem borer	5	--	--

Trade Names

Sr. No.	Trade Name	Company
1	Brego GR	RACCOLTO Agritech India Pvt. Ltd.

Quinalphos 20.00 % AF

Crop	Pests	Dose / ha		Waiting period (Days)
		Quantity of formulation (ml)	Volume of water (Lit.)	
Rice	Brown plant hopper, Green leaf hopper, Leaf folder, Stem borer	1250-1500	750-1000	40
Okra	Shoot and fruit borer	1250-1500	750-1000	07
Cotton	American bollworm, Pinlk bollworm, Spotted bollworm	1750-2500	750-1000	07
Tomato	Fruit borer	1500-1750	750-1000	07
Tea	Hopper, Caterpillar	1000	400	07
Pigeonpea	Pod borer	2500	750-1000	30
Groundnut	Tobacco caterpillar	1250-1775	750-1000	30

Trade Names

Sr. No.	Trade Name	Company
1		

Quinalphos 25.00 % EC

Crop	Pests	Dose / ha		Waiting period (Days)
		Quantity of formulation (ml)	Volume of water (Lit.)	
Rice	Brown plant hopper	1500	500-1000	40
	Hispa, Blue beetle	2000	500-1000	40
	Leaf folder	1000	500-1000	40
	Stem borer	1300	500-1000	40
Wheat	Aphid	1000	500-1000	--
	Earhead caterpillar, Mite	1600	500-1000	--
Bengal gram / Chickpea	Pod borer	1000	500-1000	--
Black gram	Bihar hairy caterpillar	1500	500-100	--
French bean	Stem fly	1000	500-1000	--

Red gram	Pod borer, Pod fly	1400	500-1000	--
Soybean	Leaf weevil	1000	500-1000	--
Groundnut	Leaf hopper, Thrips	1400	500-1000	--
	Leaf miner	1000	500-1000	--
Mustard	Sawfly	1200	500-1000	--
Okra	Fruit borer	800	500-1000	--
	Leaf hopper, Mite	1000	500-1000	--
Cauliflower	Stem borer	2000	500-1000	--
Chilli	Aphid	1000	500-1000	--
	Mite	1500	500-1000	--
Tomato	Fruit borer	1000	500-1000	--
Apple	Wooly aphid	3000-4000	500-1000	--
Banana	Tingid bug	3000-4000	500-1000	--
Citrus	Scale	4200-5600	500-1000	--
	Citrus butterfly	1500-2000	500-1000	--
Pomegranate	Scale	4800-6400	500-1000	--
Tea	Thrips	760	500-1000	--

Trade Names

Sr. No.	Trade Name	Company
1	Hilquin	HIL (India) Limited
2	Krush	Biostadt India Ltd.
3	Quinguard	Gharda Chemicals Ltd.
4	Dhanulux	Dhanuka Agritech Ltd.
5	Flash	Indofil Industries Ltd.
6	Ekalux	Syngenta India Ltd.
7	Anuphos	Anu Products Ltd.
8	Promise	Modern Insecticides Ltd.
9	Celquin	Excel Crop Care Ltd.
10	Goldlux	Crop Chemicals India Ltd.
11	Kemlox	Sumitomo Chemical India Ltd.
12	Hemquin	Hemani Industries Ltd., Ankleshwar
13	Deluxe	Tusova Agro Pvt. Ltd.
14	Rambalux	Herenba Industries Ltd.
15	Sikquin	Sikko Industries Ltd.
16	Seequeen	Silver Sine Bio Tech
17	Akkalux	Natural Crop Science
18	Alphosides	Essential Biosciences
19	Luxer-25	Synwood Agro Ltd.
20	Skill	Shivalik Crop Sciences (P) Ltd.
21	SA Flush	Saraswati Agro Life Science (India) Pvt. Ltd.
22	Agrikind Qlux	Mankind Agritech
23	Paylux	Chemley Agritech Pvt. Ltd.
24	Brego	RACCOLTO Agritech India Pvt. Ltd.

Quinalphos 01.50 % DP

Crop	Pests	Dose / ha		Waiting period (Days)
		Quantity of formulation (Kg)	Volume of water (Lit.)	
Rice	Brown plant hopper	20	--	--
Gram	Pod borer	23.30	At pod formation	
Red gram	Pod borer	23.30	--	--
Soybean	Leaf weevil	16.60	--	--
French bean	Stem fly	20	--	--
Cotton	Aphid, Jassid, Thrips	20	From square formation	
	Bollworms	30		
Groundnut	Thrips, Jassid	23.30	--	--
	Red hairy caterpillar	25	--	--
Safflower	Aphid	20	--	--
Chilli	Aphid	20	--	--

Trade Names

Sr. No.	Trade Name	Company
1	Sikquin-DP	Sikko Industries Ltd.
2	Rakshak	Global Crop Care
3	Parphos-DP	RACCOLTO Agritech India Pvt. Ltd.

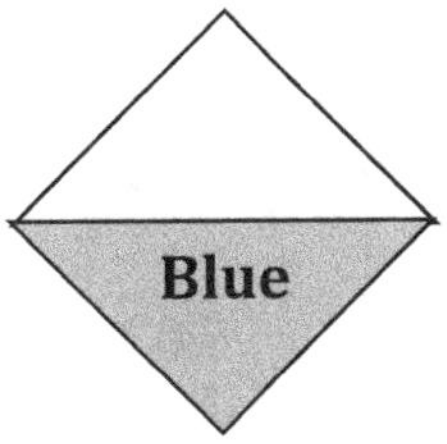

SPINETORAM

Available formulations in India : 1. **Spinetoram 11.70% SC**
2. **Spinetoram 0.10% RB**

Classification on the basis of

Target organism : Insecticide
Mode of entry : Stomach and Contact poison
Mode of action : Nerve poison
Chemical group : Spinosyns

IRAC classification

Main Group and Primary Site of Action : 5 Nicotinic acetylcholine receptor (nAChR) allosteric modulators – Site I
Sub-group or exemplifying Active Ingredient : Spinosyns

Category of toxicity : Moderately toxic
Toxicity label colour : Blue

- Spinetoram is a relatively new insecticide belonging to the spinosyn class, derived from the fermentation of the soil bacterium *Saccharopolyspora spinosa*. It's a semi-synthetic modification of spinosad.
- Spinetoram's mode of action involves the disruption of the insect nervous system. It is believed to primarily affect the nicotinic acetylcholine receptors (nAChR) and gamma-aminobutyric acid (GABA) receptors at the postsynaptic membranes of insect nerve cells. The disruption of these receptors causes overstimulation of the insect's nervous system. This overstimulation leads to involuntary muscle contractions. Eventually, the continuous and uncontrolled muscle activity leads to paralysis. The insect ultimately dies due to the disruption of its nervous system and vital functions.
- Spinetoram exhibits both contact and stomach action. It also has translaminar activity, meaning it can penetrate the leaf cuticle and move into the leaf tissue, providing control of pests on both the upper and lower leaf surfaces.

Label claim (DPPQ & S, Faridabad)

Spinetoram 11.70% SC

Crop	Pests	Dose / ha		Waiting period (Days)
		Quantity of formulation (ml)	Volume of water (Lit.)	
Cotton	Thrips	420	500-1000	30
	Tobacco caterpillar, Spotted bollworm	420-470	500-1000	30
	Pink bollworm	450	500-1000	30
Soybean	Tobacco caterpillar	450	500-625	30

Chilli	Thrips, Fruit borer, Tobacco caterpillar	470-500	400-500	07
Okra	Fruit borer, Shoot and fruit borer	375-450	500-1000	03
Brinjal	Shoot and fruit borer	375-450	500-1000	03
Chickpea	Pod borer	375-450	500-1000	20
Red gram	Pod borer, Spotted pod borer	375-450	500-1000	23
Grapes	Thrips	300	500-1000	05
Rice	Yellow stem borer	350-375	500	20
	Leaf folder	250	500	20
Tomato	Fruit borer, Tobacco caterpillar, Leaf miner, Tomato pinworm	375-450	500	03
Maize	Fall armyworm	250	500	32

Trade Names

Sr. No.	Trade Name	Company
1	Delegate	Dow AgroSciences India Pvt. Ltd.
2	Largo	Dhanuka Agritech Ltd.
3	Champion	Tropical Agrosystem India (P) Ltd.
4	Summit	Rallis India Ltd.

Spinetoram 0.10% RB

Crop	Pests	Dose / ha		Waiting period (Days)
		Quantity of formulation (g)	Volume of water (Lit.)	
Maize	Fall armyworm	5000-7500	--	21

Trade Names

Sr. No.	Trade Name	Company
1	Bexavie	Corteva Agriscience

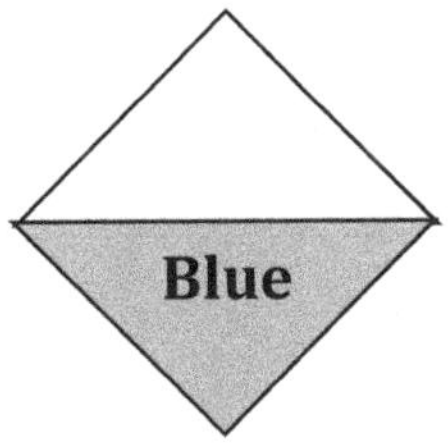

SPINOSAD

Available formulations in India : 1. **Spinosad 45.00% SC**
2. **Spinosad 02.50% SC**

Classification on the basis of

Target organism	:	Insecticide
Mode of entry	:	Stomach and Contact poison
Mode of action	:	Nerve poison
Chemical group	:	Spinosyns

IRAC classification

Main Group and Primary Site of Action	:	5 Nicotinic acetylcholine receptor (nAChR) allosteric modulators – Site I
Sub-group or exemplifying Active Ingredient	:	Spinosyns

Category of toxicity : Slightly toxic
Toxicity label colour : Blue

- Spinosad is a natural insecticide derived from the fermentation of the soil bacterium *Saccharopolyspora spinosa*. It is a mixture of two active components, spinosyn A and spinosyn D.
- Spinosad has a unique mode of action that affects the insect nervous system. It primarily acts on nicotinic acetylcholine receptors (nAChRs), causing neuronal excitation. It also has secondary effects on gamma-aminobutyric acid (GABA) receptors. This causes the insect's nerves to fire uncontrollably. The overstimulation leads to involuntary muscle contractions. Eventually, the insect becomes paralyzed. Death occurs due to the failure of the nervous system and muscle functions.
- Spinosad works through both contact and ingestion. It also exhibits translaminar activity, allowing it to move from the treated leaf surface to the other, offering some protection to pests feeding on the underside.

Label claim (DPPQ & S, Faridabad)

Spinosad 45.00 % SC

Crop	Pests	Dose / ha		Waiting period (Days)
		Quantity of formulation (ml)	Volume of water (Lit.)	
Cotton	American bollworm	165-220	500	10
Chilli	Fruit borer, Thrips	160	500	03
Red gram	Pod borer	125-162	800-1000	47
Brinjal	Shoot and fruit borer	162-187	500	03
Grapes	Thrips	250	1000	15

Trade Names

Sr. No.	Trade Name	Company
1	Tracer	Dow Agro Sciences India Pvt. Ltd.
2	Taffin	Rallis India Limited
3	One-Up	Dhanuka Agritech Ltd.
4	Spintor	Bayer Crop Science Ltd.
5	Spinner	Modern Insecticides Ltd.
6	Conserve	NACL Industries Ltd.
7	Spinosik	Sikko Industries Ltd.
8	Spiner	Tropical Agrosystem India (P) Ltd.
9	Trace One	Koronation Bio-Science
10	Spin	Global Crop Care

Spinosad 02.50 % SC

Crop	Pests	Dose / ha		Waiting period (Days)
		Quantity of formulation (ml)	Volume of water (Lit.)	
Cabbage, Cauliflower	Diamond back moth	600-700	500	03

Trade Names

Sr. No.	Trade Name	Company
1	Success	Dow Agro Sciences India Pvt. Ltd.
2	Spinokal	Kalyani Industries Ltd.

Blue

SPIROMESIFEN

Available formulations in India : **Spiromesifen 22.90% SC**

Classification on the basis of

Target organism : Insecticide, Acaricide
Mode of entry : Stomach and Contact poison
Mode of action : Lipid synthesis and growth regulator
Chemical group : Tetronic and Tetramic acid derivatives

IRAC classification

Main Group and Primary Site of Action : 23 Inhibitors of acetyl CoA carboxylase
Sub-group or exemplifying Active Ingredient : Tetronic and Tetramic acid derivatives

Category of toxicity : Moderately toxic
Toxicity label colour : Blue

- Spiromesifen is a relatively new insecticide and acaricide belonging to the tetronic acid chemical class.
- Spiromesifen works by inhibiting lipid biosynthesis in insects and mites, specifically by disrupting the acetyl CoA carboxylase enzyme. This enzyme is crucial for the production of fats and oils necessary for their growth and development. By interfering with this process, spiromesifen leads to the cessation of development and eventual death of the pest.
- It exhibits both contact and systemic (translaminar) activity. When sprayed on leaves, it can penetrate the leaf tissue and control pests on the underside as well.
- Spiromesifen is particularly effective against all stages (eggs, nymphs, and adults) of mites and whiteflies. It also reduces the fecundity of female pests and the hatching of their eggs (ovicidal effect).
- Spiromesifen is more stable in acidic and neutral conditions and degrades more rapidly in alkaline solutions.

Label claim (DPPQ & S, Faridabad)

Spiromesifen 22.90% SC

Crop	Pests	Dose / ha		Waiting period (Days)
		Quantity of formulation (ml)	Volume of water (Lit.)	
Brinjal	Red spider mite	400	500	05
Cotton	Whitefly, Mite	600	500	10
Apple	European red mite, Red spider mite	300	1000	30
Chilli	Yellow mite	400	500-750	07
Tea	Red spider mite	400	400	07

Okra	Red spider mite	400-500	500	03
Tomato	Whitefly, Mite	625	500	03

Trade Names

Sr. No.	Trade Name	Company
1	Voltage	PI Industries
2	Danfuron	Dhanuka Agritech Ltd.
3	Oberon	Bayer Crop Science Ltd.
4	Squillo	Isagro Asia Agrochemicals Pvt. Ltd.
5	Reveal	Best Agrolife Ltd.
6	Acarin	Nichino India
7	Spiro	Tropical Agrosystem India (P) Ltd.
8	Mitey	Global Crop Care
9	Kilron	JU Agri Sciences Pvt. Ltd.
10	Spirofuse	Mankind Agritech
11	Omega	Preventis Chem India (P) Ltd.

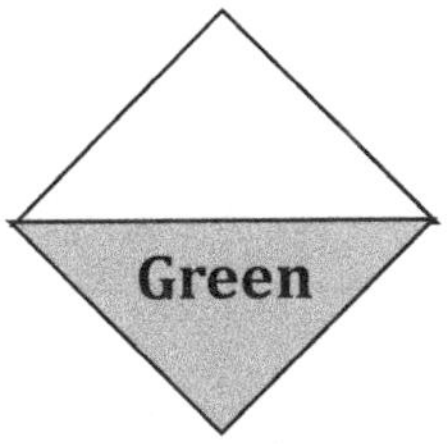

SPIROTETRAMAT

Available formulations in India : **Spirotetramat 15.31% OD**

Classification on the basis of

Target organism : Insecticide, Acaricide
Mode of entry : Systemic, Stomach and Contact poison
Mode of action : Lipid synthesis and growth regulator
Chemical group : Keto-enol

IRAC classification

Main Group and Primary Site of Action : 23 Inhibitors of acetyl CoA carboxylase
Sub-group or exemplifying Active Ingredient : Keto-enol

Category of toxicity : Slightly toxic
Toxicity label colour : Green

- Spirotetramat is a relatively new insecticide belonging to the keto-enol chemical class.
- Spirotetramat works by inhibiting lipid biosynthesis in insects and mites, specifically by disrupting the acetyl CoA carboxylase (ACC) enzyme. This enzyme is crucial for the production of fats and oils necessary for the growth and development of these pests. By interfering with this process, spirotetramat leads to the disruption of their development and eventual death.
- It exhibits two-way systemic activity, meaning it can move within the plant both upwards (via the xylem) and downwards (via the phloem). This unique characteristic allows it to reach pests feeding on various parts of the plant, including those on the underside of leaves.
- Spirotetramat is effective against various sucking pests, including mites, whiteflies, aphids, and mealybugs, at different life stages.
- Spirotetramat is rapidly hydrolyzed in alkaline conditions. It is most stable in 4-6 pH.

Label claim (DPPQ & S, Faridabad)

Spirotetramat 15.31% OD

Crop	Pests	Dose / ha		Waiting period (Days)
		Quantity of formulation (ml)	Volume of water (Lit.)	
Chilli	Thrips, Aphid	400	500	05
Okra	Aphid, Whitefly, Mite	600	500	03
Grapes	Mealybug, Mite	700	500-1000	60
Cotton	Aphid, Whitefly, Thrips	700	500	52
Citrus	Psylla, Mite	300	500	30
Cabbage	Diamond back moth	600	500	07

Trade Names

Sr. No.	Trade Name	Company
1	Movento	Bayer Crop Science Ltd.
2	Jumat	Global Crop Care

SULFOXAFLOR

Available formulations in India	:	**Sulfoxaflor 24.00 % SC**
Classification on the basis of		
Target organism	:	Insecticide
Mode of entry	:	Stomach and Contact poison
Mode of action	:	Nerve poison
Chemical group	:	Sulfoximines
IRAC classification		
Main Group and Primary Site of Action	:	4 Nicotinic acetylcholine receptor (nAChR) competitive modulators
Sub-group or exemplifying Active Ingredient	:	4C Sulfoximines
Category of toxicity	:	
Toxicity label colour	:	

- Sulfoxaflor is a relatively new insecticide belonging to the sulfoximine chemical class.
- Sulfoxaflor acts as a nicotinic acetylcholine receptor (nAChR) competitive modulator in insects. This means it binds to the same receptors in the insect's nervous system as acetylcholine, a crucial neurotransmitter. However, instead of activating the receptor normally, sulfoxaflor causes uncontrolled nerve impulses, leading to symptoms like muscle tremors, paralysis and eventually, death of the insect.
- While it affects the same general target site as neonicotinoids, sulfoxaflor binds differently to the nAChR, which is why it can be effective against neonicotinoid-resistant pests.
- It is a systemic insecticide and kills through both contact and ingestion. It also exhibits translaminar activity, allowing it to move within the leaf tissue.

Label claim (DPPQ & S, Faridabad)

Sulfoxaflor 24.00 % SC

Crop	Pests	Dose / ha		Waiting period (Days)
		Quantity of formulation (ml)	Volume of water (Lit.)	
Rice	Brown plant hopper, White backed plant hopper	375	500	14
Cotton	Jassid, Aphid	313	500	14
	Whitefly, Cotton mealybug	375	500	14

Trade Names

Sr. No.	Trade Name	Company
1	Transform	Dow AgroSciences India Pvt. Ltd.
2	D One	Dhanuka Agritech Ltd.

Blue

TETRANILIPROLE

Available formulations in India : 1. **Tetraniliprole 18.18 % SC**
2. **Tetraniliprole 40.34 % FS**

Classification on the basis of

Target organism : Insecticide
Mode of entry : Stomach and Contact poison
Mode of action : Muscle poison
Chemical group : Anthranilic diamide

IRAC classification

Main Group and Primary Site of Action : 28 Ryanodine receptor modulators
Sub-group or exemplifying Active Ingredient : Diamides

Category of toxicity : Moderately toxic
Toxicity label colour : Blue

- Tetraniliprole is a relatively new and powerful insecticide that belongs to the anthranilic diamide class of chemicals. Its main use is for controlling a wide range of insect pests in various crops.
- **Mode of Action :** Tetraniliprole works primarily through ingestion, but can also have some contact activity. Its specific mode of action is as a ryanodine receptor modulator. This means it activates the ryanodine receptors in the insect's muscle cells, leading to an uncontrolled release of calcium. This results in muscle paralysis, feeding cessation, and ultimately, the death of the insect.
- The insecticide has translaminar movement, meaning it can move through the plant's leaves to reach pests on the other side that are difficult to hit directly with a spray.
- Tetraniliprole is moderately persistent in the environment. Its degradation is influenced by factors like hydrolysis (especially under alkaline conditions) and photolysis.

Label claim (DPPQ & S, Faridabad)

Tetraniliprole 18.18 % SC

Crop	Pests	Dose / ha		Waiting period (Days)
		Quantity of formulation (ml)	Volume of water (Lit.)	
Rice	Yellow stem borer, Leaf folder	250-300	500	43
Soybean	Girdle beetle, Tobacco leaf eating caterpillar, Semilooper	250-300	500	35
Tomato	Fruit borer, Leaf miner	300	500	05

Trade Names

Sr. No.	Trade Name	Company
1	Vayego	Bayer Crop Science Ltd.

Tetraniliprole 40.34 % FS

Crop	Pests	Dose / Kg seed		Waiting period (Days)
		Quantity of formulation (ml)	Volume of water (Lit.)	
Rice	Stem borer, Leaf folder	10.0-12.5	--	--
Maize	Stem borer	5.0-7.5	--	--

Trade Names

Sr. No.	Trade Name	Company
1	Reatis	Bayer Crop Science Ltd.

Yellow

THIACLOPRID

Available formulations in India : **Thiacloprid 21.70% SC**

Classification on the basis of

Target organism : Insecticide

Mode of entry : Systemic, Stomach and Contact poison

Mode of action : Nerve poison

Chemical group : Neonicotinoid

IRAC classification

Main Group and Primary Site of Action : 4 Nicotinic acetylcholine receptor (nAchR) competitive modulators

Sub-group or exemplifying Active Ingredient : 4A Neonicotinoids

Category of toxicity : Highly toxic

Toxicity label colour : Yellow

- Thiacloprid is a neonicotinoid insecticide.
- Thiacloprid acts by binding to and blocking the nicotinic acetylcholine receptors (nAChRs) in the insect's nervous system. Acetylcholine is a crucial neurotransmitter for nerve impulse transmission. By binding to these receptors, thiacloprid causes overstimulation of the nerve cells and disrupts normal nerve function, leading to paralysis and eventual death of the insect.
- Thiacloprid is systemic, meaning it is absorbed by the plant and translocated throughout its tissues. It also exhibits some translaminar activity, allowing it to move to some extent within the leaf.
- It has both contact and stomach action.

Label claim (DPPQ & S, Faridabad)

Thiacloprid 21.70% SC

Crop	Pests	Dose / ha		Waiting period (Days)
		Quantity of formulation (ml)	Volume of water (Lit.)	
Cotton	Aphid, Thrips, Jassid	100-125	500	52
	Whitefly	500-600	500	52
Rice	Stem borer	500	500	30
Chilli	Thrips	225-300	500	05
Tea	Mosquito bug	375	400	07
Brinjal	Shoot and fruit borer	750	500	05
Soybean	Girdle beetle	750	500	017
Apple	Thrips	1.85-2.30 / lit water	As per requiremet	30

Trade Names

Sr. No.	Trade Name	Company
1	Splendour	Crystal Crop Protection Ltd.
2	Gunwaan	Anu Products Ltd.
3	Thiocel	Excel Crop Care Ltd.
4	Invest	Isagro Asia Agrochemicals Pvt. Ltd.
5	Alanto	Bayer Crop Science Ltd.
6	Expert	Global Crop Care

Blue

THIOCYCLAM HYDROGEN OXALATE

Available formulations in India : **Thiocyclam Hydrogen Oxalate 50.00 % SP**

Classification on the basis of

Target organism : Insecticide
Mode of entry : Stomach and Contact poison
Mode of action : Nerve poison
Chemical group : Nereistoxin analouge

IRAC classification

Main Group and Primary Site of Action : 14 Nicotinic acetylcholine receptor (nAChR) channel blockers
Sub-group or exemplifying Active Ingredient : Nereistoxin analogues

Category of toxicity : Moderately toxic
Toxicity label colour : Blue

- Thiocyclam hydrogen oxalate is a broad-spectrum, synthetic insecticide. It is a nereistoxin analogue, which means its chemical structure is similar to nereistoxin, a natural neurotoxin found in some marine worms.
- Thiocyclam hydrogen oxalate works by disrupting the nervous system of insects. It is a nicotinic acetylcholine receptor (nAChR) blocker. By interfering with these receptors, it causes paralysis and ultimately death in the targeted pests. It has both contact and stomach action, and some limited systemic activity, meaning it can be absorbed by the plant and translocated to other tissues, offering protection against both above-ground and below-ground pests.
- Thiocyclam hydrogen oxalate is relatively stable in acidic solution.

Label claim (DPPQ & S, Faridabad)

Thiocyclam Hydrogen Oxalate 50.00 % SP

Crop	Pests	Dose / ha		Waiting period (Days)
		Quantity of formulation (g)	Volume of water (Lit.)	
Rice	Stem borer, Leaf folder	1000	500	30

Trade Names

Sr. No.	Trade Name	Company
1	Vibrant 50 SP	PI Industries Ltd.

THIODICARB

Available formulations in India : **Thiodicarb 75.00% WP**

Classification on the basis of

Target organism : Insecticide
Mode of entry : Systemic, Stomach and Contact poison
Mode of action : Nerve poison
Chemical group : Carbamates

IRAC classification

Main Group and Primary Site of Action : 1 Acetylcholinesterase (AChE) inhibitors
Sub-group or exemplifying Active Ingredient : 1A Carbamates

Category of toxicity : Highly toxic
Toxicity label colour : Yellow

- Thiodicarb is a carbamate insecticide with primarily stomach action and some contact activity.
- Thiodicarb works by inhibiting the enzyme acetylcholinesterase (AChE) in the nervous system of insects. This enzyme is crucial for breaking down the neurotransmitter acetylcholine (ACh). When AChE is inhibited, acetylcholine accumulates at the nerve synapses, leading to overstimulation of the nervous system, followed by muscle twitching, paralysis, and ultimately, death of the insect.
- As a seed treatment, thiodicarb is rapidly translocated systemically throughout the plant. It also acts as a molluscicide, causing paralysis and death in slugs.

Label claim (DPPQ & S, Faridabad)

Thiodicarb 75.00% WP

Crop	Pests	Dose / ha		Waiting period (Days)
		Quantity of formulation (g)	Volume of water (Lit.)	
Cabbage	Diamond back moth	1000-1330	500	07
Cotton	Bollworms	1000	500	30
Brinjal	Shoot and fruit borer	625-1000	500	06
Chilli	Fruit borer	625-1000	500	06
Black gram	Pod borer, Spotted pod borer	625-750	375-500	17
Pigeonpea	Pod borer	625-1000	500	30

Trade Names

Sr. No.	Trade Name	Company
1	Larvin	Bayer Crop Science Ltd.
2	Capcadis	Syngenta India Ltd.

Blue

THIAMETHOXAM

Available formulations in India :

1. **Thiamethoxam 30.00% FS**
2. **Thiamethoxam 30.00% WS**
3. **Thiamethoxam 70.00% WS**
4. **Thiamethoxam 75.00% SG**
5. **Thiamethoxam 25.00% WG**
6. **Thiamethoxam 01.20% GR**

Classification on the basis of

Target organism : Insecticide
Mode of entry : Stomach and Contact poison
Mode of action : Nerve poison
Chemical group : Neonicotinoid

IRAC classification

Main Group and Primary Site of Action : 4 Nicotinic acetylcholine receptor (nAchR) competitive modulators
Sub-group or exemplifying Active Ingredient : 4A Neonicotinoids

Category of toxicity : Moderately toxic
Toxicity label colour : Blue

- Thiamethoxam is a neonicotinoid insecticide.
- Thiamethoxam acts as a nicotinic acetylcholine receptor (nAChR) competitive modulator in insects. It binds to the same receptors in the insect's nervous system as acetylcholine, a crucial neurotransmitter. By binding to these receptors, thiamethoxam disrupts the normal transmission of nerve impulses, leading to overstimulation of the nervous system, muscle tremors, paralysis and eventually, death of the insect.
- It is a systemic insecticide, meaning it is absorbed by the plant and transported throughout its tissues, providing protection from within.
- Thiamethoxam also exhibits contact and stomach action, killing pests that directly touch or ingest it. Furthermore, it has translaminar activity, allowing it to move within the leaf tissue, offering some control of pests on the underside of leaves.

Label claim (DPPQ & S, Faridabad)

Thiamethoxam 30.00 % FS (Seed treatment)

Crop	Pests	Dose / Kg seed		Waiting period (Days)
		Quantity of formulation (ml)	Volume of water (ml)	
Cotton	Aphid, Whitefly, Jassid, Thrips	8-10	--	--
Sorghum	Shoot fly	10	--	--
Wheat	Termite	3.30	--	--

Soybean	Stem fly, Girdle beetle, Whitefly	6	6-8	--
	Jassid	4	6-8	--
Chilli	Thrips	7	--	--
Okra	Jassid	5.70		
Sunflower	Jassid, Thrips	10		
Maize	Aphid	6	10-12	
		8	10-12	
Groundnut	Jassid	3	6-8	--
	Aphid, Thrips	5	6-8	--

Trade Names

Sr. No.	Trade Name	Company
1	Slayer Pro	GSP Crop Science Pvt. Ltd.
2	Taiyo Plus	IFFCO-MC Crop Science Private LTd.
3	Areva Super	Dhanuka Agritech Ltd.
4	Image Gold	Anu Products Ltd.
5	Bheema	Insecticides (India) Ltd.
6	Bakhtar	Crop Life Science Ltd.
7	Kaziro	Parijat Industries India Pvt. Ltd.
8	Crucial	Chemet Chemicals Pvt, Ltd.
9	Kemstar Eaasy	Sumitomo Chemical India Ltd.
10	Paras Player	Paras Crops Ltd.
11	Eco-Champ Plus	Isagro Asia Agrochemicals Pvt. Ltd.
12	Prompt Plus	Herenba Industries Ltd.
13	Sikthim	Sikko Industries Ltd.
14	Mahabali	Best Agrolife Ltd.
15	Hirota	Nichino India Pvt. Ltd.
16	Paatalok FS	Somnath Demeter Agritron Ltd.
17	Seed Cover	Silver Sine Bio Tech
18	Thiame Gold	Natural Crop Science
19	Thiala	Essential Biosciences
20	Takin Plus	Synwood Agro Ltd.
21	Maharathi	Koronation Bio-Science
22	Apterra Plus	Todays's Global Crop Protection
23	Kisan Thiam Plus	National Fertilizers Ltd.
24	Seedoxm-FS	Krishi Rasayan
25	Trot	Tata Rallis India Ltd.
26	Ikka 30	Green India Care
27	Mantra FS	JU Agri Sciences Pvt. Ltd.
28	Raxak	Linuxcrop Chemicals Pvt. Ltd.
29	SA-Glazer	Saraswati Agro Life Science (India) Pvt. Ltd.
30	Silky	Gharda Chemicals Ltd.
31	Thiakind-Strong	Mankind Agritech
32	Exylon Thiazol	Torrent Crop Science
33	Aspire FS	Preventis Chem India (P) Ltd.
34	Headlight	RACCOLTO Agritech India Pvt. Ltd.

Thiamethoxam 30.00 % WS (Seed treatment)

Crop	Pests	Dose / ha		Waiting period (Days)
		Quantity of formulation (g)	Volume of water (Lit.)	
Maize	Aphid	6	10-12	--
	Shoot fly	8	10-12	--
Cotton	Aphid, Jassid, Whitefly, Thrips	8-10	10-15	--
Soybean	Jassid	4	6-8	--
	Stem fly, Girdle beetle, Whitefly	6	6-8	--
Groundnut	Jassid	3	6-8	--
	Aphid, Thrips	5	6-8	--

Trade Names

Sr. No.	Trade Name	Company
1		

Thiamethoxam 70.00 % WS (Seed treatment)

Crop	Pests	Dose / ha		Waiting period (Days)
		Quantity of formulation (g)	Volume of water (Lit.)	
Cotton	Aphid, Thrips, Whitefly, Jassid	4.30	--	--
Okra	Aphid, Jassid	2.86	--	--
Tomato	Aphid, Thrips	6.00	--	--
Sunflower	Jassid, Thrips	4.00	--	--
Wheat	Termite, Aphid	1.75	--	--
Maize	Shoot fly, Aphid	3.50	--	--
Rice	Thrips, Green leaf hopper	1.50	--	--

Trade Names

Sr. No.	Trade Name	Company
1	Cruiser	Syngenta India Ltd.
2	Slayer	GSP Crop Science Pvt. Ltd.
3	Refuge	Riddhi Speciality Chemicals
4	Blue Tara +	ISCAR Crop Science
5	Texan PW	Nobel Crop Science
6	Swiftwind	RACCOLTO Agritech India Pvt. Ltd.

Thiamethoxam 75.00 % SG

Crop	Pests	Dose / ha		Waiting period (Days)
		Quantity of formulation (g)	Volume of water (Lit.)	
Groundnut	Termite	125	500-1000	57

Sugarcane	Termite, Early shoot borer	160	500-1000	230
Rice	Green leaf hopper, Brown plant hopper	150	Dissolve in 500 ml water and mix with 20 Kg sand / ha	60
Cotton	Jassid, Thrips	125	50-100 ml /plant	109

Trade Names

Sr. No.	Trade Name	Company
1	Neo Super	Crystal Crop Protection Ltd.
2	Ripple	Dhanuka Agritech Ltd.
3	Thiomax 75	Katyayani Organics
4	Crazy	Astha Chemicals
5	Tagxone Gold	Tropical Agrosystem India (P) Ltd.
6	Thiomax Super	Essential Biosciences
7	Takin Super 75	Synwood Agro Ltd.
8	Gajraj	Koronation Bio-Science
9	Krioxm-Gold	Krishi Rasayan
10	Ikka 75	Green India Care
11	JU-Mantra Super	JU Agri Sciences Pvt. Ltd.
12	Exylon Thia Apex	Torrent Crop Science
13	Sergeant	Preventis Chem India (P) Ltd.
14	Swiftwind Plus	RACCOLTO Agritech India Pvt. Ltd.

Thiamethoxam 25.00 % WG

Crop	Pests	Dose / ha		Waiting period (Days)
		Quantity of formulation (g)	Volume of water (Lit.)	
Rice	Stem borer, Gall midge, Leaf folder, White backed plant hopper, Brown plant hopper, Green leaf hopper, Thrips	100	500-750	14
Cotton	Jassid, Thrips, Aphid	100	500-750	21
	Whitefly	200	500-750	21
Okra	Jassid, Aphid, Whitefly	100	500-1000	05
Mango	Hoppers	100	1000	30
Wheat	Aphid	50	500	21
Mustard	Aphid	50-100	500-1000	21
Tomato	Whitefly (Foliar)	200	500	05
	Whitefly (Soil drenching)	400	500	05
Brinjal	Whitefly (Foliar)	200	500	03
Tea	Mosquito bug	100	400-500	07
Potato	Aphid (Foliar)	100	500	77
	Aphid (Soil drenching)	200	400-500	77
Citrus	Psylla	100	1000	20

Rice Nursery	Green leaf hopper, Thrips, Whorl maggot (Soil drenching)	2000	250 ml / Sq.m.	86
Cumin	Aphid	100	500	05
Grapes	Mealybug, Thrips, Flea beetle (Soil drenching)	400	200 ml / vine	10
	Mealybug, Thrips, Flea beetle (Foliar)	100	1000	15

Trade Names

Sr. No.	Trade Name	Company
1	Extra Super	Crystal Crop Protection Ltd.
2	Suckgan	ADAMA India Private Limited
3	Maxima	PI Industries
4	Renova	UPL Ltd.
5	Battalion	Swal Corporation Ltd.
6	Evident, Thiomax	Biostadt India Ltd.
7	Cover	Gharda Chemicals Ltd.
8	Areva	Dhanuka Agritech Ltd.
9	Caper	Cheminova
10	Optra	Coromandel International Ltd.
11	Chivas	Cropnosys India Pvt. Ltd.
12	Click	Indofil Industries Ltd.
13	Volt	Alfa Crop Science
14	Actara	Syngenta
15	Torpedo	Auxin Crop Care
16	Image	Anu Products Ltd.
17	Arrow	Insecticides (India) Ltd.
18	Thia-Tech WSG	Modern Insecticides Ltd.
19	Dxtar	NACL Industries Ltd.
20	Devitara	Devidayal Agro Chemicals Ltd.
21	Wonderex	Excel Crop Care Ltd.
22	Athoxam	Crop Life Science Ltd.
23	Jonga	Parijat Industries India Pvt. Ltd.
24	Kemstar	Sumitomo Chemical India Ltd.
25	Eco-Champ	Isagro Asia Agrochemicals Pvt. Ltd.
26	G-Tara	GFG Crop Science Pvt. Ltd.
27	Evader	Tusova Agro Pvt. Ltd.
28	Prompt	Herenba Industries Ltd.
29	Siktara	Sikko Industries Ltd.
30	Throne	Best Agrolife Ltd.
31	Crux	Nichino India Pvt. Ltd.
32	Pol	Somnath Demeter Agritron Ltd.
33	5-Star	Silver Sine Bio Tech
34	Thaim 25	Natural Crop Science
35	Tagxone	Tropical Agrosystem India (P) Ltd.

36	Thiomax	Essential Biosciences
37	Tudio	Synwood Agro Ltd.
38	Kisan Thiam	National Fertilizers Ltd.
39	Meghraj	Koronation Bio-Science
40	Paanch Tara	Today's Global Crop Protection
41	Kri-oxm	Krishi Rasayan
42	Act 25	Global Crop Care
43	Anant	Tata Rallis India Ltd.
44	Ikka 25	Green India Care
45	Mantra	JU Agri Sciences Pvt. Ltd.
46	Thiozim	Linuxcrop Chemicals Pvt. Ltd.
47	SA-Thiastar	Saraswati Agro Life Science (India) Pvt. Ltd.
48	Thiakind	Mankind Agritech
49	Exylon Thiapower	Torrent Crop Science
50	Aspire	Preventis Chem India (P) Ltd.
51	Thimble	RACCOLTO Agritech India Pvt. Ltd.

Thiamethoxam 01.20% GR

Crop	Pests	Dose / ha		Waiting period (Days)
		Quantity of formulation (Kg)	Volume of water (Lit.)	
Rice	Yellow stem borer, Leaf folder, Brown plant hopper	10	Broadcasting	52
Groundnut	White grub, Termite	10	Broadcasting	53
Sugarcane	Termite, Early shoot borer, White grub	10	Broadcasting	237

Trade Names

Sr. No.	Trade Name	Company
1		

Yellow

TOLFENPYRAD

Available formulations in India	:	**Tolfenpyrad 15.00% EC**
Classification on the basis of		
Target organism	:	Insecticide
Mode of entry	:	Stomach and Contact poison
Mode of action	:	Energy metabolism
Chemical group	:	Pyrazole
IRAC classification		
Main Group and Primary Site of Action	:	21 Mitochondrial complex I electron transport inhibitors
Sub-group or exemplifying Active Ingredient	:	21A Pyrazole
Category of toxicity	:	Highly toxic
Toxicity label colour	:	Yellow

- Tolfenpyrad is a relatively new insecticide and acaricide belonging to the pyrazole chemical class (IRAC Group 21A).
- Tolfenpyrad is a Mitochondrial Electron Transport Inhibitor (METI), specifically targeting Complex I of the mitochondrial respiratory chain in insects and mites. This disruption of cellular respiration leads to a rapid cessation of energy production, resulting in rapid stop of feeding, paralysis and death of the pest.
- It exhibits contact activity and is effective against all life stages of the pest, including eggs, larvae, nymphs, and adults.
- It also shows anti-feeding activity, particularly against lepidopteran larvae, and can suppress oviposition (egg-laying) in female insects.
- Some sources also indicate mild systemic action through plant uptake.

Label claim (DPPQ & S, Faridabad)

Tolfenpyrad 15.00% EC

Crop	Pests	Dose / ha		Waiting period (Days)
		Quantity of formulation (ml)	Volume of water (Lit.)	
Cabbage	Diamond back moth, Aphid	1000	500	05
Okra	Aphid, Jassid, Thrips, Whitefly	1000	500	03
Cotton	Aphid, Jassid, Thrips, Whitefly	1000	500	26
Cumin	Aphid, Thrips	1000	500	29
Chilli	Aphid, Thrips	1000	500	07
Mango	Hoppers, Thrips	1000	500	07
Onion	Thrips	1000	500	10
Pomegranate	Thrips	1000	500	07

Trade Names

Sr. No.	Trade Name	Company
1	Keefun	PI Industries
2	Tychi	Nichino India
3	Keechak	Katyayani Organics
4	Tuckzila	Parijat Industries India Pvt. Ltd.
5	Tycoon	Angel Crop Science
6	Tolfen	Windson Bio Tech
7	Aimpyrad	AIMCO Pesticides Ltd.
8	Telcy	Adicon Pesticides Pvt. Ltd.
9	Oxfen	Goldenox Crop Care
10	Tolfen	Icon Crop Care
11	Tolryd	Foreward Crop Protection Pvt. Ltd.
12	Toofani	Dr. Chemistar Agro Industries
13	Teefun	Agri Science Co.
14	Tol R	Advance Pesticides
15	Tolfetox	Mankind Agritech
16	Tender	Prisha Chemicals
17	Lause	Global Crop Care
18	Thunder	Penta Life Science
19	Patang	Rallies India Ltd.
20	Yamaraj	GSP
21	Welloen	Greenwell Biotech
22	Tollgate	Perfect Crop Science
23	H Fun	Harishree Crop Science
24	Zenith	Agrostar
25	Voltol	Volkschem Crop Science Pvt. Ltd.
26	Taifun	Neolect Crop Science
27	Tuffnut	Aramco Applomattix Chemicals
28	Torac	Unicorn Chemicals
29	PTOFE-15	Parth Chemicals
30	Tofe	Marwell Crop Science
31	Hunter	Linuxcrop Chemicals Pvt. Ltd.
32	Keyfun	Natural Bio Fuel
33	Target	Silver Sine Bio Tech
34	Typride	Ample Agrochem Industries
35	Tolins	Insectigen Crop Care
36	Tofu	Alexa Crop Science
37	Trofy	Astha Chemicals
38	Bond Plus	Indo Pacific Chemicals
39	Fun	Tropical Agrosystem India (P) Ltd.
40	Torito-15	Synwood Agro Ltd.
41	Taifen	Koronation Bio-Science
42	Taosu	JU Agri Sciences Pvt. Ltd.
43	Hunter	Linuxcrop Chemicals Pvt. Ltd.
44	Caspen	Synovita Crop science Pvt. Ltd.
45	Tolfetox	Mankind Agritech
46	Exylon Thunder	Torrent Crop Science
47	Status	Preventis Chem India (P) Ltd.
48	Taluka	Chemley Agritech Pvt. Ltd.

Blue

TRIFLUMEZOPYRIM

Available formulations in India :
1. **Triflumezopyrim 10.00 % SC**
2. **Triflumezopyrim 20.00 % WG**

Classification on the basis of

Target organism : Insecticide
Mode of entry : Systemic, Stomach poison
Mode of action : Nerve poison
Chemical group : Mesoionic

IRAC classification

Main Group and Primary Site of Action : 4 Nicotinic acetylcholine receptor (nAChR) competitive modulators
Sub-group or exemplifying Active Ingredient : 4E Mesoionics

Category of toxicity : Moderately toxic
Toxicity label colour : Blue

- Triflumezopyrim is a highly effective insecticide belonging to the novel class of mesoionic insecticides. It is particularly known for its use in controlling hoppers and planthoppers in rice crops.
- Triflumezopyrim is a systemic insecticide, which means it can be absorbed by plants and then transported throughout the plant's tissues. The primary mode of entry for an insect is ingestion.
- **Mode of action:** Unlike neonicotinoids which are agonists, triflumezopyrim acts as a competitive modulator, binding to and inhibiting the orthosteric site of the insect's nicotinic acetylcholine receptors (nAChR). This distinct mechanism of action makes it effective against pests that have developed resistance to other insecticides, including neonicotinoids.
- Triflumezopyrim is highly stable across a wide range of pH levels (4-9).

Label claim (DPPQ & S, Faridabad)

Triflumezopyrim 10.00 % SC

Crop	Pests	Dose / ha		Waiting period (Days)
		Quantity of formulation (ml)	Volume of water (Lit.)	
Rice	Brown plant hopper, White backed plant hopper	236	500	21
Mango	Hoppers	2.36 ml / lit water	1000	30

Trade Names

Sr. No.	Trade Name	Company
1	Pexalon	Corteva, Dupont
2	Hexalon	Haedway Bio Products LLP
2	Penaxa	Global Crop Care

Triflumezopyrim 20.00 % WG

Crop	Pests	Dose / ha		Waiting period (Days)
		Quantity of formulation (g)	Volume of water (Lit.)	
Rice	Brown plant hopper, White backed plant hopper	125	500	21

Trade Names

Sr. No.	Trade Name	Company
1	Vestoria	Syngenta India Ltd.

PART III

Insecticide Combinations

Acephate + Bifenthrin

Available formulations in India : Acephate 50.00% + Bifenthrin 10.00% WDG

Classification on the basis of		Acephate	Bifenthrin
Target organism	:	Insecticide	Insecticide
Mode of entry	:	Systemic, Contact and Stomach poison	Contact and Stomach poison
Mode of action	:	Nerve poison	Nerve poison
Chemical group	:	Organophosphate	Synthetic pyrethroid
Category of toxicity : Highly toxic **Toxicity label colour** : Yellow			

- The combination of these two active ingredients offers a dual mode of action, making it highly effective against a broad spectrum of insect pests. The dual action arises from the nature of its components.
- Acephate is a systemic organophosphate insecticide that provides a systemic mode of action. It is absorbed by the plant and moves through its system, offering residual control against feeding pests. Acephate works by inhibiting the acetylcholinesterase enzyme, which is crucial for nerve signal transmission in insects, leading to paralysis and death.
- On the other hand, Bifenthrin is a synthetic pyrethroid insecticide that acts primarily by contact and stomach poison. It affects the insect's nervous system by disrupting sodium channel function in nerve membranes, causing rapid knockdown and mortality, often providing a quick result against exposed pests.
- The combination leverages both systemic and contact activity, ensuring control of both internal and external pests.

Label claim (DPPQ & S, Faridabad)
Acephate 50.00% + Bifenthrin 10.00% WDG

Crop	Pests	Dose / ha		Waiting period (Days)
		Quantity of formulation (g)	Volume of water (Lit.)	
Cotton	Bollworms	500-750	500-1000	20

Trade Names

Sr. No.	Trade Name	Company
1	Acenthrin	Swal Corporation Ltd.
2	Tangent	Anu Products Ltd.
3	Super vajra	Ryzochem Crop Science
4	Mani plus	Mani Agro Chemicals
5	Commentary	Ariana Crop Science
6	Harrier	Gayatri Insecticide Pvt. Ltd.
7	Roar	Plazma Crop Science
8	Target	Sai Ram Agritech Pvt. Ltd.
9	Kilin	Universal Agro Chemical Industries
10	Lion power	Dream Field Chemical Private Limited
11	Spoodo	Zivista Crop Protection
12	Vasko	Tanvi Bio Chemical Pvt. Ltd.
13	Lowe	RACCOLTO Agritech India Pvt. Ltd.

Acephate + Cypermethrin

Available formulations in India : Acephate 45% + Cypermethrin 5% DF

Classification on the basis of		Acephate	Cypermethrin
Target organism	:	Insecticide	Insecticide
Mode of entry	:	Systemic, Contact and Stomach poison	Contact and Stomach poison
Mode of action	:	Nerve poison	Nerve poison
Chemical group	:	Organophosphate	Synthetic pyrethroid
Category of toxicity : Highly toxic **Toxicity label colour** : Yellow			

- The combination of Acephate and Cypermethrin results in a powerful, broad-spectrum insecticide that delivers dual-action control over a wide range of insect pests. The strength of this combination lies in the complementary modes of action of the two active ingredients.
- Acephate is a systemic organophosphate insecticide that provides systemic and stomach action. It is absorbed by the plant and translocated through its tissues, offering residual control against internal feeders and sucking pests by inhibiting the acetylcholinesterase enzyme in insects' nervous systems.
- Cypermethrin, on the other hand, is a non-systemic synthetic pyrethroid insecticide that works primarily by contact and stomach action. It is renowned for its quick knockdown effect, rapidly disrupting the insect's nervous system by interfering with sodium channels.
- By combining the rapid, external control of Cypermethrin with the lasting, internal protection of Acephate, the formulation achieves a synergistic effect that is often more effective than applying either chemical alone, even against pests that may have developed resistance to single-mode chemistries.

Label claim (DPPQ & S, Faridabad)
Acephate 45% + Cypermethrin 5% DF

Crop	Pests	Dose / ha		Waiting period (Days)
		Quantity of formulation (ml)	Volume of water (Lit.)	
Cotton	Aphid, Jassid, Thrips, Whitefly	850	500-600	22
Maize	Stem borer, Fall armyworm	1250	500	52

Trade Names

Sr. No.	Trade Name	Company
1	Sorter	UPL Ltd.
2	Mani gold	Mani Agro Chemicals
3	Acer 455	Tusova Agro Pvt. Ltd.
4	Lingan	RACCOLTO Agritech India Pvt. Ltd.

Acephate + Fenvalerate

Available formulations in India : Acephate 25.00% + Fenvalerate 03.00% EC

Classification on the basis of		Acephate	Fenvalerate
Target organism	:	Insecticide	Insecticide
Mode of entry	:	Systemic, Contact and Stomach poison	Contact and Stomach poison
Mode of action	:	Nerve poison	Nerve poison
Chemical group	:	Organophosphate	Synthetic pyrethroid
Category of toxicity : Highly toxic **Toxicity label colour** : Yellow			

- The product containing Acephate and Fenvalerate is a highly effective, broad-spectrum insecticide formulated to manage a wide variety of insect pests of crops.
- The combination of the two different chemical classes-Acephate being an organophosphate and Fenvalerate being a synthetic pyrethroid-provides a synergistic approach to pest control.
- Acephate is a systemic insecticide, meaning it is absorbed by the plant foliage and translocated throughout the plant tissues. This allows it to kill sucking and chewing pests that feed on the treated plant parts, providing residual protection for new growth and against internal feeders like leaf miners. Its mechanism involves inhibiting the acetylcholinesterase enzyme, which disrupts the insect's nervous system.
- Fenvalerate is a non-systemic pyrethroid known for its fast contact and stomach action. It acts as a sodium channel modulator on the insect's nervous system, causing rapid over-excitement, paralysis, and quick knockdown of pests upon direct contact or ingestion.
- The combination leverages the quick killing power of Fenvalrate with the lasting, deep-penetrating protection of Acephate.

Label claim (DPPQ & S, Faridabad)
Acephate 25.00% + Fenvalerate 03.00% EC

Crop	Pests	Dose / ha		Waiting period (Days)
		Quantity of formulation (ml)	Volume of water (Lit.)	
Cotton	Bollworms, Sucking pest	2000	500	15

Trade Names

Sr. No.	Trade Name	Company
1	Acefa	Gujarat Pestichem Industries
2	Para-Killer	Paramount Pesticides Ltd.
3	Sardar	National Pesticides and Chemicals
4	Phoenix	Aquarius Agro Chemicals
5	Acefenval	Plazma Crop Science
6	Contra	Sai Ram Agritech Pvt. Ltd.
7	V-Acefen	Varad Crop Science Pvt. Ltd.

Acephate + Fipronil

Available formulations in India : Acephate 50.00 % + Fipronil 05.00% WDG

Classification on the basis of		Acephate	Fipronil
Target organism	:	Insecticide	Insecticide
Mode of entry	:	Systemic, Contact and Stomach poison	Contact and Stomach poison
Mode of action	:	Nerve poison	Nerve poison
Chemical group	:	Organophosphate	Phenylpyrazole
Category of toxicity : Highly toxic **Toxicity label colour** : Yellow			

- The combination of Acephate and Fipronil is formulated as a powerful, dual-action insecticide, and is highly valued in agriculture for providing superior, broad-spectrum control over a wide variety of insect pests. The mixture capitalizes on the distinct yet complementary modes of action of its two active ingredients: Acephate, an organophosphate, and Fipronil, a phenylpyrazole.
- Acephate provides both systemic and contact action by acting as a cholinesterase inhibitor, which disrupts nerve impulse transmission and leads to the paralysis and death of the insect. Its systemic nature allows it to be absorbed and translocated throughout the plant tissue, offering residual protection against sucking and chewing pests.
- Fipronil adds a strong contact and stomach action, disrupting the insect's central nervous system by blocking GABA-gated chloride channels. This mechanism ensures rapid knockdown and long-lasting residual control against a diverse range of pests, including difficult-to-manage insects.
- The primary benefit of combining Acephate and Fipronil is the synergistic effect and the dual mode of action, which provides enhanced efficacy compared to either insecticide used alone. This combination results in a quick knockdown effect followed by extended residual activity.

Label claim (DPPQ & S, Faridabad)
Acephate 50.00 % + Fipronil 05.00% WDG

Crop	Pests	Dose / ha		Waiting period (Days)
		Quantity of formulation (g)	Volume of water (Lit.)	
Rice	Stem borer, Leaf folder, Brown planthopper	1000	500	27

Trade Names

Sr. No.	Trade Name	Company
1	Leneer	Agri Science Co.
2	Acefipro	Natural Crop Science
3	Acenil	Prisha Chemicals
4	Acefro	Rumikon Crop Scuence

Acephate + Imidacloprid

Available formulations in India : **1. Acephate 50.00% + Imidacloprid 01.80% SP**
2. Acephate 50.00% + Imidacloprid 05.00% WG

Classification on the basis of		Acephate	Imidacloprid
Target organism	:	Insecticide	Insecticide
Mode of entry	:	Systemic, Contact and Stomach poison	Systemic, Contact and Stomach poison
Mode of action	:	Nerve poison	Nerve poison
Chemical group	:	Organophosphate	Neonicotinoid
Category of toxicity : Highly toxic **Toxicity label colour** : Yellow			

- Acephate + Imidacloprid is a dual-action systemic and contact insecticide combination that provides broad-spectrum control over both sucking and chewing insect pests.
- Acephate, a high-concentration organophosphate, provides rapid contact and systemic activity by inhibiting the acetylcholinesterase enzyme, leading to paralysis and quick knockdown of pests.
- Complementing this is the low-dose Imidacloprid, a neonicotinoid that is powerfully systemic and works by blocking nicotinic acetylcholine receptors. This action is particularly effective against sucking pests, ensuring long-lasting protection even on new growth.

Label claim (DPPQ & S, Faridabad)
Acephate 50.00% + Imidacloprid 01.80% SP

Crop	Pests	Dose / ha		Waiting period (Days)
		Quantity of formulation (g)	Volume of water (Lit.)	
Cotton	Aphid, Jassid, Thrips, Whitefly, Bollworms	1000	500	40
Rice	Brown planthopper, Green leafhopper, Stem borer, Leaf folder	1000	500	--
Sugarcane	Termite, White grub, Early shoot borer, Black bug, Pyrilla	2500	500	123
Chilli	Jassid, Aphid, Whitefly, Thrips. Fruit borer, Tobacco caterpillar	1000	500	03

Trade Names

Sr. No.	Trade Name	Company
1	Lancer Gold	UPL Ltd.
2	Starthene Power	Swal Corporation Ltd.

3	Unicorn Gold	Auxin Crop Care
4	Mantra Gold	Herenba Industries Ltd.
5	Silver Gold	Silver Sine Bio Tech
6	Acemida	Natural Crop Science
7	Syn Rider	Synwood Agro Ltd.
8	Lexus Gold	Preventis Chem India (P) Ltd.
9	Chemlance	Chemley Agritech Pvt. Ltd.
10	Zefiro	RACCOLTO Agritech India Pvt. Ltd.

Acephate 50.00% + Imidacloprid 05.00% WG

Crop	Pests	Dose / ha		Waiting period (Days)
		Quantity of formulation (g)	Volume of water (Lit.)	
Cotton	Aphid, Jassid, Thrips, Whitefly, Bollworms	1000	500	53

Trade Names

Sr. No.	Trade Name	Company
1		

Acetamiprid + Bifenthrin

Available formulations in India : Acetamiprid 25.00% + Bifenthrin 25.00% WG

Classification on the basis of		Acetamiprid	Bifenthrin
Target organism	:	Insecticide	Insecticide
Mode of entry	:	Systemic, Contact poison	Stomach and Contact poison
Mode of action	:	Nerve poison	Nerve poison
Chemical group	:	Neonicotinoid	Synthetic pyrethroid
Category of toxicity : Highly toxic **Toxicity label colour** : Yellow			

- This insecticide is designed for broad-spectrum pest control, particularly in soybean and cotton crops. It is effective against a wide range of pests.
- This combination of active ingredients provides a dual-action approach to pest control, offering both rapid knockdown and long-lasting residual protection.
- Acetamiprid: It is a systemic insecticide, which means it is absorbed by the plant tissues and transported throughout the plant. It works by disrupting the nervous system of sucking pests, leading to paralysis and death.
- Bifenthrin: It works through contact and stomach poisoning. It acts on the nervous system of insects by interacting with their sodium channels, causing paralysis and quick death.
- The combination of these two modes of action makes the product effective against a broad spectrum of pests, including both sucking and chewing insects.

Label claim (DPPQ & S, Faridabad)
Acetamiprid 25.00% + Bifenthrin 25.00% WG

Crop	Pests	Dose / ha		Waiting period (Days)
		Quantity of formulation (g)	Volume of water (Lit.)	
Cotton	Jassid, Aphid, Thrips, Whitefly, Pink bollworm, American bollworm, Spotted bollworm	160	500	33
Soybean	Whitefly, Girdle beetle, Semilooper, Tobacco caterpillar	250	500	28

Trade Names

Sr. No.	Trade Name	Company
1	Sperto	Swal Corporation Ltd.
2	Argyle	UPL Ltd.

Acetamiprid + Chlorantraniliprole

Available formulations in India : 1. Acetamiprid 20.00% + Chlorantraniliprole 20.00% WG
2. Acetamiprid 19.50% + Chlorantraniliprole 19.50% FS

Classification on the basis of		Acetamiprid	Chlorantraniliprole
Target organism	:	Insecticide	Insecticide
Mode of entry	:	Systemic, Contact poison	Systemic, Contact and Stomach poison
Mode of action	:	Nerve poison	Muscle poison
Chemical group	:	Neonicotinoid	Anthranilic diamide
Category of toxicity : Moderately toxic **Toxicity label colour** : Blue			

- It is a combination insecticide that brings together two powerful active ingredients to provide broad-spectrum control of various insect pests.
- Acetamiprid is a systemic insecticide belonging to the neonicotinoid class. It works by acting on the insect's nervous system, specifically by disrupting the nicotinic acetylcholine receptors. This leads to the paralysis and death of the insect. Acetamiprid provides a systemic and translaminar effect, meaning it is absorbed by the plant and moves throughout its tissues, controlling pests that feed on the underside of leaves or in other hidden areas.
- Chlorantraniliprole is a selective insecticide from the diamide class. It is a ryanodine receptor modulator, which means it targets the muscles of the insect. By causing an uncontrolled release of calcium ions from the muscles, it leads to paralysis and, ultimately, death. This mode of action is particularly effective against lepidopteran pests (caterpillars).
- The two ingredients act on different physiological systems in the insect, making the product highly effective and helping to manage insecticide resistance.

Label claim (DPPQ & S, Faridabad)

Acetamiprid 20.00% + Chlorantraniliprole 20.00% WG

Crop	Pests	Dose / ha		Waiting period (Days)
		Quantity of formulation (g)	Volume of water (Lit.)	
Cabbage	Diamond back moth, Tobacco caterpillar, Mustard aphid	75	500	03
Tomato	Fruit borer, Tobacco caterpillar, Aphid, Jassid, Whitefly	150	500	07

Trade Names

Sr. No.	Trade Name	Company
1	Lexicon	Swal Corporation Ltd.
2	Sekito	UPL Ltd.

Acetamiprid 19.50% + Chlorantraniliprole 19.50% FS

Crop	Pests	Dose / Kg Seed		Waiting period (Days)
		Quantity of formulation (ml)	Volume of water (Lit.)	
Maize	Shoot fly, Fall armyworm, Stem borer, Aphid	10	--	--

Trade Names

Sr. No.	Trade Name	Company
1		

Acetamiprid + Chlorpyriphos

Available formulations in India : Acetamiprid 00.40%+Chlorpyriphos 20.00% EC

Classification on the basis of		Acetamiprid	Chlorpyrifos
Target organism	:	Insecticide	Insecticide
Mode of entry	:	Systemic, Contact poison	Stomach, Contact and Respiratory poison
Mode of action	:	Nerve poison	Nerve poison
Chemical group	:	Neonicotinoid	Organophosphates
Category of toxicity : Highly toxic **Toxicity label colour** : Yellow			

- It is a combination product designed for broad-spectrum pest control. This combination is effective because it targets insects in two different ways, which helps manage resistance and provides fast, long-lasting results.
- Chlorpyrifos is a well-known insecticide from the organophosphate class. It acts as a nerve poison, disrupting the insect's nervous system by inhibiting an enzyme called acetylcholinesterase. This leads to the buildup of a neurotransmitter, causing overstimulation of the nerves, followed by paralysis and death. Chlorpyrifos works on contact, stomach, and fumigant action, making it effective against a wide variety of pests.
- Acetamiprid is a newer-generation, systemic insecticide from the neonicotinoid class. It also attacks the insect's nervous system, but it works differently from chlorpyrifos. Acetamiprid disrupts nicotinic acetylcholine receptors, causing paralysis and death. Its systemic action means it is absorbed by the plant and moves through its tissues, providing protection to new growth and controlling pests that feed on parts of the plant not directly hit by the spray.
- The formulation provides both immediate knockdown (from chlorpyrifos) and long-term, systemic protection (from acetamiprid), ensuring the crop remains protected for an extended period.

Label claim (DPPQ & S, Faridabad)
Acetamiprid 00.40%+Chlorpyriphos 20.00% EC

Crop	Pests	Dose / ha		Waiting period (Days)
		Quantity of formulation (ml)	Volume of water (Lit.)	
Rice	Stem borer, Brown plant hopper, White backed plant hopper	2500	500-800	10

Trade Names

Sr. No.	Trade Name	Company
1	Trump	Sydney Formulation & Chemicals
2	Filtamol	Bhumi Agro Center
3	Aristo	Aquarius Agro Chemicals
4	Acephos	Volkschem Crop Science Pvt. Ltd.
5	Priban	RACCOLTO Agritech India Pvt. Ltd.

Acetamiprid + Cypermethrin

Available formulations in India : Acetamiprid 01.10% + Cypermethrin 05.50% EC

Classification on the basis of		Acetamiprid	Cypermethrin
Target organism	:	Insecticide	Insecticide
Mode of entry	:	Systemic, Contact poison	Stomach and Contact poison
Mode of action	:	Nerve poison	Nerve poison
Chemical group	:	Neonicotinoid	Synthetic pyrethroid
Category of toxicity : Highly toxic **Toxicity label colour** : Yellow			

- Acetamiprid is a systemic insecticide that provides excellent, long-lasting residual control, especially against sucking pests like aphids and whiteflies. Its mode of action is to act as an agonist (mimic) on the insect's nicotinic acetylcholine receptors (nAChR) in the central nervous system. This causes uncontrolled overstimulation, leading to paralysis and death. Because it is absorbed and translocated throughout the plant, it protects new growth. It is highly effective against pests that have developed resistance to older chemistries such as organophosphates, carbamates, and pyrethroids.
- Cypermethrin is a powerful insecticide known for its fast knockdown action. It works primarily through contact with the insect. As a synthetic pyrethroid, it attacks the insect's nervous system by disrupting the function of voltage-gated sodium channels in the nerve cell membranes. This forces the nerves to fire repeatedly, resulting in rapid neurotoxicity, incoordination, paralysis, and swift mortality across a wide range of insect pests.
- The combination of these two chemicals provides a highly potent insecticide mixture: the fast contact knockdown of Cypermethrin paired with the systemic, long-residual action of Acetamiprid.

Label claim (DPPQ & S, Faridabad)
Acetamiprid 01.10% + Cypermethrin 05.50% EC

Crop	Pests	Dose / ha		Waiting period (Days)
		Quantity of formulation (ml)	Volume of water (Lit.)	
Cotton	Aphid, Jassid, Thrips, Bollworms	1000	400-1000	30

Trade Names

Sr. No.	Trade Name	Company
1	Aceper	Sikko Industries Ltd.
2	Tadache	Greenovate AGROTECH Pvt. Ltd.
3	Tryzer	Aquarius Agro Chemicals

Benzpyrimoxam + Pymetrozine

Available formulations in India : Benzpyrimoxam 10% + Pymetrozine 20% WG

Classification on the basis of		Benzpyrimoxam	Pymetrozine
Target organism	:	Insecticide	Insecticide
Mode of entry	:	Contact and Stomach poison	Systemic, Stomach poison
Mode of action	:	Insect Growth Regulator (IGR)	Nerve poison
Chemical group	:	Benzyloxypyrimidines	Pyridine Azomethine
Category of toxicity :			
Toxicity label colour :			

- Benzpyrimoxam is a novel insecticide primarily targeting the nymphal stages of pests, especially planthoppers in rice (including resistant strains). Its unique mode of action is to disrupt the insect's molting process. It works by interfering with the natural concentration (titer) of ecdysteroids (the insect's molting hormones), causing the nymphs to undergo abnormal and ultimately fatal molting.
- Pymetrozine is a highly selective, systemic insecticide known for being friendly to beneficial insects. Its action is rapid and distinct: it targets the insect's nervous system (specifically the chordotonal organs/TRPV channels) to immediately inhibit feeding behavior. Pests, such as aphids, whiteflies, and planthoppers, cease feeding almost instantly upon contact or ingestion, leading to starvation and death over a few days. It has good systemic and translaminar activity, meaning it moves throughout the plant tissue for residual protection.
- The combination of these two active ingredients provides an excellent dual-action strategy for resistance management, offering both immediate feeding shutdown and long-term control by blocking pest development.

Label claim (DPPQ & S, Faridabad)

Benzpyrimoxam 10% + Pymetrozine 20% WG

Crop	Pests	Dose / ha		Waiting period (Days)
		Quantity of formulation (g)	Volume of water (Lit.)	
Rice	Brown plant hopper, White backed plant hopper, Green leaf hopper	500-700	500	36

Trade Names

Sr. No.	Trade Name	Company
1	Orchestra Duet	Nichino India Private Ltd.

Benzpyrimoxam + Thiamethoxam

Available formulations in India : Benzpyrimoxam 10.00% + Thiamethoxam 3.30% SC

Classification on the basis of		Benzpyrimoxam	Thiamethoxam
Target organism	:	Insecticide	Insecticide
Mode of entry	:	Contact and Stomach poison	Systemic, Contact and Stomach poison
Mode of action	:	Insect Growth Regulator (IGR)	Nerve poison
Chemical group	:	Benzyloxypyrimidines	Neonicotinoid
Category of toxicity : **Toxicity label colour :**			

- Benzpyrimoxam is a novel insecticide that targets the insect's endocrine system. It works by severely disrupting the concentration (titer) of ecdysteroids (molting hormones), which causes abnormal and fatal molting primarily in the nymphal stages of the pest. It's especially effective against difficult pests like the Brown Planthopper and is utilized as a crucial tool in resistance management strategies.
- Thiamethoxam is a systemic insecticide known for providing reliable, long-lasting control of both sucking and certain chewing pests. It acts directly on the insect's central nervous system as an agonist (mimic) of acetylcholine, binding to the nicotinic acetylcholine receptors (nAChR). This binding overstimulates the nerve cells, rapidly causing paralysis, feeding cessation, and ultimately, death. Its systemic action allows it to be absorbed and translocated throughout the entire plant, protecting new growth.
- When combined, these two chemicals provide a powerful strategy: Thiamethoxam offers quick, systemic control and immediate feeding shutdown, while Benzpyrimoxam offers long-term control by disrupting the growth and development of new generations.

Label claim (DPPQ & S, Faridabad)
Benzpyrimoxam 10.00% + Thiamethoxam 3.30% SC

Crop	Pests	Dose / ha		Waiting period (Days)
		Quantity of formulation (ml)	Volume of water (Lit.)	
Rice	Brown plant hopper, White backed plant hopper	750	500	30

Trade Names

Sr. No.	Trade Name	Company
1		

Benzpyrimoxam + Fipronil

Available formulations in India : Benzpyrimoxam 15% + Fipronil 10% SC

Classification on the basis of		Benzpyrimoxam	Fipronil
Target organism	:	Insecticide	Insecticide
Mode of entry	:	Contact and Stomach poison	Contact and Stomach poison
Mode of action	:	Insect Growth Regulator (IGR)	Nerve poison
Chemical group	:	Benzyloxypyrimidines	Phenylpyrazole
Category of toxicity : **Toxicity label colour :**			

- Benzpyrimoxam is a novel, selective insecticide that targets the insect's endocrine system. Its mechanism of action involves disrupting the concentration (titer) of ecdysteroids (molting hormones). Instead of stopping hormone production entirely, Benzpyrimoxam prevents the essential drop in hormone levels required for a successful molt. This results in abnormal, fatal molting and is particularly effective against the nymphal stages of pests. It provides crucial control against major rice pests like the Brown Planthopper and Whitebacked Planthopper.
- Fipronil is a well-established, broad-spectrum insecticide that acts on the insect's central nervous system. It works by blocking the GABA-gated chloride channels in nerve cells. This inhibition leads to the over-excitation of nerves, quickly causing paralysis, feeding cessation, and death. Fipronil is effective via both contact and ingestion. It is known for its residual action and its ability to control a wide range of pests, including chewing, sucking, and soil insects. A key feature is its non-repellent nature, which, in social pests like ants or termites, allows the chemical to be transferred throughout the colony, creating a powerful domino effect.
- The combination of these two compounds provides a comprehensive pest control strategy: Fipronil offers rapid, broad-spectrum neurotoxic control with long residual action, while Benzpyrimoxam provides targeted, future-generation control by disrupting the development of younger nymphs.

Label claim (DPPQ & S, Faridabad)
Benzpyrimoxam 15% + Fipronil 10% SC

Crop	Pests	Dose / ha		Waiting period (Days)
		Quantity of formulation (ml)	Volume of water (Lit.)	
Rice	Brown plant hopper, Whitebacked plant hopper, Leaf folder	400-500	500	22

Trade Names

Sr. No.	Trade Name	Company
1	Benzilla	Tata Rallies India Ltd.

Beta-cyfluthrin + Imidacloprid

Available formulations in India : Beta-cyfluthrin 08.49% + Imidacloprid 19.81% OD

Classification on the basis of		Beta-cyfluthrin	Imidacloprid
Target organism	:	Insecticide	Insecticide
Mode of entry	:	Contact and Stomach poison	Systemic, Contact poison
Mode of action	:	Nerve poison	Nerve poison
Chemical group	:	Synthetic pyrethroid	Neonicotinoid
Category of toxicity : Highly toxic **Toxicity label colour** : Yellow			

- The combination of Beta-cyfluthrin and Imidacloprid creates a powerful, broad-spectrum insecticide, leveraging two distinct methods for killing pests.
- Beta-cyfluthrin provides rapid knockdown by acting as a contact and ingestive poison. As a sodium channel blocker, it overstimulates the insect's nervous system, causing rapid paralysis and quick death.
- Imidacloprid delivers long-lasting, systemic protection. Once absorbed and moved throughout the plant, this insecticide disrupts the insect's central nervous system, causing nerve failure and death when pests feed on the foliage. It also provides anti-feeding effects, helping to stop crop damage immediately.
- In short, Beta-cyfluthrin ensures the fast initial kill, while Imidacloprid guarantees extended residual control.

Label claim (DPPQ & S, Faridabad)
Beta-cyfluthrin 08.49% + Imidacloprid 19.81% OD

Crop	Pests	Dose / ha		Waiting period (Days)
		Quantity of formulation (ml)	Volume of water (Lit.)	
Brinjal	Aphid, Jassid, Shoot and fruit borer	175-200	500	07
Soybean	Girdle beetle, Semilooper	350	500	17
Cotton	Jassid, Whitefly	200	500	21
Chilli	Thrips, Aphid, Whitefly	310	500	05
Cucumber	Whitefly	500	500	05
Wheat	Aphid	400	500	31
Cashew	Tea mosquito bug, Thrips	250	500-1000	45
Tea	Mosquito bug, Looper	300	400	07

Trade Names

Sr. No.	Trade Name	Company
1	Solomon	Bayer Crop Science Ltd.
2	Insectora	Katyayani Organics
3	Solo	Global Crop Care

Bifenthrin + Chlorpyrifos

Available formulations in India : Bifenthrin 03.00% + Chlorpyriphos 30.00% EC

Classification on the basis of		Bifenthrin	Chlorpyrifos
Target organism	:	Insecticide	Insecticide, Acaricide
Mode of entry	:	Contact and Stomach poison	Contact, Stomach and Respiratory poison
Mode of action	:	Nerve poison	Nerve poison
Chemical group	:	Synthetic pyrethroid	Organophosphate
Category of toxicity : Highly toxic **Toxicity label colour** : Yellow			

- This dual-action formula targets the insect nervous system in two distinct ways to deliver comprehensive pest control:
- Bifenthrin (a pyrethroid) provides fast knockdown and extended residual activity. It works by interfering with the insect's sodium channels, forcing nerve cells to fire rapidly and repeatedly. This immediate overstimulation quickly results in paralysis and death upon contact or ingestion.
- Chlorpyrifos (an organophosphate) ensures a deep, fatal action by acting as an irreversible acetylcholinesterase inhibitor. By blocking this critical enzyme, it causes a toxic buildup of the neurotransmitter acetylcholine, leading to uncontrollable nerve cell firing, convulsions, and ultimate death. It is effective through contact, ingestion, and sometimes fumigation, offering good soil residual protection.
- The combined use of these two chemicals provides a highly effective, broad-spectrum solution with both rapid surface action and long-lasting residual control.

Label claim (DPPQ & S, Faridabad)
Bifenthrin 03.00% + Chlorpyriphos 30.00% EC

Crop	Pests	Dose / ha		Waiting period (Days)
		Quantity of formulation (ml)	Volume of water (Lit.)	
Rice	Stem borer, Leaf folder	800-1000	500	21

Trade Names

Sr. No.	Trade Name	Company
1	Lethal Gold	Insecticides (India) Ltd.
2	Bifenos	Aquarius Agro Chemicals
3	Chloro Gold	Saurabh Agro Industries
4	Lansu	Global Crop care
5	Hiro	Grow Safe Chemicals Pvt. Ltd.
6	Bifenac	Agri India Science
7	Multion	Maziq Farm Technologies Pvt. Ltd.

Bifenthrin + Clothianidin

Available formulations in India : Bifenthrin 08.00% + Clothianidin 10.00% SC

Classification on the basis of		Bifenthrin	Clothianidin
Target organism	:	Insecticide	Insecticide
Mode of entry	:	Contact and Stomach poison	Systemic, Contact and Stomach poison
Mode of action	:	Nerve poison	Nerve poison
Chemical group	:	Synthetic pyrethroid	Neonicotinoid
Category of toxicity : Highly toxic **Toxicity label colour** : Yellow			

- The combination of Bifenthrin and Clothianidin is a highly effective, dual-action broad-spectrum insecticide. This product is extensively used to manage complex infestations of both sucking and chewing pests. The mixture capitalizes on the complementary strengths of its two active ingredients: Bifenthrin, a synthetic pyrethroid, and Clothianidin, a neonicotinoid.
- Bifenthrin is known for its fast contact and stomach action, delivering rapid knockdown by interfering with the insect's nervous system through the sodium channels. As a non-systemic pyrethroid, it provides immediate control of surface-feeding and chewing pests. It is also valued for its long residual effect in the soil, which makes it particularly effective against soil-dwelling pests.
- Clothianidin provides the crucial systemic component, a characteristic of neonicotinoids, which allows it to be absorbed by plant roots and leaves and translocated throughout the plant tissue. It acts as an agonist of the nicotinic acetylcholine receptors (nAChR) in the insect's central nervous system, leading to paralysis and death. This mode of action ensures long-lasting control of difficult sucking pests.
- Due to its dual nature, it is highly recommended for controlling mixed populations of pests from different orders, offering both immediate curative and extended preventive effects.

Label claim (DPPQ & S, Faridabad)
Bifenthrin 8% + Clothianidin 10% SC

Crop	Pests	Dose / ha		Waiting period (Days)
		Quantity of formulation (ml)	Volume of water (Lit.)	
Groundnut	White grub, Thrips, Aphid	1000	1000	83
Cotton	Grey weevil, Mealybug, Jassid, Whitefly, Aphid, Thrips	1000	1000	73
Sugarcane	Termite, Early shoot borer	1000	1000	300

Trade Names

Sr. No.	Trade Name	Company
1	Talstar Plus	FMC Corporation
2	Batter	GSP Crop Science Pvt. Ltd.
3	X-Ron	Chemley Agritech Pvt. Ltd.
4	Third	Ayushi Crop Science

Bifenthrin + Chlorantraniliprole

Available formulations in India : Bifenthrin 32.00% + Chlorantraniliprole 12.00% WG

Classification on the basis of		Bifenthrin	Chlorantraniliprole
Target organism	:	Insecticide	Insecticide
Mode of entry	:	Contact and Stomach poison	Systemic, Contact and Stomach poison
Mode of action	:	Nerve poison	Muscle poison
Chemical group	:	Synthetic pyrethroid	Anthranilic diamide
Category of toxicity : **Toxicity label colour :**			

- Bifenthrin and Chlorantraniliprole are a potent insecticidal duo, each contributing a distinct mode of action.
- Bifenthrin is a broad-spectrum neurotoxin that acts quickly. It interferes with the insect's nervous system by targeting the sodium channels, causing rapid paralysis, a quick "knockdown", and death upon contact or ingestion. It provides fast-acting control with good residual activity, meaning it remains effective on treated surfaces for a period of time.
- In contrast, Chlorantraniliprole is a newer, more selective insecticide. It targets the insect's muscle function by activating the ryanodine receptors, leading to uncoordinated muscle contraction, paralysis, and eventual death. This action is typically slower than bifenthrin, but it is known for providing long-lasting residual control. A key benefit of Chlorantraniliprole is its generally lower toxicity to many beneficial insects, like bees, especially when applied as a soil treatment or on non-flowering crops according to label instructions.

Label claim (DPPQ & S, Faridabad)
Bifenthrin 32.00% + Chlorantraniliprole 12.00% WG

Crop	Pests	Dose / ha		Waiting period (Days)
		Quantity of formulation (g)	Volume of water (Lit.)	
Soybean	Whitefly, Girdle beetle, Green semilooper, Stem fly	250	500	40

Trade Names

Sr. No.	Trade Name	Company
1		

Bifenthrin + Thiamethoxam

Available formulations in India :

1. **Bifenthrin 10.00% + Thiamethoxam 05.00% SE**
2. **Bifenthrin 25.00% + Thiamethoxam 25.00% WG**
3. **Thiamethoxam 0.40% + Bifenthrin 0.80% GR**

Classification on the basis of		**Bifenthrin**	**Thiamethoxam**
Target organism	:	Insecticide	Insecticide
Mode of entry	:	Contact and Stomach poison	Systemic, Contact and Stomach poison
Mode of action	:	Nerve poison	Nerve poison
Chemical group	:	Synthetic pyrethroid	Neonicotinoid
Category of toxicity : **Toxicity label colour :**			

- The combination of Bifenthrin and Thiamethoxam provides a powerful two-pronged defence against a broad spectrum of insect pests.
- Bifenthrin is a contact and stomach poison that belongs to the pyrethroid chemical class. It works by interfering with the insect's central nervous system, specifically by disrupting the sodium channels. This results in rapid paralysis and a quick knockdown or death upon contact or ingestion. It provides excellent residual control, remaining effective on treated surfaces for a significant period. Bifenthrin is highly effective against chewing and crawling pests.
- Thiamethoxam is a newer, systemic insecticide from the neonicotinoid class. It offers longer-term internal protection for the plant. Once applied, it is readily absorbed by the plant and circulated throughout the tissues. It targets the insect's nervous system by acting as an agonist on the nicotinic acetylcholine receptors (nAChRs), leading to irreversible nerve overstimulation, paralysis, and death. Its systemic action makes it particularly effective against sucking pests that feed on plant sap.
- By combining the fast-acting surface control of Bifenthrin with the long-lasting internal protection of Thiamethoxam, this mixture offers a comprehensive and effective pest management strategy.

Label claim (DPPQ & S, Faridabad)
Bifenthrin 10.00% + Thiamethoxam 05.00% SE

Crop	**Pests**	**Dose / ha**		**Waiting period (Days)**
		Quantity of formulation (ml)	**Volume of water (Lit.)**	
Rice	Stem borer, Leaf folder, Gall midge, Brown plant hopper	500	500	39

Trade Names

Sr. No.	**Trade Name**	**Company**
1	Blend	Tata Rallis India Ltd.

Bifenthrin 25.00% + Thiamethoxam 25.00% WG

Crop	Pests	Dose / ha		Waiting period (Days)
		Quantity of formulation (g)	Volume of water (Lit.)	
Cotton	Jassid, Aphid, Thrips, Whitefly, American bollworm, Pink bollworm, Spotted bollworm	200	500	42
Rice	Stem borer, Leaf folder, Hispa, Brown plant hopper, Green leaf hopper	200	500	25

Trade Names

Sr. No.	Trade Name	Company
1		

Bifenthrin 25.00% + Thiamethoxam 25.00% WG

Crop	Pests	Dose / ha		Waiting period (Days)
		Quantity of formulation (ml)	Volume of water (Lit.)	
Cotton	Jassid, Thrips, Aphid, Whitefly, American bollworm, Pink bollworm, Spotted bollworm	200	500	42
Rice	Stem borer, Leaf folder, Hispa, Brown plant hopper, Green leaf hopper	200	500	25

Trade Names

Sr. No.	Trade Name	Company
1		

Thiamethoxam 0.40% + Bifenthrin 0.80% GR

Crop	Pests	Dose / ha		Waiting period (Days)
		Quantity of formulation (Kg)	Volume of water (Lit.)	
Groundnut	White grub, Termite	12	--	105

Trade Names

Sr. No.	Trade Name	Company
1		

Buprofezin + Acephate

Available formulations in India :

1. **Buprofezin 09.00% + Acephate 24.00% WP**
2. **Buprofezin 15.00% + Acephate 35.00% WP**
3. **Buprofezin 20.00% + Acephate 50.00% WP**

Classification on the basis of		Buprofezin	Acephate
Target organism	:	Insecticide	Insecticide
Mode of entry	:	Contact and Stomach poison	Systemic, Contact and Stomach poison
Mode of action	:	Insect Growth Regulator (IGR)	Nerve poison
Chemical group	:	Thiadiazine	Organophosphate
Category of toxicity : Moderately toxic **Toxicity label colour** : Blue			

- The exceptional effectiveness of this combination comes from its two distinct modes of action, a strategy that is highly valuable for preventing and managing pesticide resistance.
- Buprofezin works as an Insect Growth Regulator (IGR). Its primary role is to stop the pest life cycle by inhibiting chitin biosynthesis, the essential process insects use to build their new exoskeleton. This mechanism leads to the death of nymphs and larvae because it prevents them from successfully molting. Furthermore, Buprofezin suppresses egg-laying by adult females, and the few eggs that are laid are often sterile. This provides a long-lasting, residual effect that controls the pest population over time and is generally regarded as safer for some beneficial insects.
- Acephate provides the fast, neurotoxic element. It functions as a cholinesterase inhibitor, specifically targeting the acetylcholinesterase (AChE) enzyme in the insect's nervous system. By blocking this enzyme, it causes the neurotransmitter acetylcholine to build up, leading to continuous and over-excited nerve signaling. This results in the paralysis and rapid death of the pest. Acephate is a systemic insecticide, meaning it's absorbed by the plant and moves through the tissues to kill sucking pests that feed on the internal sap.
- In essence, this mixture is a potent tool that combines the quick knockdown of Acephate on adult pests with the long-term life-cycle disruption of Buprofezin on the immature stages.

Label claim (DPPQ & S, Faridabad)
Buprofezin 09.00% + Acephate 24.00% WP

Crop	Pests	Dose / ha		Waiting period (Days)
		Quantity of formulation (g)	Volume of water (Lit.)	
Rice	Brown planthopper	600	500	20

Trade Names

Sr. No.	Trade Name	Company
1	Nova Topper Plus	Nova Agri Sciences Pvt. Ltd.
2	Prophate	Plantix Crop Care

Buprofezin 15.00% + Acephate 35.00% WP

Crop	Pests	Dose / ha		Waiting period (Days)
		Quantity of formulation (g)	Volume of water (Lit.)	
Cotton	Jassid, Thrips, Whitefly	1250	500	48
Okra	Jassid, Whitefly	750	500	07
Rice	Brown plant hopper, White backed plant hopper	1250	500	20

Trade Names

Sr. No.	Trade Name	Company
1	Tapuz	ADAMA India Private Limited
2	Superstar Gold	Anu Products Ltd.
3	Sazedar	Crop Life Science Ltd.
4	Bazooka	Sydney Formulation & Chemicals
5	Bace-50	Parijat Industries India Pvt. Ltd.
6	Reker	Sumitomo Chemical India Ltd.
7	Inova	Paras Crops Ltd.
8	Buprostar	Herenba Industries Ltd.
9	Unione	Nichino India
10	Kpmire	KP Agro Industries
11	Piranha	Coromandel International
12	Papro	HPM Chemicals and Fertilizers Ltd.
13	Unione	Nichino India Pvt. Ltd.
14	Teproz	Agri Science Co.
15	Inova	Paras Crop Ltd.
16	Kabucee	KR LifeSciences Pvt. Ltd.
17	Tadaka	Bhaveesh Agency
18	Bace-50	Parijat Industries India Pvt. Ltd.
19	Founder	Essential Biosciences
20	Acebu	Goodweel Crop Care
21	Goly	Bharat Certis Agriscience Ltd.
22	Elephant	Silver Sine Bio Tech
23	Topaz	Natural Crop Science
24	Buphate	Tropical Agrosystem India (P) Ltd.
25	AB 1535	Global Crop Care
26	Gunship	RACCOLTO Agritech India Pvt. Ltd.

Buprofezin 20.00% + Acephate 50.00% WP

Crop	Pests	Dose / ha		Waiting period (Days)
		Quantity of formulation (g)	Volume of water (Lit.)	
Rice	Stem borer, Leaf folder, Brown plant hopper	1000	500	20
Cotton	Thrips, Jassid, Mealybug	1250	500	15
Chilli	Thrips, Aphid, Yellow mite	1250	500	07

Trade Names

Sr. No.	Trade Name	Company
1	Record	Crystal Crop Protection Ltd.
2	Bazooka Plus	Sydney Formulation & Chemicals
3	Odis	Tata Rallis India Ltd.
4	Veer	Angel Crop Science
5	Synprotect	Synwood Agro Ltd.

Buprofezin + Acetamiprid

Available formulations in India : **Buprofezin 20.00% + Acetamiprid 02.00% WP**

Classification on the basis of		Buprofezin	Acetamiprid
Target organism	:	Insecticide	Insecticide
Mode of entry	:	Contact and Stomach poison	Systemic, Contact and Stomach poison
Mode of action	:	Insect Growth Regulator (IGR)	Nerve poison
Chemical group	:	Thiadiazine	Neonicotinoid
Category of toxicity : **Toxicity label colour :**			

- The combination of Buprofezin and Acetamiprid is a powerful insecticide blend widely used in agriculture to manage sucking pests. Its effectiveness stems from a dual-mode of action that targets insects across their entire life cycle, making it an excellent tool for resistance management and achieving broad-spectrum control.
- Buprofezin is an Insect Growth Regulator (IGR). It primarily affects the nymphal and larval stages by acting as a chitin biosynthesis inhibitor, disrupting the formation of the insect's protective exoskeleton. It prevents successful molting, and also reduces the viability of eggs laid by adult females, effectively halting the pest's reproductive cycle. It is not a fast-acting killer.
- Acetamiprid is a systemic neonicotinoid that acts as a nerve poison. It binds to receptors in the insect's central nervous system, causing rapid overstimulation, paralysis, and a quick "knockdown" effect, particularly on adults. Because it is systemic, the plant absorbs it and moves it to all tissues, providing thorough, long-lasting protection against sap-feeding pests, even those on the underside of leaves or inside the plant.
- Acetamiprid provides immediate control and kills existing adult insects, while Buprofezin stops the next generation from developing, breaking the pest lifecycle and preventing populations from rebounding.

Label claim (DPPQ & S, Faridabad)
Buprofezin 20.00% + Acetamiprid 02.00% WP

Crop	Pests	Dose / ha		Waiting period (Days)
		Quantity of formulation (g)	Volume of water (Lit.)	
Rice	Brown plant hopper, White backed plant hopper, Green leaf hopper, Leaf folder, Stem borer	800	400	15

Trade Names

Sr. No.	Trade Name	Company
1		

Buprofezin + Fipronil

Available formulations in India	:	1. **Buprofezin 22.00% + Fipronil 03.00% SC** 2. **Buprofezin 23.10%+Fipronil 03.85% SC**

Classification on the basis of		**Buprofezin**	**Fipronil**
Target organism	:	Insecticide	Insecticide
Mode of entry	:	Contact and Stomach poison	Contact and Stomach poison
Mode of action	:	Insect Growth Regulator (IGR)	Nerve poison
Chemical group	:	Thiadiazine	Phenylpyrazole
Category of toxicity : Moderately toxic **Toxicity label colour** : Blue			

- The combination of Buprofezin and Fipronil is a popular and highly effective dual-action insecticide mixture.
- Buprofezin is an Insect Growth Regulator (IGR), it interferes with the formation of the insect's exoskeleton (chitin). This primarily affects the nymphal and larval stages (immature insects), preventing them from successfully molting into the next stage, leading to slow death. It also has a suppressive effect on adult egg-laying (oviposition) and reduces the viability of eggs. It provides long-term, residual control by breaking the pest life cycle and controlling the next generation. It is not a fast-acting adulticide.
- Fipronil is a broad-spectrum insecticide belonging to the Phenylpyrazole chemical family. It acts as a nerve poison by blocking the passage of chloride ions through GABA-gated chloride channels in the insect's central nervous system. This results in hyperexcitation of the nerves and muscles, leading to paralysis and rapid death. It works by contact and stomach action and often has systemic properties (it can be absorbed by the plant and translocated). It provides a quick "knockdown" effect, effectively controlling both adults and immature stages with immediate action.

Label claim (DPPQ & S, Faridabad)
Buprofezin 22.00% + Fipronil 03.00% SC

Crop	**Pests**	**Dose / ha**		**Waiting period (Days)**
		Quantity of formulation (ml)	**Volume of water (Lit.)**	
Rice	Brown plant hopper	500	400-500	32
Cotton	Mealybug, Thrips	1000	500	17

Trade Names

Sr. No.	**Trade Name**	**Company**
1	Aikido	Insecticides (India) Ltd.
2	Bufisik	Sikko Industries Ltd.

3	Trianmore	Synergy
4	Thripisa	Silver Sine Bio Tech
5	Debu Pro	Somnath Demeter Agritron Ltd.
6	Kraftfull	RACCOLTO Agritech India Pvt. Ltd.

Buprofezin 23.10% + Fipronil 03.85% SC

Crop	Pest	Dose / ha		Waiting period (Days)
		Quantity of formulation (ml)	Volume of water (Lit.)	
Rice	Brown plant hopper	750	500	30

Trade Names

Sr. No.	Trade Name	Company
1	Zohar	ADAMA India Private Limited
2	Bupronil	Natural Crop Science

Cartap hydrochloride + Buprofezin

Available formulations in India : **Cartap Hydrochloride 50.00% + Buprofezin 10.00% WP**

Classification on the basis of		Cartap hydrochloride	Buprofezin
Target organism	:	Insecticide	Insecticide
Mode of entry	:	Systemic, Contact and Stomach poison	Contact and Stomach poison
Mode of action	:	Nerve poison	Insect Growth Regulator
Chemical group	:	Thiocarbamate	Thiadiazine
Category of toxicity : Highly toxic **Toxicity label colour** : Yellow			

- Cartap Hydrochloride is a fast-acting systemic insecticide that works by mimicking the natural toxin nereis toxin. It attacks the insect's nervous system (as a nicotinic acetylcholine receptor agonist/antagonist), causing immediate paralysis and death upon contact and ingestion. It provides the initial, rapid control to quickly reduce the adult pest population.
- Buprofezin is an Insect Growth Regulator (IGR), meaning it targets pest development rather than the adult nervous system. It works by inhibiting chitin synthesis, which prevents the pest from forming a new cuticle (shell) after molting. Buprofezin is highly effective on nymphs and immature stages, preventing their development and providing long-lasting residual control to stop population resurgence over time.
- The combination delivers Dual-Action Protection: Cartap provides the rapid knockdown, while Buprofezin ensures extended persistence, making it a highly effective strategy for managing pest resistance and controlling a wide range of insects.

Label claim (DPPQ & S, Faridabad)
Cartap Hydrochloride 50.00% + Buprofezin 10.00% WP

Crop	Pests	Dose / ha		Waiting period (Days)
		Quantity of formulation (g)	Volume of water (Lit.)	
Rice	Yellow stem borer, Leaf folder, Brown plant hopper, Green leaf hopper, White backed plant hopper	800	500	20

Trade Names

Sr. No.	Trade Name	Company
1	Kritap Gold	Krishi Rasayan Exports Private Limited
2	MeLeaf	Global Crop Care
3	Cartazen	Katyayani Organics
4	Cartomax	Essential Biosciences
5	Kritap-Gold	Krishi Rasayan

Cartap hydrochloride + Chlorantraniliprole

Available formulations in India : **Cartap hydrochloride 07.50% + Chlorantraniliprole 0.40% GR**

Classification on the basis of		Cartap hydrochloride	Chlorantraniliprole
Target organism	:	Insecticide	Insecticide
Mode of entry	:	Systemic, Contact and Stomach poison	Systemic, Contact and Stomach poison
Mode of action	:	Nerve poison	Muscle poison
Chemical group	:	Thiocarbamate	Anthranilic diamide
Category of toxicity : Highly toxic **Toxicity label colour** : Yellow			

- The combination of Cartap hydrochloride and Chlorantraniliprole is a widely used dual-action insecticide, particularly in the cultivation of paddy (rice).
- Cartap hydrochloride acts as a Nicotinic Acetylcholine Receptor Agonist/Antagonist. It blocks nerve impulse transmission at the nerve cell, causing insect paralysis. It is a contact, systemic, and stomach poison.
- Chlorantraniliprole acts as a Ryanodine Receptor Activator. It causes an abnormal release of calcium ions from the insect's muscle cells, leading to muscle spasms, paralysis, and eventually death. It is primarily a stomach poison but also has contact and systemic activity.
- It provides prolonged efficacy against targeted pests due to the systemic nature of both components. Using two different modes of action makes it much harder for pests to develop resistance, compared to using a single-molecule product.

Label claim (DPPQ & S, Faridabad)
Cartap hydrochloride 07.50% + Chlorantraniliprole 0.40% GR

Crop	Pests	Dose / ha		Waiting period (Days)
		Quantity of formulation (Kg)	Volume of water (Lit.)	
Rice	Yellow stem borer, Leaf folder, Whorl maggot	8.75	--	59

Trade Names

Sr. No.	Trade Name	Company
1	Convay	Swal Corporation Ltd.
2	Zevigo	UPL Ltd.

Cartap hydrochloride + Emamectin benzoate

Available formulations in India : **Cartap Hydrochloride 07.50% + Emamectin benzoate 0.25% GR**

Classification on the basis of		Cartap hydrochloride	Emamectin benzoate
Target organism	:	Insecticide	Insecticide, Acaricide
Mode of entry	:	Systemic, Contact and Stomach poison	Systemic, Contact and Stomach poison
Mode of action	:	Nerve poison	Nerve poison
Chemical group	:	Thiocarbamate	Avermectin (Macrocyclic lactone)
Category of toxicity : Moderately toxic **Toxicity label colour** : Blue			

- This insecticide blend is an excellent strategy for resistance management because it delivers control through two distinctly different modes of action.
- Cartap Hydrochloride acts as a Nicotinic Acetylcholine Receptor Agonist/Antagonist, swiftly disrupting the insect's central nervous system by blocking nerve impulse transmission. This provides a fast, potent knockdown effect—via contact, stomach, and systemic action—to immediately paralyze and kill existing pests.
- Complementing this is Emamectin Benzoate, which works as a Chloride Channel Activator. It binds to specific receptors (GABA and H-Glutamate), causing an irreversible flow of chloride ions into the nerve cells, also resulting in paralysis and death. Emamectin Benzoate is especially effective against Lepidopteran larvae (caterpillars) at very low dosages and is prized for providing long-lasting residual control to prevent future generations of pests from establishing.
- In short, the combination achieves both a rapid kill and extended protection.

Label claim (DPPQ & S, Faridabad)
Cartap Hydrochloride 07.50% + Emamectin benzoate 0.25% GR

Crop	Pests	Dose / ha		Waiting period (Days)
		Quantity of formulation (Kg)	Volume of water (Lit.)	
Rice	Yellow stem borer	7.50	--	35

Trade Names

Sr. No.	Trade Name	Company
1	Barroz	Adama India Pvt. Ltd.
2	Samharak	Katyayani Organics

Cartap hydrochloride + Fipronil

Available formulations in India : Cartap Hydrochloride 04.00% + Fipronil 0.50% CG

Classification on the basis of		Cartap hydrochloride	Fipronil
Target organism	:	Insecticide	Insecticide
Mode of entry	:	Systemic, Contact poison	Contact and Stomach poison
Mode of action	:	Nerve poison	Nerve poison
Chemical group	:	Neonicotinoid	Phenylpyrazole
Category of toxicity : Moderately toxic **Toxicity label colour** : Blue			

- Cartap hydrochloride acts as a neurotoxic insecticide. It is a propesticide of nereistoxin, which causes paralysis by blocking cholinergic transmission (inhibiting the activity of the enzyme acetylcholinesterase) in the insect's central nervous system. This leads to overstimulation, paralysis, and death. It's effective by contact, stomach, and systemic action.
- Fipronil acts as a neurotoxic insecticide. It targets the GABA-gated chloride channels in the insect's central nervous system. By disrupting this channel, it prevents the uptake of chloride ions, resulting in hyperexcitation of the nerves, leading to paralysis and death. It is highly effective with contact and stomach action, and often provides residual control.
- The combination of these two different modes of action provides synergistic control, helping to manage a broader range of pests and also helping with Insecticide Resistance Management (IRM).

Label claim (DPPQ & S, Faridabad)
Cartap Hydrochloride 04.00% + Fipronil 0.50% CG

Crop	Pests	Dose / ha		Waiting period (Days)
		Quantity of formulation (Kg)	Volume of water (Lit.)	
Rice	Stem borer, Leaf folder	15-20	--	27
Soybean	Leaf worm, Girdle beetle, Semilooper, Stem fly	0.20	500	41

Trade Names

Sr. No.	Trade Name	Company
1	Exotica	Willowood
2	Excellent	Silver Sine Bio Tech
3	Thavma	RACCOLTO Agritech India Pvt. Ltd.

Chlorantraniliprole + Abamectin

Available formulations in India : **Chlorantraniliprole 04.30% + Abamectin 01.70% SC**

Classification on the basis of		Chlorantraniliprole	Abamectin
Target organism	:	Insecticide	Acaricide, Insecticide
Mode of entry	:	Systemic, Contact and Stomach poison	Contact and Stomach poison
Mode of action	:	Muscle poison	Nerve poison
Chemical group	:	Anthranilic diamide	Avermectin (Macrocyclic Lactone)
Category of toxicity : Moderately toxic **Toxicity label colour** : Blue			

- The typical formulation of Chlorantraniliprole and Abamectin is designed as a broad-spectrum insecticide mixture, offering a powerful, dual-action approach for pest management, especially within agricultural environments. Each component targets insects through a distinct mechanism, making the combination highly effective.
- Chlorantraniliprole functions as a Ryanodine Receptor Modulator. It targets the ryanodine receptors located in insect muscle cells, triggering an uncontrolled, massive release of calcium from the internal storage sites. This over-release of calcium causes rapid muscle dysfunction, leading to paralysis, an immediate halt in feeding, and ultimately, death. While primarily a stomach poison, it also offers good systemic and contact activity. It is particularly valued for its high effectiveness against the larval stage (caterpillars) of Lepidoptera (moths and butterflies).
- Abamectin acts as a GABA (Gamma-Aminobutyric Acid) agonist and a chloride channel activator. It achieves its lethal effect by binding to glutamate-gated chloride channels in the insect's nerve and muscle cells. This binding increases the permeability of chloride ions across the cell membrane, severely disrupting neurotransmission. The result is an irreversible, flaccid paralysis, cessation of feeding, and death. Abamectin is non-systemic, relying on contact and stomach action, but it exhibits strong translaminar movement, meaning it can penetrate the leaf surface to reach pests feeding on the underside. It is highly effective as an acaricide (against mites like spider mites) and an insecticide (against thrips, leaf miners, and some sucking/burrowing pests).
- The key advantage of this mixture lies in its dual mode of action, which simultaneously targets the insect's muscular system (via Chlorantraniliprole) and its nervous system (via Abamectin). This combined assault is highly effective at quickly controlling a wide range of pests and is also a strong strategy for mitigating the development of pesticide resistance in a pest population.

Label claim (DPPQ & S, Faridabad)

Chlorantraniliprole 04.30% + Abamectin 01.70% SC

Crop	Pests	Dose / ha		Waiting period (Days)
		Quantity of formulation (ml)	Volume of water (Lit.)	
Chilli	Thrips, Mites, Fruit borer	625	500	05
Tomato	Fruit borer, Leaf miner, Mites	500	500	05

Trade Names

Sr. No.	Trade Name	Company
1	Voliam Targo	Syngenta India Ltd.
2	Army	Anmol Biotech Private Limited
3	Hunter	Growtix Agrotech
4	Trygun	Koronation Bio-Science
5	Hum Rahi	Global Crop Care
6	Abacat	Chemley Agritech Pvt. Ltd.

Chlorantraniliprole + Clothianidin

Available formulations in India : **Chlorantraniliprole 15.75% + Clothianidin 26.25% SC**

Classification on the basis of		Chlorantraniliprole	Clothianidin
Target organism	:	Insecticide	Insecticide
Mode of entry	:	Systemic, Contact and Stomach poison	Systemic, Contact and Stomach poison
Mode of action	:	Muscle poison	Nerve poison
Chemical group	:	Anthranilic diamide	Neonicotinoid
Category of toxicity : **Toxicity label colour :**			

- Chlorantraniliprole 15.75% + Clothianidin 26.25% SC is a powerful, broad-spectrum premix insecticide.
- This "dual-action" formula attacks the insect's system from two angles.
- Chlorantraniliprole targets the ryanodine receptors in insect muscles. It causes a continuous release of calcium, leading to muscle paralysis, immediate cessation of feeding, and eventual death.
- Clothianidin acts on the central nervous system as an agonist of nicotinic acetylcholine receptors (nAChR). It disrupts nerve signal transmission, causing hyper-excitation and lethargy.
- Primarily applied via soil drenching (especially in sugarcane) or foliar spray depending on the target pest. For sugarcane, it is often applied as a drench over the cane setts at the time of planting or during the early growth stage.

Label claim (DPPQ & S, Faridabad)
Chlorantraniliprole 15.75% + Clothianidin 26.25% SC

Crop	Pests	Dose / ha		Waiting period (Days)
		Quantity of formulation (ml)	Volume of water (Lit.)	
Sugarcane	Early shoot borer, Termite, White grub	400	1000	258

Trade Names

Sr. No.	Trade Name	Company
1	Tirracto	FMC India Pvt. Ltd.

Chlorantraniliprole + Fipronil

Available formulations in India :

1. **Chlorantraniliprole 0.53% + Fipronil 0.80% GR**
2. **Chlorantraniliprole 0.35% + Fipronil 0.35% GR**
3. **Chlorantraniliprole 12.00% + Fipronil 30.00% GR**
4. **Chlorantraniliprole 0.83% + Fipronil 01.33% GR**

Classification on the basis of		**Chlorantraniliprole**	**Fipronil**
Target organism	:	Insecticide	Insecticide
Mode of entry	:	Systemic, Contact and Stomach poison	Contact and Stomach poison (Systemic)
Mode of action	:	Muscle poison	Nerve poison
Chemical group	:	Anthranilic diamide	Phenylpyrazole
Category of toxicity :			
Toxicity label colour :			

- The combination of Chlorantraniliprole and Fipronil is a popular broad-spectrum insecticide product in agriculture, particularly for crops like rice.
- Chlorantraniliprole activates insect ryanodine receptors, causing an uncontrolled release of calcium ions in muscle cells. This leads to muscle paralysis and rapid cessation of feeding.
- Fipronil blocks the GABA-gated chloride channels in the insect's central nervous system. This overstimulates the nerves, leading to hyperexcitation, paralysis, and death. There's some debate in scientific literature over how systemic fipronil is, especially when compared to highly systemic products like neonicotinoids. Nonetheless, it's frequently used in ways that rely on its movement within plants.
- Fipronil often provides a faster knockdown effect, while Chlorantraniliprole offers a prolonged residual activity, protecting the crop for an extended period.

Label claim (DPPQ & S, Faridabad)

Chlorantraniliprole 0.53% + Fipronil 0.80% GR

Crop	**Pest**	**Dose / ha**		**Waiting period (Days)**
		Quantity of formulation (Kg)	**Volume of water (Lit.)**	
Rice	Yellow stem borer	7.50	Broadcasting mixed with sand	53

Trade Names

Sr. No.	**Trade Name**	**Company**
1	Forpido	Adama India Pvt. Ltd.

Chlorantraniliprole 0.35% + Fipronil 0.35% GR

Crop	Pests	Dose / ha		Waiting period (Days)
		Quantity of formulation (Kg)	Volume of water (Lit.)	
Rice	Stem borer, Leaf folder	10	Broadcasting mixed with sand	65

Trade Names

Sr. No.	Trade Name	Company
1	Centran GR	Insecticide India Ltd.

Chlorantraniliprole 12.00% + Fipronil 30.00% GR

Crop	Pests	Dose / ha		Waiting period (Days)
		Quantity of formulation (g)	Volume of water (Lit.)	
Chilli	Thrips, Aphid, Fruit borer, Tobacco caterpillar	250	500	10
Cotton	Thrips, Aphid, American bollworm, Pink bollworm	250	500	41

Trade Names

Sr. No.	Trade Name	Company
1		

Chlorantraniliprole 0.83% + Fipronil 01.33% GR

Crop	Pests	Dose / ha		Waiting period (Days)
		Quantity of formulation (Kg)	Volume of water (Lit.)	
Rice	Stem borer, Leaf folder, Hispa	5	--	49

Trade Names

Sr. No.	Trade Name	Company
1	Balveer	Swal Corporation Ltd.

Chlorantraniliprole + Fipronil + Emamectin benzoate

Available formulations in India : **Chlorantraniliprole 0.45% + Fipronil 0.50% + Emamectin benzoate 0.18% GR**

Classification on the basis of		Chlorantraniliprole	Fipronil	Emamectin benzoate
Target organism	:	Insecticide	Insecticide	Insecticide, Acaricide
Mode of entry	:	Systemic, Contact and Stomach poison	Contact and Stomach poison	Systemic, Contact and Stomach poison
Mode of action	:	Muscle poison	Nerve poison	Nerve poison
Chemical group	:	Anthranilic diamide	Phenylpyrazole	Avermectin (Macrocyclic lactone)
Category of toxicity : Moderately toxic **Toxicity label colour** : Blue				

- This formulation (Granular) is an advanced, triple-action insecticide designed for soil application, which offers a powerful and comprehensive strategy for pest control, particularly in crops like tomato. This unique Granular (GR) form is typically applied by broadcasting it into the soil during early crop stages, allowing the active ingredients to be absorbed by the plant roots, providing systemic and long-lasting protection against both soil-borne and foliar-feeding pests.
- Chlorantraniliprole (Diamide class) acts primarily via the ryanodine receptors, causing uncontrolled release of calcium ions from the muscle cells. This leads to rapid muscle paralysis and death, and is highly effective against chewing pests.
- Fipronil (Phenylpyrazole) disrupts the insect's central nervous system by blocking GABA-regulated chloride channels. This causes hyperexcitation, ensuring a strong contact, stomach, and systemic action that provides excellent residual control, particularly against sucking pests, root-feeding insects, and certain borers.
- Emamectin benzoate (Avermectin) works by binding to chloride ion channels, causing irreversible paralysis and death. It is very potent against Lepidopteran (caterpillar) pests, such as borers, by causing immediate feeding inhibition.

Label claim (DPPQ & S, Faridabad)

Chlorantraniliprole 0.45% + Fipronil 0.50% + Emamectin benzoate 0.18% GR

Crop	Pests	Dose / ha		Waiting period (Days)
		Quantity of formulation (Kg)	Volume of water (Lit.)	
Tomato	Whitefly, Fruit borer	10	--	

Trade Names

Sr. No.	Trade Name	Company
1	Fetagan	Best Agrolife Ltd.

Chlorantraniliprole + Lambda-cyhalothrin

Available formulations in India : **Chlorantraniliprole 09.30% + Lambda-cyhalothrin 04.60% ZC**

Classification on the basis of		Chlorantraniliprole	Lambda-cyhalothrin
Target organism	:	Insecticide	Insecticide
Mode of entry	:	Systemic, Contact and Stomach poison	Contact and Stomach poison
Mode of action	:	Muscle poison	Nerve poison
Chemical group	:	Anthranilic diamide	Synthetic pyrethroid
Category of toxicity : Highly toxic **Toxicity label colour** : Yellow			

- This combination provides powerful control through two distinct modes of action:
- Chlorantraniliprole is Ryanodine Receptor Modulator. It activates ryanodine receptors in insect muscles, causing uncontrolled calcium release, leading to impaired muscle regulation, paralysis, and ultimately death. It is systemic (absorbed by the plant) and provides long-lasting residual control (up to 14-21 days). It stops insect feeding very quickly (1-2 hours).
- Lambda-cyhalothrin is Sodium Channel Modulator. It disrupts the insect's central nervous system by interfering with nerve signal transmission. It has contact and stomach action with a fast knockdown effect (paralyzing pests within minutes). It repels and kills adult pests.
- The synergistic action of both components ensures rapid knockdown of adult and contact-sensitive pests, along with prolonged, systemic control of chewing pests and larvae, even targeting egg stages.

Label claim (DPPQ & S, Faridabad)
Chlorantraniliprole 09.30% + Lambda-cyhalothrin 04.60% ZC

Crop	Pests	Dose / ha		Waiting period (Days)
		Quantity of formulation (ml)	Volume of water (Lit.)	
Pigeonpea	Pod borer	200	500	18
Cotton	Bollworms complex	250	500	200
Brinjal	Shoot and fruit borer, Jassid	200	500	05
Okra	Shoot and fruit borer, Jassid	200	500	03
Rice	Stem borer, Leaf folder, Green leaf hopper	200-250	500	53
Soybean	Leaf worm, Girdle beetle, Semilooper, Stem fly	200	500	41

Cabbage	Tobacco caterpillar, Diamond back moth	200	500	05
Groundnut	Leaf miner, Leaf feeder, Thrips	200	500	20
Black gram	Pod borers (Tobacco caterpillar, Spotted pod borer, Pod borer)	200	500	24
Pomegranate	Fruit borer, Thrips	200	1000	07
Maize	Fall armyworm, Stem borer	250	500	36

Trade Names

Sr. No.	Trade Name	Company
1	Ampligo	Syngenta India Ltd.
2	Sil Vigo	Silver Sine Bio Tech
3	Volium Star	Natural Crop Science
4	Chloda	Katyayani Organics
5	Gammon	Tropical Agrosystem India (P) Ltd.
6	Helipower	GSP Crop Science Pvt. Ltd.
7	Topligo	Thakar Chemicals
8	Lamgo	Sahib Pesticides and Seeds Ltd.
9	Alishan Gold	Ichiban Crop Science
10	Emmigo	Shivalik Crop Sciences Pvt. Ltd.
11	Cuba	Essential Biosciences
12	Simigo	Synwood Agro Ltd.
13	Ampy	Koronation Bio-Science
14	Power	Vajra Crop Science
15	Amplify	Global Crop Care
16	Zeuesbrut	Mankind Agritech
17	Lodigo	Chemley Agritech Pvt. Ltd.

Chlorantraniliprole + Novaluron

Available formulations in India :

1. **Chlorantraniliprole 05.20% + Novaluron 08.50% SC**
2. **Chlorantraniliprole 10.00% + Novaluron 20.00% SC**

Classification on the basis of		**Chlorantraniliprole**	**Novaluron**
Target organism	:	Insecticide	Insecticide
Mode of entry	:	Systemic, Contact and Stomach poison	Contact and Stomach poison
Mode of action	:	Muscle poison	Insect Growth Regulator (IGR)
Chemical group	:	Anthranilic diamide	Benzoylurea
Category of toxicity : **Toxicity label colour** :			

- The combination of Chlorantraniliprole and Novaluron is a mixture of insecticides often used in crop protection to control a broad spectrum of insect pests, particularly those from the order Lepidoptera (moths and butterflies).
- It is Ryanodine Receptor Modulator. It causes the uncontrolled release of calcium from the internal stores of insect muscle cells, leading to muscle paralysis, rapid cessation of feeding, lethargy, and ultimately death. It is primarily toxic by ingestion and secondarily by contact, with ovicidal and larvicidal activity.
- Novaluron is Inhibitor of Chitin Biosynthesis. It acts as an Insect Growth Regulator (IGR). It disrupts the molting process in the larval and nymphal stages by inhibiting the synthesis of chitin, a key component of the insect exoskeleton. This prevents the pests from successfully completing their development and molting into the next stage or adult, leading to death.
- Chlorantraniliprole provides quick cessation of feeding and knockdown, while Novaluron offers a long residual effect, particularly against immature stages, due to its IGR activity.

Label claim (DPPQ & S, Faridabad)
Chlorantraniliprole 05.20% + Novaluron 08.50% SC

Crop	**Pests**	**Dose / ha**		**Waiting period (Days)**
		Quantity of formulation (ml)	**Volume of water (Lit.)**	
Chickpea	Pod borer	400	500	07
Chilli	Fruit borer, Tobacco caterpillar	500	500	07
Tomato	Fruit borer, Tobacco caterpillar	500	500	07

Trade Names

Sr. No.	Trade Name	Company
1	Tewlis	Swal Corporation Ltd.

Chlorantraniliprole 10.00% + Novaluron 20.00% SC

Crop	Pests	Dose / ha		Waiting period (Days)
		Quantity of formulation (ml)	Volume of water (Lit.)	
Cabbage	Diamond back moth	300	500	10
Chilli	Fruit borer, Tobacco caterpillar	300	500	10
Maize	Fall armyworm	300	500	21

Trade Names

Sr. No.	Trade Name	Company
1		

Chlorantraniliprole + Novaluron + Emamectin benzoate

Available formulations in India : **Chlorantraniliprole 04.50% + Novaluron 11.50% + Emamectin benzoate 01.50% SE**

Classification on the basis of		Chlorantraniliprole	Novaluron	Emamectin benzoate
Target organism	:	Insecticide	Insecticide	Insecticide, Acaricide
Mode of entry	:	Systemic, Contact and Stomach poison	Contact and Stomach poison	Systemic, Contact and Stomach poison
Mode of action	:	Muscle poison	Insect Growth Regulator (IGR)	Nerve poison
Chemical group	:	Anthranilic diamide	Benzoylurea	Avermectin (Macrocyclic lactone)
Category of toxicity : Moderately toxic **Toxicity label colour** : Blue				

- The combination of Chlorantraniliprole + Novaluron + Emamectin benzoate is an insecticide mixture that combines three active ingredients with different modes of action to provide broad-spectrum control, particularly against certain types of insect pests.
- Chlorantraniliprole is a Ryanodine Receptor Modulator. It disrupts the calcium balance in insect muscles, causing rapid paralysis, cessation of feeding, and ultimately, death. It has systemic properties.
- Novaluron is an Insect Growth Regulator (IGR) and a Chitin Synthesis Inhibitor. It prevents the insect from molting (shedding its exoskeleton), mainly targeting the larval stages and leading to "abortive molting" and death.
- Emamectin Benzoate is a Glutamate-gated Chloride Channel (GluCl) Allosteric Modulator. It is a neurotoxin that inhibits muscle contraction, leading to paralysis and death. It works mainly by ingestion but also has some contact action and good translaminar activity (penetrates the leaf tissue).
- The combination of three different modes of action offers a powerful solution for immediate kill and extended residual control, which is essential for managing tough pests and preventing/delaying the development of insecticide resistance.

Label claim (DPPQ & S, Faridabad)
Chlorantraniliprole 04.50% + Novaluron 11.50% + Emamectin benzoate 01.50% SE

Crop	Pests	Dose / ha		Waiting period (Days)
		Quantity of formulation (ml)	Volume of water (Lit.)	
Brinjal	Shoot and fruit borer	625	500	16

Trade Names

Sr. No.	Trade Name	Company
1	Nemagen	Best Agrolife Ltd.
2	Akhand	Sudarshan Farm Chemicals India Pvt. Ltd.

Chlorantraniliprole + Thiamethoxam

Available formulations in India	:	**1. Chlorantraniliprole 0.50% + Thiamethoxam 01.00% GR** **2. Chlorantraniliprole 08.80% + Thiamethoxam 17.50% SC** **3. Chlorantraniliprole 07.50% + Thiamethoxam 07.50% OD** **4. Chlorantraniliprole 20.00% + Thiamethoxam 20% GR**

Classification on the basis of		**Chlorantraniliprole**	**Thiamethoxam**
Target organism	:	Insecticide	Insecticide
Mode of entry	:	Systemic, Contact and Stomach poison	Systemic, Contact and Stomach poison
Mode of action	:	Muscle poison	Nerve poison
Chemical group	:	Anthranilic diamide	Neonicotinoid
Category of toxicity : Moderately toxic **Toxicity label colour** : Blue			

- The mixture is effective due to the complementary action of its two active ingredients, which target different systems in the insect's body.
- Chlorantraniliprole is primarily a ryanodine receptor modulator. It binds to and activates ryanodine receptors in insect muscle cells, causing an uncontrolled release of calcium ions from internal stores. This leads to impaired muscle contraction, feeding cessation, paralysis, and eventual death. It is effective against primarily chewing pests, especially Lepidopteran larvae (caterpillars, loopers, borers) and some Coleopteran (beetles).
- Thiamethoxam is a neurotoxin that acts as an agonist on the nicotinic acetylcholine receptors (nAChRs) in the insect's central nervous system. This blocks the normal transfer of nerve impulses, leading to rapid paralysis and death. It is a systemic insecticide. It is effective against primarily sucking pests (aphids, whiteflies, hoppers) and some chewing pests.
- It provides control over a wide range of both sucking and chewing pests (Lepidopteran, Hemipteran, and others).
- Thiamethoxam offers systemic action (absorbed by the plant and transported throughout), while both components work via ingestion and contact, providing comprehensive and long-lasting protection.

Label claim (DPPQ & S, Faridabad)
Chlorantraniliprole 0.50% + Thiamethoxam 01.00% GR

Crop	**Pests**	**Dose / ha**		**Waiting period (Days)**
		Quantity of formulation (Kg)	**Volume of water (Lit.)**	
Rice	Stem borer, Leaf folder, Brown plant hopper, Green leaf hopper	6	--	60

Trade Names

Sr. No.	Trade Name	Company
1	Virtako	Syngenta India Ltd.
2	Power GR	Goldenox Crop Care
3	Zeuesforce	Mankind Agritech

Chlorantraniliprole 08.80% + Thiamethoxam 17.50% SC

Crop	Pests	Dose / ha		Waiting period (Days)
		Quantity of formulation (ml)	Volume of water (ml)	
Tomato	Leaf miner, Whitefly, Fruit borer	▪ 500 ml ▪ Soil drench, ▪ Single application, ▪ Application time : 8-10 days after transplanting)	50-100 ml/plant	36
Rice nursery	Stem borer, Leaf folder, Green leaf hopper	▪ 600 ml ▪ Soil drench, ▪ Single application, ▪ Application time : At the time of sowing to before transplanting)	100 lit/ha	116

Trade Names

Sr. No.	Trade Name	Company
1	Voliam Flexi	Syngenta India Pvt. Ltd.
2	Vijeta	Silver Sine Bio Tech
3	Volium Plus	Natural Crop Science
4	Chakryaan	Katyayani Organics
5	Kalia	Vajra Crop Science
6	KTM	Koronation Bio-Science
7	Flexon	Global Crop Care
8	Ambivi	Parijat Industries India Pvt. Ltd.
9	Vantor-GR	Synwood Agro Ltd.
10	Tora Tora	Koronation Bio-Science
11	Koka	Green India Care
12	Tajchem	Chemley Agritech Pvt. Ltd.

Chlorantraniliprole 07.50% + Thiamethoxam 07.50% OD

Crop	Pests	Dose / ha		Waiting period (Days)
		Quantity of formulation (ml)	Volume of water (Lit.)	
Transplanted rice	Stem borer, Whorl maggot	480	Splash inti water at 15-20 days after transplanting	95

Trade Names

Sr. No.	Trade Name	Company

Chlorantraniliprole 20.00% + Thiamethoxam 20.00% GR

Crop	Pests	Dose / ha		Waiting period (Days)
		Quantity of formulation (Kg)	Volume of water (Lit.)	

Trade Names

Sr. No.	Trade Name	Company
1	Mironto	Tropical Agrosystem India (P) Ltd.

Clothianidin + Pyriproxifen

Available formulations in India : **Clothianidin 03.50% + Pyriproxyfen 08.00% SE**

Classification on the basis of		**Clothianidin**	**Pyriproxyfen**
Target organism	:	Insecticide	Insecticide
Mode of entry	:	Systemic, Contact and Stomach poison	Contact and Stomach poison
Mode of action	:	Nerve poison	Insect Growth Regulator (IGR)
Chemical group	:	Neonicotinoid	Pyridine ether
Category of toxicity : **Toxicity label colour :**			

- Clothianidin is a nicotinic acetylcholine receptor (nAChR) competitive modulator. It acts on the insect's central nervous system, causing overstimulation, paralysis, and death. It provides a rapid "knockdown" effect and is systemic (can be absorbed by the plant and translocated throughout), controlling both sucking and chewing pests.
- Pyriproxyfen mimics the insect's natural juvenile hormone. It disrupts the metamorphosis (development) of the insect, preventing larvae from developing into viable, reproducing adults. It also often exhibits an ovicidal (egg-killing) effect. It is effective against the young, immature stages of pests and has a longer-lasting residual effect.
- By combining a neurotoxic insecticide (Clothianidin) with an insect growth regulator (Pyriproxyfen), the mixture provides control over both the adult stage (fast control by Clothianidin) and the immature stages (long-term control and reproductive disruption by Pyriproxyfen).

Label claim (DPPQ & S, Faridabad)
Clothianidin 03.50% + Pyriproxyfen 08.00% SE

Crop	Pests	Dose / ha		Waiting period (Days)
		Quantity of formulation (ml)	Volume of water (Lit.)	
Brinjal	Whitefly, Jassid, Thrips, Aphid	1250	500	03
Cotton	Whitefly, Jassid, Thrips, Aphid	1250-1500	500	60

Trade Names

Sr. No.	Trade Name	Company
1	Pyclome	Sumitomo Chemical India Ltd.

Chlorpyrifos + Alphacypermethrin

Available formulations in India : **Chlorpyrofos160.00% + Alphacypermethrin 01.00% EC**

Classification on the basis of		Chlorpyrifos	Alphacypermethrin
Target organism	:	Insecticide, Acaricide	Insecticide
Mode of entry	:	Contact, Stomach and Respiratory poison	Contact and Stomach poison
Mode of action	:	Nerve poison	Nerve poison
Chemical group	:	Organophosphate	Synthetic pyrethroid
Category of toxicity : Highly toxic **Toxicity label colour** : Yellow			

- Chlorpyrifos 16% + Alphacypermethrin 1% EC is a broad-spectrum, dual-action insecticide formulation widely used in India for controlling bollworms in cotton.
- Chlorpyrifos is non-systemic with contact, stomach, and respiratory (fumigant) action. It is an Acetylcholinesterase inhibitor, which affects the insect's nervous system leading to paralysis and death.
- Alphacypermethrin is non-systemic with contact and stomach action. It is a Sodium Channel Modulator, disrupting nerve function, causing hyperexcitation, paralysis, and rapid knockdown.
- Combining two different chemical classes with different mechanisms of action helps to provide rapid knockdown (from Alphacypermethrin) and residual/longer-lasting control, including vapor action (from Chlorpyrifos).

Label claim (DPPQ & S, Faridabad)
Chlorpyrofos 16.00% + Alphacypermethrin 01.00% EC

Crop	Pests	Dose / ha		Waiting period (Days)
		Quantity of formulation (ml)	Volume of water (Lit.)	
Cotton	Spotted bollworm, Pink bollworm, American bollworm	2500	500-750	15

Trade Names

Sr. No.	Trade Name	Company
1	Chlorthrin	Arysta Lifescience
2	AC-116	Parijat Industries India Pvt. Ltd.
3	Raffle	Crop Chemicals India Ltd.
4	Shark	Tusova Agro Pvt. Ltd.
5	Herashakti	Herenba Industries Ltd.
6	Chlothrin	Sikko Industries Ltd.

7	Chlorin	Silver Sine Bio Tech
8	Alpha Cyper 161	Essential Biosciences
9	Syn Plus	Synwood Agro Ltd.
10	Anth Super	Krishi Rasayan
11	Takara	Global Crop Care
12	Mamaki	RACCOLTO Agritech India Pvt. Ltd.

Chlorpyrifos + Cypermethrin

Available formulations in India : **Chlorpyrofos 50.00% + Cypermethrin 05.00% EC**

Classification on the basis of		Chlorpyrifos	Cypermethrin
Target organism	:	Insecticide, Acaricide	Insecticide
Mode of entry	:	Contact, Stomach and Respiratory poison	Contact and Stomach poison
Mode of action	:	Nerve poison	Nerve poison
Chemical group	:	Organophosphate	Synthetic pyrethroid
Category of toxicity : Highly toxic **Toxicity label colour** : Yellow			

- Chlorpyrifos is non-systemic with contact, stomach, and respiratory (fumigant) action. It is an Acetylcholinesterase inhibitor, which affects the insect's nervous system leading to paralysis and death.
- Cypermethrin is non-systemic with contact and stomach action. It is a Sodium Channel Modulator, disrupting nerve function, causing hyperexcitation, paralysis, and rapid knockdown.
- Combining two different chemical classes with different mechanisms of action helps to provide rapid knockdown (from Cypermethrin) and residual/longer-lasting control, including vapor action (from Chlorpyrifos).
- It is effective against a wide range of both sucking and chewing/lepidopteran pests.

Label claim (DPPQ & S, Faridabad)
Chlorpyrofos 50.00% + Cypermethrin 05.00% EC

Crop	Pests	Dose / ha		Waiting period (Days)
		Quantity of formulation (ml)	Volume of water (Lit.)	
Cotton	Aphid, Jassid, Thrips, Whitefly, Spotted bollworm, Pink bollworm, American bollworm, Tobacco caterpillar	1000	500-1000	15
Rice	Stem borer, Leaf folder	625-750	500-700	15
Brinjal	Shoot and fruit borer	1000	500	07
Okra	Shoot and fruit borer	1000	500	10
Cabbage	Diamond back moth	750	500	05

Trade Names

Sr. No.	Trade Name	Company
1	Nurelle D -505, Lara-909	Crystal Crop Protection Ltd.
2	Premain Super	ADAMA India Private Limited
3	Koranda 505	Rallis India Limited

4	Hilhunter	HIL (India) Limited
5	Yujo	IFFCO-MC Crop Science Private Ltd.
6	Double Star	Swal Corporation Ltd.
7	Ulka 505	Biostadt India Ltd.
8	Hamla 550	Gharda Chemicals Ltd.
9	Catchh	Coromandel International Ltd.
10	Bhishma-505	Auxin Crop Care
11	Noorani 505	Anu Products Ltd.
12	Lethal Super 505	Insecticides (India) Ltd.
13	Cannon	NACL Industries Ltd.
14	Bancyp 505	Devidayal Agro Chemicals Ltd.
15	Bang X	Excel Crop Care Ltd.
16	Karan-Arjun	Sydney Formulation & Chemicals
17	Shatrujeet 505	Parijat Industries India Pvt. Ltd.
18	Nagraj-505	Crop Chemicals India Ltd.
19	Combi-X-Super	Sumitomo Chemical India Ltd.
20	Plus-500	Paras Crops Ltd.
21	Hemthrin Duo	Hemani Industries Ltd., Ankleshwar
22	Super 505	Super Crop Safe Ltd., Ahemadabad
23	Clorex Plus	GFG Crop Science Pvt. Ltd.
24	Param	Herenba Industries Ltd.
25	Cyclone	Mangroove Laboratory Pvt. Ltd.
26	Lenocet-505	Sikko Industries Ltd.
27	Chandika	Nichino India
28	Kaichi-505	Somnath Demeter Agritron Ltd.
29	Nuclear 505	Silver Sine Bio Tech
30	Clormex Gold 505	Natural Crop Science
31	Docter 505	Katyayani Organics
32	Action-505	Tropical Agrosystem India (P) Ltd.
33	German 505	Essential Biosciences
34	Syn-505	Synwood Agro Ltd.
35	Akhanda 505	Koronation Bio-Science
36	Qairawan 505	Today's Global Crop Protection
37	Super-D	Dhanuka Agrotech Ltd.
38	Clorin	Rain Bio Tech Industries
39	Naag 505	Safex Chemicals (India) Ltd.
40	Seturn 505	ATPL
41	Anth-505	Krishi Rasayan
42	Impact-505	Global Crop Care
43	Uri 505	Green India Care
44	Jupiter-505	JU Agrisciences Pvt. Ltd.
45	Run 505	Linuxcrop Chemicals Pvt. Ltd.
46	SA-Rainbow	Saraswati Agro Life Science (India) Pvt. Ltd.
47	Cairo	Synovita Crop science Pvt. Ltd.
48	Mahasodi	Mankind Agritech
49	Rambo 505	Preventis Chem India (P) Ltd.
50	Telpher-505	RACCOLTO Agritech India Pvt. Ltd.

Chlorpyrifos + Fipronil

Available formulations in India : **Chlorpyrifos 35.00% + Fipronil 03.50% EC**

Classification on the basis of		Chlorpyrifos	Fipronil
Target organism	:	Insecticide, Acaricide	Insecticide
Mode of entry	:	Contact, Stomach and Respiratory poison	Contact and Stomach poison (Systemic)
Mode of action	:	Nerve poison	Nerve poison
Chemical group	:	Organophosphate	Phenylpyrazole
Category of toxicity : Highly toxic **Toxicity label colour** : Yellow			

- Chlorpyrifos, an organophosphate, is a non-systemic insecticide that works primarily by contact, ingestion, and respiratory action, inhibiting the enzyme acetylcholinesterase, which leads to overstimulation and death of the insect nervous system.
- Fipronil, a phenylpyrazole, provides some systemic action and works by blocking GABA-gated chloride channels in the central nervous system, causing hyperexcitation and death.
- The combination is recommended for broad-spectrum control of pests, often exhibiting a quick knock-down effect and extended residual protection due to the combined non-systemic and systemic activity.
- However, like both individual compounds, the mixture requires careful use due to potential toxicity to non-target organisms like bees and aquatic life.

Label claim (DPPQ & S, Faridabad)
Chlorpyrofos 35.00% + Fipronil 03.50% EC

Crop	Pests	Dose / ha		Waiting period (Days)
		Quantity of formulation (ml)	Volume of water (Lit.)	
Rice	Stem borer, Plant hopper	1250-1500	500-625	
Cabbage	Diamondback moth	1125-1375	500-625	

Trade Names

Sr. No.	Trade Name	Company
1	Rustam	Global Crop Care
2	Chip	Green India Care

0

Cyantraniliprole + Diafenthiuron

Available formulations in India : **Cyantraniliprole 08.00% + Diafenthiuron 40.00% SC**

Classification on the basis of		Cyantraniliprole	Diafenthiuron
Target organism	:	Insecticide	Insecticide, Acaricide
Mode of entry	:	Systemic, Contact and Stomach poison	Contact and Stomach poison
Mode of action	:	Muscle poison	Energy metabolism
Chemical group	:	Anthranilic Diamide	Thiourea
Category of toxicity : Moderately toxic **Toxicity label colour** : Blue			

- Cyantraniliprole acts by modulating insect ryanodine receptors (RyRs) in muscle cells. This leads to the uncontrolled release of calcium ions, causing muscle contraction, paralysis, rapid feeding cessation, and eventual death of the insect. It exhibits systemic and translaminar activity, meaning it is absorbed by the plant and translocated through the xylem, controlling pests that feed on the roots, stems, or undersides of leaves.
- Diafenthiuron is a pro-insecticide that is metabolically converted into its active form (a carbodiimide) in the insect. The active metabolite inhibits mitochondrial ATPase, disrupting the insect's energy production and leading to metabolic disruption, paralysis, and death. It is a non-systemic insecticide with strong contact and translaminar action. It also has some acaricidal (mite-killing) activity.
- It is effective against both chewing pests (like caterpillars) and a wide range of sucking pests (like whiteflies, aphids, and thrips), as well as mites.
- The two different modes of action provide a faster knockdown effect and excellent residual (long-lasting) control. Combining two active ingredients with different modes of action helps in an anti-resistance strategy by reducing the likelihood of pests developing resistance to a single compound.

Label claim (DPPQ & S, Faridabad)
Cyantraniliprole 08.00% + Diafenthiuron 40.00% SC

Crop	Pests	Dose / ha		Waiting period (Days)
		Quantity of formulation (ml)	Volume of water (Lit.)	
Cotton	Whitefly, Jassid, Thrips, Aphid, Pink bollworm	625	500	29
Chilli	Thrips, Mites, Whitefly, Fruit borer	625	500	05

Trade Names

Sr. No.	Trade Name	Company
1	Plesiva	Syngenta India Ltd.

Cyantraniliprole + Lufenuron

Available formulations in India : Cyantraniliprole 16.90% + Lufenuron 16.90% SC

Classification on the basis of		Cyantraniliprole	Lufenuron
Target organism	:	Insecticide	Insecticide
Mode of entry	:	Systemic, Contact and Stomach poison	Contact and Stomach poison
Mode of action	:	Muscle poison	Insect Growth Regulator (IGR)
Chemical group	:	Anthranilic Diamide	Benzoylurea
Category of toxicity : Moderately toxic **Toxicity label colour** : Blue			

- Cyantraniliprole is Ryanodine Receptor Modulator. It activates the ryanodine receptors in the insect's muscle cells, causing an uncontrolled release of calcium ions. This leads to muscle contraction, paralysis, rapid feeding cessation, and ultimately, death of the pest. It is primarily systemic (absorbed and distributed within the plant) and has strong stomach and good contact activity. It provides quick control and immediate feeding stop. It is effective against a cross-spectrum of chewing and sucking pests, including Lepidopteran larvae (caterpillars/bollworms), leaf miners, and certain sucking pests like aphids and thrips.
- Lufenuron is an Insect Growth Regulator (IGR). It inhibits the synthesis of chitin, which is essential for the formation of the insect's cuticle (exoskeleton). This prevents the insect larvae from molting properly, disrupting their development and killing them. It is primarily non-systemic with stomach and contact activity. It provides long residual control and is highly effective against eggs and larvae (ovicidal and larvicidal activity). It is especially effective against Lepidopteran larvae (caterpillars) and also controls pests like whiteflies and rust mites.
- The combination provides a dual mode of action that results in broad-spectrum control and fast knockdown and long-lasting control.

Label claim (DPPQ & S, Faridabad)
Cyantraniliprole 16.90% + Lufenuron 16.90% SC

Crop	Pests	Dose / ha		Waiting period (Days)
		Quantity of formulation (ml)	Volume of water (Lit.)	
Rice	Stem borer, Leaf folder	50	500	39
Tomato	Fruit borer, Leaf miner, Thrips	150	500	05
Maize	Fall armyworm, Stem borer	150	500	44
Cotton	Pink & American bollworm, Tobacco caterpillar, Thrips	250	500	25

Trade Names

Sr. No.	Trade Name	Company
1	Minecto Xtra	Syngenta India Ltd.

Cyantraniliprole + Pymetrozine

Available formulations in India : **Cyantraniliprole 10.00% + Pymetrozine 50.00% WG**

Classification on the basis of		Cyantraniliprole	Pymetrozine
Target organism	:	Insecticide	Insecticide
Mode of entry	:	Systemic, Contact and Stomach poison	Systemic, Stomach poison
Mode of action	:	Muscle poison	Nerve poison
Chemical group	:	Anthranilic Diamide	Pyridine azomethine
Category of toxicity : **Toxicity label colour :**			

- Cyantraniliprole is Ryanodine Receptor Modulator. It disrupts calcium ion regulation in insect muscles, leading to muscle contraction, rapid feeding cessation, paralysis, and death. It provides systemic and translaminar action.
- Pymetrozine is Chordotonal Organ TRPV Channel Modulator (Anti-feeding). It selectively affects the nervous system of sap-feeding insects, causing a loss of coordination and a rapid and irreversible cessation of feeding (insects starve). It acts as a nutritional deterrent and has systemic activity.
- It effectively controls a wider range of pests, including both chewing (Lepidoptera) and sucking (Hemiptera) insects.
- Both components exhibit systemic and/or translaminar movement within the plant, providing long-lasting control and protecting parts of the plant not directly hit by the spray.
- The Pymetrozine component quickly stops sap-feeding pests like aphids and whiteflies from feeding, limiting crop damage and virus transmission.

Label claim (DPPQ & S, Faridabad)

Cyantraniliprole 10.00% + Pymetrozine 50.00% WG

Crop	Pests	Dose / ha		Waiting period (Days)
		Quantity of formulation (g)	Volume of water (Lit.)	
Rice	Stem borer, Leaf folder, Brown plant hopper	300	500	43

Trade Names

Sr. No.	Trade Name	Company
1	Synero	Syngenta India Pvt. Ltd.

Cyantraniliprole + Thiamethoxam

Available formulations in India : **Cyantraniliprole 19.80% + Thiamethoxam 19.80% FS**

Classification on the basis of		Cyantraniliprole	Thiamethoxam
Target organism	:	Insecticide	Insecticide
Mode of entry	:	Systemic, Contact and Stomach poison	Systemic, Contact and Stomach poison
Mode of action	:	Muscle poison	Nerve poison
Chemical group	:	Anthranilic Diamide	Neonicotinoid
Category of toxicity : Moderately toxic **Toxicity label colour** : Blue			

- he combination of Cyantraniliprole 19.80% + Thiamethoxam 19.80% FS is a high-performance, dual-action seed treatment insecticide.
- Cyantraniliprole targets the muscles of chewing pests (like caterpillars), causing immediate paralysis.
- Thiamethoxam is a systemic nerve poison that is highly effective against sucking pests.
- Application Tip: Mix the required dose with about 10–15 ml of water per kg of seed to create a slurry. Ensure every seed is evenly coated and let them dry in the shade before sowing.
- Seed treatment protects the crop for the first 30–40 days, reducing the need for early-stage foliar sprays. The chemical is absorbed by the roots and moves upward as the plant grows, protecting new leaves. It promotes better seedling vigour and uniform germination.

Label claim (DPPQ & S, Faridabad)
Cyantraniliprole 19.80% + Thiamethoxam 19.80% FS

Crop	Pests	Dose / Kg Seed		Waiting period (Days)
		Quantity of formulation (ml)	Volume of water (ml)	
Maize	Shoot fly, Stem borer, Cutworm, Aphid	4	10-12	Seed treatment
	Fall armyworm	6		

Trade Names

Sr. No.	Trade Name	Company
1	Fortenza Duo	Syngenta India Pvt. Ltd.

Cypermethrin + Indoxacarb

Available formulations in India : **Cypermethrin 10.00% + Indoxacarb 10.00% SC**

Classification on the basis of		Cypermethrin	Indoxacarb
Target organism	:	Insecticide	Insecticide
Mode of entry	:	Contact and Stomach poison	Contact and Stomach poison
Mode of action	:	Nerve poison	Nerve poison
Chemical group	:	Synthetic pyrethroid	Oxidiazine
Category of toxicity : Highly toxic **Toxicity label colour** : Yellow			

- Cypermethrin acts as a fast-acting neurotoxin. It targets the voltage-gated sodium channels in the insect's nervous system, disrupting the normal function of nerve signaling, which leads to paralysis and rapid death (knockdown effect). It is non-systemic, acting primarily by contact and ingestion.
- Indoxacarb is a pro-insecticide that is bioactivated within the insect's body. The active metabolite blocks the voltage-gated sodium channels in a different way than pyrethroids, inhibiting nerve function. This leads to cessation of feeding, paralysis, and eventual death. It primarily acts by ingestion, but also has contact activity.
- The mixture of Cypermethrin and Indoxacarb offers several advantages. Cypermethrin is effective against a wide range of chewing and sucking pests, and Indoxacarb provides strong control, particularly against Lepidopteran (moth and butterfly) larvae. This combination targets a greater variety of insects.
- Using two insecticides with different modes of action can help prevent or delay the development of insecticide resistance in pest populations, which is a major concern when using a single class of chemistry repeatedly.

Label claim (DPPQ & S, Faridabad)
Cypermethrin 10.00% + Indoxacarb 10.00% SC

Crop	Pests	Dose / ha		Waiting period (Days)
		Quantity of formulation (ml)	Volume of water (Lit.)	
Cotton	Jassid, Thrips, Bollworms	500	400-1000	07
Rice	Yellow stem borer	250	500	37
	Leaf folder	375	500	37

Trade Names

Sr. No.	Trade Name	Company
1	Auxi-Carb	Auxin Crop Care
2	Demonexia 110	Somnath Demeter Agritron Ltd.
3	Intra	Aquarius Agro Chemicals
4	Thunderhead	RACCOLTO Agritech India Pvt. Ltd.

Cypermethrin + Quinalphos

Available formulations in India : **Cypermethrin 03.00% + Quinalphos 20.00% EC**

Classification on the basis of		Cypermethrin	Quinalphos
Target organism	:	Insecticide	Insecticide
Mode of entry	:	Contact and Stomach poison	Stomach and Contact poison
Mode of action	:	Nerve poison	Nerve poison
Chemical group	:	Synthetic pyrethroid	Organophosphates
Category of toxicity : Highly toxic **Toxicity label colour** : Yellow			

- The formulation of Cypermethrin 03.00% + Quinalphos 20.00% EC (Emulsifiable Concentrate) is a widely used, powerful, dual-action insecticide designed for the effective control of a broad spectrum of pests, especially in major crops. This product leverages the strengths of two distinct chemical classes: Cypermethrin, a synthetic pyrethroid, and Quinalphos, an organophosphate.
- The efficacy of this combination relies on its dual mode of action, which provides both rapid knockdown and long-lasting control.
- Cypermethrin acts primarily as a contact and stomach insecticide, working immediately by interfering with the insect's nervous system through modulation of the sodium channels, leading to rapid paralysis and death. This component is highly effective against chewing pests.
- In contrast, Quinalphos, at a higher concentration, provides robust contact, stomach, and systemic activity by acting as a cholinesterase inhibitor, disrupting the nerve signals and ensuring a more sustained residual kill against both chewing and sucking pests.
- By simultaneously attacking the pest's nervous system via two different biochemical pathways, the combination reduces the likelihood of insects developing resistance to either chemical alone.

Label claim (DPPQ & S, Faridabad)
Cypermethrin 03.00% + Quinalphos 20.00% EC

Crop	Pests	Dose / ha		Waiting period (Days)
		Quantity of formulation (ml)	Volume of water (Lit.)	
Brinjal	Shoot and fruit borer	350-400	500-600	07
Cotton	American, bollworm, Spotted bollworm, Jassid	100-1250	500-600	15

Trade Names

Sr. No.	Trade Name	Company
1	Viraat	UPL Ltd.
2	Relay-101	Swal Corporation Ltd.
3	Samrat	Anu Products Ltd.
4	Alert	Arysta Lifescience
5	German Plus	Essential Biosciences
6	Quincyper	Synwood Agro Ltd.
7	Topthron	RACCOLTO Agritech India Pvt. Ltd.
8	Cyqueen	Eagle Plant Protection Pvt. Ltd.
9	Cykill	Azeel Cropscience Ltd.
10	Lebero	Aquarius Agro Chemicals
11	Kwincyp	Krishi Rasayan
12	Cyko	Hindustan Agro Life
13	Cyphos-320	Angel Crop Science
14	Sonic Plus	Swastik Agri Science

Deltamethrin + Buprofezin

Available formulations in India : **Deltamethrin 0.72% + Buprofezin 05.65% EC**

Classification on the basis of		Deltamethrin	Buprofezin
Target organism	:	Insecticide	Insecticide
Mode of entry	:	Contact and Stomach poison	Contact and Stomach poison
Mode of action	:	Nerve poison	Insect Growth Regulator (IGR)
Chemical group	:	Synthetic pyrethroid	Thiadiazine
Category of toxicity : Highly toxic **Toxicity label colour** : Yellow			

- Deltamethrin is non-systemic, fast-acting insecticide with contact and stomach action. It is a synthetic pyrethroid that disrupts the normal functioning of the insect's nervous system by preventing sodium channels from closing, leading to rapid knockdown and death of pests, especially adults. It is broad-spectrum, effective against a wide range of pests. In this combination, it provides quick action against adult insects.
- Buprofezin is a persistent insecticide and acaricide with contact and stomach action. It is not translocated in the plant. It acts as an Insect Growth Regulator by inhibiting chitin biosynthesis, which is necessary for the formation of the insect's cuticle. This leads to the death of nymphs and larvae during the molting process. It also suppresses egg-laying by adults, and any eggs laid are often sterile. It is highly effective against Homoptera (such as Planthoppers and Leafhoppers), certain Coleoptera, and Acarina, with a persistent effect.
- The mixture combines the fast-acting knockdown effect of Deltamethrin (pyrethroid) against adult insects with the persistent, growth-regulating action of Buprofezin against nymphs and larvae. This helps in controlling the pest population across different life stages and managing insects that may have developed tolerance to single-action insecticides.

Label claim (DPPQ & S, Faridabad)
Deltamethrin 0.72% + Buprofezin 05.65% EC

Crop	Pests	Dose / ha		Waiting period (Days)
		Quantity of formulation (ml)	Volume of water (Lit.)	
Rice	Brown plant hopper, Leaf folder	1250-1500	500	30

Trade Names

Sr. No.	Trade Name	Company
1	Astu	Crop Life Science Ltd.
2	Dell Plus	Synergy

3	Dettol	Silver Sine Bio Tech
4	Dhadaka	Volkschem Crop Science Pvt. Ltd.
5	Buprothrin	Skyn Crop care Pvt. Ltd.
6	Black Apollo	HPM Chemicals and Fertilizers Ltd.

Diafenthiuron + Acetamiprid

Available formulations in India : **Diafenthiuron 40.10% + Acetamiprid 03.90% WP**

Classification on the basis of		**Diafenthiuron**	**Acetamiprid**
Target organism	:	Insecticide, Acaricide	Insecticide
Mode of entry	:	Contact and Stomach poison	Systemic, Contact and Stomach poison
Mode of action	:	Energy metabolism	Nerve poison
Chemical group	:	Thiourea	Neonicotinoid
Category of toxicity : Moderately toxic **Toxicity label colour** : Blue			

- The insecticide formulation Diafenthiuron 40.10% + Acetamiprid 03.90% WP (Wettable Powder) is a potent, dual-action product specifically engineered to combat a wide spectrum of difficult sucking pests and mites in crops.
- The combination of its two distinct active ingredients, Diafenthiuron and Acetamiprid, provides a critical dual mode of action, which is a key strategy for effective pest resistance management.
- Diafenthiuron : As a pro-insecticide (a compound that converts to its active form after application), Diafenthiuron provides contact, stomach, translaminar, and vapor action. Its primary mode of action is the inhibition of mitochondrial respiration in the pest, leading to a severe depletion of energy, paralysis, and eventual death. This component is highly effective against mites and various sucking pests, offering rapid control and a powerful residual effect through its ability to penetrate the leaf tissue (translaminar action) and reach pests on the underside of leaves via vapor action.
- Acetamiprid : As a neonicotinoid, Acetamiprid provides strong systemic and translaminar activity, ensuring the chemical is absorbed and translocated throughout the plant. Its mode of action involves interfering with the insect's central nervous system by binding to nicotinic acetylcholine receptors (nAChR), resulting in paralysis. This is particularly effective against tenacious sucking pests, providing protection to new growth and hidden parts of the plant.
- This combination not only achieves a fast kill but also offers a longer duration of control.

Label claim (DPPQ & S, Faridabad)
Diafenthiuron 40.10% + Acetamiprid 03.90% WP

Crop	Pests	Dose / ha		Waiting period (Days)
		Quantity of formulation (g)	Volume of water (Lit.)	
Cotton	Thrips, Whitefly, Mites	500-600	500	21
Chilli	Thrips, Whitefly, Mites	500-600	500	21

Trade Names

Sr. No.	Trade Name	Company
1	Herculus	Insecticides (India) Ltd.
2	Dicet	Sikko Industries Ltd.
3	Ashwamedh Plus	Katyayani Organics
4	EBS Auron-44	Essential Biosciences

Diafenthiuron + Bifenthrin

Available formulations in India : **Diafenthiuron 47.00% + Bifenthrin 09.40% SC**

Classification on the basis of		Diafenthiuron	Bifenthrin
Target organism	:	Insecticide, Acaricide	Insecticide
Mode of entry	:	Contact and Stomach poison	Contact and Stomach poison
Mode of action	:	Energy metabolism	Nerve poison
Chemical group	:	Thiourea	Synthetic pyrethroid
Category of toxicity : Moderately toxic **Toxicity label colour** : Blue			

- Diafenthiuron, belonging to the thiourea chemical class, is a pro-insecticide that converts to its active metabolite. This metabolite works by inhibiting mitochondrial ATP synthase, thereby disrupting the pest's energy production, which leads to paralysis and eventual mortality. This component provides long-lasting residual control, especially against mites and whiteflies.
- Conversely, Bifenthrin, a synthetic pyrethroid, provides a rapid knockdown effect by interfering with the insect's central and peripheral nervous system, specifically by disrupting the sodium channel gating, resulting in quick paralysis and death.
- The combined product is highly effective against a broad spectrum of insect pests, including key sucking pests such as whiteflies, aphids, thrips, and leafhoppers. By integrating a respiration inhibitor (Diafenthiuron) with a sodium channel modulator (Bifenthrin), the formulation provides both immediate cessation of pest feeding and prolonged protection, ensuring pest-free crop growth and ultimately leading to enhanced yields and improved crop health.

Label claim (DPPQ & S, Faridabad)
Diafenthiuron 47.00% + Bifenthrin 09.40% SC

Crop	Pests	Dose / ha		Waiting period (Days)
		Quantity of formulation (ml)	Volume of water (Lit.)	
Cotton	Thrips, Leafhopper, Whitefly, Aphid	625	500	30
Chilli	Thrips, Aphid	625	500	07

Trade Names

Sr. No.	Trade Name	Company
1	Takaf	Adama India Private Limited
2	Super Zeep	Synergy
3	Josh	Silver Sine Bio Tech
4	Taqat	Natural Crop Science

5	Taqila	Angel Crop Science
6	Dualfen	Katyayani Organics
7	Takata	Global Crop Care
8	Bifenforce Xtra	Mankind Agritech
9	Takara	Preventis Chem India (P) Ltd.
10	Polon	RACCOLTO Agritech India Pvt. Ltd.
11	Dosser	Zivista Crop Protection

Diafenthiuron + Dinotefuran

Available formulations in India : **Diafenthiuron 48.00% + Dinotefuran 08.00% WG**

Classification on the basis of		**Diafenthiuron**	**Dinotefuran**
Target organism	:	Insecticide, Acaricide	Insecticide
Mode of entry	:	Contact and Stomach poison	Systemic, Contact and Stomach poison
Mode of action	:	Energy metabolism	Nerve poison
Chemical group	:	Thiourea	Neonicotinoid
Category of toxicity : **Toxicity label colour :**			

- Diafenthiuron and Dinotefuran are active ingredients combined in certain insecticides to provide dual-action control against a wide range of agricultural pests, primarily sucking pests.
- Diafenthiuron is inhibitor of mitochondrial respiration. It acts as a pro-insecticide, converting to its active form (carbodiimide) upon light exposure or metabolism. The active metabolite inhibits ATP synthesis in mitochondria, disrupting the insect's energy production, leading to paralysis and death. It provides contact and stomach action, and is also effective against mites (acaricide).
- Dinotefuran is Nicotinic Acetylcholine Receptor (nAChR) competitive modulator. It acts as a systemic and contact insecticide with rapid knockdown. It disrupts the insect's nervous system by binding to and stimulating the nAChR, leading to over-excitation of nerves, paralysis, and death. It is absorbed by the plant and moves throughout the tissues (systemic action).

Label claim (DPPQ & S, Faridabad)
Diafenthiuron 48.00% + Dinotefuran 08.00% WG

Crop	**Pests**	**Dose / ha**		**Waiting period (Days)**
		Quantity of formulation (g)	**Volume of water (Lit.)**	
Cotton	Thrips, Jassid, Whitefly, Mites	625	500	43
Brinjal	Thrips, Jassid, Whitefly, Mites	625	500	03

Trade Names

Sr. No.	**Trade Name**	**Company**
1	Kuebiko	Indofil Industries Ltd.
2	Dorito	PI Industries Ltd.

Diafenthiuron + Fipronil + Dinotefuran

Available formulations in India : **Diafenthiuron 18% + Fipronil 8% + Dinotefuran 5% SC**

Classification on the basis of		Diafenthiuron	Fipronil	Dinotefuran
Target organism	:	Insecticide, Acaricide	Insecticide	Insecticide
Mode of entry	:	Contact and Stomach poison	Contact and Stomach poison	Systemic, Contact and Stomach poison
Mode of action	:	Energy metabolism	Nerve poison	Nerve poison
Chemical group	:	Thiourea	Phenylpyrazole	Neonicotinoid
Category of toxicity : Moderately toxic **Toxicity label colour** : Blue				

- This specific combination—Diafenthiuron 18% + Fipronil 8% + Dinotefuran 5% SC—is a powerful, triple-action insecticide formulation.
- It offers a "three-way" attack strategy:
- Diafenthiuron acts as a pro-insecticide. It is converted by sunlight (or metabolism) into a form that inhibits the insect's energy production (ATP synthesis). It also has excellent miticidal (mite-killing) properties.
- Fipronil is a broad-spectrum insecticide that disrupts the insect central nervous system. It provides both contact and stomach action and is particularly effective against Thrips and soil pests.
- Dinotefuran is a third-generation neonicotinoid. It is highly systemic, meaning the plant absorbs it, protecting new growth and killing pests that hide under leaves.
- Dinotefuran provides the "instant" kill, while Diafenthiuron and Fipronil provide long-lasting protection (often up to 15–20 days). Diafenthiuron has a slight vapor phase, helping it reach pests hiding in dense foliage.

Label claim (DPPQ & S, Faridabad)
Diafenthiuron 18% + Fipronil 8% + Dinotefuran 5% SC

Crop	Pests	Dose / ha		Waiting period (Days)
		Quantity of formulation (ml)	Volume of water (Lit.)	
Chilli	Thrips, Fruit borers (*Helicoverpa armigera, Spodoptera litura*)	875	500	28

Trade Names

Sr. No.	Trade Name	Company
1	Tag Fly Gold	Tropical Agroscience (India) Pvt. Ltd.

Diafenthiuron + Pyriproxyfen

Available formulations in India :

1. **Diafenthiuron 30.00% + Pyriproxyfen 08.00% SC**
2. **Pyriproxyfen 05.00% + Diafenthiuron 25.00% SE**
3. **Pyriproxifen 08.00% + Diafenthiuron 30.00% SE**

Classification on the basis of		**Diafenthiuron**	**Pyriproxyfen**
Target organism	:	Insecticide, Acaricide	Insecticide
Mode of entry	:	Contact and Stomach poison	Contact and Stomach poison
Mode of action	:	Energy metabolism	Insect Growth Regulator (IGR)
Chemical group	:	Thiourea	Pyridine ether
Category of toxicity : Moderately toxic **Toxicity label colour** : Blue			

- Diafenthiuron and Pyriproxyfen are a common combination in insecticide products, offering a dual mode of action for effective pest control, particularly against sucking pests.
- Diafenthiuron is Thio-urea derivative (Inhibitor of oxidative phosphorylation; ATP disruption). It provides quick knockdown effect by contact, ingestion, and vapor action, leading to immediate paralysis and death of the pest.
- Pyriproxifen is Juvenile Hormone Mimic / Insect Growth Regulator (IGR). It disrupts the insect's life cycle by preventing egg hatching, inhibiting the metamorphosis from nymph to adult, and causing sterility in adult pests.
- This dual action provides a quick knockdown of adults (from Diafenthiuron) and long-lasting residual control by managing all three life stages: eggs, nymphs, and adults (from Pyriproxyfen).

Label claim (DPPQ & S, Faridabad)

Diafenthiuron 30.00% + Pyriproxyfen 08.00% SC

Crop	**Pests**	**Dose / ha**		**Waiting period (Days)**
		Quantity of formulation (ml)	**Volume of water (Lit.)**	
Chilli	Mites, Whitefly	1000	500	03
Cotton	Whitefly	1000	500	35

Trade Names

Sr. No.	**Trade Name**	**Company**
1	Finio	Coromandel International Ltd.
2	Dipye	Global Crop Care

Pyriproxyfen 05.00% + Diafenthiuron 25.00% SE

Crop	Pests	Dose / ha		Waiting period (Days)
		Quantity of formulation (ml)	Volume of water (Lit.)	
Cotton	Whitefly, Thrips, Jassid, Aphid	1000-1250	500	35
Chilli	Whitefly, Thrips, Yellow mite	750-1000	500	10

Trade Names

Sr. No.	Trade Name	Company
1	Tecax	Swal Corporation Ltd.
2	Xyfen	Parijat Industries India Pvt. Ltd.
3	Pydon	Best Agrolife Ltd.
4	Finio	Coromandel International Ltd.
5	SLR 525	GSP Crop Science Pvt. Ltd.
6	Lapolo	Agrosis
7	Cimax	UPL Ltd.
8	HRL 525	Hydrabad Crop Science
9	LMG	Safex
10	Pyron	Katyayani Organics
11	SLR-525	GSP Crop Science Pvt. Ltd.
12	Derby Duo	Biostadt India Ltd.
13	Dyna	Mikado Crop Science
14	Apache 525	Sahib Pecticide and Seeds Ltd.
15	Para-525	

Pyriproxifen 08.00% + Diafenthiuron 30.00% SE

Crop	Pests	Dose / ha		Waiting period (Days)
		Quantity of formulation (ml)	Volume of water (Lit.)	
Chilli	Mites, Whitefly	1000	500	03
Cotton	Whitefly	1000	500	

Trade Names

Sr. No.	Trade Name	Company
.0	Xyfen Ultra	Parijat Industries India Pvt. Ltd.
2	Difenhuge-P	Mankind Agritech
3	Boris Super	Tata Rallis India Ltd.

Dimethoate + Cypermethrin

Available formulations in India : Dimethoate 20.00% + Cypermethrin 03.00% EC

Classification on the basis of		Dimethoate	Cypermethrin
Target organism	:	Insecticide, Acaricide	Insecticide
Mode of entry	:	Systemic, Contact and Stomach poison	Contact and Stomach poison
Mode of action	:	Nerve poison	Nerve poison
Chemical group	:	Organophosphate	Synthetic pyrethroid
Category of toxicity : Highly toxic **Toxicity label colour** : Yellow			

- Dimethoate is an organophosphate insecticide with both systemic and contact action. It acts as an acetylcholinesterase inhibitor, which blocks the enzyme responsible for breaking down acetylcholine in the nervous system. This leads to the accumulation of acetylcholine, causing continuous nerve impulses, overstimulation, paralysis, and eventually death of the insect. Dimethoate is effective against a wide range of sucking and chewing pests, and is absorbed and transported throughout plant tissues.
- Cypermethrin is a synthetic pyrethroid insecticide that primarily acts by contact and ingestion. It targets voltage-gated sodium channels in nerve cell membranes, causing them to remain open longer and resulting in excessive nerve stimulation. This leads to symptoms such as hyper-excitation, tremors, paralysis, and death in insects. Cypermethrin remains on plant surfaces and provides fast knockdown action against various insect pests.
- The mixture of dimethoate and cypermethrin provides a dual action against insects by combining a systemic organophosphate (disrupting acetylcholinesterase) with a non-systemic pyrethroid (modifying sodium channel activity). This synergy overwhelms insect detoxification systems, reduces the likelihood of resistance, and enhances efficacy against a broad range of pests.

Label claim (DPPQ & S, Faridabad)
Dimethoate 20.00% + Cypermethrin 03.00% EC

Crop	Pests	Dose / ha		Waiting period (Days)
		Quantity of formulation (ml)	Volume of water (Lit.)	
Brinjal	Shoot and fruit borer, Jassid, Epilachna beetle	800	500	07

Trade Names

Sr. No.	Trade Name	Company
1	Kri Kreate	Krishi Rasayan

Dinotefuran + Acephate

Available formulations in India : **Dinotefuran 04.00% + Acephate 50.00% SG**

Classification on the basis of		Dinotefuran	Acephate
Target organism	:	Insecticide	Insecticide
Mode of entry	:	Systemic, Contact and Stomach poison	Systemic, Contact and Stomach poison
Mode of action	:	Nerve poison	Nerve poison
Chemical group	:	Neonicotinoid	Organophosphate
Category of toxicity : Highly toxic **Toxicity label colour** : Yellow			

- Dinotefuran is a neonicotinoid insecticide that acts by blocking the nicotinic acetylcholine receptors in the insect nervous system, causing paralysis and eventual death of the pest.
- Acephate is an organophosphate insecticide that works by inhibiting acetylcholinesterase, disrupting nerve transmission and leading to the pest's death.
- The combination delivers dual action on pests, helping in resistance management due to these different modes of action.
- This combination is preferred for its rapid knockdown, systemic action, and broad-spectrum efficacy against sucking pests

Label claim (DPPQ & S, Faridabad)
Dinotefuran 04.00% + Acephate 50.00% SG

Crop	Pests	Dose / ha		Waiting period (Days)
		Quantity of formulation (g)	Volume of water (Lit.)	
Rice	Brown plant hopper, White backed plant hopper	500	500	28
Cotton	Aphid, Jassid, Thrips, Whitefly	880	500	10

Trade Names

Sr. No.	Trade Name	Company
1	Dinoace	PI Industries Ltd.
2	Sapper	Indofil Industries Ltd.
3	Dinosar	Silver Sine Bio Tech
4	Zapper	Global Crop Care

Dinotefuran + Ethion

Available formulations in India : **Dinotefuran 05.00% + Ethion 50.00% EC**

Classification on the basis of		Dinotefuran	Ethion
Target organism	:	Insecticide	Insecticide, Acaricide
Mode of entry	:	Systemic, Contact and Stomach poison	Contact and Stomach poison
Mode of action	:	Nerve poison	Nerve poison
Chemical group	:	Neonicotinoid	Organophosphate
Category of toxicity : Highly toxic **Toxicity label colour** : Yellow			

- Dinotefuran is a neonicotinoid insecticide, acting as a systemic, contact, and stomach poison. It selectively binds to nicotinic acetylcholine receptors (nAChRs) in the insect nervous system, causing continuous nerve stimulation, paralysis, and death. As a systemic agent, it is absorbed and translocated through plant tissues, protecting even unsprayed parts.
- Ethion is an organophosphorus insecticide with chiefly contact, but also some systemic and stomach action. It inhibits the enzyme acetylcholinesterase, leading to the accumulation of acetylcholine in synapses, which disrupts nerve function and causes paralysis and death in insects.
- When formulated together (such as Dinotefuran 5% + Ethion 50% EC), the insecticide provides rapid knockdown and extended protection due to the dual mode of action. Dinotefuran's systemic and translaminar activity is complemented by Ethion's strong contact and stomach effect, making the combination effective against both sucking and chewing pests across various crop stages.

Label claim (DPPQ & S, Faridabad)
Dinotefuran 05.00% + Ethion 50.00% EC

Crop	Pests	Dose / ha		Waiting period (Days)
		Quantity of formulation (ml)	Volume of water (Lit.)	
Rice	Brown plant hopper, White backed plant hopper, Green leaf hopper, Leaf folder	1000	500	33

Trade Names

Sr. No.	Trade Name	Company
1	Wash	Green India Care
2	Claret	PI Industries Ltd.

Dinotefuran + Pymetrozine

Available formulations in India	:	1. **Dinotefuran 11.00% + Pymetrozine 36.00% WG** 2. **Dinotefuran 15.00% + Pymetrozine 45.00% WG** 3. **Pymetrozine 29.20% + Dinotefuran 11.70% WDG**

Classification on the basis of		**Dinotefuran**	**Pymetrozine**
Target organism	:	Insecticide	Insecticide
Mode of entry	:	Systemic, Contact and Stomach poison	Systemic, Stomach poison
Mode of action	:	Nerve poison	Nerve poison
Chemical group	:	Neonicotinoid	Pyridine azomethine
Category of toxicity : Moderately toxic **Toxicity label colour** : Blue			

- Dinotefuran is a neonicotinoid insecticide that acts by binding to acetylcholine receptors in the insect's nervous system, causing disrupted nerve transmission, abnormal behavior, paralysis, and death. It enters insects both orally (when the pest feeds on treated plant sap) and through contact (dermal absorption).
- Pymetrozine is a selective feeding blocker insecticide. Its unique mode of action involves disrupting the nervous system of sap-feeding insects. It is a neuroactive compound that acts on the chordotonal mechanoreceptors in the insect's legs by modulating TRPV channels. This immediately stops the insect from feeding by preventing it from inserting its stylet into the plant tissue. While the insect dies from starvation in a few days, crop damage ceases almost immediately. Pymetrozine is a systemic and translaminar product, providing protection throughout the plant.
- The combination provides contact and systemic activity throughout the plant, fast coverage and rainfastness, reduced egg-laying in female pestsand long-lasting pest control and improved crop health at reproductive stages. This dual action gives both immediate mortality and suppression of future pest generations, making it a robust solution for paddy pest management.

Label claim (DPPQ & S, Faridabad)
Dinotefuran 11.00% + Pymetrozine 36.00% WG

Crop	Pests	Dose / ha		Waiting period (Days)
		Quantity of formulation (g)	Volume of water (Lit.)	
Rice	Brown plant hopper, White backed plant hopper	350	500	36

Trade Names

Sr. No.	Trade Name	Company
1	Tandav	SP Agrochemical Industries

Dinotefuran 15.00% + Pymetrozine 45.00% WG

Crop	Pests	Dose / ha		Waiting period (Days)
		Quantity of formulation (g)	Volume of water (Lit.)	
Rice	Brown plant hopper, White backed plant hopper, Green leaf hopper, Rice earhead bug	333	500	24

Trade Names

Sr. No.	Trade Name	Company
1	Axeman	Best Agrolife Ltd.
2	Polish	Era Agri Tech Co.
3	BPH Super +	Katyayani Organics
4	Dolby	Today's Global Crop Protection
5	Skystar	Indofill Industries Ltd.
6	Dixon	Gujrat Pesticides
7	Star	Global Crop Care
8	Enconter	Silver Sine Bio Tech
9	Distrozin	Natural Crop Science
10	Warrior	Tropical Agrosystem India (P) Ltd.
11	Aura Plus	Essential Biosciences
12	Dolby	Today's Global Crop Protection
13	Distruptor	PI Industries
14	Tagawa	Green India Care
15	Dinopy	R. K. Chemicals
16	Dynemo	Torrent Crop Science
17	Turban	JU Agri Sciences Pvt. Ltd.
18	Raynox	Synovita Crop science Pvt. Ltd.
19	Pymeneed	Mankind Agritech
20	Magnus	Preventis Chem India (P) Ltd.
21	Expostar	Chemley Agritech Pvt. Ltd.

Pymetrozine 29.20% + Dinotefuran 11.70% WDG

Crop	Pests	Dose / ha		Waiting period (Days)
		Quantity of formulation (g)	Volume of water (Lit.)	
Rice	Brown plant hopper	500	500	22

Trade Names

Sr. No.	Trade Name	Company
1	Bazak	Adama India Private Limited

Dinotefuran + Thiamethoxam + Pymetrozine

Available formulations in India : **Dinotefuran 12.00% + Thiamethoxam 22.50% + Pymetrozine 30.00% WG**

Classification on the basis of		**Dinotefuran**	**Thiamethoxam**	**Pymetrozine**
Target organism	:	Insecticide	Insecticide	Insecticide
Mode of entry	:	Systemic, Contact and Stomach poison	Systemic, Contact and Stomach poison	Systemic, Stomach poison
Mode of action	:	Nerve poison	Nerve poison	Nerve poison
Chemical group	:	Neonicotinoid	Neonicotinoid	Pyridine azomethine
Category of toxicity : **Toxicity label colour** :				

- This specific triple-combination—Dinotefuran 12% + Thiamethoxam 22.5% + Pymetrozine 30% w/w WG—is a highly potent, synergistic insecticide formulation. This formulation combines three distinct active ingredients to attack pests at multiple biological levels.
- Thiamethoxam provides broad-spectrum, long-lasting systemic protection.
- Pymetrozine paralyzes the pest's mouthparts instantly, causing starvation.
- Dinotefuran has rapid "knockdown" effect; kills pests within hours of contact or ingestion.

Label claim (DPPQ & S, Faridabad)

Dinotefuran 12.00% + Thiamethoxam 22.50% + Pymetrozine 30.00% WG

Crop	**Pests**	**Dose / ha**		**Waiting period (Days)**
		Quantity of formulation (g)	**Volume of water (Lit.)**	
Rice	Brown plant hopper, White backed plant hopper	350	500	18

Trade Names

Sr. No.	**Trade Name**	**Company**
1	Triston	JU Agri Sciences Pvt. Ltd.

Emamectin benzoate + Chlorantraniliprole + Deltamethrin

Available formulations in India : **Emamectin benzoate 01.90% + Chlorantraniliprole 05.00% + Deltamethrin 01.00% EC**

Classification on the basis of		Emamectin benzoate	Chlorantraniliprole	Deltamethrin
Target organism	:	Insecticide, Acaricide	Insecticide	Insecticide
Mode of entry	:	Systemic, Contact and Stomach poison	Systemic, Contact and Stomach poison	Contact and Stomach poison
Mode of action	:	Nerve poison	Muscle poison	Nerve poison
Chemical group	:	Avermectin (Macrocyclic lactones)	Anthranilic diamide	Synthetic pyrethroid
Category of toxicity : **Toxicity label colour :**				

- This emulsifiable concentrate (EC) combines semi-synthetic avermectin Emamectin benzoate for stomach and contact action, ryanodine receptor modulator Chlorantraniliprole for muscle paralysis in lepidopterans, and pyrethroid Deltamethrin for rapid knockdown.
- The synergy provides quick knockdown, broad-spectrum control, and resistance management, rainfast within hours, safe for beneficials when used as directed.

Label claim (DPPQ & S, Faridabad)
Emamectin benzoate 01.90% + Chlorantraniliprole 05.00% + Deltamethrin 01.00% EC

Crop	Pests	Dose / ha		Waiting period (Days)
		Quantity of formulation (ml)	Volume of water (Lit.)	
Chilli	Thrips, Mites, Fruit borers (*Helicoverpa armigera*, *Spodoptera litura*)	375	500	28

Trade Names

Sr. No.	Trade Name	Company
1	Tag Hikima	Tropical Agriscience (India) Pvt. Ltd.

Emamectin benzoate + Diafenthiuron

Available formulations in India : **Emamectin benzoate 01.10% + Diafenthiuron 30.00% SC**

Classification on the basis of		Emamectin benzoate	Diafenthiuron
Target organism	:	Insecticide, Acaricide	Insecticide, Acaricide
Mode of entry	:	Systemic, Contact and Stomach poison	Contact and Stomach poison
Mode of action	:	Nerve poison	Energy metabolism
Chemical group	:	Avermectin (Macrocyclic lactones)	Thiourea
Category of toxicity : Highly toxic **Toxicity label colour** : Yellow			

- Emamectin Benzoate belongs to the avermectin group and is derived from *Streptomyces avermitilis*. It acts primarily as a neurotoxic agent by binding to glutamate-gated chloride channels and GABA (gamma-aminobutyric acid) receptors in insect nerve cells, causing hyperpolarization, disruption in nerve signal transmission, paralysis, and death. It shows stomach and translaminar activity, effective when pests ingest treated plant material or come into contact with treated surfaces.
- Diafenthiuron is classified as a thiourea insecticide and acaricide with a unique mode of inhibiting mitochondrial function. It is metabolized into a carbodiimide compound within the pest, which irreversibly binds to the glucose-6-phosphate translocase in the mitochondria, inhibiting ATP production and disrupting energy metabolism, leading to cell dysfunction and death. It shows activity by contact, ingestion (stomach action), and has some ovicidal effect (kills eggs as well as mobile stages).
- The combination expands insecticidal spectrum by combining neurotoxic action (Emamectin benzoate) with energetic metabolism inhibition (Diafenthiuron), synergy delays resistance buildup, controls overlapping pest populations (e.g., both caterpillars and sucking pests).
- This combination provides both quick knockdown and long-lasting residual control, making it effective for integrated pest management in multiple crops.

Label claim (DPPQ & S, Faridabad)
Emamectin benzoate 01.10% + Diafenthiuron 30.00% SC

Crop	Pests	Dose / ha		Waiting period (Days)
		Quantity of formulation (ml)	Volume of water (Lit.)	
Cotton	Thrips, Leafhopper, Whitefly, Aphid, Pink bollworm	1000	500	38
Brinjal	Thrips, Aphid, Whitefly, red spider mite, Shoot and fruit borer	1000	500	08

Trade Names

Sr. No.	Trade Name	Company
1	Trassid	Adama India Private Limited
2	ED-Max	Willwood Chemicals Ltd.

Emamectin benzoate + Fipronil

Available formulations in India : **Emamectin benzoate 01.50% + Fipronil 03.50% SC**

Classification on the basis of		Emamectin benzoate	Fipronil
Target organism	:	Insecticide, Acaricide	Insecticide
Mode of entry	:	Systemic, Contact and Stomach poison	Contact and Stomach poison
Mode of action	:	Nerve poison	Nerve poison
Chemical group	:	Avermectin (Macrocyclic lactones)	Phenylpyrazole
Category of toxicity : Moderately toxic **Toxicity label colour** : Blue			

- Emamectin Benzoate is a semi-synthetic avermectin-derived insecticide. It paralyzes and kills insects by binding to and opening glutamate-gated chloride channels in their nerve cells. This hyperpolarization is especially effective against Lepidopteran pests like caterpillars and borers
- Fipronil, a phenylpyrazole, works by blocking GABA-gated chloride channels. This interference with chloride ion regulation leads to the hyperexcitation, paralysis, and death of the pest. It offers broad-spectrum control, targeting soil insects, ants, termites, thrips, and beetles.
- The combination delivers systemic and contact action, ensuring rapid knockdown and extended protection. Its ovilarvicidal action kills larvae immediately post-hatching, providing broad-spectrum control against both sucking and chewing pests to maximize crop yields.

Label claim (DPPQ & S, Faridabad)
Emamectin benzoate 01.50% + Fipronil 03.50% SC

Crop	Pests	Dose / ha		Waiting period (Days)
		Quantity of formulation (ml)	Volume of water (Lit.)	
Chilli	Thrips, Fruit borer	500-750	500	03

Trade Names

Sr. No.	Trade Name	Company
1	Apex-50	Crystal Crop Protection Ltd.
2	Zordar	Anu Products Ltd.
3	Femasik	Sikko Industries Ltd.
4	Don	KP Agro Industries
5	Result	Somnath Demeter Agritron Ltd.
6	Kranti	Essential Biosciences

7	Leaf	Global Crop Care
8	Elaan	Green India Care
9	Sipahi	Katyayani Organics
10	Epee	Bharat Agro Chemical and Fertilizer
11	Licence	Advance Pesticides
12	Exylon Elance	Torrent Crop Science
13	Full Power	Kisan Pesticides
14	Empala	ATPL

Emamectin benzoate + Lambda-cyhalothrin

Available formulations in India : **Emamectin benzoate 03.50% + Lambda-cyhalothrin 05.00% WP**

Classification on the basis of		Emamectin benzoate	Lambda-cyhalothrin
Target organism	:	Insecticide, Acaricide	Insecticide
Mode of entry	:	Systemic, Contact and Stomach poison	Contact and Stomach poison
Mode of action	:	Nerve poison	Nerve poison
Chemical group	:	Avermectin (Macrocyclic lactones)	Synthetic pyrethroid
Category of toxicity : **Toxicity label colour :**			

- Emamectin Benzoate is derived from avermectins, acts primarily as a nerve poison by binding to glutamate-gated chloride channels within the nerve cells of insects, causing hyperpolarization, paralysis, and death. It is effective against Lepidopteran larvae and certain chewing pests.
- Lambda-cyhalothrin is a highly active pyrethroid insecticide that works by disrupting the function of voltage-gated sodium channels in nerve membranes, resulting in paralysis and death of the pest. It has both contact and stomach action, with repellent properties for rapid knockdown and long residual activity. It controls a wide spectrum of pests, including lepidopteran and some sucking pests.
- The combination offers quick knockdown and extended residual activity for both chewing (caterpillars) and some sucking pests. It provides rapid pest suppression with long-lasting protection.

Label claim (DPPQ & S, Faridabad)
Emamectin benzoate 03.50% + Lambda-cyhalothrin 05.00% WP

Crop	Pests	Dose / ha		Waiting period (Days)
		Quantity of formulation (g)	Volume of water (Lit.)	
Chilli	Fruit borer, Thrips, Mites	250	500-700	07

Trade Names

Sr. No.	Trade Name	Company
1	Krill	Krishi Rasayan Exports Pvt. Ltd.

Emamectin benzoate + Lambda-cyhalothrin + Bifenthrin

Available formulations in India : **Emamectin benzoate 01.25% + Lambda-cyhalothrin 01.50% + Bifenthrin 04.50% EC**

Classification on the basis of		Emamectin benzoate	Lambda-cyhalothrin	Bifenthrin
Target organism	:	Insecticide, Acaricide	Insecticide	Insecticide
Mode of entry	:	Systemic, Contact and Stomach poison	Contact and Stomach poison	Contact and Stomach poison
Mode of action	:	Nerve poison	Nerve poison	Nerve poison
Chemical group	:	Avermectin (Macrocyclic lactones)	Synthetic pyrethroid	Synthetic pyrethroid
Category of toxicity :				
Toxicity label colour :				

- Emamectin benzoate activates Glutamate-Gated Chloride Channels (GluCl). It disrupts nerve signal transmission, causing flaccid paralysis and cessation of feeding.
- Lambda-cyhalothrin modulates Voltage-Gated Sodium Channels. It disrupts the insect's nervous system by causing overexcitement (repeated firing of nerves).
- Bifenthrin modulates Voltage-Gated Sodium Channels and similar to Lambda-cyhalothrin, causing nerve overexcitement, spasms, and paralysis.
- This triple combination is a broad-spectrum insecticide primarily aimed at chewing pests, with the dual Pyrethroids also offering control over some sucking pests and mites.
- The pyrethroids (Lambda-cyhalothrin and Bifenthrin) provide fast knockdown, while Emamectin Benzoate offers a longer residual effect.

Label claim (DPPQ & S, Faridabad)
Emamectin benzoate 01.25% + Lambda-cyhalothrin 01.50% + Bifenthrin 04.50% EC

Crop	Pests	Dose / ha		Waiting period (Days)
		Quantity of formulation (ml)	Volume of water (Lit.)	
Rice	Stem borer, Leaf folder	750	500	17

Trade Names

Sr. No.	Trade Name	Company
1	Akaya	JU Agri Sciences Pvt. Ltd.

Emamectin benzoate + Lufenuron

Available formulations in India	:	1. **Emamectin benzoate 05.00% + Lufenuron 40.00% WG** 2. **Lufenuron 04.00% + Emamectin benzoate 01.50% EC**

Classification on the basis of		**Emamectin benzoate**	**Lufenuron**
Target organism	:	Insecticide, Acaricide	Insecticide
Mode of entry	:	Systemic, Contact and Stomach poison	Contact and Stomach poison
Mode of action	:	Nerve poison	Insect Growth Regulator (IGR)
Chemical group	:	Avermectin (Macrocyclic lactones)	Benzoylurea
Category of toxicity : Moderately toxic **Toxicity label colour** : Blue			

- Emamectin benzoate plus lufenuron is a modern insecticidal combination that delivers dual action for controlling lepidopteran and other resistant pests. Their combined effect targets both the nervous system (paralysis and rapid feeding cessation) and the molting process (disrupting chitin synthesis), making them highly effective in crop protection.
- Emamectin benzoate is a semi-synthetic compound derived from avermectin, used extensively against caterpillars, thrips, and other chewing insects in a variety of crops. It acts by binding to GABA and glutamate-gated chloride channels in the insect nervous system, causing increased chloride ion influx, leading to paralysis and eventual death. The primary mode is stomach and some contact action; targeted pests usually stop feeding within hours and die within a day or two. Emamectin benzoate exhibits translaminar movement in plants, offering extended residual action.
- Lufenuron is a benzoylurea class insect growth regulator (IGR) widely used for lepidopteran and coleopteran pests. Its key mode of action is inhibiting chitin synthesis, a critical component of the insect exoskeleton. Without chitin, larvae cannot successfully molt, leading to death. Lufenuron targets immature stages (larvae and eggs), directly affecting larvae on foliage and also preventing egg hatch when applied ovicidally. While it does not kill adults, it reduces their reproductive capacity and lowers subsequent pest populations.
- This dual action provides both immediate knockdown (from Emamectin Benzoate) and extended control by breaking the pest's life cycle (from Lufenuron). The combination is highly valued in resistance management programs because the two chemicals target different physiological pathways, thus reducing the likelihood of cross-resistance development in target pests.

Label claim (DPPQ & S, Faridabad)

Emamectin benzoate 05.00% + Lufenuron 40.00% WG

Crop	Pests	Dose / ha		Waiting period (Days)
		Quantity of formulation (g)	Volume of water (Lit.)	
Cauliflower	Diamond back moth	60	500	03
Chilli	Fruit borer, Tobacco caterpillar, Thrips, Yellow mite	60	500	03
Maize	Fall armyworm, Stem borer	80	500	36
Tea	Looper, Thrips, Leaf roller	60	450	07
Tomato	Fruit borer, Thrips	60	500	05
Brinjal	Shoot and fruit borer	60	500	05

Trade Names

Sr. No.	Trade Name	Company
1	Evicent	Syngenta India Pvt. Ltd.
2	Cent	Tropical Agrosystem India (P) Ltd.

Lufenuron 04.00% + Emamectin benzoate 01.50% EC

Crop	Pests	Dose / ha		Waiting period (Days)
		Quantity of formulation (ml)	Volume of water (Lit.)	
Maize	Fall armyworm	500	500	27

Trade Names

Sr. No.	Trade Name	Company
1		

Emamectin benzoate + Permethrin

Available formulations in India : **Emamectin benzoate 02.20% + Permethrin 15.30% EC**

Classification on the basis of		Emamectin benzoate	Permethrin
Target organism	:	Insecticide, Acaricide	Insecticide, Acaricide
Mode of entry	:	Systemic, Contact and Stomach poison	Contact and Stomach poison
Mode of action	:	Nerve poison	Nerve poison
Chemical group	:	Avermectin (Macrocyclic lactones)	Synthetic pyrethroid
Category of toxicity : Highly toxic **Toxicity label colour** : Yellow			

- Emamectin benzoate combined with permethrin is a dual-action insecticidal formulation designed primarily for robust management of crop pests. This combination harnesses both systemic and contact activity, offering a quick knockdown of pests as well as extended residual protection.
- Emamectin Benzoate, a macrocyclic lactone insecticide derived from avermectin, it mainly acts by causing paralysis in pests through stimulation of GABA and glutamate-gated chloride channels in their nervous system. It primarily works as a stomach poison and also exhibits some contact activity, resulting in rapid feeding cessation and eventual death of the insect.
- Permethrin is a synthetic pyrethroid that disrupts sodium ion channels in nerve membranes, leading to knockdown, paralysis, and speedy death of pests. Permethrin kills through both contact and stomach action, and also has a mild fumigant effect. It is highly stable and leaves minimal residue when used as directed.
- When combined, there is a notable synergistic action: increased insecticidal activity, a broader spectrum of action (targets eggs, larvae, adults), and improved resistance management compared to single agents. The formulation is effective against eggs, larvae, and adults, thereby giving sustainable crop protection

Label claim (DPPQ & S, Faridabad)
Emamectin benzoate 02.20% + Permethrin 15.30% EC

Crop	Pests	Dose / ha		Waiting period (Days)
		Quantity of formulation (ml)	Volume of water (Lit.)	
Maize	Fall armyworm	750	500	--

Trade Names

Sr. No.	Trade Name	Company
1	Marvex	Coromandel International Limited

Emamectin benzoate + Profenofos

Available formulations in India : **Emamectin benzoate 01.50% + Profenofos 35.00% WDG**

Classification on the basis of		Emamectin benzoate	Profenofos
Target organism	:	Insecticide, Acaricide	Insecticide, Acaricide
Mode of entry	:	Systemic, Contact and Stomach poison	Contact and Stomach poison
Mode of action	:	Nerve poison	Nerve poison
Chemical group	:	Avermectin (Macrocyclic lactones)	Organophosphate
Category of toxicity : Moderately toxic **Toxicity label colour** : Blue			

- Emamectin Benzoate disrupts insect nervous system by increasing chloride ion influx via GABA/glutamate pathways, causing paralysis, rapid feeding cessation, and death. Acts mostly by ingestion (stomach action) and also penetrates leaf tissue (translaminar activity).
- Profenofos is an organophosphate insecticide that inhibits acetylcholinesterase, blocking nerve impulse transmission, leading to paralysis and death. Acts both by contact and ingestion, and is ovicidal (kills eggs), providing translaminar effects and residual control.
- The combination provides both contact and systemic action for broader pest control, including eggs and larvae (ovi-larvicidal). Pests stop feeding quickly after application, with visible results within 24 hours.

Label claim (DPPQ & S, Faridabad)
Emamectin benzoate 01.50% + Profenofos 35.00% WDG

Crop	Pests	Dose / ha		Waiting period (Days)
		Quantity of formulation (ml)	Volume of water (Lit.)	
Cotton	Thrips, Leafhopper, Whitefly, Aphid, Pink bollworm	700	500	15
Chilli	Thrips, Fruit borer, Yellow mite	700	500	07
Maize	Fall armyworm	750	500	35

Trade Names

Sr. No.	Trade Name	Company
1	Velecktin	Parijat Industries India Pvt. Ltd.
2	Apex 50	Crystal Crop Protection Ltd.
3	Korko	Sumitomo Chemical India Ltd.
4	Ant	Global Crop Care
5	Miraya	Tata Rallies India Limited
6	Emaneed Prophos	Mankind Agritech

Emamectin benzoate + Thiamethoxam

Available formulations in India :

1. **Emamectin benzoate 03.80% + Thiamethoxam 20.00% WDG**
2. **Emamectin benzoate 03.00% + Thiamethoxam 12.00% WG**

Classification on the basis of		Emamectin benzoate	Thiamethoxam
Target organism	:	Insecticide, Acaricide	Insecticide
Mode of entry	:	Systemic, Contact and Stomach poison	Systemic, Contact and Stomach poison
Mode of action	:	Nerve poison	Nerve poison
Chemical group	:	Avermectin (Macrocyclic lactones)	Neonicotinoid
Category of toxicity : Moderately toxic **Toxicity label colour** : Blue			

- Emamectin benzoate blocks neurotransmission via GABA/glutamate-gated chloride channels in pest nervous and muscular systems. This leads to paralysis, rapid feeding cessation, and ultimately death. It acts mostly by stomach (ingestion) and contact, with quick translaminar movement within leaves.
- Thiamethoxam is a systemic neonicotinoid that binds to nicotinic acetylcholine receptors in the pest CNS, causing nerve overstimulation, paralysis, and death. It has strong contact and systemic action control sucking pests with fast mobility and spread throughout plant tissues.
- The combination of two different modes of action enhances efficacy, provides a quick knockdown effect, and helps in managing or delaying insect resistance development.

Label claim (DPPQ & S, Faridabad)

Emamectin benzoate 03.80% + Thiamethoxam 20.00% WDG

Crop	Pests	Dose / ha		Waiting period (Days)
		Quantity of formulation (g)	Volume of water (Lit.)	
Okra	Jassid, Whitefly, Aphid, Shoot and fruit borer	100	500-700	05

Trade Names

Sr. No.	Trade Name	Company
1	Kougar	Krishi Rasayan

Emamectin benzoate 03.00% + Thiamethoxam 12.00% WG

Crop	Pests	Dose / ha		Waiting period (Days)
		Quantity of formulation (g)	Volume of water (Lit.)	
Chilli	Thrips, Fruit borer	200	500	03
Green gram	Pod borer, Leafhopper	200	500	15
Tea	Tea mosquito bug, Tea looper	150-200	400-500	07

Trade Names

Sr. No.	Trade Name	Company
1	Jury	Anu Products Ltd.
2	Encounter	Insecticides (India) Ltd.
3	Benzomax	Sikko Industries Ltd.
4	Pro Thaio	Natural Crop Science
5	Emathio	Katyayani Organics
6	Emaan Plus	Essential Biosciences
7	Counter	Global Crop Care
8	Emathia Max	Bhumi Growth Solution
9	Emethron	Linuxcrop Chemicals Pvt. Ltd.
10	Emoxam	ATPL

Ethion + Cypermethrin

Available formulations in India : **Ethion 40.00% + Cypermethrin 05.00% EC**

Classification on the basis of		Ethion	Cypermethrin
Target organism	:	Insecticide, Acaricide	Insecticide
Mode of entry	:	Contact and Stomach poison	Contact and Stomach poison
Mode of action	:	Nerve poison	Nerve poison
Chemical group	:	Organophosphate	Synthetic pyrethroid
Category of toxicity : Highly toxic **Toxicity label colour** : Yellow			

- Ethion acts as a cholinesterase inhibitor (neurotoxic action). It disrupts the nervous system by inhibiting the enzyme acetylcholinesterase (AChE), leading to an accumulation of acetylcholine and causing paralysis and death in the insects. It is primarily a non-systemic acaricide/insecticide with contact action.
- Cypermethrin acts as a sodium channel modulator (neurotoxic action). It interferes with the transmission of nerve signals by binding to sodium channels in the nerve membranes, causing continuous nerve impulses, rapid knockdown, and eventually death. It has non-systemic contact and stomach action, and provides a residual effect.
- The dual mode of action helps to provide a broader spectrum of control.

Label claim (DPPQ & S, Faridabad)
Ethion 40.00% + Cypermethrin 05.00% EC

Crop	Pests	Dose / ha		Waiting period (Days)
		Quantity of formulation (ml)	Volume of water (Lit.)	
Cotton	American bollworm	1000	500	15
Chilli	Mites, Thrips	1500	500	10

Trade Names

Sr. No.	Trade Name	Company
1	Colfos	PI Industries
2	Nagata	Tata Rallis India Limited
3	Ananda	Anu Products Ltd.
4	Nasa	Crop Chemicals India Ltd.
5	Colonel	Tusova Agro Pvt. Ltd.
6	Heron Plus	Herenba Industries Ltd.
7	Samrat King	Gujarat Crop Care
8	Bifors	Nichino India
9	Ethion Super	Synergy
10	E-Cyper	Silver Sine Bio Tech

11	Ethion Gold	Natural Crop Science
12	Sonata	Tropical Agro
13	Ethon Cyper	Essential Biosciences
14	MIT-405	Synwood Agro Ltd.
15	Mig 405	Koronation Bioscience
16	Sabu	Green India Care
17	Exylon Eclipse	Torrent Crop Science
18	Regal	Preventis Chem India (P) Ltd.
19	Sifang	RACCOLTO Agritech India Pvt. Ltd.
20	Ethiocyper 45	Katyayani Organics
21	Zapat-X	Agrostar

Ethiprole + Imidacloprid

Available formulations in India : **Ethiprole 40.00% + Imidacloprid 40.00% WG**

Classification on the basis of		Ethiprole	Imidacloprid
Target organism	:	Insecticide	Insecticide
Mode of entry	:	Contact and Stomach poison	Systemic, Contact and Stomach poison
Mode of action	:	Nerve poison	Nerve poison
Chemical group	:	Phenylpyrazole	Neonicotinoid
Category of toxicity : **Toxicity label colour :**			

- Ethiprole is a phenylpyrazole insecticide that acts on gamma-aminobutyric acid (GABA) regulated chloride channels in the nervous system of insects, causing disruption of central nervous system activity. This leads to paralysis and death, providing high selective toxicity and lack of cross-resistance with other hopper insecticides, making it instrumental for resistance management.
- Imidacloprid is a neonicotinoid insecticide that acts as an antagonist of nicotinic acetylcholine receptors in the insect central nervous system. This blocks signal transmission by causing sustained excitation and eventual paralysis of nerve cells, quickly disrupting insect feeding and leading to death.
- This combination gives proactive and broad-spectrum control of hoppers, including all pest life stages.

Label claim (DPPQ & S, Faridabad)

Ethiprole 40.00% + Imidacloprid 40.00% WG

Crop	Pests	Dose / ha		Waiting period (Days)
		Quantity of formulation (g)	Volume of water (Lit.)	
Rice	Brown plant hopper,	93.75	375	15
	White backed plant hopper	125	375	15

Trade Names

Sr. No.	Trade Name	Company
1	Glamore	Bayer Crop Science Ltd.

Ethiprole + Pymetrozine

Available formulations in India : **Ethiprole 10.70% + Pymetrozine 40.00% WG**

Classification on the basis of		Ethiprole	Pymetrozine
Target organism	:	Insecticide	Insecticide
Mode of entry	:	Contact and Stomach poison	Systemic, Stomach poison
Mode of action	:	Nerve poison	Nerve poison
Chemical group	:	Phenylpyrazole	Pyridine azomethine
Category of toxicity : Moderately toxic **Toxicity label colour** : Blue			

- Ethiprole acts by inhibiting gamma-aminobutyric acid (GABA)-gated chloride channels in the insect nervous system, causing uncontrolled nerve impulses, paralysis, and eventual death. This neurotoxic action provides fast knockdown of pests and effective resistance management.
- Pymetrozine disrupts the feeding mechanism of sucking pests. By interfering with the chordotonal organs and other internal sensory mechanisms, it stops pests from feeding almost immediately after contact or ingestion. Affected insects stop feeding, become inactive, and die from starvation, offering long-lasting protection.
- The formulated product is described as having dual action i.e. contact and stomach action and systemic / translaminar action.
- Ethiprole + Pymetrozine offers both rapid knockdown and sustained suppression of rice hoppers.

Label claim (DPPQ & S, Faridabad)
Ethiprole 10.70% + Pymetrozine 40.00% WG

Crop	Pests	Dose / ha		Waiting period (Days)
		Quantity of formulation (g)	Volume of water (Lit.)	
Rice	Brown plant hopper, White backed plant hopper	375-425	375	27

Trade Names

Sr. No.	Trade Name	Company
1	Curbix Pro	Bayer Crop Science Ltd.
2	Combi	AMICO Pesticides Ltd.
3	Duranga	Saffire

Fenazaquin + Bifenthrin

Available formulations in India : **Fenazaquin 10.00% + Bifenthrin 04.00% EC**

Classification on the basis of		Fenazaquin	Bifenthrin
Target organism	:	Insecticide, Acaricide	Insecticide
Mode of entry	:	Contact and Stomach poison	Contact and Stomach poison
Mode of action	:	Mitochondrial complex I electron transport inhibitor	Nerve poison
Chemical group	:	Quinazoline	Synthetic pyrethroid
Category of toxicity : Highly toxic **Toxicity label colour** : Yellow			

- Fenazaquin is a non-systemic acaricide and insecticide that acts by inhibiting mitochondrial electron transport in mites, specifically by blocking the function of complex I within the mitochondria. This disruption leads to reduced energy production, immobilization, and death of the mite. Fenazaquin has both contact and stomach action, highly effective for quick knockdown and long-lasting residual control.
- Bifenthrin is a synthetic pyrethroid with contact and stomach poison activity. It acts on the nervous system by modifying sodium channel gating, causing repeated and prolonged nerve impulses, rapid paralysis, and death in insects and mites. Bifenthrin provides rapid adulticidal effects and broad-spectrum activity.
- Synergistic formulation offers both quick suppression of existing pest populations (adulticidal, knockdown) and residual control of developing stages.
- It controls both sucking pests and mites, providing wide crop protection.

Label claim (DPPQ & S, Faridabad)
Fenazaquin 10.00% + Bifenthrin 04.00% EC

Crop	Pests	Dose / ha		Waiting period (Days)
		Quantity of formulation (ml)	Volume of water (Lit.)	
Tomato	Red spider mite, Whitefly	1250	500	05
Tea	Ted spider mite, Mosquito bug	1000	400	07
Cotton	Whitefly, Mite	1250	500	74

Trade Names

Sr. No.	Trade Name	Company
1	Avanger	Global Crop Care
2	SA-Magnitude	Saraswati Agro Life Science (India) Pvt. Ltd.

Fenobucarb + Buprofezin

Available formulations in India : Fenobucarb 20.00% + Buprofezin 05.00% SE

Classification on the basis of		Fenobucarb	Buprofezin
Target organism	:	Insecticide	Insecticide
Mode of entry	:	Contact and Stomach poison	Contact and Stomach poison
Mode of action	:	Nerve poison	Insect Growth Regulator (IGR)
Chemical group	:	Carbamate	Thiadiazine
Category of toxicity : Moderately toxic **Toxicity label colour** : Blue			

- Fenobucarb (BPMC) acts as a reversible inhibitor of acetylcholinesterase (AChE) in the nervous system, causing accumulation of acetylcholine, disrupting nerve signal transmission, leading to paralysis and quick knockdown of adult insects. It is broad-spectrum with contact and stomach activity, primarily adulticidal action.
- Buprofezin inhibits chitin biosynthesis in immature stages, blocking molting and development of nymphs/larvae. It has ovicidal and nymphicidal actions, disrupting the life cycle and preventing hoppers from reaching adulthood.
- This dual action ensures the control of both the adult pests (quick kill by Fenobucarb) and the development of the next generation (long-term control by Buprofezin).

Label claim (DPPQ & S, Faridabad)

Fenobucarb 20.00% + Buprofezin 05.00% SE

Crop	Pests	Dose / ha		Waiting period (Days)
		Quantity of formulation (ml)	Volume of water (Lit.)	
Rice	Brown plant hopper, Green leaf hopper	2000	500	30

Trade Names

Sr. No.	Trade Name	Company
1	Snaper	Crystal Crop Protection Ltd.
2	Zorba	Anu Products Ltd.
3	Silective	Synergy
4	Iffective	Silver Sine Bio Tech
5	Super Knock	Krishi Rasayan
6	Curat	RACCOLTO Agritech India Pvt. Ltd.

Fenobucarb + Buprofezin + Acephate

Available formulations in India : **Fenobucarb 22.50% + Buprofezin11.25% + Acephate 02.50% ME**

Classification on the basis of		Fenobucarb	Buprofezin	Acephate
Target organism	:	Insecticide	Insecticide	Insecticide
Mode of entry	:	Contact and Stomach poison	Contact and Stomach poison	Systemic, Contact and Stomach poison
Mode of action	:	Nerve poison	Insect Growth Regulator (IGR)	Nerve poison
Chemical group	:	Carbamate	Thiadiazine	Organophosphate
Category of toxicity :				
Toxicity label colour :				

- Fenobucarb 22.5% + Buprofezin 11.25% + Acephate 2.5% ME is a novel, multi-action insecticide mixture formulated as a microemulsion (ME). This 3-way combination provides contact, systemic, and insect growth regulator activities for the control of a broad spectrum of sucking and chewing pests.
- Fenobucarb is a carbamate insecticide, known for quick contact action and rapid knockdown of pests. Its mode of action is generally through acetylcholinesterase (AChE) inhibition, which disrupts the insect's nervous system.
- Buprofezin is an Insect Growth Regulator (IGR). Its primary mode of action is to interfere with the molting process (chitin synthesis inhibitor) and can also suppress egg-laying, offering long-term control by targeting the nymphal stages.
- Acephate is an organophosphate insecticide, known for its systemic and contact/stomach action. As a systemic insecticide, it is absorbed by the plant and moves throughout the tissue, providing residual protection. Its mode of action is also through AChE inhibition.

Label claim (DPPQ & S, Faridabad)

Fenobucarb 22.50% + Buprofezin11.25% + Acephate 02.50% ME

Crop	Pests	Dose / ha		Waiting period (Days)
		Quantity of formulation (ml)	Volume of water (Lit.)	
Rice	Green leaf hopper, Brown plant hopper, White backed plant hopper	1750	500	25

Trade Names

Sr. No.	Trade Name	Company
1	Insta-X	Willwood Chemicals Pvt. Ltd.

Fipronil + Acetamiprid

Available formulations in India : **Fipronil 04.00% + Acetamiprid 04.00% SC**

Classification on the basis of		Fipronil	Acetamiprid
Target organism	:	Insecticide	Insecticide
Mode of entry	:	Contact and Stomach poison	Systemic, Contact and Stomach poison
Mode of action	:	Nerve poison	Nerve poison
Chemical group	:	Phenylpyrazole	Neonicotinoid
Category of toxicity : Highly toxic **Toxicity label colour** : Yellow			

- Fipronil belongs to the Phenylpyrazole chemical group. It is a broad-spectrum insecticide that disrupts the insect central nervous system by blocking GABA-gated chloride channels. This leads to hyperexcitation, paralysis, and death. It works via contact and ingestion.
- Acetamiprid belongs to the Neonicotinoid chemical group. It acts as a systemic insecticide that targets the insect central nervous system by binding to nicotinic acetylcholine receptors (nAChR). This interferes with nerve impulse transmission, causing overstimulation, paralysis, and death. It has systemic and translaminar activity (moves across the leaf surface).
- The combination provides a synergistic effect and multiple benefits. By attacking two different sites in the insect's nervous system, it provides a quick knockdown effect and sustained control. Acetamiprid provides systemic control (moving throughout the plant to kill hidden pests), while Fipronil offers quick contact and stomach action. The combination is often noted for controlling all stages of the targeted sucking pests (egg, nymph, and adult).

Label claim (DPPQ & S, Faridabad)

Fipronil 04.00% + Acetamiprid 04.00% SC

Crop	Pests	Dose / ha		Waiting period (Days)
		Quantity of formulation (ml)	Volume of water (Lit.)	
Cotton	Aphid, Jassid, Whitefly	1000	500	30

Trade Names

Sr. No.	Trade Name	Company
1	Supranil	Paraijat Industries (India) Ltd.
2	Metabu	Advance Pesticides
3	Fantasy Plus	Katyayani Organics
4	Typhon	HPM Chemicals and Fertilizers Ltd.
5	Mahaviz	Maharashtra Insecticide Ltd.

6	Rangeela	Gharda Chemicals Ltd.
7	Master	Shrinathji Enterprise
8	Dahak	Oswal Crop Protection Pvt. Ltd.
9	Sesh	Veeaar Industries
10	Parviz	ADAMA India Private Limited
11	Trillion	Anu Products Ltd.
12	Flyvosh	Crop Life Science Ltd.
13	Hello	Somnath Demeter Agritron Ltd.
14	Toofan	Global Crop Care
15	Acenilchem	Creative Unichem Pvt. Ltd.
16	Agenda	Preventis Chem India (P) Ltd.

Fipronil + Buprofezin

Available formulations in India : **Fipronil 05.00% + Buprofezin 20.00% SC**

Classification on the basis of		Fipronil	Buprofezin
Target organism	:	Insecticide	Insecticide
Mode of entry	:	Contact and Stomach poison	Contact and Stomach poison
Mode of action	:	Nerve poison	Insect Growth Regulator (IGR)
Chemical group	:	Phenylpyrazole	Thiadiazine
Category of toxicity : **Toxicity label colour** :			

- Fipronil acts by blocking GABA-gated chloride channels, causing hyperexcitation and paralysis in insect pests.
- Buprofezin works by inhibiting chitin synthesis, disrupting molting, and causing nymphal deformities and death.
- Combination achieves quick knockdown (fipronil) along with persistent suppression of immature stages (buprofezin).
- Fipronil offers contact, stomach, and moderate systemic action, while Buprofezin primarily works as a contact and stomach poison, providing comprehensive coverage.

Label claim (DPPQ & S, Faridabad)
Fipronil 05.00% + Buprofezin 20.00% SC

Crop	Pests	Dose / ha		Waiting period (Days)
		Quantity of formulation (ml)	Volume of water (Lit.)	
Chilli	Thrips	750	500	05
	Fruit borer	1000	500	05
Cotton	Jassid, Thrips, Aphid, Whitefly	1000	500	06
Rice	Brown plant hopper	500	500	20
	Yellow stem borer, Leaf folder	1000	500	20

Trade Names

Sr. No.	Trade Name	Company
1		

Fipronil + Chlorantraniliprole

Available formulations in India : **Fipronil 15.00% + Chlorantraniliprole 05.00% SC**

Classification on the basis of		Fipronil	Chlorantraniliprole
Target organism	:	Insecticide	Insecticide
Mode of entry	:	Contact and Stomach poison	Systemic, Contact and Stomach poison
Mode of action	:	Nerve poison	Muscle poison
Chemical group	:	Phenylpyrazole	Anthranilic diamide
Category of toxicity : Highly toxic **Toxicity label colour** : Yellow			

- The combination is effective due to the two insecticides targeting different sites in the insect's body, which provides a dual mode of action.
- Fipronil is a phenylpyrazole insecticide. It is a GABA-gated chloride channel blocker. It disrupts the central nervous system of insects, leading to hyperexcitation, paralysis, and ultimately death. Primarily it is an ingestion toxicant with contact and moderate systemic action.
- Chlorantraniliprole is a diamide insecticide. It is a ryanodine receptor activator. It causes the uncontrolled release of calcium ions from the muscles, leading to muscle paralysis, rapid feeding cessation, and death of the pest. Primarily it acts via ingestion, with secondary contact action and systemic movement in the plant.
- The dual mode of action offers several advantages such as broad-spectrum control, resistance management, systemic and contact action, quick knockdown and long-lasting control.

Label claim (DPPQ & S, Faridabad)
Fipronil 15.00% + Chlorantraniliprole 05.00% SC

Crop	Pests	Dose / ha		Waiting period (Days)
		Quantity of formulation (ml)	Volume of water (Lit.)	
Rice	Stem borer, Leaf folder	500	500	49

Trade Names

Sr. No.	Trade Name	Company
1	XFactor	Gharada Chemicals Ltd.
2	Centran SC	Insecticide India Ltd.
3	Escort	Synovita Crop science Pvt. Ltd.

Fipronil + Deltamethrin

Available formulations in India : **Fipronil 15.00% + Deltamethrin 02.50% SC**

Classification on the basis of		Fipronil	Deltamethrin
Target organism	:	Insecticide	Insecticide
Mode of entry	:	Contact and Stomach poison	Contact and Stomach poison
Mode of action	:	Nerve poison	Nerve poison
Chemical group	:	Phenylpyrazole	Synthetic pyrethroid
Category of toxicity : Highly toxic **Toxicity label colour** : Yellow			

- Fipronil is a GABA-gated chloride channel blocker. It disrupts the insect's central nervous system by blocking the entry of chloride ions, which leads to hyperexcitation, paralysis, and death. It is generally more toxic to insects due to its higher binding affinity for insect nerve receptors than for mammalian ones. It works via both contact and stomach action, and often has some systemic activity (absorbed by the plant).
- Deltamethrin is a sodium channel modulator. It acts on nerve cell membranes by delaying the closing of the sodium ion channels, leading to prolonged nerve impulses, hyperexcitation, tremors, and paralysis. Primarily it works via contact and stomach action, often providing a quick "knockdown" effect.
- The combination delivers fast knockdown and longer residual control due to these complementary modes of action.

Label claim (DPPQ & S, Faridabad)
Fipronil 15.00% + Deltamethrin 02.50% SC

Crop	Pests	Dose / ha		Waiting period (Days)
		Quantity of formulation (ml)	Volume of water (Lit.)	
Tomato	Fruit borer	500	500	07

Trade Names

Sr. No.	Trade Name	Company
1	Etherium	Synovita Crop Science Pvt. Ltd.
2	Vajra	Gharda Chemicals Ltd.

Fipronil + Diafenthiuron

Available formulations in India : **Fipronil 10.00% + Diafenthiuron 30.00% WG**

Classification on the basis of		Fipronil	Diafenthiuron
Target organism	:	Insecticide	Insecticide, Acaricide
Mode of entry	:	Contact and Stomach poison	Contact and Stomach poison
Mode of action	:	Nerve poison	Energy metabolism
Chemical group	:	Phenylpyrazole	Thiourea
Category of toxicity : Highly toxic **Toxicity label colour** : Yellow			

- Fipronil primarily acts as a GABA-gated chloride channel blocker in the central nervous system of insects. This disrupts the normal function of the central nervous system, leading to hyper-excitation, paralysis, and eventual death. It has both contact and stomach action, and often exhibits systemic and translaminar (moves from the top surface of the leaf to the bottom) properties.
- Diafenthiuron is a pro-insecticide, meaning it is converted into its active form (Diafenthiuron carbodiimide) after entering the pest. The active metabolite acts as a mitochondrial electron transport inhibitor (specifically, it inhibits ATP synthesis by binding to mitochondrial ATPase/porin). This disrupts the pest's energy production, leading to paralysis and death. It exhibits contact, stomach, translaminar, and vapor action. Mortality is often delayed (hidden knock-down effect) for about 3-4 days, but pests stop feeding immediately upon intake.
- Fipronil offers quick initial control, while Diafenthiuron provides a powerful, prolonged effect with its contact, translaminar, and vapor actions.

Label claim (DPPQ & S, Faridabad)
Fipronil 10.00% + Diafenthiuron 30.00% WG

Crop	Pests	Dose / ha		Waiting period (Days)
		Quantity of formulation (g)	Volume of water (Lit.)	
Cotton	Jassid, Thrips, Pink bollworm	750	500	21

Trade Names

Sr. No.	Trade Name	Company
1	Guru	Gharada Chemicals Ltd.
2	Neuton	Synovita Crop Science Pvt. Ltd.
3	Dictator	Gujarat Insecticides Ltd.

Fipronil + Lambda-cyhalothrin

Available formulations in India : Fipronil 12.50% + Lambda-cyhalothrin 05.00% ZC

Classification on the basis of		Fipronil	Lambda-cyhalothrin
Target organism	:	Insecticide	Insecticide
Mode of entry	:	Contact and Stomach poison	Contact and Stomach poison
Mode of action	:	Nerve poison	Nerve poison
Chemical group	:	Phenylpyrazole	Synthetic pyrethroid
Category of toxicity : **Toxicity label colour :**			

- The combination is designed to leverage the distinct modes of action of the two active ingredients for enhanced efficacy, quicker knockdown, and potential management of insecticide resistance.
- Fipronil is a GABA-gated chloride channel blocker (a neurotoxin). It disrupts the central nervous system of insects by blocking the passage of chloride ions into the nerve cells, leading to uncontrolled neural activity, hyperexcitation, paralysis, and death. Primarily it works by contact and ingestion, and typically has a systemic and long-lasting residual effect, especially effective against soil-dwelling insects and certain foliar pests.
- Lambda-cyhalothrin is a sodium channel modulator (a neurotoxin). It disrupts the nervous system by blocking sodium channels in the nerve cells, which prevents the normal transmission of nerve impulses. This results in nerve hyperexcitation, paralysis, and a rapid knockdown effect. Primarily it works by contact and stomach action, and is known for its quick knockdown activity and repellent properties.
- The combination offers both rapid knockdown (via Lambda-cyhalothrin) and prolonged residual activity (from Fipronil), ensuring broad-spectrum, long-lasting pest control.

Label claim (DPPQ & S, Faridabad)
Fipronil 12.50% + Lambda-cyhalothrin 05.00% ZC

Crop	Pests	Dose / ha		Waiting period (Days)
		Quantity of formulation (ml)	Volume of water (Lit.)	
Cotton	Thrips, Leafhopper, Aphid, Pink bollworm	400	500	35
Chilli	Thrips, Aphid, Fruit borer	400	500	07
Onion	Thrips	400	500	07
Tomato	Thrips, Whitefly, Fruit borer	400	500	07

Trade Names

Sr. No.	Trade Name	Company
1		

Fipronil + Propargite

Available formulations in India : **Fipronil 02.50% + Propargite 35.00% SE**

Classification on the basis of		Fipronil	Propargite
Target organism	:	Insecticide	Acaricide
Mode of entry	:	Contact and Stomach poison	Contact poison
Mode of action	:	Nerve poison	Affect energy metabolism
Chemical group	:	Phenylpyrazole	Sulfite ester
Category of toxicity : **Toxicity label colour** :			

- Fipronil belongs to the phenylpyrazole class. It acts on the insect nervous system by blocking GABA-gated and glutamate-gated chloride channels, leading to hyperexcitation, paralysis, and death. It is effective against a broad spectrum of soil and foliar pests.
- Propargite is a non-systemic contact acaricide (from the sulfite ester group) with and vapor action. It inhibits mitochondrial oxidative phosphorylation (ATP synthesis), causing energy loss and desiccation in mites. It is selective for mite species such as red spider mite, two-spotted spider mite, and scarlet mite.
- This combination provides a dual-action solution that controls both damaging insect pests and mites simultaneously in a single application.
- The combination of two different modes of action (nervous system disruption by Fipronil and energy production inhibition by Propargite) is valuable for managing pest resistance that can develop when using a single class of pesticide repeatedly.

Label claim (DPPQ & S, Faridabad)
Fipronil 02.50% + Propargite 35.00% SE

Crop	Pests	Dose / ha		Waiting period (Days)
		Quantity of formulation (ml)	Volume of water (Lit.)	
Chilli	Thrips, Aphid, Mites, Fruit borer	2000	500	20

Trade Names

Sr. No.	Trade Name	Company
1	Fiprogite	Mankind Agritech
2	Ju Tako	

Fipronil + Imidacloprid

Available formulations in India :

1. **Fipronil 15.00% + Imidacloprid 05.00% SC**
2. **Fipronil 40.00% + Imidacloprid 40.00% WG**

Classification on the basis of		**Fipronil**	**Imidacloprid**
Target organism	:	Insecticide	Insecticide
Mode of entry	:	Contact and Stomach poison	Systemic, Contact and Stomach poison
Mode of action	:	Nerve poison	Nerve poison
Chemical group	:	Phenylpyrazole	Neonicotinoid
Category of toxicity : Highly toxic **Toxicity label colour** : Yellow			

- The combination Fipronil + Imidacloprid is a broad-spectrum, dual-action insecticide designed to control both chewing and sucking pests.
- Fipronil (Phenylpyrazole insecticide) acts on GABA-regulated chloride channels, blocking neural transmission in insects. It causes hyperexcitation, paralysis, and death of target pests. It provides both contact and ingestion (stomach) action and long residual effect.
- Imidacloprid (Neonicotinoid insecticide) targets nicotinic acetylcholine receptors (nAChRs) in the insect nervous system, disrupting signal transduction. It works primarily through systemic and contact activity, leading to paralysis and death of pests.
- The combination provides contact/stomach action (Fipronil) and systemic action (Imidacloprid).

Label claim (DPPQ & S, Faridabad)

Fipronil 15.00% + Imidacloprid 05.00% SC

Crop	**Pests**	**Dose / ha**		**Waiting period (Days)**
		Quantity of formulation (ml)	**Volume of water (Lit.)**	
Cotton	Aphid, Jassid, Thrips, Pink bollworm	500	500	06

Trade Names

Sr. No.	**Trade Name**	**Company**
1	Bakeel	Gharda Chemicals Ltd.

Fipronil 40.00% + Imidacloprid 40.00% WG

Crop	**Pests**	**Dose / ha**		**Waiting period (Days)**
		Quantity of formulation (g)	**Volume of water (Lit.)**	
Sugarcane	White grub	437.5-500	1000-1250	296

Groundnut	White grub	250-300	1000	106
Cotton	Jassid, Thrips	125	500	36
Rice	Yellow stem borer, Leaf folder, Brown plant hopper	100-125	500	07
Chilli	Thrips, Fruit borer	75-100	500	07

Trade Names

Sr. No.	Trade Name	Company
1	Lesenta	Bayer Crop Science Ltd.
2	Shirasagi	IFFCO-MC Crop Science Private LTd.
3	Police	Gharda Chemicals Ltd.
4	Umpire	Crystal Crop Protection Ltd.
5	Panther	Auxin Crop Care
6	Akira	Anu Products Ltd.
7	imFi	Excel Crop Care Ltd.
8	Shamsher	Crop Life Science Ltd.
9	Fiptor 80	Parijat Industries India Pvt. Ltd.
10	Fida	Crop Chemicals India Ltd.
11	Enova	Sumitomo Chemical India Ltd.
12	Fida	Herenba Industries Ltd.
13	Congo	Sikko Industries Ltd.
14	Yuvan	Nichino India
15	Fimida	Silver Sine Bio Tech
16	Recenta	Natural Crop Science
17	Poll	Tropical Agrosystem India (P) Ltd.
18	F+imida	Essential Biosciences
19	Kill Plus	Synwood Agro Ltd.
20	Black Cat	Frontline Agri Science
21	Pluto	Koronation Bio-Science
22	Kri-Max 80	Krishi Rasayan
23	Policy	Global Crop Care
24	Mahakal	JU Agri Sciences Pvt. Ltd.
25	Profin	Synovita Crop science Pvt. Ltd.
26	Fiprokind-Mida	Mankind Agritech
27	Exylon Flybird	Torrent Crop Science
28	Nashak	Katyayani
29	Ranger	Preventis Chem India (P) Ltd.
30	Farydor	Chemley Agritech Pvt. Ltd.
31	Deputy-40	RACCOLTO Agritech India Pvt. Ltd.

Fipronil + Thiamethoxam

Available formulations in India :

1. **Fipronil 04.00% + Thiamethoxam 04.00% SC**
2. **Thiamethoxam 0.90% + Fipronil 0.20% GR**

Classification on the basis of		**Fipronil**	**Thiamethoxam**
Target organism	:	Insecticide	Insecticide
Mode of entry	:	Contact and Stomach poison	Systemic, Contact and Stomach poison
Mode of action	:	Nerve poison	Nerve poison
Chemical group	:	Phenylpyrazole	Neonicotinoid
Category of toxicity : Highly toxic **Toxicity label colour** : Yellow			

- Thiamethoxam acts as an agonist of the nicotinic acetylcholine receptors (nAChR) in the insect's central nervous system. It disrupts the normal transmission of nerve impulses, leading to paralysis and eventual death. Its systemic action means it is absorbed by the plant and translocated throughout the tissues, protecting the entire plant from feeding insects.
- Fipronil acts as a GABA-gated chloride channel antagonist. It blocks the passage of chloride ions through the GABA receptor, leading to the hyperexcitation of the insect's central nervous system, followed by paralysis and death. It provides a fast knockdown effect through contact or ingestion.
- This combination is highly effective against a wide range of pests, particularly those that suck plant sap and soil-dwelling insects.
- It combines contact/stomach action (Fipronil) and systemic action (Thiamethoxam).
- The dual action helps in delaying the development of resistance in pest populations.
- Fipronil offers rapid initial control, while Thiamethoxam generally provides longer-lasting systemic protection.

Label claim (DPPQ & S, Faridabad)

Fipronil 04.00% + Thiamethoxam 04.00% SC

Crop	**Pests**	**Dose / ha**		**Waiting period (Days)**
		Quantity of formulation (ml)	**Volume of water (Lit.)**	
Rice	Brown plant hopper, White backed plant hopper, Green Leaf hopper	1100	500	45

Trade Names

Sr. No.	**Trade Name**	**Company**
1	Mojati	Parijat Industries India Pvt. Ltd.
2	Karrif	KP Agro Industries

3	Firse Pro	Somnath Demeter Agritron Ltd.
4	Profit	Silver Sine Bio Tech
5	Origen	Essential Biosciences
6	Jatayu	Global Crop Care
7	Jhumru	Gharda Chemicals Ltd.
8	Foxy	Katyayani Organics
9	Canoza	Advance Pesticides
10	Fifthi	R. K. Chemicals
11	Full Down	Greenwell Biotech
12	Thianil	Plazma Crop Science
13	Fiprozam	Maxwell Biotech
14	Aatank	Rumicon Crop Science
15	Trigent+	Green Life Save & Protect Pvt. Ltd.
16	Thianilchem	Creative Unichem Pvt. Ltd.

Thiamethoxam 0.90% + Fipronil 0.20% GR

Crop	Pests	Dose / ha		Waiting period (Days)
		Quantity of formulation (Kg)	Volume of water (Lit.)	
Groundnut	White grub, Termite	12.15	--	106
Sugarcane	White grub, Termite	15	--	296

Trade Names

Sr. No.	Trade Name	Company
1	Voltax	Crystal Crop Protection Ltd.
2	Vibgyor	Best Agrolife Ltd.
3	Thionil	Katyayani Organics Silver Sine Bio Tech
4	Thiakind-F	Mankind Agritech

Fipronil + Hexythiazox

Available formulations in India : **Fipronil 07.00% + Hexythiazox 02.00% SC**

Classification on the basis of		Fipronil	Hexythiazox
Target organism	:	Insecticide	Acaricide
Mode of entry	:	Contact and Stomach poison	Contact and Stomach poison
Mode of action	:	Nerve poison	Insect Growth Regulator (IGR)
Chemical group	:	Phenylpyrazole	Thiazolidinone
Category of toxicity : Moderately toxic **Toxicity label colour** : Blue			

- Fipronil + Hexythiazox is a modern dual-action insecticide–acaricide formulation widely used in Indian agriculture, particularly for managing thrips and mite complexes in chilli and vegetable crops.
- Fipronil acts on the central nervous system by blocking GABA-gated chloride channels, causing neuronal hyperexcitation, paralysis, and death. It exhibits both contact and ingestion toxicity, offering some systemic and translaminar activity in plants.
- Hexythiazox functions as a mite growth regulator, inhibiting chitin synthesis and disrupting molting in eggs, larvae, and nymphs. It possesses strong ovicidal and nymphicidal properties, ensuring long-lasting control of mite populations.
- The mixture provides broad-spectrum control through complementary actions. Fipronil gives rapid knockdown of adult thrips and mites. Hexythiazox provides prolonged suppression by targeting immature mite stages. Together, they deliver both immediate and residual protection, reducing the need for repeated sprays.

Label claim (DPPQ & S, Faridabad)
Fipronil 07.00% + Hexythiazox 02.00% SC

Crop	Pests	Dose / ha		Waiting period (Days)
		Quantity of formulation (ml)	Volume of water (Lit.)	
Chilli	Mites, Thrips	1000	500	07

Trade Names

Sr. No.	Trade Name	Company
1	Mythri	Coromandel International Ltd.
2	Cubax power	Best Agrolife Ltd.
3	Pestop	JU Agri Sciences Pvt. Ltd.
4	Fiprokind-H	Mankind Agritech
5	Kleen Mite	Tropical Agrosystem India (P) Ltd.
6	Matri	Global Crop Care
7	Mantle	Nagarjuna Agrochemicals Pvt. Ltd.
8	Dominate	Preventis Chem India (P) Ltd.

Fipronil + Tolfenpyrad + Abamectin

Available formulations in India : **Fipronil 07.00% + Tolfenpyrad 15.00% + Abamectin 01.25% SC**

Classification on the basis of		Fipronil	Tolfenpyrad	Abamectin
Target organism	:	Insecticide	Insecticide	Acaricide, Insecticide
Mode of entry	:	Contact and Stomach poison	Stomach and Contact poison	Stomach and Contact poison
Mode of action	:	Nerve poison	Energy metabolism	Nerve poison
Chemical group	:	Phenylpyrazole	Pyrazole	Avermectin
Category of toxicity : Highly toxic **Toxicity label colour** : Yellow				

- This triple-combination insecticide—Fipronil 7% + Tolfenpyrad 15% + Abamectin 1.25% SC—is a specialized, broad-spectrum formulation primarily used for managing severe infestations of sucking and chewing pests.
- This SC formulation combines phenylpyrazole Fipronil for GABA-gated channel disruption, METI acaricide Tolfenpyrad for mitochondrial electron transport inhibition, and avermectin Abamectin for chloride channel modulation, offering contact, stomach, and translaminar action.

Label claim (DPPQ & S, Faridabad)
Fipronil 07.00% + Tolfenpyrad 15.00% + Abamectin 01.25% SC

Crop	Pests	Dose / ha		Waiting period (Days)
		Quantity of formulation (ml)	Volume of water (Lit.)	
Okra	Aphid, Jassid, Thrips, Whitefly, Mites, Fruit borer	750	500	03
Chilli	Aphid, Thrips, Mites, Fruit borer	750	500	10

Trade Names

Sr. No.	Trade Name	Company
1	Best Man	Best Agrolife Ltd.

Fipronil + Flonicamid

Available formulations in India :

1. **Fipronil 15.00% + Flonicamid 15.00% WDG**
2. **Flonicamid 20.00% + Fipronil 08.00% SC**

Classification on the basis of		**Fipronil**	**Flonicamid**
Target organism	:	Insecticide	Insecticide
Mode of entry	:	Contact and Stomach poison	Systemic, Contact and Stomach poison
Mode of action	:	Nerve poison	Affect on chordotonal organ function
Chemical group	:	Phenylpyrazole	Pyridincarboxamide
Category of toxicity : Moderately toxic **Toxicity label colour** : Blue			

- Fipronil acts on the insect's nervous system by blocking gamma-aminobutyric acid (GABA)-gated chloride channels, causing paralysis and death. It provides both contact and ingestion activity, with translaminar and residual action.
- Flonicamid is a selective feeding blocker that acts rapidly to inhibit the feeding behavior of piercing and sucking pests. The exact target site is the Chordotonal organ nicotinamidase inhibitors. Pests stop feeding minutes after exposure, leading to death.
- Fipronil offers quick knockdown and a broad-spectrum nerve-disrupting effect. Flonicamid offers rapid cessation of feeding, systemic action, and a unique mode of action, which is especially good against resistant sucking pests.
- The combination is highly effective against the sucking pest complex and certain borers.

Label claim (DPPQ & S, Faridabad)
Fipronil 15.00% + Flonicamid 15.00% WDG

Crop	**Pests**	**Dose / ha**		**Waiting period (Days)**
		Quantity of formulation (g)	**Volume of water (Lit.)**	
Cotton	Aphid, Jassid, Thrips, Whitefly, Mealybug, Bollworm	400	500	33
Rice	Brown plant hopper, Green leaf hopper, Stem borer, Leaf folder	400	500	30

Trade Names

Sr. No.	**Trade Name**	**Company**
1	Apache	UPL Ltd.
2	Oxalis	Swal Corporation Ltd.

Flonicamid 20.00% + Fipronil 08.00% SC

Crop	Pests	Dose / ha		Waiting period (Days)
		Quantity of formulation (ml)	Volume of water (Lit.)	
Cotton	Aphid, Jassid, Thrips, Whitefly	375	500	35

Trade Names

Sr. No.	Trade Name	Company
1	Iconiq	Adarsh Agro seeds

Flubendiamide + Buprofezin

Available formulations in India : **Flubendiamide 04.00% + Buprofezin 20.00% SC**

Classification on the basis of		Flubendiamide	Buprofezin
Target organism	:	Insecticide	Insecticide
Mode of entry	:	Contact and Stomach poison	Contact and Stomach poison
Mode of action	:	Muscle poison	Insect Growth Regulator (IGR)
Chemical group	:	Diamide	Thiadiazine
Category of toxicity : Moderately toxic **Toxicity label colour** : Blue			

- Flubendiamide + Buprofezin is a broad-spectrum pre-mix insecticide designed primarily for controlling both lepidopteran (chewing) and sucking pests in crops.
- Flubendiamide acts by activating the ryanodine receptors in insect muscle cells, which causes the uncontrolled release of calcium ions from internal stores. This leads to immediate muscle paralysis, rapid cessation of feeding, and ultimately, death. It is highly effective against Lepidopteran larvae (caterpillars), which are chewing pests (e.g., stem borers, leaf folders, bollworms).
- Buprofezin is an Insect Growth Regulator (IGR). It primarily works by inhibiting the biosynthesis of chitin, a critical component of the insect's exoskeleton. This prevents the insect from molting properly, leading to their death during the molting process (ecdysis). It also reduces the egg-laying capacity of female pests. It is highly effective against Homoptera/Hemiptera (sucking pests), particularly those that are immature (nymphs), such as leafhoppers and planthoppers.
- The pre-mix is designed to provide comprehensive control over both major chewing and sucking pests simultaneously, offering a dual-mode of action strategy. Flubendiamide provides rapid control of caterpillars (borers/folders). Buprofezin provides residual control of sucking pest nymphs (hoppers/flies) by disrupting their life cycle.

Label claim (DPPQ & S, Faridabad)
Flubendiamide 04.00% + Buprofezin 20.00% SC

Crop	Pests	Dose / ha		Waiting period (Days)
		Quantity of formulation (ml)	Volume of water (Lit.)	
Rice	Yellow stem borer, Leaf folder, Brown plant hopper	875	500	30

Trade Names

Sr. No.	Trade Name	Company
1	Rizostar	NACL Industries Ltd.
2	Rapid	Global Crop Care

Flubendiamide + Deltamethrin

Available formulations in India : **Flubendiamide 08.33% + Deltamethrin 05.56% SC**

Classification on the basis of		Flubendiamide	Deltamethrin
Target organism	:	Insecticide	Insecticide
Mode of entry	:	Contact and Stomach poison	Contact and Stomach poison
Mode of action	:	Muscle poison	Nerve poison
Chemical group	:	Diamide	Synthetic pyrethroid
Category of toxicity : Highly toxic **Toxicity label colour** : Yellow			

- The effectiveness of this combination comes from the two active ingredients targeting different physiological systems in insects, providing both quick knockdown and long-lasting protection, while also aiding in resistance management.
- Flubendiamide acts as a ryanodine receptor modulator, disrupting calcium ion release in insect muscle cells. This causes muscular paralysis and death due to loss of control over muscular movements.
- Deltamethrin is a synthetic pyrethroid that acts as a sodium channel modulator, rapidly disrupting nerve signal transmission leading to knockdown paralysis and mortality.
- Deltamethrin provides the quick knockdown (immediate paralysis/mortality) upon contact or ingestion. Flubendiamide provides the long residual control (extended period of protection) and immediate feeding cessation.

Label claim (DPPQ & S, Faridabad)
Flubendiamide 08.33% + Deltamethrin 05.56% SC

Crop	Pests	Dose / ha		Waiting period (Days)
		Quantity of formulation (ml)	Volume of water (Lit.)	
Chickpea	Pod borer	250	500	07
Cucumber	Cucumber beetle, Fruit fly	200-250	500	05
Rice	Leaf folder, Hispa	300	500	35
Chilli	Thrips. Fruit borer	400	500	05
Tomato	Thrips, Fruit borer	400	500	05

Trade Names

Sr. No.	Trade Name	Company
1	Fenos Quick	Bayer Crop Science Ltd.
2	Sine Quick	Silver Sine Bio Tech
3	Lehomo	Global Crop Care
4	Forum	Prisha Chemicals
5	Flern	Samridhdhii Agritech Ltd.

Flubendiamide + Thiacloprid

Available formulations in India : **Flubendiamide 19.92% + Thiacloprid 19.92% SC**

Classification on the basis of		Flubendiamide	Thiacloprid
Target organism	:	Insecticide	Insecticide
Mode of entry	:	Contact and Stomach poison	Systemic, Contact and Stomach poison
Mode of action	:	Muscle poison	Nerve poison
Chemical group	:	Diamide	Neonicotinoid
Category of toxicity : Highly toxic **Toxicity label colour** : Yellow			

- Flubendiamide sisrupts the ryanodine receptors (RyR) in insect muscle cells, leading to an uncontrolled release of calcium ions. This causes lethargy, paralysis, and rapid cessation of feeding, leading to the insect's death. It acts primarily via ingestion/contact and is effective against chewing pests.
- Thiacloprid acts as an agonist on the postsynaptic nicotinic acetylcholine receptors (nAChR) in the insect's central nervous system. This causes over-stimulation of the nervous system, ultimately killing the insect. It is generally systemic and highly effective against sucking pests.
- This combination offers broad-spectrum control over both chewing and sucking insects.
- The combination of the two distinct chemical groups provides a powerful "knockdown" and systemic effect.

Label claim (DPPQ & S, Faridabad)
Flubendiamide 19.92% + Thiacloprid 19.92% SC

Crop	Pests	Dose / ha		Waiting period (Days)
		Quantity of formulation (ml)	Volume of water (Lit.)	
Chilli	Thrips, Fruit borer	200-250	500	05
Rice	Yellow stem borer, Leaf folder	250	500	33
Tea	Mosquito bug, Semilooper	250	400	07
Brinjal	Shoot and fruit borer, Aphid, Thrips, Jassid	350	500	05

Trade Names

Sr. No.	Trade Name	Company
1	Belt Expert	Bayer Crop Science Ltd.

Flonicamid + Diafenthiuron

Available formulations in India : **Flonicamid 11.70% + Diafenthiuron 36.00% WG**

Classification on the basis of		Flonicamid	Diafenthiuron
Target organism	:	Insecticide	Insecticide, Acaricide
Mode of entry	:	Systemic, Contact and Stomach poison	Contact and Stomach poison
Mode of action	:	Affect on chordotonal organ function	Energy metabolism
Chemical group	:	Pyridincarboxamide	Thiourea
Category of toxicity :			
Toxicity label colour :			

- The combination of Flonicamid and Diafenthiuron is an insecticide mixture designed to control a broad spectrum of insect pests, particularly those with piercing-sucking mouthparts.
- Flonicamid is a systemic and selective insecticide that disrupts the feeding behavior of insects. It causes rapid antifeeding effect (starvation) by inhibiting the stylet penetration into plant tissues, and is thought to modulate chordotonal organ function.
- Diafenthiuron is a pro-insecticide, meaning it is converted into its active form inside the insect. The active compound acts as a mitochondrial electron transport inhibitor, disrupting the insect's energy production (ATP synthesis). This results in metabolic dysfunction, quick paralysis, and ultimately, death. It has contact and stomach action, and translaminar and vapor action.
- Diafenthiuron often controls both nymphs/larvae and adults, while Flonicamid provides rapid anti-feeding action. Diafenthiuron's mode of action often provides control of hidden pests within the plant canopy.

Label claim (DPPQ & S, Faridabad)
Flonicamid 11.70% + Diafenthiuron 36.00% WG

Crop	Pests	Dose / ha		Waiting period (Days)
		Quantity of formulation (g)	Volume of water (Lit.)	
Cotton	Whitefly, Aphid, Jassid	625	500	80
Tomato	Whitefly, Aphid, Jassid	625	500	05

Trade Names

Sr. No.	Trade Name	Company
1		

Fluxametamide + Bifenthrin

Available formulations in India : **Fluxametamide 05.81% + Bifenthrin 05.81% EC**

Classification on the basis of		Fluxametamide	Bifenthrin
Target organism	:	Insecticide	Insecticide
Mode of entry	:	Contact and Stomach poison	Contact and Stomach poison
Mode of action	:	Nerve poison	Nerve poison
Chemical group	:	Isoxazoline	Synthetic pyrethroid
Category of toxicity : Moderately toxic **Toxicity label colour** : Blue			

- This combination is used for broad-spectrum protection against a wide range of both sucking and chewing pests in various crops. The combination is designed to provide dual-action control by targeting the insect nervous system in two different ways, which also helps in pest resistance management.
- Fluxametamide is a newer insecticide belonging to the isoxazoline class. It works as a GABA-gated chloride channel blocker, disrupting the inhibitory signals in the nervous system, which leads to over-excitation, stopping of feeding, and death. It provides both contact and translaminar activity. It is particularly effective against sucking pests.
- Bifenthrin is a broad-spectrum pyrethroid insecticide. It acts on the sodium channels in the pest's nervous system, causing rapid over-excitation, paralysis (quick knockdown), and eventual death, primarily through contact and ingestion. It is effective against both chewing and some sucking pests.
- This formulation integrates the novel isoxazoline chemistry of Fluxametamide with the pyrethroid power of Bifenthrin, providing fast knockdown and long residual protection.

Label claim (DPPQ & S, Faridabad)
Fluxametamide 05.81% + Bifenthrin 05.81% EC

Crop	Pests	Dose / ha		Waiting period (Days)
		Quantity of formulation (ml)	Volume of water (Lit.)	
Chilli	Fruit borer	625	375	07
	Thrips, Whitefly	500	375	07
Tomato	Fruit borer	625	375	07
	Leaf miner	500	375	07
Brinjal	Shoot and fruit borer	625	375	07
	Whitefly, Jassid	500	375	07

Trade Names

Sr. No.	Trade Name	Company
1	LaNevo	Dhanuka Agritech

Fluxametamide + Pyridaben

Available formulations in India : **Fluxametamide 03.80% + Pyridaben 09.50% SC**

Classification on the basis of		Fluxametamide	Pyridaben
Target organism	:	Insecticide	Insecticide, Acaricide
Mode of entry	:	Contact and Stomach poison	Contact and Stomach poison
Mode of action	:	Nerve poison	Energy metabolism
Chemical group	:	Isoxazoline	Pyridazinone
Category of toxicity :			
Toxicity label colour :			

- Fluxametamide + Pyridaben is an innovative dual-action insecticide and miticide combination recently introduced in India for broad-spectrum pest management. This mix targets both sucking and chewing pest complexes, providing rapid knockdown and long residual control through distinct mechanisms of action.
- Fluxametamide acts as a GABA-gated chloride channel antagonist, blocking chloride ion flow across nerve membranes. This leads to paralysis and death of insects. It provides contact, stomach, and translaminar activity. It is effective against thrips, lepidopteran larvae, and resistant pest strains.
- Pyridaben is a mitochondrial complex I electron transport inhibitor that prevents energy (ATP) production. It blocks cellular respiration, rapidly stopping energy production in the pest, which results in quick knockdown. It is primarily a fast-acting contact poison. It is highly effective against mites, whiteflies, leafhoppers, and aphids by contact and stomach action.
- Pyridaben ensures rapid mortality, while Fluxametamide offers more residual activity. The Fluxametamide component helps the product penetrate the leaf tissue to reach pests feeding on the underside.

Label claim (DPPQ & S, Faridabad)
Fluxametamide 03.80% + Pyridaben 09.50% SC

Crop	Pests	Dose / ha		Waiting period (Days)
		Quantity of formulation (ml)	Volume of water (Lit.)	
Brinjal	Leafhopper, Thrips, Whitefly, Shoot and fruit borer, Red spider mite	500	375	05
Chilli	Thrips, Fruit borer, Tobacco caterpillar, Yellow mite	1000	500	05

Trade Names

Sr. No.	Trade Name	Company
1	Rashinban	Nissan Chemical Corporation , Godrej Agrovet

Hexythiazox + Diafenthiuron

Available formulations in India : Hexythiazox 03.5% + Diafenthiuron 42.00% WDG

Classification on the basis of		Hexythiazox	Diafenthiuron
Target organism	:	Acaricide	Insecticide, Acaricide
Mode of entry	:	Contact and Stomach poison	Contact and Stomach poison
Mode of action	:	Mite Growth Regulator	Energy metabolism
Chemical group	:	Thiazolidinone	Thiourea
Category of toxicity : Moderately toxic **Toxicity label colour** : Blue			

- The combination of Hexythiazox and Diafenthiuron is designed to provide broad-spectrum control over various pests, especially mites and certain sucking insects.
- Hexythiazox primarily acts as a miticide/acaricide and is classified as a Mite Growth Regulator. It controls all stages of the mite lifecycle, particularly eggs, larvae, and nymphs. It is generally non-systemic but has good translaminar activity (moves from the treated leaf surface to the opposite side).
- Diafenthiuron primarily acts as an insecticide and acaricide. It is a mitochondrial respiration inhibitor. It controls nymphs and adults of many sucking pests and mites. It is known for its contact and stomach action, and often possesses translaminar and vapor action.
- The two different modes of action help prevent the development of pest resistance.
- It offers long-duration control, reducing the frequency of sprays.
- Diafenthiuron and its metabolites may sometimes provide a positive effect on plant health and vigor.

Label claim (DPPQ & S, Faridabad)
Hexythiazox 03.5% + Diafenthiuron 42.00% WDG

Crop	Pests	Dose / ha		Waiting period (Days)
		Quantity of formulation (g)	Volume of water (Lit.)	
Chilli	Mites, Thrips, Jassid, Aphid, Whitefly	650	500	07

Trade Names

Sr. No.	Trade Name	Company
1	Mighty	Global Crop Care
2	Aster	Synovita Crop Science Pvt. Ltd.
3	Canister	Coromamdel International
4	Debate	Preventis Chem India (P) Ltd.

Imidacloprid + Bifenthrin

Available formulations in India : **Imidacloprid 20.00% + Bifenthrin 08.00% SL**

Classification on the basis of		Imidacloprid	Bifenthrin
Target organism	:	Insecticide	Insecticide
Mode of entry	:	Systemic, Contact and Stomach poison	Contact and Stomach poison
Mode of action	:	Nerve poison	Nerve poison
Chemical group	:	Neonicotinoid	Synthetic pyrethroid
Category of toxicity : Highly toxic **Toxicity label colour** : Yellow			

- The combination product containing Imidacloprid 20.00% + Bifenthrin 08.00% SL is a dual-action insecticide designed for broad-spectrum pest control in agriculture. This specific mixture combines two different classes of insecticides, providing quick knockdown and long-lasting residual control.
- Imidacloprid is a systemic insecticide belonging to the neonicotinoid chemical group. It works by acting as an agonist on the postsynaptic nicotinic acetylcholine receptors (nAChR) in the central nervous system of insects, leading to their irreversible blockage, paralysis, and death. Because it is systemic, it is absorbed by the plant and translocated throughout the tissues, providing protection against sucking pests.
- Bifenthrin is a synthetic pyrethroid insecticide and acaricide. Its mode of action involves disrupting the nervous system of insects by modulating the sodium channel, which causes rapid paralysis and death (known as a "knockdown" effect). Bifenthrin is a contact and stomach poison with some residual activity on the plant surface. It is primarily effective against chewing pests such as caterpillars, borers, and certain mites and can also control some sucking pests.
- The combination of Imidacloprid's systemic action and Bifenthrin's contact/stomach action ensures that pests are controlled both from the inside and outside of the plant.

Label claim (DPPQ & S, Faridabad)
Imidacloprid 20.00% + Bifenthrin 08.00% SL

Crop	Pests	Dose / ha		Waiting period (Days)
		Quantity of formulation (ml)	Volume of water (Lit.)	
Rice	Stem borer, Hispa, Plant hopper, Gundhi bug	300	500	10

Trade Names

Sr. No.	Trade Name	Company
1	Talextra	Krishi Rasayan Export Pvt. Ltd.

Imidacloprid + Lambda-cyhalothrin

Available formulations in India : **Imidacloprid 06.00% + Lambda-cyhalothrin 04.00% SL**

Classification on the basis of		Imidacloprid	Lambda-cyhalothrin
Target organism	:	Insecticide	Insecticide
Mode of entry	:	Systemic, Contact and Stomach poison	Contact and Stomach poison
Mode of action	:	Nerve poison	Nerve poison
Chemical group	:	Neonicotinoid	Synthetic pyrethroid
Category of toxicity : Highly toxic **Toxicity label colour** : Yellow			

- Imidacloprid has systemic (absorbed by the plant and moves throughout) and contact/stomach action. It affects the insect's central nervous system (nicotinic acetylcholine receptor). Primarily sucking pests (e.g., aphids, jassids, whiteflies, thrips, leafhoppers) are target.
- Lambda-cyhalothrin has contact and stomach action (non-systemic). It affects the insect's central and peripheral nervous systems by interfering with sodium channel function. It is a broad-spectrum killer effective against a wide range of pests, including chewing pests (e.g. caterpillars/larvae, beetles) and sucking pests (e.g., thrips, mites, jassids).
- The mixture is designed to provide comprehensive control by combining the systemic and long-residual effect of Imidacloprid with the quick knockdown and broad-spectrum contact effect of Lambda-cyhalothrin.

Label claim (DPPQ & S, Faridabad)
Imidacloprid 06.00% + Lambda-cyhalothrin 04.00% SL

Crop	Pests	Dose / ha		Waiting period (Days)
		Quantity of formulation (ml)	Volume of water (Lit.)	
Rice	Stem borer, Hispa, Plant hopper, Gundhi bug	300	500	10

Trade Names

Sr. No.	Trade Name	Company
1	Emidalem	Natural Crop Science
2	Inovexia	Willowood Crop Science Pvt. Ltd.
3	Volcano	Angel Crop Science
4	Kiken	RACCOLTO Agritech India Pvt. Ltd.
5	Leimidex	Era Agri Tech Co.
6	Indra	Royal Agro Tech
7	Dhamaka	Rain Bio Tech Industries

Indoxacarb + Acetamiprid

Available formulations in India : **Indoxacarb 14.50% + Acetamiprid 07.70% SC**

Classification on the basis of		Indoxacarb	Acetamiprid
Target organism	:	Insecticide	Insecticide
Mode of entry	:	Contact and Stomach poison	Systemic, Contact and Stomach poison
Mode of action	:	Nerve poison	Nerve poison
Chemical group	:	Oxadiazine	Neonicotinoid
Category of toxicity : Highly toxic **Toxicity label colour** : Yellow			

- The combination of Indoxacarb and Acetamiprid is a modern insecticide formulation designed for broad-spectrum pest control in agriculture.
- Indoxacarb belongs to the oxadiazine chemical class and works as a sodium channel blocker. It is often referred to as a pro-insecticide because it must be metabolized by the insect to become fully active. Once ingested or absorbed, it leads to rapid cessation of feeding, paralysis, and eventual death. Indoxacarb is particularly effective against Lepidopteran pests.
- Acetamiprid is a neonicotinoid insecticide. Its mode of action involves interfering with the transmission of nerve impulses by acting on the nicotinic acetylcholine receptors in the insect's central nervous system. Acetamiprid provides powerful systemic action, meaning it is absorbed by the plant and moves through its tissues, providing protection to new growth and targeting hidden, piercing-sucking pests that feed on plant sap.
- Indoxacarb provides rapid, robust control over chewing pests, while the systemic and fast-acting nature of Acetamiprid tackles the difficult-to-manage sucking pests.

Label claim (DPPQ & S, Faridabad)
Indoxacarb 14.50% + Acetamiprid 07.70% SC

Crop	Pests	Dose / ha		Waiting period (Days)
		Quantity of formulation (ml)	Volume of water (Lit.)	
Cotton	Jassid, Whitefly, Bollworms	400-500	500	30
Chilli	Thrips, Fruit borer	400-500	500	05

Trade Names

Sr. No.	Trade Name	Company
1	Kite	Gharda Chemicals Ltd.
2	Quba	Anu Products Ltd.
3	Kalash	KP Agro Industries
4	Ace-Indox	Katyayani Organics
5	Moola	Tropical Agrosystem India (P) Ltd.
6	Bajirao	HPM Chemicals and Fertilizers Ltd.
7	Topmost	RACCOLTO Agritech India Pvt. Ltd.

Indoxacarb + Fipronil

Available formulations in India : **Indoxacarb 05.00% + Fipronil 05.00% SC**

Classification on the basis of		**Indoxacarb**	**Fipronil**
Target organism	:	Insecticide	Insecticide
Mode of entry	:	Contact and Stomach poison	Contact and Stomach poison
Mode of action	:	Nerve poison	Nerve poison
Chemical group	:	Oxadiazine	Phenylpyrazole
Category of toxicity : Moderately toxic **Toxicity label colour** : Blue			

- This combination insecticide works by attacking the insect's nervous system in two different ways.
- Indoxacarb belongs to the Oxadiazine chemical class. It is a pro-insecticide that is bioactivated within the insect. It works primarily by blocking sodium channels in the nerve cells, causing rapid cessation of feeding, paralysis, and eventual death. It acts mainly by contact and stomach (ingestion) action and is especially effective against chewing pests (Lepidoptera larvae).
- Fipronil belongs to the Phenylpyrazole chemical class. It is a broad-spectrum insecticide that disrupts the insect's central nervous system by blocking GABA-gated chloride channels. It exhibits contact and some systemic action and is effective against both chewing and sucking pests and some soil pests.
- The dual mode of action and broad spectrum of the combination make it effective against a variety of pests.

Label claim (DPPQ & S, Faridabad)
Indoxacarb 05.00% + Fipronil 05.00% SC

Crop	Pests	Dose / ha		Waiting period (Days)
		Quantity of formulation (ml)	Volume of water (Lit.)	
Chilli	Thrips	750-1000	500	05
Cabbage	Diamond back moth	750-1000	500	07

Trade Names

Sr. No.	Trade Name	Company
1	Indic	Synovita Crop Science Pvt. Ltd.
2	Black	Gharda Chemicals Ltd.

Indoxacarb + Thiamethoxam

Available formulations in India : **Indoxacarb 10.00% + Thiamethoxam 10.00% WG**

Classification on the basis of		Indoxacarb	Thiamethoxam
Target organism	:	Insecticide	Insecticide
Mode of entry	:	Contact and Stomach poison	Systemic, Contact and Stomach poison
Mode of action	:	Nerve poison	Nerve poison
Chemical group	:	Oxadiazine	Neonicotinoid
Category of toxicity : Moderately toxic **Toxicity label colour** : Blue			

- The effectiveness of the Indoxacarb + Thiamethoxam combination is rooted in its dual mode of action, which provides comprehensive pest control and is a strong tool for resistance management.
- Indoxacarb belongs to the Oxadiazine class (IRAC Group 22A) and acts as a voltage-dependent sodium channel blocker. This mechanism disrupts nerve signal transmission, quickly causing the insect to cease feeding, leading to paralysis and eventual death, making it highly effective against chewing pests, such as lepidopteran larvae.
- In contrast, Thiamethoxam is a neonicotinoid (IRAC Group 4A) that works by binding to nicotinic acetylcholine receptors in the insect's central nervous system. This results in nerve hyperexcitation, leading to rapid knockdown and paralysis, and it is particularly effective against sucking pests. As a highly systemic insecticide, Thiamethoxam provides protection via both contact and ingestion, delivering long-lasting residual activity.
- By pairing these two distinct modes of action, the combination achieves broad-spectrum control over both chewing and sap-feeding insects in a single application, significantly reducing the risk of pests developing cross-resistance.

Label claim (DPPQ & S, Faridabad)
Indoxacarb 10.00% + Thiamethoxam 10.00% WG

Crop	Pests	Dose / ha		Waiting period (Days)
		Quantity of formulation (g)	Volume of water (Lit.)	
Tomato	Whitefly, Fruit borer	750	500	05
Rice	Brown plant hopper, Yellow stem borer, Leaf folder	500	500	14

Trade Names

Sr. No.	Trade Name	Company
1	Firrangi	Gharda Chemicals Ltd.

Lufenuron + Chlorantraniliprole + Pyriproxyfen

Available formulations in India : **Lufenuron 20.00% + Chlorantraniliprole 10.00% + Pyriproxyfen 05.00% WG**

Classification on the basis of		Lufenuron	Chlorantraniliprole	Pyriproxyfen
Target organism	:	Insecticide	Insecticide	Insecticide
Mode of entry	:	Contact and Stomach poison	Systemic, Contact and Stomach poison	Contact and Stomach poison
Mode of action	:	Insect Growth Regulator	Muscle poison	Insect Growth Regulator (IGR)
Chemical group	:	Benzoylurea	Anthranilic diamide	Pyridine ether
Category of toxicity : **Toxicity label colour** :				

- This triple-combination insecticide—Lufenuron 20% + Chlorantraniliprole 10% + Pyriproxyfen 5% WG—is a high-strength, water-dispersible granule (WG) formulation designed for advanced pest management.
- Lufenuron prevents larvae from molting and forming new skin; also has strong ovicidal (egg-killing) properties.
- Chlorantraniliprole paralyzes the insect's muscles, causing them to stop feeding within minutes and highly effective against caterpillars.
- Pyriproxyfen is an Insect Growth Regulator (IGR) that prevents larvae from maturing into adults and sterilizes adult females.
- While Chlorantraniliprole kills the feeding larvae, Lufenuron and Pyriproxyfen ensure that any surviving eggs or nymphs never reach adulthood.

Label claim (DPPQ & S, Faridabad)

Lufenuron 20.00% + Chlorantraniliprole 10.00% + Pyriproxyfen 05.00% WG

Crop	Pests	Dose / ha		Waiting period (Days)
		Quantity of formulation (g)	Volume of water (Lit.)	
Chilli	Thrips, Mites, Fruit borers (*Helicoverpa armigera, Spodoptera litura*)	112.50	500	19

Trade Names

Sr. No.	Trade Name	Company
1		Parijat Industries (India) Ltd.

Lufenuron + Emamectin benzoate

Available formulations in India : **Lufenuron 04.00% + Emamectin benzoate 01.50% EC**

Classification on the basis of		Lufenuron	Emamectin benzoate
Target organism	:	Insecticide	Insecticide, Acaricide
Mode of entry	:	Contact and Stomach poison	Systemic, Contact and Stomach poison
Mode of action	:	Insect Growth Regulator	Nerve poison
Chemical group	:	Benzoylurea	Avermectin (Macrocyclic lactones)
Category of toxicity : **Toxicity label colour :**			

- The Lufenuron and Emamectin Benzoate combination is a powerful insecticide mixture that provides broad-spectrum control, primarily targeting sLepidopteran pests. The formulation's efficacy stems from the synergy of its two distinct modes of action.
- Emamectin benzoate (an Avermectin derivative, IRAC Group 6) acts as a potent non-systemic insecticide with primarily stomach action and secondary contact activity. It works by interfering with the insect's nervous system, leading to rapid paralysis, causing the pest to stop feeding within 1-2 hours, and resulting in death within 2-4 days. It is known for its quick knockdown and strong efficacy against caterpillars.
- Lufenuron (a Benzoylurea Insect Growth Regulator or IGR, IRAC Group 15) provides long-lasting residual control by disrupting the insect's life cycle. It acts mostly by ingestion, inhibiting the synthesis of chitin, a vital component of the insect exoskeleton. This prevents larvae from successfully molting and developing, and it also exhibits ovicidal activity, helping to sterilize eggs and adults to halt population growth.
- This dual action provides both immediate knockdown (from Emamectin Benzoate) and extended control by breaking the pest's life cycle (from Lufenuron).

Label claim (DPPQ & S, Faridabad)
Lufenuron 04.00% + Emamectin benzoate 01.50% EC

Crop	Pests	Dose / ha		Waiting period (Days)
		Quantity of formulation (ml)	Volume of water (Lit.)	
Maize	Fall armyworm	750-1150	400-500	35
Soybean	Tobacco caterpillar	625	375	37

Trade Names

Sr. No.	Trade Name	Company
1		

Methoxyfenozide + Chlorantraniliprole

Available formulations in India : **Methoxyfenozide 20.00% + Chlorantraniliprole 05.00% SC**

Classification on the basis of		Methoxyfenozide	Chlorantraniliprole
Target organism	:	Insecticide	Insecticide
Mode of entry	:	Stomach and Contact poison	Systemic, Contact and Stomach poison
Mode of action	:	Insect Growth regulator (IGR)	Muscle poison
Chemical group	:	Diacylhydrazines	Anthranilic diamide
Category of toxicity : **Toxicity label colour** :			

- The Methoxyfenozide + Chlorantraniliprole combination is a highly effective, dual-action insecticide primarily used to control a wide range of Lepidopteran pests (caterpillars and borers) in different crops.
- Methoxyfenozide acts as an Insect Growth Regulator (IGR), disrupting the molting process and preventing the pest from maturing, thereby providing long-lasting control. It mimics the natural insect molting hormone, causing the larvae to attempt a premature, lethal molt. It stops the insect life cycle.
- Concurrently, Chlorantraniliprole works as a Ryanodine Receptor Modulator, causing immediate muscle paralysis and rapid cessation of feeding, which protects the crop from instantaneous damage.
- This synergy ensures both a quick knockdown and robust, long-term residual activity, while also serving as a crucial tool in resistance management strategies by employing two different IRAC groups (18 and 28).

Label claim (DPPQ & S, Faridabad)
Methoxyfenozide 20.00% + Chlorantraniliprole 05.00% SC

Crop	Pests	Dose / ha		Waiting period (Days)
		Quantity of formulation (ml)	Volume of water (Lit.)	
Groundnut	Tobacco caterpillar, American bollworm, Leaf miner	500	500	27

Trade Names

Sr. No.	Trade Name	Company
1	Brizesta	Mankind Agritech
2	Liger	GSP Crop Science, Gagan Pesticides
3	Armda	Agrostar

Methoxyfenozide + Chlorpyrifos

Available formulations in India : **Methoxyfenozide 05.00% + Chlorpyrifos 25.00% SE**

Classification on the basis of		Methoxyfenozide	Chlorpyrifos
Target organism	:	Insecticide	Insecticide, Acaricide
Mode of entry	:	Stomach and Contact poison	Contact, Stomach and Respiratory poison
Mode of action	:	Insect Growth regulator (IGR)	Nerve poison
Chemical group	:	Diacylhydrazines	Organophosphate
Category of toxicity : Highly toxic **Toxicity label colour** : Yellow			

- The combination of Methoxyfenozide and Chlorpyrifos is an insecticide mixture designed to provide broad-spectrum control by leveraging two distinct modes of action.
- Methoxyfenozide belongs to the Diacylhydrazine class (IRAC Group 18) and functions as an Insect Growth Regulator (IGR), specifically an ecdysone agonist. It mimics the insect's natural molting hormone, leading to a premature and lethal molt, thus primarily targeting the larval stages of Lepidopteran pests (caterpillars) and offering long-term residual activity.
- In contrast, Chlorpyrifos is a broad-spectrum organophosphate compound (IRAC Group 1B) that works as an Acetylcholinesterase (AChE) inhibitor. It affects the nervous system of insects by preventing the breakdown of the neurotransmitter acetylcholine, leading to the rapid overstimulation of nerve cells, paralysis, and swift knockdown of a wide range of pests, including Homoptera (e.g., aphids, leafhoppers), Coleoptera, and Diptera.
- Combining these two active ingredients aims to achieve both quick pest elimination and prolonged control, while utilizing different modes of action is a critical strategy in insecticide resistance management (IRM) to delay the development of resistance in pest populations.

Label claim (DPPQ & S, Faridabad)
Methoxyfenozide 05.00% + Chlorpyrifos 25.00% SE

Crop	Pests	Dose / ha		Waiting period (Days)
		Quantity of formulation (ml)	Volume of water (Lit.)	
Chickpea	Pod borer	2000	500	15

Trade Names

Sr. No.	Trade Name	Company
1		

Methoxyfenozide + Emamectin benzoate

Available formulations in India : **Methoxyfenozide 18.00% + Emamectin benzoate 01.80% SC**

Classification on the basis of		Methoxyfenozide	Emamectin benzoate
Target organism	:	Insecticide	Insecticide, Acaricide
Mode of entry	:	Stomach and Contact poison	Systemic, Contact and Stomach poison
Mode of action	:	Insect Growth regulator (IGR)	Nerve poison
Chemical group	:	Diacylhydrazines	Avermectin (Macrocyclic lactones)
Category of toxicity : **Toxicity label colour :**			

- The combination of Methoxyfenozide and Emamectin benzoate forms a highly effective insecticide that provides a powerful dual mode of action against a broad spectrum of insect pests.
- Emamectin benzoate acts as a potent, fast-acting neurotoxin with contact and translaminar activity, causing rapid paralysis and cessation of feeding almost immediately upon ingestion or contact, leading to quick knockdown of the target pests.
- Complementing this, Methoxyfenozide functions as an insect growth regulator (IGR) from the diacylhydrazine family, which disrupts the larval molting process by mimicking the insect's natural molting hormone, thereby offering prolonged residual control by halting the development of larvae and preventing them from reaching the adult stage.
- Methoxyfenozide offers prolonged residual activity by interfering with insect development. Emamectin benzoate provides rapid knockdown via neurotoxicity.
- Combination of growth regulation and neurotoxic action offers broad and effective pest control, reducing chances of resistance development.

Label claim (DPPQ & S, Faridabad)
Methoxyfenozide 18.00% + Emamectin benzoate 01.80% SC

Crop	Pests	Dose / ha		Waiting period (Days)
		Quantity of formulation (ml)	Volume of water (Lit.)	
Chilli	Fruit borer, Thrips, Yellow mite	500	500	03

Trade Names

Sr. No.	Trade Name	Company
1	Fighter	GSP Crop Science

Nitenpyram + Fipronil

Available formulations in India : **Nitenpyram 15.00% + Fipronil 20.00% WP**

Classification on the basis of		**Nitenpyram**	**Fipronil**
Target organism	:	Insecticide	Insecticide
Mode of entry	:	Systemic, Contact and Stomach poison	Contact and Stomach poison (Systemic)
Mode of action	:	Nerve poison	Nerve poison
Chemical group	:	Neonicotinoid	Phenylpyrazole
Category of toxicity : **Toxicity label colour :**			

- Nitenpyram 15.0% + Fipronil 20.0% WP is a synergistic insecticide mixture primarily used in agricultural settings for broad-spectrum pest control. It combines a fast-acting neonicotinoid with a long-lasting phenylpyrazole to target both sucking and chewing pests.
- This formulation utilizes two distinct modes of action to ensure high mortality rates and reduced resistance development.
- Nitenpyram is a systemic neonicotinoid that acts as an agonist of nicotinic acetylcholine receptors (nAChR). It disrupts the insect's central nervous system, causing rapid paralysis and knockdown, often within minutes to hours.
- Fipronil is a phenylpyrazole that blocks GABA-gated chloride channels. This causes hyperexcitation of the nerves and muscles, leading to death. Fipronil provides strong contact and stomach action with significant residual activity.

Label claim (DPPQ & S, Faridabad)

Nitenpyram 15.00% + Fipronil 20.00% WP

Crop	**Pests**	**Dose / ha**		**Waiting period (Days)**
		Quantity of formulation (g)	**Volume of water (Lit.)**	
Cotton	Thrips, Leafhopper	400	500	70

Trade Names

Sr. No.	**Trade Name**	**Company**
1		

Novaluron + Emamectin benzoate

Available formulations in India : **Novaluron 05.25% + Emamectin benzoate 0.90% SC**

Classification on the basis of		Novaluron	Emamectin benzoate
Target organism	:	Insecticide	Insecticide, Acaricide
Mode of entry	:	Stomach and Contact poison	Systemic, Contact and Stomach poison
Mode of action	:	Insect Growth Regulator (IGR)	Nerve poison
Chemical group	:	Benzoylureas	Avermectin (Macrocyclic lactones)
Category of toxicity : Moderately toxic **Toxicity label colour** : Blue			

- This combination is highly valued because it incorporates two active ingredients with different modes of action (Dual-Action Formula), which is crucial for effective resistance management and comprehensive control.
- Novaluron is an insect growth regulator (IGR) that works by inhibiting chitin synthesis in immature insects. Chitin is an essential component for the molting process. By disrupting this process, Novaluron prevents insect larvae from progressing to later life stages, causing mortality during development. This affects mainly the larval stages of pests.
- Emamectin Benzoate, a semi-synthetic derivative of avermectin, acts on the nervous system of insects. It specifically binds to glutamate-gated chloride channels in the insect's nerve cells, causing overstimulation of the nervous system which leads to paralysis and eventual death. It acts quickly as a neurotoxin, providing contact and stomach poisoning effects against pests. It also has systemic and translaminar action, meaning it is absorbed and moves within the plant to protect even hidden pests.
- Together, Novaluron + Emamectin Benzoate provides broad-spectrum pest control by targeting pest development through growth inhibition (Novaluron) and immediate pest knockdown via neurotoxicity (Emamectin).

Label claim (DPPQ & S, Faridabad)
Novaluron 05.25% + Emamectin benzoate 0.90% SC

Crop	Pests	Dose / ha		Waiting period (Days)
		Quantity of formulation (ml)	Volume of water (Lit.)	
Cabbage	Diamond back moth, Tobacco caterpillar	875	500	03
Chilli	Gram pod borer, Tobacco caterpillar	875	500	03
Pigeonpea	Pod borer	875	500	25

Rice	Stem borer	1500	500	32
Brinjal	Shoot and fruit borer	1500	500	05
Chickpea	Pod borer	875	500	11
Maize	Fall armyworm, Stem borer	1500	500	27

Trade Names

Sr. No.	Trade Name	Company
1	Barazide	ADAMA India Private Limited
2	Gunther	UPL Ltd.
3	Hamor	Swal Corporation Ltd.
4	Comex	Anu Products Ltd.
5	Medad	Parijat Industries India Pvt. Ltd.
6	Mastery	Best Agrolife Ltd.
7	Bushido	Nichino India
8	Novadem	Somnath Demeter Agritron Ltd.
9	Novabenzo	Natural Crop Science
10	Aakramak	Katyayani Organics
11	Novatin	Tropical Agrosystem India (P) Ltd.
12	Ghaatak	Essential Biosciences
13	Novatin	Synwood Agro Ltd.
14	Wood Guard	Krishi Rasyan
15	Spodo	Global Crop Care
16	Kakuna	Tata Rallis India Ltd.
17	Novazide	JU Agri Sciences Pvt. Ltd.
18	Nexor	Synovita Crop science Pvt. Ltd.
19	Emaneed-Plus	Mankind Agritech
20	Exylon Novamax	Torrent Crop Science
21	Parade	Preventis Chem India (P) Ltd.
22	Vultopia	Chemley Agritech Pvt. Ltd.

Novaluron + Indoxacarb

Available formulations in India : **Novaluron 05.25% + Indoxacarb 04.50% SC**

Classification on the basis of		Novaluron	Indoxacarb
Target organism	:	Insecticide	Insecticide
Mode of entry	:	Stomach and Contact poison	Contact and Stomach poison
Mode of action	:	Insect Growth Regulator (IGR)	Nerve poison
Chemical group	:	Benzoylureas	Oxadiazine
Category of toxicity : Highly toxic **Toxicity label colour** : Yellow			

- Novaluron + Indoxacarb is a broad-spectrum insecticide combination widely used to manage lepidopteran pests in various crops.
- Novaluron acts as an insect growth regulator (IGR) by inhibiting chitin synthesis, which is essential for the molting process. This disruption prevents larvae from molting properly, leading to death in immature stages.
- Indoxacarb is a neurotoxin that targets the insect nervous system by blocking sodium channels in nerve cells, causing paralysis and death. It acts through contact and stomach poisoning, providing fast knockdown effects.
- Together, this combination provides dual action. Novaluron disrupts pest development, while Indoxacarb delivers rapid control by affecting the nervous system.

Label claim (DPPQ & S, Faridabad)
Novaluron 05.25% + Indoxacarb 04.50% SC

Crop	Pests	Dose / ha		Waiting period (Days)
		Quantity of formulation (ml)	Volume of water (Lit.)	
Tomato	Fruit borer, Tobacco caterpillar	825-875	500	05
Chickpea	Gram pod borer	825-875	500	09
Soybean	Tobacco caterpillar, Gram pod borer, Semilooper	825-875	500	14
Pigeonpea	Pod borer, Pod fly	825-875	500	25
Chilli	Fruit borer, Tobacco caterpillar	825-875	500	07
Black gram	Black gram pod borer complex (*Etiella zinckenella, Spodoptera litura, Maruca vitrata*)	825-875	500	14
Rice	Leaf folder	437.5	500	40
Groundnut	Gram pod borer, Tobacco caterpillar	875	500	34

Trade Names

Sr. No.	Trade Name	Company
1	Plethora	ADAMA India Private Limited
2	Novacarb	UPL Ltd.
3	Novacode	Swal Corporation Ltd.
4	Senora	Anu Products Ltd.
5	Kizomba	Parijat Industries India Pvt. Ltd.
6	Fighter	Paras Crops Ltd.
7	Novaxa	Best Agrolife Ltd.
8	Indoclub	Somnath Demeter Agritron Ltd.
9	Aakramak Plus	Katyayani Organics
10	Novadox	Tropical Agrosystem India (P) Ltd.
11	Rakshak	Essential Biosciences
12	Hydra	Today's Global Crop Protection
13	Cabo	Global Crop Care
14	Blower	Gharda Chemicals Ltd.
15	Pledge	Preventis Chem India (P) Ltd.
16	Wingman	Mahindra Summit

Novaluron + Lambda-cyhalothrin

Available formulations in India : **Novaluron 09.45% + Lambda-cyhalothrin 01.90% ZC**

Classification on the basis of		Novaluron	Lambda-cyhalothrin
Target organism	:	Insecticide	Insecticide
Mode of entry	:	Stomach and Contact poison	Contact and Stomach poison
Mode of action	:	Insect Growth Regulator (IGR)	Nerve poison
Chemical group	:	Benzoylureas	Synthetic pyrethroid
Category of toxicity : **Toxicity label colour** :			

- The combination of Novaluron and Lambda-cyhalothrin is a popular mixed-formulation insecticide used to provide broad-spectrum control over a wide range of agricultural and sometimes public health pests.
- Novaluron is an insect growth regulator that disrupts chitin synthesis, which is vital for insect molting. By preventing molting, it causes death in immature stages of pests like caterpillars.
- Lambda-cyhalothrin is a synthetic pyrethroid insecticide that acts as a neurotoxin by targeting the sodium channels in insect nerve cells, causing paralysis and rapid death. It has a fast knockdown effect and provides effective contact insecticidal action.
- The combination thus provides both long-term pest control through disruption of development by Novaluron and immediate knockdown of pests through neurotoxicity by Lambda-cyhalothrin.
- This dual mode helps manage resistant pest populations and offers broad-spectrum control of pests.

Label claim (DPPQ & S, Faridabad)
Novaluron 09.45% + Lambda-cyhalothrin 01.90% ZC

Crop	Pests	Dose / ha		Waiting period (Days)
		Quantity of formulation (ml)	Volume of water (Lit.)	
Pigeonpea	Pod borer, Leafhopper	750	500	39
Tomato	Fruit borer, Tobacco caterpillar	750	500	07
Cabbage	Diamond back moth, Tobacco caterpillar	750	500	07

Trade Names

Sr. No.	Trade Name	Company
1		

Phenthoate + Cypermethrin

Available formulations in India : **Phenthoate 45.00% + Cypermethrin 06.00% EC**

Classification on the basis of		Phenthoate	Cypermethrin
Target organism	:	Insecticide	Insecticide
Mode of entry	:	Contact and Stomach poison	Contact and Stomach poison
Mode of action	:	Nerve poison	Nerve poison
Chemical group	:	Organophosphate	Synthetic pyrethroid
Category of toxicity : Highly toxic **Toxicity label colour** : Yellow			

- Phenthoate + Cypermethrin is a combination insecticide formulation that provides broad-spectrum control of both chewing and sucking insect pests in various crops.
- Phenthoate is an organophosphate insecticide with contact, stomach, and ovicidal action. It works by inhibiting cholinesterase enzyme in the insect nervous system, disrupting nerve impulses and causing paralysis and death. It also acts as a repellent and has ovicidal effects that prevent egg hatching.
- Cypermethrin is a synthetic pyrethroid insecticide with contact and stomach poison action. It targets the sodium channels in insect nerve cells, causing rapid paralysis and death, providing quick knockdown effects.
- Phenthoate often has a strong, pungent odor which acts as a repellent, discouraging adult moths from laying eggs.

Label claim (DPPQ & S, Faridabad)
Phenthoate 45.00% + Cypermethrin 06.00% EC

Crop	Pests	Dose / ha		Waiting period (Days)
		Quantity of formulation (ml)	Volume of water (Lit.)	
Rice	Yellow stem borer, Leaf folder, Brown plant hopper	1000	500	At the end of harvest

Trade Names

Sr. No.	Trade Name	Company
1	Phendal Plus	Coromandel International Ltd.

Phenthoate + Emamectin benzoate

Available formulations in India : **Phenthoate 35.00% + Emamectin benzoate 03.50% EC**

Classification on the basis of		Phenthoate	Emamectin benzoate
Target organism	:	Insecticide	Insecticide, Acaricide
Mode of entry	:	Contact and Stomach poison	Systemic, Contact and Stomach poison
Mode of action	:	Nerve poison	Nerve poison
Chemical group	:	Organophosphate	Avermectin (Macrocyclic lactones)
Category of toxicity : **Toxicity label colour** :			

- Phenthoate 35.00% + Emamectin benzoate 3.50% EC is a high-potency, dual-action insecticide mixture designed to control a broad range of lepidopteran (caterpillar) and sucking pests.
- The combination of an organophosphate (Phenthoate) and an avermectin (Emamectin benzoate) provides both a fast knockdown effect and long-lasting residual control, making it particularly effective against pests that have developed resistance to single-ingredient insecticides.
- Phenthoate acts as an acetylcholinesterase (AChE) inhibitor. It disrupts the nervous system by causing an accumulation of acetylcholine, leading to rapid paralysis and death of the insect. It provides strong contact and stomach action.
- Emamectin benzoate is a semi-synthetic derivative of abamectin that acts as a chloride channel activator. It causes an influx of chloride ions into nerve cells, leading to irreversible muscle contraction and paralysis. It has excellent translaminar activity, meaning it penetrates leaf tissues to kill pests feeding on the underside.

Label claim (DPPQ & S, Faridabad)
Phenthoate 35.00% + Emamectin benzoate 03.50% EC

Crop	Pests	Dose / ha		Waiting period (Days)
		Quantity of formulation (ml)	Volume of water (Lit.)	
Chilli	Mites, Thrips, Fruit borer	750	500	05

Trade Names

Sr. No.	Trade Name	Company
1		

Profenofos + Cypermethrin

Available formulations in India :

1. **Profenofos 40.00% + Cypermethrin 04.00% EC**
2. **Profenofos 40.00% + Cypermethrin 04.00% EW**

Classification on the basis of		**Profenofos**	**Cypermethrin**
Target organism	:	Insecticide	Insecticide
Mode of entry	:	Contact and Stomach poison	Contact and Stomach poison
Mode of action	:	Nerve poison	Nerve poison
Chemical group	:	Organophosphate	Synthetic pyrethroid
Category of toxicity : Highly toxic **Toxicity label colour** : Yellow			

- This combination is a highly effective, broad-spectrum insecticide widely used in agriculture to control a variety of insect pests.
- Profenofos is an organophosphate insecticide that acts as a potent inhibitor of the acetylcholinesterase enzyme, essential for nerve signal transmission. This inhibition disrupts nerve impulses causing paralysis and death in insects.
- Cypermethrin is a synthetic pyrethroid insecticide that acts by modulating sodium channels in insect nerve cells. It causes nerve hyperexcitation, convulsions, paralysis, and rapid death.
- The combination provides dast knockdown effect due to Cypermethrin; strong ovicidal, larvicidal, and acaricidal activities; excellent translaminar action, killing pests on both upper and lower surfaces of leaves and long residual control with strong contact and stomach action.

Label claim (DPPQ & S, Faridabad)

Profenofos 40.00% + Cypermethrin 04.00% EC

Crop	Pests	Dose / ha		Waiting period (Days)
		Quantity of formulation (ml)	Volume of water (Lit.)	
Cotton	Bollworm complex	1000-1500	500-1000	14
Sugarcane	Early shoot borer	1000	500	138

Trade Names

Sr. No.	Trade Name	Company
1	Polytrin	Syngenta India Ltd.
2	Profigan Super	ADAMA India Private Limited
3	Roket	PI Industries
4	Jashn Super	Tata Rallis India Limited
5	Himawari	IFFCO-MC Crop Science Private Ltd.
6	Ajanta Super	Coromandel International Ltd.
7	Kilcron Plus	Crystal Crop Protection Ltd.

8	Super Mario	Auxin Crop Care
9	Googly	Anu Products Ltd.
10	Profex Super	NACL Industries Ltd.
11	Hitcel	Excel Crop Care Ltd.
12	Profkel Super	Crop Life Science Ltd.
13	Take 44	Parijat Industries India Pvt. Ltd.
14	Legend Super	Crop Chemicals India Ltd.
15	Kemcron Plus	Sumitomo Chemical India Ltd.
16	Blast 404	Paras Crops Ltd.
17	Hempro	Hemani Industries Ltd., Ankleshwar
18	Terror Super	Isagro Asia Agrochemicals Pvt. Ltd.
19	Protector 404	Gujarat Pesti-Chem Industries
20	Profit Super	GFG Crop Science Pvt. Ltd.
21	Score	Tusova Agro Pvt. Ltd.
22	Progress Plus	Herenba Industries Ltd.
23	Fire	Mangroove Laboratory Pvt. Ltd.
24	Cyprosik	Sikko Industries Ltd.
25	Getrid Plus	
26	Mor	Somnath Demeter Agritron Ltd.
27	Polister 44 EC	Silver Sine Bio Tech
28	Profmaxy	Natural Crop Science
29	Pataka	Tropical Agrosystem India (P) Ltd.
30	Proper 404	Essential Biosciences
31	Professor Plus	Synwood Agro Ltd.
32	Profas	Frontline Agri Science
33	Krifos Super	Krishi Rasayan
34	Rocker	Global Crop Care
35	Egg Kill 404	Green India Care
36	Minister	JU Agri Sciences Pvt. Ltd.
37	Proper Super	Linuxcrop Chemicals Pvt. Ltd.
38	Cyrex Super	Synovita Crop science Pvt. Ltd.
39	Jugaad	Gharda Chemicals Ltd.
40	Propibrut	Mankind Agritech
41	Procyp	Preventis Chem India (P) Ltd.
42	Saylaab	Chemley Agritech Pvt. Ltd.
43	Kombat Super	RACCOLTO Agritech India Pvt. Ltd.

Profenofos 40.00% + Cypermethrin 04.00% EW

Crop	Pests	Dose / ha		Waiting period (Days)
		Quantity of formulation (ml)	Volume of water (Lit.)	
Rice	Yellow stem borer, Leaf folder	1000	500	25

Trade Names

Sr. No.	Trade Name	Company
1		

Profenofos + Fenpropathrin

Available formulations in India : **Profenofos 50.00% + Fenpropathrin 05.00% EC**

Classification on the basis of		Profenofos	Fenpropathrin
Target organism	:	Insecticide	Insecticide
Mode of entry	:	Contact and Stomach poison	Contact and Stomach poison
Mode of action	:	Nerve poison	Nerve poison
Chemical group	:	Organophosphate	Synthetic pyrethroid
Category of toxicity : Highly toxic **Toxicity label colour** : Yellow			

- The combination of Profenofos + Fenpropathrin is a highly effective, broad-spectrum insecticide used in agriculture. It's designed to provide control over a wide range of sucking and chewing pests.
- Profenofos is an organophosphate insecticide that acts by inhibiting the enzyme acetylcholinesterase in insects. This causes accumulation of acetylcholine in nerve synapses, disrupting nerve impulse transmission, leading to paralysis and death.
- Fenpropathrin is a synthetic pyrethroid insecticide that acts on the sodium channels of insect nerve cells. It prolongs the opening of these channels, causing overstimulation of nerves, paralysis, and rapid insect death.
- The combination of these two distinct modes of action helps in achieving fast knockdown (quick killing of pests) and extended residual control.

Label claim (DPPQ & S, Faridabad)

Profenofos 50.00% + Fenpropathrin 05.00% EC

Crop	Pests	Dose / ha		Waiting period (Days)
		Quantity of formulation (ml)	Volume of water (Lit.)	
Cotton	Pink bollworm, Thrips, Aphid, Jassid, Whitefly	1500	500	22

Trade Names

Sr. No.	Trade Name	Company
1	Meshi	Sumitomo Chemical India Ltd.

Profenofos + Fenpyroximate

Available formulations in India : **Profenofos 40.00% + Fenpyroximate 02.50% EC**

Classification on the basis of		Profenofos	Fenpyroximate
Target organism	:	Insecticide	Insecticide, Acaricide
Mode of entry	:	Contact and Stomach poison	Stomach and Contact poison
Mode of action	:	Nerve poison	Energy metabolism
Chemical group	:	Organophosphate	Pyrazole
Category of toxicity : Highly toxic **Toxicity label colour** : Yellow			

- This combination is designed to offer a broad spectrum of control, targeting both insect pests and mites simultaneously, making it a valuable tool in integrated pest management strategies. The dual-action formula helps in managing pest populations, including those with developing resistance to single-active-ingredient products.
- The effectiveness of this combination stems from the distinct actions of its two components.
- Profenofos is an organophosphate insecticide that acts primarily as a contact and stomach poison. Its mode of action involves potent inhibition of the enzyme acetylcholinesterase, which is crucial for nerve impulse transmission in insects. This disruption leads to neurophysiological impairments, paralysis, and ultimately death.
- Fenpyroximate, on the other hand, is a specific acaricide (miticide). It belongs to the Mitochondrion Electron Transport Inhibition (METI) group, blocking cellular respiration in mites by inhibiting mitochondrial complex I electron transport. This action causes a quick cessation of feeding and reproduction, leading to paralysis and death of the mites.
- The combination provides rapid "quick knockdown" action along with residual activity for prolonged crop protection.

Label claim (DPPQ & S, Faridabad)
Profenofos 40.00% + Fenpyroximate 02.50% EC

Crop	Pests	Dose / ha		Waiting period (Days)
		Quantity of formulation (ml)	Volume of water (Lit.)	
Chilli	Thrips, Mites, Fruit borer	1000	500	07
Brinjal	Whitefly, Mites, Shoot and fruit borer	1000	500	07

Trade Names

Sr. No.	Trade Name	Company
1	Etna	Sumitomo Chemical India Ltd., Excel Crop Care Ltd.
2	Galwan	Krishi Rasayan

3	Altair	Synovita Crop science Pvt. Ltd.
4	Promate	BHL Rasayan
5	Orax Gold	Atul
6	Fosmite	Sudarshan Bio-Crop Science
7	Proximat	Rumicon Crop Science
8	Promate 425	Volkschem Crop Science Pvt. Ltd.

Profenofos + Lambda-cyhalothrin

Available formulations in India : **Profenofos 40.00% + Lambda-cyhalothrin 01.50% EC**

Classification on the basis of		Profenofos	Lambda-cyhalothrin
Target organism	:	Insecticide	Insecticide
Mode of entry	:	Contact and Stomach poison	Contact and Stomach poison
Mode of action	:	Nerve poison	Nerve poison
Chemical group	:	Organophosphate	Synthetic pyrethroid
Category of toxicity : Highly toxic **Toxicity label colour** : Yellow			

- The combination of Profenofos (an organophosphate) and Lambda-cyhalothrin (a synthetic pyrethroid) is formulated as a powerful, broad-spectrum insecticide.
- Profenofos works primarily as a cholinesterase inhibitor (IRAC Group 1B), disrupting the nervous system function of the insect, and it exhibits translaminar movement, allowing it to penetrate the leaf tissue for extended control, especially against Lepidopteran pests, aphids, and mites.
- In contrast, Lambda-cyhalothrin is a sodium channel modulator (IRAC Group 3) that acts through contact and stomach action, causing rapid paralysis and a fast knockdown effect against a broad spectrum of chewing and sucking insects, including bollworms, whiteflies, thrips, and leafhoppers.
- The combination of these two modes of action helps in managing populations that may have developed resistance to single-ingredient insecticides, offering both quick elimination and residual protection.

Label claim (DPPQ & S, Faridabad)
Profenofos 40.00% + Lambda-cyhalothrin 01.50% EC

Crop	Pests	Dose / ha		Waiting period (Days)
		Quantity of formulation (ml)	Volume of water (Lit.)	
Cotton	Aphid, Thrips, Leafhopper, Whitefly, Bollworm	1000	500-1000	15

Trade Names

Sr. No.	Trade Name	Company
1	Krifos Gold	Krishi Rasayan

Propargite + Bifenthrin

Available formulations in India : **Propargite 50.00% + Bifenthrin 05.00% SE**

Classification on the basis of		**Propargite**	**Bifenthrin**
Target organism	:	Acaricide	Insecticide
Mode of entry	:	Stomach and Contact poison	Contact and Stomach poison
Mode of action	:	Energy metabolism	Nerve poison
Chemical group	:	Sulfite ester	Synthetic pyrethroid
Category of toxicity : Moderately toxic **Toxicity label colour** : Blue			

- The two active ingredients target pests through different mechanisms, leading to dual action.
- Propargite is primarily an acaricide/miticide with contact and vapor action. It belongs to the sulfite ester chemical class (IRAC Group 12) and functions by disrupting the normal respiratory system and vital functions of mites, leading to their suffocation and death. It is highly effective against all mobile stages of various mites, such as the red spider mite.
- On the other hand, Bifenthrin is a synthetic pyrethroid insecticide (IRAC Group 3A) that works through contact and stomach action. Its mode of action involves interfering with the sodium channels in the nervous system of insects, causing hyperexcitation, paralysis, and rapid knockdown. This makes it effective against a broad array of insect pests.
- This dual-component formulation provides superior control over both mites and numerous key insect pests.
- This dual chemistry offers rapid cessation of pest feeding, minimizing crop damage, and provides extended residual protection.

Label claim (DPPQ & S, Faridabad)
Propargite 50.00% + Bifenthrin 05.00% SE

Crop	**Pests**	**Dose / ha**		**Waiting period (Days)**
		Quantity of formulation (ml)	**Volume of water (Lit.)**	
Okra	Mite, Whitefly, Jassid	1100-1150	500	05
Tomato	Mite, Whitefly, Jassid	1100-1150	500	05

Trade Names

Sr. No.	**Trade Name**	**Company**
1	Dammu	Indofil Industries Ltd.
2	Brigade Power	FMC India Pvt. Ltd.

Pymetrozine + Fipronil + Dinotefuran

Available formulations in India : **Pymetrozine 14.80% + Fipronil 07.50% + Dinotefuran 04.80% SC**

Classification on the basis of		Lufenuron	Chlorantraniliprole	Pyriproxyfen
Target organism	:	Insecticide	Insecticide	Insecticide
Mode of entry	:	Contact and Stomach poison	Systemic, Contact and Stomach poison	Contact and Stomach poison
Mode of action	:	Insect Growth Regulator	Muscle poison	Insect Growth Regulator (IGR)
Chemical group	:	Benzoylurea	Anthranilic diamide	Pyridine ether
Category of toxicity :				
Toxicity label colour :				

- This unique combination of Pymetrozine, Fipronil, and Dinotefuran in a Suspension Concentrate (SC) form is a powerhouse "three-way" insecticide.
- By combining three different chemical classes, this formulation provides systemic, contact, and stomach action with both immediate and long-lasting effects.
- This mixture targets the nervous and feeding systems of insects through three distinct pathways:
- Pymetrozine is a selective feeding blocker. It doesn't kill the insect instantly but causes it to stop feeding immediately. The insect remains on the plant but eventually dies of starvation within a few days.
- Fipronil is a phenylpyrazole that disrupts the central nervous system by blocking GABA-gated chloride channels. It provides excellent contact and residual control, ensuring pests that touch the treated surface are eliminated.
- Dinotefuran is a third-generation neonicotinoid known for its extreme solubility and systemic speed. It is rapidly absorbed by the plant and moves to new growth, providing a quick "knockdown" of sucking pests.

Label claim (DPPQ & S, Faridabad)
Pymetrozine 14.80% + Fipronil 07.50% + Dinotefuran 04.80% SC

Crop	Pests	Dose / ha		Waiting period (Days)
		Quantity of formulation (ml)	Volume of water (Lit.)	
Paddy (Rice)	Stem borer, Brown plant hopper, White backed plant hopper, Green leaf hopper	875	500	25

Trade Names

Sr. No.	Trade Name	Company
1		

Pymetrozine + Triflumezopyrim

Available formulations in India : **Pymetrozine 11.30% +Triflumezopyrim 01.90% SC**

Classification on the basis of		Pymetrozine	Triflumezopyrim
Target organism	:	Insecticide	Insecticide
Mode of entry	:	Systemic, Stomach poison	Systemic, Stomach poison
Mode of action	:	Nerve poison	Nerve poison
Chemical group	:	Pyridine azomethine	Mesoionic
Category of toxicity : Moderately toxic **Toxicity label colour** : Blue			

- The combined insecticide leverages two distinct modes of action for robust pest control and resistance management.
- Pymetrozine is classified in the IRAC Group 9B (Chordotonal Organ TRPV Channel Modulators, Pyridine Azomethine Derivatives). Its unique mode of action involves selectively affecting the chordotonal mechanoreceptors in the insect's nervous system, which are crucial for sensing functions like gravity and balance. Physiologically, it prevents sap-feeding insects, such as BPH, from inserting their stylet into the plant tissue, leading to a rapid cessation of feeding.
- Triflumezopyrim, on the other hand, is a relatively newer molecule belonging to the novel mesoionic chemical class (IRAC Group 4E, Nicotinic acetylcholine receptor (nAChR) Inhibitors). It works by binding to and inhibiting the orthosteric site of the nicotinic acetylcholine receptors (nAChRs) in the insect's nervous system. This inhibitory action is distinct from neonicotinoids (Group 4A, which are nAChR agonists) and provides a valuable tool for managing pests that have developed resistance to other common insecticides.
- The combination of Pymetrozine and Triflumezopyrim provides both quick knockdown (due to feeding cessation) and long-lasting control against target pests like BPH.

Label claim (DPPQ & S, Faridabad)
Pymetrozine 11.30% + Triflumezopyrim 01.90% SC

Crop	Pests	Dose / ha		Waiting period (Days)
		Quantity of formulation (ml)	Volume of water (Lit.)	
Rice	Brown plant hopper, White backed plant hopper, Green leaf hopper	1000	500	31

Trade Names

Sr. No.	Trade Name	Company
1	Vestoria Pro	Syngenta India Ltd.

Pyriproxyfen + Fenpropathrin

Available formulations in India : Pyriproxyfen 05.00% + Fenpropathrin 15.00% EC

Classification on the basis of		Pyriproxyfen	Fenpropathrin
Target organism	:	Insecticide	Insecticide
Mode of entry	:	Contact and Stomach poison	Contact and Stomach poison
Mode of action	:	Insect Growth Regulator (IGR)	Nerve poison
Chemical group	:	Pyridine ether	Synthetic pyrethroid
Category of toxicity : Highly toxic **Toxicity label colour** : Yellow			

- The mixture combines two distinct modes of action for enhanced efficacy and resistance management.
- Pyriproxyfen is an Insect Growth Regulator (IGR), which acts by mimicking juvenile hormones, thereby disrupting the life cycle of the pest, particularly preventing the metamorphosis of larvae into adults and reducing the egg-laying capacity of adults.
- Complementing this, Fenpropathrin is a fast-acting pyrethroid class insecticide that works via contact and ingestion, causing hyper-excitation of the insect's nervous system by affecting sodium channels, leading to a quick knockdown and kill of adult pests.
- This combination provides both immediate relief from active infestations and long-term control by breaking the pest's reproductive cycle, making it effective against a broad spectrum of chewing and sucking pests.

Label claim (DPPQ & S, Faridabad)
Pyriproxyfen 05.00% + Fenpropathrin 15.00% EC

Crop	Pests	Dose / ha		Waiting period (Days)
		Quantity of formulation (ml)	Volume of water (Lit.)	
Cotton	Bollworms	500-750	500-750	14
	Whitefly	600	500	19
Brinjal	Whitefly, Shoot and fruit borer	500-750	500-750	07
Okra	Whitefly, Fruit borer	500-750	500-750	07
Chilli	Whitefly, Fruit borer	500-750	500-750	07

Trade Names

Sr. No.	Trade Name	Company
1	Sumiprempt	Sumitomo Chemical India Ltd.
2	Fenox	Global Crop Care
3	Rudram	Sahib Pesticides Pvt. Ltd.

Pyriproxyfen + Bifenthrin

Available formulations in India : **Pyriproxyfen 10.00% + Bifenthrin 10.00% EC**

Classification on the basis of		Pyriproxyfen	Bifenthrin
Target organism	:	Insecticide	Insecticide
Mode of entry	:	Contact and Stomach poison	Contact and Stomach poison
Mode of action	:	Insect Growth Regulator (IGR)	Nerve poison
Chemical group	:	Pyridine ether	Synthetic pyrethroid
Category of toxicity : Moderately toxic **Toxicity label colour** : Blue			

- This combination is highly effective due to its dual mode of action that targets different life stages of pests, providing a strong synergistic effect.
- Pyriproxyfen acts as an Insect Growth Regulator (IGR), mimicking the insect's juvenile hormone. Its action is slower but critical, preventing the hatching of eggs, inhibiting the metamorphosis of nymphs into adults, and causing sterility in adult insects, thus controlling the next generation.
- In contrast, Bifenthrin is a fast-acting pyrethroid insecticide that provides a rapid knockdown effect on adult insects primarily through contact and stomach action. It works by interfering with the insect's central and peripheral nervous system, specifically by disrupting the sodium channels, which leads to paralysis and death.
- This dual action ensures complete control over all stages (eggs, nymphs, and adults) and offers a longer-lasting residual activity.
- The combination is particularly recommended for controlling whiteflies in crops like cotton, but is also effective against other sucking pests.

Label claim (DPPQ & S, Faridabad)
Pyriproxyfen 10.00% + Bifenthrin 10.00% EC

Crop	Pests	Dose / ha		Waiting period (Days)
		Quantity of formulation (ml)	Volume of water (Lit.)	
Cotton	Whitefly	600	500	19
Chilli	Yellow mite, Thrips, Whitefly, Fruit borer	1250	500	15
Brinjal	Whitefly, Shoot and fruit borer	1250	500	15
Soybean	Whitefly	1000	500	35
Green gram	Whitefly	1000	500	25

Trade Names

Sr. No.	Trade Name	Company
1	Stoxx Super	Anu Products Ltd.
2	Leonis	Parijat Industries India Pvt. Ltd.
3	Prudens	Godrej Agrovet
4	Egonis	Eagle Plant Protect Pvt. Ltd.
5	Antkal	Cropgate Chemicals Pvt. Ltd.
6	Final Call	R K Chemicals
7	Raftaar	Essential Biosciences
8	Pyrinox	Silver Sine Bio Tech
9	Demand	Somnath Demeter Agritron Ltd.
10	Pyriseen	Synergy
11	Proxyfen	Natural Crop Science
12	Hathiyar	Tropical Agrosystem India (P) Ltd.
13	K-10 Super	Koronation Bio-Science
14	Sota	Green India Care
15	Exylon Pyrithrin	Torrent Crop Science
16	Trophy	Preventis Chem India (P) Ltd.
17	Cease	Gujrat Pesticides
18	Pyfengold	Star Chemicals

Pyriproxyfen + Dinotefuran + Diafenthiuron

Available formulations in India : **Pyriproxyfen 08.00% + Dinotefuran 05.00% + Diafenthiuron 18.00% EC**

Classification on the basis of		Pyriproxyfen	Dinotefuran	Diafenthiuron
Target organism	:	Insecticide	Insecticide	Insecticide, Acaricide
Mode of entry	:	Contact and Stomach poison	Systemic, Contact and Stomach poison	Contact and Stomach poison
Mode of action	:	Insect Growth Regulator (IGR)	Nerve poison	Energy metabolism
Chemical group	:	Pyridine ether	Neonicotinoid	Thiourea
Category of toxicity : Moderately toxic **Toxicity label colour** : Blue				

- The combination product containing Pyriproxyfen, Dinotefuran, and Diafenthiuron is a highly effective, triple-action insecticide often formulated as a Suspension Concentrate (SC). This ternary mixture is specifically designed to provide comprehensive and long-lasting control of a broad spectrum of persistent sucking pests, including whiteflies, jassids, aphids, and thrips, and often demonstrates acaricidal activity against mites. The combination of three distinct modes of action is a critical strategy for managing pest resistance and ensuring the suppression of all life stages of the target insects.
- Each active ingredient contributes a unique mechanism.
- Pyriproxyfen is an Insect Growth Regulator (IGR) that mimics the natural juvenile hormone of insects, primarily acting as an ovicidal (prevents egg hatching) and larvicidal agent (inhibits metamorphosis from nymph/larva to adult), thus disrupting the pest's life cycle and reproductive capacity.
- Dinotefuran is a third-generation neonicotinoid insecticide with strong systemic action, meaning it is absorbed by the plant and moves throughout its tissues, killing pests that feed on the treated foliage. It acts rapidly on the insect's nervous system, providing a quick knockdown effect.
- Diafenthiuron is a pro-insecticide that converts into an active compound upon exposure to light or inside the insect, acting as a mitochondrial respiration inhibitor to disrupt ATP synthesis. It offers contact, stomach, and often a vapor action, providing long-lasting residual activity and control over mobile stages (nymphs and adults), and is also effective against mites.
- Commonly used on crops like cotton, brinjal (eggplant), and other vegetables, this combination provides a preventive and curative solution that targets eggs, nymphs, and adults, leading to a healthier crop and better yield by mitigating damage from hard-to-kill and resistant pest populations.

Label claim (DPPQ & S, Faridabad)

Pyriproxyfen 08.00% + Dinotefuran 05.00% + Diafenthiuron 18.00% EC

Crop	Pests	Dose / ha		Waiting period (Days)
		Quantity of formulation (ml)	Volume of water (Lit.)	
Brinjal	Whitefly, Jassid, Thrips, Aphid	600-825	500	8-10
Cotton	Whitefly, Thrips, Jassid, Aphid	600	500	48
Chilli	Whitefly, Jassid, Thrips, Aphid	825	500	05

Trade Names

Sr. No.	Trade Name	Company
1	Ronfen	Best Agrolife Ltd.
2	Tridev	JU Agri Sciences Pvt. Ltd.
3	Triveni	Natural Crop Science
4	Fli	Tropical Agrosystem India (P) Ltd.
5	TrianT	Global Crop Care
6	Caliber	Preventis Chem India (P) Ltd.
7	Sutathion	Sudarshan Farm Chemical India Pvt. Ltd.
8	Tag Fli	Tropical Agrosystem India (P) Ltd.
9	Peri Peri	Praveka Life Science

Spinetoram + Methoxyfenozide

Available formulations in India : Spinetoram 05.66% +Methoxyfenozide 28.30% SC

Classification on the basis of		Spinetoram	Methoxyfenozide
Target organism	:	Insecticide	Insecticide
Mode of entry	:	Stomach and Contact poison	Stomach and Contact poison
Mode of action	:	Nerve poison	Insect Growth regulator (IGR)
Chemical group	:	Spinosyns	Diacylhydrazines
Category of toxicity : Moderately toxic **Toxicity label colour** : Blue			

- Spinetoram + Methoxyfenozide is a specialized insecticide mixture of two powerful active ingredients to provide a comprehensive solution for managing tough lepidopteran (caterpillar) pests.
- This mixture is highly effective because it attacks pests using two different biological pathways, which also helps in managing insecticide resistance.
- Spinetoram is a member of the Spinosyn class (IRAC Group 5). It acts as an allosteric modulator of nicotinic acetylcholine receptors, leading to rapid nervous system disruption, paralysis, and death. It offers fast knockdown through contact and ingestion.
- Methoxyfenozide is an Insect Growth Regulator (IGR) belonging to the diacylhydrazine class (IRAC Group 18). It mimics the molting hormone (ecdysone), causing the larvae to undergo a lethal, premature molt. It is highly specific to lepidopteran pests and provides long-lasting residual control.

Label claim (DPPQ & S, Faridabad)
Spinetoram 05.66% + Methoxyfenozide 28.30% SC

Crop	Pests	Dose / ha		Waiting period (Days)
		Quantity of formulation (ml)	Volume of water (Lit.)	
Rice	Stem borer, Leaf folder	375-400	500	56
Cabbage, Cauliflower	Diamond back moth, Leaf eating caterpillar	375-400	375-500	10
Black gram	Pod borer, Spotted pod borer	375-400	375-500	48

Trade Names

Sr. No.	Trade Name	Company
1	Engage	Corteva Agriscience

Spiromesifen + Hexythiazox + Abamectin

Available formulations in India : **Spiromesifen 15.00% + Hexythiazox 02.20% + Abamectin 01.15% SC**

Classification on the basis of		Spiromesifen	Hexythiazox	Abamectin
Target organism	:	Insecticide, Acaricide	Acaricide	Acaricide, Insecticide
Mode of entry	:	Stomach and Contact poison	Contact and Stomach poison	Stomach and Contact poison
Mode of action	:	Lipid synthesis and growth regulator	Growth regulator	Nerve poison
Chemical group	:	Tetronic and Tetramic acid derivatives	Thiazolidinecarboxamide	Avermectin
Category of toxicity : Moderately toxic **Toxicity label colour** : Blue				

- This potent "Triple Mix" in a Suspension Concentrate (SC) form is a specialized acaricide (miticide) designed for total management of mites at all life stages.
- This formulation is highly effective because it attacks the mite's biological processes from three different angles.
- Spiromesifen (15.00%): Ais a quaternary sugar derivative (lipid biosynthesis inhibitor). It primarily targets the nymph stages and female fertility. It prevents mites from creating the lipids necessary for growth and egg production, effectively "sterilizing" the population.
- Hexythiazox is a specialized ovicide and larvicide. It is highly effective at killing mite eggs and early-stage nymphs. It has excellent residual activity, staying on the leaf surface to catch new hatchlings.
- Abamectin is a chloride channel activator that provides the immediate knockdown. It is highly toxic to adult mites upon contact or ingestion and has translaminar activity, meaning it moves through the leaf to kill mites hiding on the underside.

Label claim (DPPQ & S, Faridabad)
Spiromesifen 15.00% + Hexythiazox 02.20% + Abamectin 01.15% SC

Crop	Pests	Dose / ha		Waiting period (Days)
		Quantity of formulation (ml)	Volume of water (Lit.)	
Chilli	Yellow mite, Black thrips	625	500	16

Trade Names

Sr. No.	Trade Name	Company
1	Cubax Power Extra	Best Agrolife Ltd.

Spirotetramat + Imidacloprid

Available formulations in India : Spirotetramat 11.01% + Imidacloprid 11.01% SC

Classification on the basis of		Spirotetramat	Imidacloprid
Target organism	:	Insecticide, Acaricide	Insecticide
Mode of entry	:	Systemic, Contact poison	Systemic, Contact and Stomach poison
Mode of action	:	Nerve poison	Nerve poison
Chemical group	:	Tetronic / Tetramic acid derivatives (Ketoenol)	Neonicotinoid
Category of toxicity : Highly toxic **Toxicity label colour** : Yellow			

- Spirotetramat belongs to the keto-enol group and works by inhibiting lipid biosynthesis. It mainly affects the immature stages of sucking pests and provides prolonged control due to its capability for two-way systemic translocation within the plant (xylem and phloem), ensuring "shoot to root" protection.
- Imidacloprid is a neonicotinoid insecticide that acts as a nicotinic acetylcholine receptor (nAChR) inhibitor, disrupting the insect's nervous system and causing rapid knockdown. It is especially effective on adult sucking pests and eggs.
- The dual systemicity and different targets (nervous system and lipid synthesis/developmental stages) ensure complete protection of the entire plant.
- It offers extended suppression of pest populations, leading to improved crop vigor and higher yields.

Label claim (DPPQ & S, Faridabad)
Spirotetramat 11.01% + Imidacloprid 11.01% SC

Crop	Pests	Dose / ha		Waiting period (Days)
		Quantity of formulation (ml)	Volume of water (Lit.)	
Okra	Red spider mite	500	500	03
Brinjal	Whitefly, Red spider mite	500	500	05
Mango	Mealybug	0.075%	As required	15
Cotton	Mealybug	625	500	22
Grapes	Thrips, Mites, Mealybug	750	1000	60

Trade Names

Sr. No.	Trade Name	Company
1	Movento Energy	Bayer Crop Science Ltd.
2	MiteOff	Global Crop Care
3	Thadam	Iscar Crop Science
4	Tetramichem	Creative Unichem Pvt. Ltd.

Spiropidon + Acetamiprid

Available formulations in India : Spiropidon 30.00% + Acetamiprid 24.00% WG

Classification on the basis of		Spiropidon	Acetamiprid
Target organism	:	Insecticide	Insecticide
Mode of entry	:	Systemic, Contact poison	Systemic, Contact and Stomach poison
Mode of action	:	Nerve poison	Nerve poison
Chemical group	:	Tetronic / Tetramic acid derivatives (Ketoenol)	Neonicotinoid
Category of toxicity : Moderately toxic **Toxicity label colour** : Blue			

- Spiropidion is a novel molecule from the spirocyclic tetramic acid chemical class. It targets the insect's lipid synthesis pathway, effectively disrupting energy metabolism and leading to feeding inhibition, paralysis, and death. It acts on both nymphs and adult stages by contact and systemic movement within the plant.
- Acetamiprid is a neonicotinoid insecticide acting on nicotinic acetylcholine receptors (nAChRs), which causes continuous nerve excitation, paralysis, and death of the pest. It has strong systemic and translaminar movement, protecting new growth.
- The combination offers dual-site activity with novel chemistry plus neonicotinoid synergy, providing comprehensive pest control and delaying resistance development.
- It delivers fast knockdown and long residual control against key sucking pests.

Label claim (DPPQ & S, Faridabad)
Spiropidon 30.00% + Acetamiprid 24.00% WG

Crop	Pests	Dose / ha		Waiting period (Days)
		Quantity of formulation (g)	Volume of water (Lit.)	
Tomato	Thrips, Aphid, Whitefly	200-250	500	05
Cotton	Mealybug, Thrips, Jassid, Aphid, Whitefly	200-250	500	85

Trade Names

Sr. No.	Trade Name	Company
1	Elestal Neo	Syngenta India Pvt. Ltd.

Sulfoxaflor + Buprofezin

Available formulations in India : **Sulfoxaflor 07.50% + Buprofezin 15.00% SC**

Classification on the basis of		Sulfoxaflor	Buprofezin
Target organism	:	Insecticide	Insecticide
Mode of entry	:	Systemic, Contact poison	Contact and Stomach poison
Mode of action	:	Nerve poison	Insect Growth Regulator (IGR)
Chemical group	:	Sulfoximine	Thiadiazin
Category of toxicity : **Toxicity label colour** :			

- Sulfoxaflor belongs to Sulfoximine (a newer class of insecticide). It acts as an agonist at the insect's nicotinic acetylcholine receptors (nAChR) in the central nervous system, leading to uncontrolled nerve impulses, tremors, paralysis, and ultimately, death. Its action is distinct from older neonicotinoids (Group 4A), making it effective against pests resistant to carbamates, neonicotinoids, organophosphates, and pyrethroids. It is effective against a broad range of sap-feeding pests.
- Buprofezin belongs to Thiadiazin (or Buprofezin) group. It is an Insect Growth Regulator (IGR). It acts by inhibiting the synthesis of chitin, a key component of the insect exoskeleton. This disrupts the molting process, causing the death of nymphs and larvae, and often reduces the egg-laying capacity of adult females (ovicidal action). It has primarily a contact and stomach action. It is used for controlling Homopteran pests.
- The mixture can target both adults and nymphs/larvae (due to Buprofezin's IGR activity) and cover a wider range of pests or life stages.
- It provides a more robust control, often with quicker knockdown (from Sulfoxaflor) and longer residual control (due to Buprofezin's IGR effect on future generations).

Label claim (DPPQ & S, Faridabad)
Sulfoxaflor 07.50% + Buprofezin 15.00% SC

Crop	Pests	Dose / ha		Waiting period (Days)
		Quantity of formulation (ml)	Volume of water (Lit.)	
Rice	Brown plant hopper	800-1000	500	26

Trade Names

Sr. No.	Trade Name	Company
1		

Tetraniliprole + Fipronil

Available formulations in India : **Tetraniliprole 0.40% + Fipronil 0.60% GR**

Classification on the basis of		Tetraniliprole	Fipronil
Target organism	:	Insecticide	Insecticide
Mode of entry	:	Systemic, Stomach poison	Contact and Stomach poison
Mode of action	:	Nerve poison	Nerve poison
Chemical group	:	Anthranilic diamide	Phenylpyrazole
Category of toxicity : Moderately toxic **Toxicity label colour** : Blue			

- Tetraniliprole 0.40% + Fipronil 0.60% GR is a modern, granular insecticide mixture designed for soil application or broadcasting. This combination is particularly effective in high-value cereal and plantation crops, providing a "shield" against both surface-dwelling and soil-borne pests.
- This formulation combines an anthranilic diamide with a phenylpyrazole to deliver comprehensive protection.
- Tetraniliprole targets the ryanodine receptors in insect muscles, causing immediate cessation of feeding, muscle paralysis, and death. It is highly systemic, moving from the roots to the foliage.
- Fipronil is a central nervous system disruptor that blocks GABA-gated chloride channels. It provides excellent contact and stomach action with a strong "root growth stimulant" effect often observed in crops like rice and sugarcane.
- The granular form ensures a slow, sustained release of active ingredients, providing long-lasting protection compared to liquid sprays.

Label claim (DPPQ & S, Faridabad)
Tetraniliprole 0.40% + Fipronil 0.60% GR

Crop	Pests	Dose / ha		Waiting period (Days)
		Quantity of formulation (Kg)	Volume of water (Lit.)	
Rice	Stem borer, Leaf folder	10.0-12.5	--	65
Sugarcane	Early shoot borer, White grub	12.5-15.0	--	314

Trade Names

Sr. No.	Trade Name	Company
1	Bicota	Bayer CropScience Ltd.

Tetraniliprole + Spirotetramat

Available formulations in India : **Tetraniliprole 10.81% + Spirotetramat 21.62% SC**

Classification on the basis of		Tetraniliprole	Spirotetramat
Target organism	:	Insecticide	Insecticide, Acaricide
Mode of entry	:	Systemic, Stomach poison	Systemic, Contact poison
Mode of action	:	Nerve poison	Nerve poison
Chemical group	:	Anthranilic diamide	Tetronic / Tetramic acid derivatives (Ketoenol)
Category of toxicity : **Toxicity label colour :**			

- Tetraniliprole 10.81% + Spirotetramat 21.62% SC is a premium, dual-systemic insecticide mixture primarily used for high-value horticultural and field crops. This formulation combines a diamide and a ketoenol to provide complete plant protection from both chewing and sucking pests.
- This Suspension Concentrate (SC) works through two distinct, highly advanced pathways.
- Tetraniliprole is an anthranilic diamide (Group 28) that targets insect ryanodine receptors. It causes immediate cessation of feeding and muscle paralysis. It is highly effective against caterpillars and beetles.
- Spirotetramat is a ketoenol (Group 23) known for its "two-way" systemicity (ambimobile). Once absorbed, it moves both upward to new growth and downward to the roots. It inhibits lipid biosynthesis, preventing sucking pests from maturing or reproducing.

Label claim (DPPQ & S, Faridabad)
Tetraniliprole 10.81% + Spirotetramat 21.62% SC

Crop	Pests	Dose / ha		Waiting period (Days)
		Quantity of formulation (ml)	Volume of water (Lit.)	
Okra	Shoot and fruit borer, Aphid, Whitefly, Mite	375	500	03
Cabbage	Diamond back moth, Tobacco caterpillar, Aphid	375	500	07
Chilli	Fruit borer, Thrips, Whitefly, Mite	375	500	05
Brinjal	Shoot and fruit borer, Whitefly, Mite	375	500	05
Tomato	Fruit borer, Leaf miner, Thrips, Whitefly	375	500	05

Trade Names

Sr. No.	Trade Name	Company
1		

Tetraniliprole + Thiacloprid

Available formulations in India : **Tetraniliprole 10.08% + Thiacloprid 30.25% SC**

Classification on the basis of		Tetraniliprole	Thiacloprid
Target organism	:	Insecticide	Insecticide
Mode of entry	:	Systemic, Stomach poison	Systemic, Contact and Stomach poison
Mode of action	:	Nerve poison	Nerve poison
Chemical group	:	Anthranilic diamide	Neonicotinoid
Category of toxicity : Moderately toxic **Toxicity label colour** : Blue			

- The mixture of Tetraniliprole and Thiacloprid is an insecticide combination that provides a broad spectrum of control against different types of insect pests.
- Tetraniliprole belongs to anthranilic diamide group. It is primarily a ryanodine receptor modulator. It acts by interfering with the ryanodine-sensitive calcium release channels in the insect's muscles. This leads to uncontrolled release of calcium, causing muscle paralysis, cessation of feeding almost immediately, and ultimately, death of the pest. Primarily it is effective against chewing pests, especially Lepidopteran larvae (caterpillars/borers), as well as some Coleopteran (beetles) and Dipteran pests. It provides systemic and translaminar activity.
- Thiacloprid belongs to neonicotinoid group. It acts on the insect's central nervous system. It is an agonist (mimic) of the nicotinic acetylcholine receptors (nAChR), disrupting signal transmission and leading to paralysis and death. Primarily it is effective against sucking and piercing-sucking pests such as aphids, whiteflies, and jassids, but also against some chewing pests. It has systemic and contact action.
- Tetraniliprole moves via xylem; Thiacloprid translocate through both xylem and phloem.
- Thiacloprid ensures a quick knockdown of adult insects, while Tetraniliprole provides longer-term residual protection through larval control.

Label claim (DPPQ & S, Faridabad)
Tetraniliprole 10.08% + Thiacloprid 30.25% SC

Crop	Pests	Dose / ha		Waiting period (Days)
		Quantity of formulation (ml)	Volume of water (Lit.)	
Rice	Stem borer, Leaf folder	312.5	500	43

Trade Names

Sr. No.	Trade Name	Company
1	Shriram Chrone	DCM Shriram Ltd.

Thiamethoxam + Lambda-cyhalothrin

Available formulations in India : **Thiamethoxam 12.60% + Lambda-cyhalothrin 09.50% ZC**

Classification on the basis of		Thiamethoxam	Lambda-cyhalothrin
Target organism	:	Insecticide	Insecticide
Mode of entry	:	Systemic, Contact and Stomach poison	Contact and Stomach poison
Mode of action	:	Nerve poison	Nerve poison
Chemical group	:	Neonicotinoid	Synthetic pyrethroid
Category of toxicity : Highly toxic **Toxicity label colour** : Yellow			

- The combination of Thiamethoxam and Lambda-cyhalothrin is formulated as a powerful, broad-spectrum insecticide, often available as a ZC (a mixed formulation of Capsule Suspension and Suspension Concentrate). This product is highly valued in agriculture for offering both quick knockdown and long residual control against a wide array of insect pests.
- The dual mode of action is key to its effectiveness.
- Thiamethoxam belongs to the neonicotinoid chemical class and is a systemic insecticide. Once applied, it is rapidly absorbed by the plant's roots and foliage and is translocated throughout the plant via the xylem. It works by irreversibly blocking postsynaptic nicotinic acetylcholine receptors in the insect's central nervous system, leading to hyperexcitation, paralysis, and eventual death. This systemic activity is particularly effective against sucking pests like aphids, jassids, thrips, and whiteflies, and offers protection from within the plant tissues, which also helps in controlling insect vectors that transmit viral diseases.
- Conversely, Lambda-cyhalothrin is a synthetic pyrethroid insecticide that acts primarily as a contact and stomach poison. It works rapidly by targeting the voltage-gated sodium channels in the nerve cells of insects, leading to immediate hyperexcitation, convulsions, and rapid paralysis, providing a fast knockdown effect. This component is highly effective against chewing pests such as various species of bollworms, stem borers, fruit borers, and leaf-eating caterpillars.
- By combining the immediate action of Lambda-cyhalothrin with the prolonged systemic protection of Thiamethoxam, the formulation provides comprehensive control against multiple pest species and life stages, helping to manage pest populations below economic threshold levels and potentially reducing the total number of sprays required during a crop season.

Label claim (DPPQ & S, Faridabad)

Thiamethoxam 12.60% + Lambda-cyhalothrin 09.50% ZC

Crop	Pests	Dose / ha		Waiting period (Days)
		Quantity of formulation (ml)	Volume of water (Lit.)	
Cotton	Jassid, Aphid, Thrips, Bollworms	200	500	26
Maize	Aphid, Shoot fly, Stem borer	125	500	42
Groundnut	Leafhopper, Leaf eating caterpillar	150	500	28
Soybean	Stem fly, Semilooper, Girdle beetle	125	500	48
Chilli	Thrips, Fruit borer	150	500	03
Tea	Tea mosquito bug, Thrips, Semilooper	150	400	01
Tomato	Thrips, Whitefly, Fruit borer	125	500	05

Trade Names

Sr. No.	Trade Name	Company
1	Iruka	IFFCO-MC Crop Science Private LTd.
2	Alika	Syngenta India Pvt. Ltd.
3	Cocoa	Anu Products Ltd.
4	Cropdon	Crop Life Science Ltd.
5	Tegata	Parijat Industries India Pvt. Ltd.
6	Cresta	Paras Crops Ltd.
7	Thialem	Sikko Industries Ltd.
8	Zapac	Dhanuka Agrotech Ltd.
9	Thorlet	ATPL
10	BACF Hover	Bharat Agro Chemicals & Fertilizers
11	Arion	Rain Bio Tech Industries
12	Chakrawarti	Katyayani Organics
13	Anolik	Adama India Pvt. Ltd.
14	Thilamcy	R K Chemicals
15	Thioshield	Essential Biosciences
16	Antkaal	HPM Chemicals and Fertilizers Ltd.
17	Tegata	Parijat Industries India Pvt. Ltd.
18	Hakura	Shivalik Crop Science Pvt. Ltd.
19	Thiola	JU Agri Science Pvt. Ltd.
20	Lumina	Torrent Crop Science
21	Astron	India Pesticides Ltd.
22	2-In-One Ultra	Somnath Demeter Agritron Ltd.
23	Challange	Silver Sine Bio Tech
24	Zepok Thaio	Natural Crop Science
25	Ekka	Tropical Agrosystem India (P) Ltd.
26	Double Hit	Synwood Agro Ltd.
27	Thiano	Frontline Agri Science

28	Dhumketu	Koronation Bio-Science
29	Korect	Krishi Rasayan
30	Alik	Global Crop Care
31	Panther	Green India Care
32	JU-Thiola	JU Agri Sciences Pvt. Ltd.
33	X-Man	Linuxcrop Chemicals Pvt. Ltd.
34	Talagold	Mankind Agritech
35	Exylon Lumina	Torrent Crop Science
36	Nato	Preventis Chem India (P) Ltd.
37	Loadout	Chemley Agritech Pvt. Ltd.
38	Paheredar	RACCOLTO Agritech India Pvt. Ltd.

Thiamethoxam + Fipronil + Chlorantraniliprole

Available formulations in India : **Thiamethoxam 01.25% + Fipronil 01.25% + Chlorantraniliprole 0.60 GR**

Classification on the basis of		Thiamethoxam	Fipronil	Chlorantraniliprole
Target organism	:	Insecticide	Insecticide	Insecticide
Mode of entry	:	Systemic, Contact and Stomach poison	Contact and Stomach poison	Systemic, Contact and Stomach poison
Mode of action	:	Nerve poison	Nerve poison	Muscle poison
Chemical group	:	Neonicotinoid	Phenylpyrazole	Anthranilic diamide
Category of toxicity :				
Toxicity label colour :				

- A product combining Thiamethoxam, Fipronil, and Chlorantraniliprole would represent an extremely potent, triple-action insecticide designed to combat a very wide spectrum of agricultural pests.
- This combination brings together three powerful and chemically distinct insecticide classes.
- Thiamethoxam belongs to the neonicotinoid family (IRAC Group 4A), acting systemically as a neurotoxin that disrupts the insect's central nervous system by binding to nicotinic acetylcholine receptors, primarily targeting sucking pests like aphids, whiteflies, and leafhoppers.
- Fipronil is a phenylpyrazole insecticide (IRAC Group 2B) with contact, stomach, and moderate systemic action; it blocks GABA-gated chloride channels in the central nervous system, causing over-excitation and rapid death, making it highly effective against soil pests (like grubs and termites) and certain chewing and sucking insects.
- Finally, Chlorantraniliprole is an anthranilic diamide (IRAC Group 28) that is systemic and primarily targets chewing pests (like borers and caterpillars) by activating insect ryanodine receptors, leading to calcium ion release, muscle paralysis, and quick feeding cessation.
- The primary benefit of combining these three active ingredients is the delivery of unparalleled broad-spectrum efficacy against the two main categories of agricultural pests: sucking pests (Thiamethoxam, Fipronil) and chewing/borer pests (Chlorantraniliprole, Fipronil).
- The triple action provides systemic protection (Thiamethoxam, Chlorantraniliprole) by being absorbed and translocated throughout the plant, as well as rapid contact and stomach action (Fipronil, Chlorantraniliprole).

Label claim (DPPQ & S, Faridabad)

Thiamethoxam 01.25% + Fipronil 01.25% + Chlorantraniliprole 0.60 GR

Crop	Pests	Dose / ha		Waiting period (Days)
		Quantity of formulation (Kg)	Volume of water (Lit.)	
Rice	Yellow stem borer, Leaf folder	4.50	--	77

Trade Names

Sr. No.	Trade Name	Company
1	Stem-Lee	Tropical Agrosystem India (P) Ltd.

Thiocyclam hydrogen oxalate + Clothianidin

Available formulations in India : **Thiocyclam hydrogen oxalate 03.00% + Clothianidin 01.20% GR**

Classification on the basis of		Thiocyclam hydrogen oxalate	Clothianidin
Target organism	:	Insecticide	Insecticide
Mode of entry	:	Stomach and Contact poison	Systemic, Contact and Stomach poison
Mode of action	:	Nerve poison	Nerve poison
Chemical group	:	Nereistoxin analouge	Neonicotinoid
Category of toxicity : **Toxicity label colour :**			

- This combination leverages the distinct modes of action of the two active ingredients to deliver both rapid knockdown and long-lasting systemic protection.
- This insecticide utilizes a dual mode of action to achieve comprehensive pest control.
- Thiocyclam hydrogen oxalate acts as a neurotoxin with primary contact and stomach action. It disrupts the insect's nervous system, leading to rapid cessation of feeding, paralysis, and death.
- Conversely, Clothianidin is a systemic neonicotinoid insecticide. Once applied, it is absorbed by the plant and translocated throughout the tissues, providing protection from within. Clothianidin also acts as a neurotoxin by interfering with insect nerve signals, ultimately causing paralysis and death.
- The synergism between the fast-acting, contact/stomach poison of Thiocyclam and the systemic, residual activity of Clothianidin ensures superior control over both visible and hidden pests, and prevents pests from developing resistance as quickly as they might to a single-ingredient product.

Label claim (DPPQ & S, Faridabad)
Thiocyclam hydrogen oxalate 03.00% + Clothianidin 01.20% GR

Crop	Pests	Dose / ha		Waiting period (Days)
		Quantity of formulation (Kg)	Volume of water (Lit.)	
Rice	Yellow stem borer, Leaf folder, Brown plant hopper, White backed plant hopper	10.00-12.50	--	56

Trade Names

Sr. No.	Trade Name	Company
1	Aurthor	GPS Crop Science Pvt. Ltd.

Tolfenpyrad + Bifenthrin

Available formulations in India : **Tolfenpyrad 15.00% + Bifenthrin 07.50% SE**

Classification on the basis of		Tolfenpyrad	Bifenthrin
Target organism	:	Insecticide	Insecticide
Mode of entry	:	Stomach and Contact poison	Contact and Stomach poison
Mode of action	:	Energy metabolism	Nerve poison
Chemical group	:	Pyrazole	Synthetic pyrethroid
Category of toxicity : Highly toxic **Toxicity label colour** : Yellow			

- Tolfenpyrad 15.00% + Bifenthrin 7.50% SE is a high-performance, broad-spectrum insecticide and miticide formulation.
- The combination provides both respiratory inhibition and nervous system disruption.
- Tolfenpyrad is a Mitochondrial Complex I Electron Transport Inhibitor (METI). It blocks the energy production cells of the insect, leading to rapid cessation of feeding and death. It is highly effective against a wide range of sucking and chewing pests, including those resistant to other chemicals.
- Bifenthrin is a potent pyrethroid that targets the nervous system by keeping sodium channels open. It provides a fast knockdown and has excellent residual activity on the leaf surface. It is also an effective miticide.

Label claim (DPPQ & S, Faridabad)
Tolfenpyrad 15.00% + Bifenthrin 07.50% SE

Crop	Pests	Dose / ha		Waiting period (Days)
		Quantity of formulation (ml)	Volume of water (Lit.)	
Cotton	Leafhopper, Aphid, Thrips, Pink bollworm	750	500	54
Transplanted Paddy	Stem borer, Leaf folder, Brown plant hopper, Green leaf hopper	750	500	35

Trade Names

Sr. No.	Trade Name	Company
1	Treza	Agrostar
2	Raavan	GPS Crop Science Pvt. Ltd.

Tolfenpyrad + Emamectin benzoate

Available formulations in India : **Tolfenpyrad 18.75% + Emamectin benzoate 0.94% SC**

Classification on the basis of		Tolfenpyrad	Emamectin benzoate
Target organism	:	Insecticide	Insecticide, Acaricide
Mode of entry	:	Stomach and Contact poison	Systemic, Contact and Stomach poison
Mode of action	:	Energy metabolism	Nerve poison
Chemical group	:	Pyrazole	Avermectin (Macrocyclic lactones)
Category of toxicity : **Toxicity label colour :**			

- Tolfenpyrad 18.75% + Emamectin benzoate 0.94% SC is a high-performance insecticide mixture designed to control a wide range of chewing and sucking pests. This Suspension Concentrate (SC) formulation combines the energy-disrupting power of Tolfenpyrad with the paralytic action of Emamectin benzoate.
- Tolfenpyrad is a Mitochondrial Complex I Electron Transport Inhibitor (METI). It blocks energy production at the cellular level, causing the insect to stop feeding almost immediately after exposure.
- Emamectin benzoate is a chloride channel activator derived from the avermectin family. It causes irreversible muscle contraction and paralysis in larvae. It also has excellent translaminar activity, meaning it moves through the leaf to kill pests hiding on the underside.

Label claim (DPPQ & S, Faridabad)
Tolfenpyrad 18.75% + Emamectin benzoate 0.94% SC

Crop	Pests	Dose / ha		Waiting period (Days)
		Quantity of formulation (ml)	Volume of water (Lit.)	
Cauliflower	Diamond back moth, Tobacco caterpillar	700	500	07
Brinjal	Shoot and fruit borer, Jassid	700	500	03

Trade Names

Sr. No.	Trade Name	Company
1		

Tolfenpyrad + Pyriproxyfen + Acetamiprid

Available formulations in India : **Tolfenpyrad 30% + Pyriproxyfen 10% + Acetamiprid 4% EC**

Classification on the basis of		Tolfenpyrad	Pyriproxyfen	Acetamiprid
Target organism	:	Insecticide	Insecticide	Insecticide
Mode of entry	:	Stomach and Contact poison	Contact and Stomach poison	Systemic, Contact and Stomach poison
Mode of action	:	Energy metabolism	Insect Growth Regulator (IGR)	Nerve poison
Chemical group	:	Pyrazole	Pyridine ether	Neonicotinoid
Category of toxicity : Moderately toxic **Toxicity label colour** : Blue				

- This unique combination of Tolfenpyrad (30%), Pyriproxyfen (10%), and Acetamiprid is a high-potency insecticide mixture designed for comprehensive pest management. By combining three different chemical classes, it provides adulticidal, ovicidal, and larvicidal activity, making it a "total lifecycle" solution for difficult-to-control sucking and chewing pests.
- This formulation attacks pests through three distinct biological pathways.
- Tolfenpyrad is a Mitochondrial Complex I Electron Transport Inhibitor (METI). It blocks energy production at the cellular level, causing pests to stop feeding immediately and die within hours. It is effective against a broad spectrum of insects and mites.
- Pyriproxyfen is an Insect Growth Regulator (IGR) that mimics juvenile hormones. It does not kill adults immediately but prevents eggs from hatching and larvae from molting into adults, effectively breaking the pest's life cycle.
- Acetamiprid is a systemic neonicotinoid that targets the nicotinic acetylcholine receptors in the nervous system. It provides rapid knockdown and has excellent translaminar movement, protecting the undersides of leaves.

Label claim (DPPQ & S, Faridabad)
Tolfenpyrad 30% + Pyriproxyfen 10% + Acetamiprid 4% EC

Crop	Pests	Dose / ha		Waiting period (Days)
		Quantity of formulation (ml)	Volume of water (Lit.)	
Chilli	Aphid, Black thrips, Whitefly, Jassid	500	500	07

Trade Names

Sr. No.	Trade Name	Company
1	Trishanku	Best Agrolife Ltd.

Triflumezopyrim + Spinetoram

Available formulations in India	:	1. **Triflumezopyrim 04.79% + Spinetoram 08.62% SC** 2. **Triflumezopyrim 10.00% + Spinetoram 12.00% WG**

Classification on the basis of		**Triflumezopyrim**	**Spinetoram**
Target organism	:	Insecticide	Insecticide
Mode of entry	:	Systemic, Stomach poison	Stomach and Contact poison
Mode of action	:	Nerve poison	Nerve poison
Chemical group	:	Mesoionic	Spinosyns
Category of toxicity : Moderately toxic **Toxicity label colour** : Blue			

- The combination of Triflumezopyrim and Spinetoram is a powerful insecticide formulation designed for broad-spectrum pest control, particularly in major crops like rice.
- Triflumezopyrim is a new mesoionic insecticide, representing a distinct chemical class. Its mode of action is classified by the Insecticide Resistance Action Committee (IRAC) as a mesoionic nAChR inhibitor (Group 4D). This means it targets the insect's nervous system by binding to the orthosteric site of the nicotinic acetylcholine receptors (nAChR) and acting as an antagonist, or inhibitor. This disruption of normal neural signaling leads to the insect's paralysis and eventual death. Triflumezopyrim is particularly effective against sucking pests, such as the Brown Plant Hopper (BPH) and White Backed Plant Hopper (WBPH) in rice, including strains that have developed resistance to older chemistries like neonicotinoids. It is a systemic insecticide, providing long-lasting protection.
- Spinetoram belongs to the spinosyn chemical group (IRAC Group 5) and is a semi-synthetic derivative. It is characterized by its neurotoxic action, which involves interference with both the nicotinic acetylcholine receptors (nAChR) and the GABA (gamma-aminobutyric acid) receptors. Its action results in rapid excitation of the insect nervous system, leading to involuntary muscle contractions, prostration, and death. Spinetoram exhibits contact and translaminar activity (moving through the leaf tissue) and is highly effective against a wide range of Lepidopteran pests (caterpillars) like Yellow Stem Borer and Leaf Folder, as well as other pests such as thrips and leaf miners.
- The formulation provides strong initial control and significantly reduces the probability of resistance development in target pests. This dual-action insecticide offers robust efficacy against both sucking pests (like hoppers) and chewing pests (like borers and leaf folders), making it highly valuable in rice and other crops.

Label claim (DPPQ & S, Faridabad)

Triflumezopyrim 04.79% + Spinetoram 08.62% SC

Crop	Pests	Dose / ha		Waiting period (Days)
		Quantity of formulation (ml)	Volume of water (Lit.)	
Rice	Brown plant hopper, White backed plant hopper, Stem borer	500	500	21

Trade Names

Sr. No.	Trade Name	Company
1	Deladaxin	Corteva Agriscience

Triflumezopyrim 10.00% + Spinetoram 12.00% WG

Crop	Pests	Dose / ha		Waiting period (Days)
		Quantity of formulation (g)	Volume of water (Lit.)	
Rice	Brown plant hopper, White backed plant hopper, Leaf folder	250	500	21

Trade Names

Sr. No.	Trade Name	Company
1		

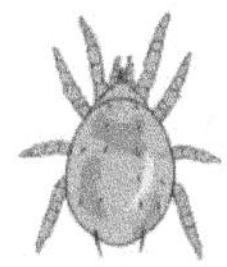

PART IV

Acaricides (Sole and Combinations)

Blue

ABAMECTIN

Available formulations in India : **Abamectin 1.90 % EC**

Classification on the basis of

Target organism : Acaricide, Insecticide
Mode of entry : Stomach and Contact poison
Mode of action : Nerve poison
Chemical group : Avermectin

IRAC classification

Main Group and Primary Site of Action : 6 Glutamate-gated chloride channel (GluCl) allosteric modulators
Sub-group or exemplifying Active Ingredient : Avermectins

Category of toxicity : Moderately toxic
Toxicity label colour : Blue

- Abamectin (Avermectin B1) is a natural fermentation product of the soil-dwelling actinomycete *Streptomyces avermitilis*, demonstrating significant acaricidal and insecticidal activity.
- Its environmental persistence is limited due to rapid photodegradation on plant and soil matrices, as well as in aquatic systems following agricultural application. Biodegradation by soil microorganisms also contributes to its relatively short half-life.
- While exhibiting lower toxicity to many non-target species, abamectin presents a high acute toxicity risk to *Apis mellifera* via direct contact or ingestion.
- **Mode of Action:** Abamectin acts as a nerve poison. It binds to glutamate-gated chloride channels in invertebrate nerve and muscle cells, causing hyperpolarization, paralysis, and eventual death of the pest. It also affects the gamma-aminobutyric acid (GABA) system, inhibiting nerve-to-nerve and nerve-to-muscle communication.
- Abamectin is stable to hydrolysis across a wide range of pH values, specifically at pH 5, 7, and 9. This means it doesn't break down easily in neutral or slightly acidic and alkaline water.

Label claim (DPPQ & S, Faridabad)

Abamectin 1.90 % EC

Crop	Pests	Dose / ha		Waiting period (Days)
		Quantity of formulation (ml)	Volume of water (Lit.)	
Rose	Red spider mites	400-665	500	03
Grape	Mites	375-750	500-1000	03
Apple	European red mite, Spotted red spider mite	1.6 ml / lit.	6-7 lit. water / tree	07
Tomato	Red spider mite, Leaf miner	450-600	500	03

Trade Names

Sr. No.	Trade Name	Company
1	Abacin	Crystal Crop Protection Ltd.
2	Tagmec	Tropical Agrosystem India (P) Ltd.
3	Chilli Max / Chilly Power	Silver Sine Bio Tech
4	Abaxin	Natural Crop Science
5	Kacin	Katyayani Organics
6	Tagmec	Tropical Agrosystem India (P) Ltd.
7	Antigo	Essential Biosciences
8	Abawin	Synwood Agro Ltd.
9	Abamek	Koronation Bio-Science
10	Mecin	Global Crop Care
11	Actin	Synovita Crop science Pvt. Ltd.
12	Exylon Azura	Torrent Crop Science
13	Abactin	R. K. Chemicals
14	Miticide	Katra Fertilizers and Chemicals Pvt. Ltd.

ACEQUINOCYL

Available formulations in India	:	**Acequinocyl 15.00 % SC**
Classification on the basis of		
Target organism	:	Acaricide
Mode of entry	:	Contact poison
Mode of action	:	Disruption of energy metabolism
Chemical group	:	Quinoline
IRAC classification		
Main Group and Primary Site of Action	:	20 Mitochondrial complex III electron transport inhibitors – Qo site
Sub-group or exemplifying Active Ingredient	:	Acequinocyl
Category of toxicity	:	
Toxicity label colour	:	

- Acequinocyl is a chemical compound primarily used as an acaricide.
- Acequinocyl is a member of the quinoline class of miticides.
- **Mode of action :** Its primary mode of action is as a mitochondrial electron transport inhibitor. Specifically, it binds to the Qo center of Complex III in the mitochondria of mites. This action prevents the transfer of electrons, which disrupts the mite's energy production and leads to its death. It is effective against all immature life stages of mites (eggs, larvae, nymphs) and may have indirect effects on adults. Its action is mainly through contact and some via ingestion; it has no systemic activity.
- Acequinocyl is most stable in acidic solutions. As the pH becomes more neutral, it starts to break down more quickly, although its half-life remains reasonable. However, in alkaline water, the compound degrades very rapidly.

Label claim (DPPQ & S, Faridabad)

Acequinocyl 15.00 % SC

Crop	Pests	Dose / ha		Waiting period (Days)
		Quantity of formulation (ml)	Volume of water (Lit.)	
Rose	Two spotted mites	500	500	10

Trade Names

Sr. No.	Trade Name	Company
1	Kanemite	Mahindra Summit

ACYNONAPYR

Available formulations in India : **Acynonapyr 18.19 % SC**

Classification on the basis of

Target organism : Acaricide
Mode of entry : Contact poison
Mode of action : Nerve poison
Chemical group : Azabicycle

IRAC classification

Main Group and Primary Site of Action : 33 Calcium-activated potassium channel (KCa2) modulators
Sub-group or exemplifying Active Ingredient : Acynonapyr

Category of toxicity :
Toxicity label colour :

- Acequinocyl is a chemical compound primarily used as an acaricide.
- Acynonapyr is a cutting-edge, novel acaricide (miticide) developed by Nippon Soda Co., Ltd. It is particularly notable for being the first agricultural chemical to feature a completely new mode of action, making it a powerful tool for managing mite populations that have grown resistant to older pesticides.
- **Mode of action** : By targeting the calcium-activated potassium (KCa2) channels, it causes neurological symptoms like convulsions and impaired mobility, eventually leading to death. Because no other pesticide uses this "lock," it is highly effective against mites that are resistant to other chemical groups.
- It works against mites at all life stages, including eggs, larvae, nymphs, and adults.
- It is highly selective. While it is lethal to spider mites, it has a very low impact on beneficial insects (like predatory mites and bees) and natural enemies, making it ideal for Integrated Pest Management (IPM) programs.
- It has excellent photostability and residual efficacy, meaning it stays active on the leaf surface for a significant period after application.

Label claim (DPPQ & S, Faridabad)

Acynonapyr 18.19 % SC

Crop	Pests	Dose / ha		Waiting period (Days)
		Quantity of formulation (ml)	Volume of water (Lit.)	
Tea	Red spider mite	500-750	500	07

Trade Names

Sr. No.	Trade Name	Company
1		

BIFENAZATE

Available formulations in India : 1. **Bifenazate 22.60 % SC**
2. **Bifenzate 50.00 % WP**

Classification on the basis of

Target organism : Acaricide
Mode of entry : Stomach and Contact poison
Mode of action : Affect Energy metabolism
Chemical group : Carbazate

IRAC classification

Main Group and Primary Site of Action : 20 Mitochondrial complex III electron transport inhibitors - Qo site
Sub-group or exemplifying Active Ingredient : 20D Bifenazate

Category of toxicity : --
Toxicity label colour : --

- Bifenazate is a selective carbazate acaricide, primarily employed in agricultural and ornamental crops to control a broad spectrum of mite species across all life stages, including eggs, larvae, nymphs, and adults.
- **Mode of action :** Its non-systemic mode of action combines contact and residual activity. Notably, bifenazate functions as a mitochondrial complex III electron transport inhibitor at the Qo site in mites. Additionally, it is reported to act as a GABA (gamma-aminobutyric acid) antagonist in the peripheral nervous system of insects at the neuromuscular synapse.
- This unique mode of action among acaricides makes bifenazate a valuable tool for resistance management.
- It provides a rapid knockdown effect, typically within 48-72 hours, and offers residual control lasting up to 4 weeks.
- To maintain the efficacy and residual control of bifenazate, it is recommended to keep the spray mixture within a slightly acidic to neutral pH range. Many product labels suggest a range of pH 5.5 to 6.5.

Label claim (DPPQ & S, Faridabad)

Bifenazate 22.60 % SC

Crop	Pests	Dose / ha		Waiting period (Days)
		Quantity of formulation (ml)	Volume of water (Lit.)	
Rose	Two spotted mite	500	2000	--
Grapes	Mites	500	1000	15

Trade Names

Sr. No.	Trade Name	Company
1	Floramite	UPL Ltd.
2	Starmite	Swal Corporation Ltd.

Bifenazate 50.00 % WP

Crop	Pests	Dose / ha		Waiting period (Days)
		Quantity of formulation (g)	Volume of water (Lit.)	
Rose	Two spotted mite	750	3000	--

Trade Names

Sr. No.	Trade Name	Company
1		

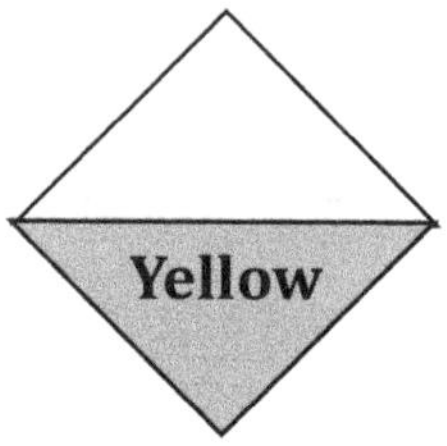

BIFENTHRIN

Available formulations in India : **Bifenthrin 08.00 % SC**

Classification on the basis of

Target organism : Insecticide, Acaricide

Mode of entry : Stomach and Contact poison

Mode of action : Nerve poison

Chemical group : Pyrethroid

IRAC classification

Main Group and Primary Site of Action : 3 Sodium channel modulators

Sub-group or exemplifying Active Ingredient : 3A Pyrethroid

Category of toxicity : Highly toxic

Toxicity label colour : Yellow

Label claim (DPPQ & S, Faridabad)

Bifenthrin 08.00 % SC

Crop	Pests	Dose / ha		Waiting period (Days)
		Quantity of formulation (ml)	Volume of water (Lit.)	
Rose	Red spider mite, Tea mosquito bug	500	400	11
Apple	Mite	7.5 ml / tree	10 lit/ tree	21

Brief information, mode of action, and trade names for this acaricide are mentioned in Part II: Sole Insecticides.

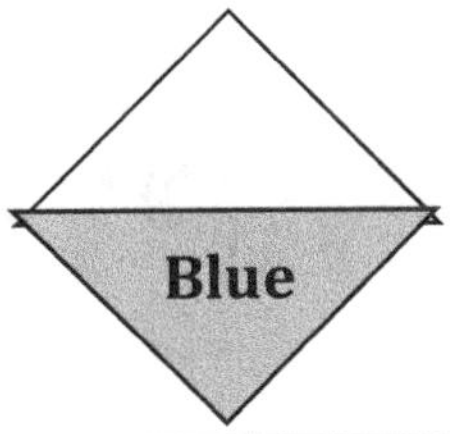

BUPROFEZIN

Available formulations in India : **Buprofezin 25.00 % SC**

Classification on the basis of

Target organism : Insecticide
Mode of entry : Stomach and contact poison
Mode of action : Growth regulator (Inhibition of chitin synthesis)
Chemical group : Thiadiazine

IRAC classification

Main Group and Primary Site of Action : 16 Inhibitors of chitin biosynthesis, type 1
Sub-group or exemplifying Active Ingredient : Buprofezin

Category of toxicity : Moderately toxic
Toxicity label colour : Blue

Label claim (DPPQ & S, Faridabad)

Buprofezin 25 % SC

Crop	Pests	Dose / ha		Waiting period (Days)
		Quantity of formulation (ml)	Volume of water (Lit.)	
Chilli	Yellow mite	300-600	500-750	05

<u>Brief information, mode of action, and trade names for this acaricide are mentioned in Part II: Sole Insecticides.</u>

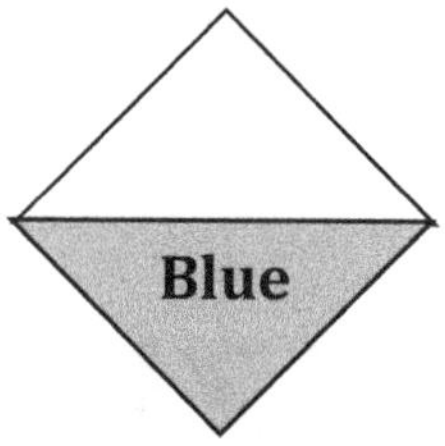

CHLORFENAPYR

Available formulations in India : 1. **Chlorfenapyr 10.00 % SC**
2. **Chlorfenapyr 240 g/L SC**

Classification on the basis of

Target organism : Insecticide, Acaricide
Mode of entry : Stomach and Contact poison
Mode of action : Energy metabolism
Chemical group : Pyrroles

IRAC classification

Main Group and Primary Site of Action : 13 Uncouplers of oxidative phosphorylation via disruption of the proton gradient
Sub-group or exemplifying Active Ingredient : Pyrroles

Category of toxicity : Moderately toxic
Toxicity label colour : Blue

Label claim (DPPQ & S, Faridabad)

Chlorfenapyr 10.00 % SC

Crop	Pests	Dose / ha		Waiting period (Days)
		Quantity of formulation (ml)	Volume of water (Lit.)	
Cabbage	Diamondback moth	750-1000	500	07
Chilli	Yellow mite	750-1000	500	05

Chlorfenapyr 240 g/L SC

Crop	Pests	Dose / ha		Waiting period (Days)
		Quantity of formulation (ml)	Volume of water (Lit.)	
Brinjal	Shoot and fruit borer, Mite	800-1000	500	10
Chilli	Mite, Thrips, Tobacco caterpillar	312.5-425	500	07
	Fruit borer	800-1000	500	07

Brief information, mode of action, and trade names for this acaricide are mentioned in Part II: Sole Insecticides.

Blue

CYENOPYRAFEN

Available formulations in India : **Cyflumetofen 30.00 % SC**

Classification on the basis of

Target organism : Acaricide
Mode of entry : Stomach and Contact poison
Mode of action : Energy metabolism
Chemical group : Pyrazole

IRAC classification

Main Group and Primary Site of Action : 25 Mitochondrial complex II electron transport inhibitors
Sub-group or exemplifying Active Ingredient : 25A *Beta*-ketonitrile derivatives

Category of toxicity : Moderately toxic
Toxicity label colour : Blue

- Cyenopyrafen is an acaricide belonging to the pyrazole chemical class.
- It is used for controlling mites across a range of crops. Notably effective against all mite developmental stages (eggs, nymphs, and adults).
- **Mode of action :** Its mode of action involved the disruption of the mite's energy production. Specifically, cyenopyrafen acted as a Mitochondrial Electron Transport Inhibitor (METI), targeting complex II (succinate dehydrogenase) within the respiratory electron transport chain. As a pro-acaricide, it required metabolic conversion within the mite to its active hydroxyl form. This activation led to the rapid inhibition of cellular respiration, resulting in paralysis and ultimately, the death of the mite.

Label claim (DPPQ & S, Faridabad)

Cyflumetofen 30.00 % SC

Crop	Pests	Dose / ha		Waiting period (Days)
		Quantity of formulation (ml)	Volume of water (Lit.)	
Apple	Mite	200-300	1000	15
Chilli	Mite	200-300	400-600	07
Rose	Mite	300	500	10
Brinjal	Mite	300	500	10

Trade Names

Sr. No.	Trade Name	Company
1	Kunoichi	Insecticides (India) Ltd.

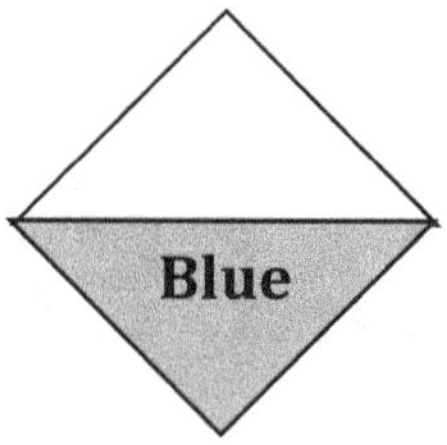

CYFLUMETOFEN

Available formulations in India : **Cyflumetofen 20.00 % SC**

Classification on the basis of

Target organism : Acaricide /Insecticide
Mode of entry : Stomach and Contact poison
Mode of action : Energy metabolism
Chemical group : Benzoylacetonitrile

IRAC classification

Main Group and Primary Site of Action : 25 Mitochondrial complex II electron transport inhibitors
Sub-group or exemplifying Active Ingredient : 25A *Beta*-ketonitrile derivatives

Category of toxicity : Moderately toxic
Toxicity label colour : Blue

- Cyflumetofen, a more recent acaricide (sometimes classified as an insecticide), belongs to the benzoylacetonitrile chemical class.
- It effectively controls all developmental stages of mites, including eggs, nymphs, and adults, and provides a good knockdown effect.
- **Mode of action :** Its mode of action involves the disruption of the mite's energy production by acting as a Mitochondrial Electron Transport Inhibitor (METI), specifically targeting complex II (succinate dehydrogenase) in the respiratory electron transport chain. Cyflumetofen inhibits ATP production, leading to energy loss, paralysis, and ultimately, mite death. Notably, cyflumetofen itself exhibits relatively weak inhibitory activity and functions as a pro-acaricide. It requires metabolism within the mite to its active form, the de-esterified metabolite AB-1, which demonstrates significantly stronger inhibition of complex II.
- Its primary activity occurs through contact and ingestion, with no significant systemic, translaminar, or vapor action.

Label claim (DPPQ & S, Faridabad)

Cyflumetofen 20.00 % SC

Crop	Pests	Dose / ha		Waiting period (Days)
		Quantity of formulation (g)	Volume of water (Lit.)	
Tea	Red spider mite	625-750	400-500	05

Trade Names

Sr. No.	Trade Name	Company
1	Foster	Dhanuka Agritech Ltd.

Blue

DIAFENTHIURON

Available formulations in India : 1. **Diafenthiuron 50.00 % WP**
2. **Diafenthiuron 47.80 % SC**

Classification on the basis of

Target organism : Insecticide, Acaricide
Mode of entry : Stomach and Contact poison
Mode of action : Energy metabolism
Chemical group : Thiourea

IRAC classification

Main Group and Primary Site of Action : 12 Inhibitors of mitochondrial ATP synthase
Sub-group or exemplifying Active Ingredient : 12A Diafenthiuron

Category of toxicity : Moderately toxic
Toxicity label colour : Blue

Label claim (DPPQ & S, Faridabad)

Diafenthiuron 50.00% WP

Crop	Pests	Dose / ha		Waiting period (Days)
		Quantity of formulation (g)	Volume of water (Lit.)	
Okra	Whitefly, Red spider mite	600	500	05
Tomato	Whitefly, Red spider mite	600	500	05
Chilli	Mites	600	500-750	03
Citrus	Mite	2 g / lit.	2-3 lit. / tree	30
Watermelon	Whitefly, Red spider mite	600	500	05

Diafenthiuron 47.80% SC

Crop	Pests	Dose / ha		Waiting period (Days)
		Quantity of formulation (ml)	Volume of water (Lit.)	
Potato	Whitefly, Leafhopper, Mite	500	500	44
Soybean	Whitefly, Mite	500	500	44

Brief information, mode of action, and trade names for this acaricide are mentioned in Part II: Sole Insecticides.

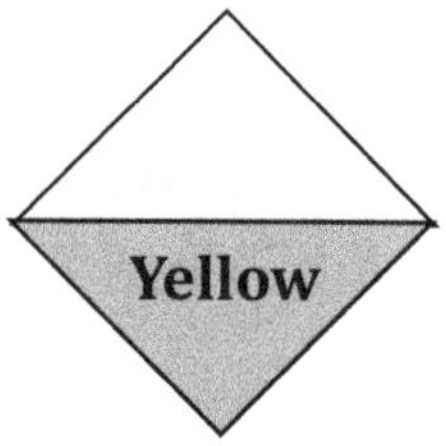

EMAMECTIN BENZOATE

Available formulations in India :
1. **Emamectin benzoate 05.00 % SG**
2. **Emamectin benzoate 01.90 % EC**

Classification on the basis of

Target organism : Insecticide
Mode of entry : Stomach and Contact poison
Mode of action : Nerve and muscle poison
Chemical group : Avermectins (Macrocyclic lactones)

IRAC classification

Main Group and Primary Site of Action : 6 Glutamate-gated chloride channel (GluCl) allosteric modulators
Sub-group or exemplifying Active Ingredient : Avermectins

Category of toxicity : Highly toxic
Toxicity label colour : Yellow

Label claim (DPPQ & S, Faridabad)

Emamectin benzoate 05.00% SG

Crop	Pests	Dose / ha		Waiting period (Days)
		Quantity of formulation (g)	Volume of water (Lit.)	
Chilli	Fruit borer, Thrips, Mite	200	500	03

<u>Brief information, mode of action, and trade names for this acaricide are mentioned in Part II: Sole Insecticides.</u>

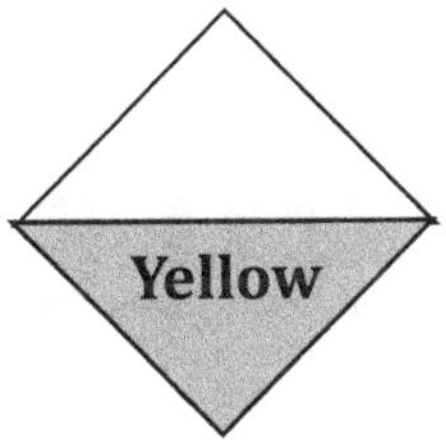

ETHION

Available formulations in India : **Ethion 50.00 % EC**

Classification on the basis of

Target organism : Insecticide, Acaricide
Mode of entry : Stomach and Contact poison
Mode of action : Nerve poison
Chemical group : Organophosphates

IRAC classification

Main Group and Primary Site of Action : 1 Acetylcholinesterase (AChE) inhibitors
Sub-group or exemplifying Active Ingredient : 1B Organophosphates

Category of toxicity : Highly toxic
Toxicity label colour : Yellow

Label claim (DPPQ & S, Faridabad)

Ethion 50.00 % EC

Crop	Pests	Dose / ha		Waiting period (Days)
		Quantity of formulation (ml)	Volume of water (Lit.)	
Chilli	Mites, Thrips	1500-2000	500-1000	05
Tea	Red spider mites, Purple mites, Yellow mite, Thrips, Scale	500	500-1000	03

Brief information, mode of action, and trade names for this acaricide are mentioned in Part II: Sole Insecticides.

Green

ETOXAZOLE

Available formulations in India	:	**Etoxazole 10.00 % SC**
Classification on the basis of		
Target organism	:	Acaricide
Mode of entry	:	Contact poison
Mode of action	:	Growth regulator
Chemical group	:	Diphenyl oxazoline
IRAC classification		
Main Group and Primary Site of Action	:	10 Mite growth inhibitors affecting CHS1
Sub-group or exemplifying Active Ingredient	:	10B Etoxazole
Category of toxicity	:	Slightly toxic
Toxicity label colour	:	Green

- Etoxazole, a selective acaricide from the diphenyl oxazoline group, is mainly employed to control various plant-feeding mites in agriculture and horticulture.
- Characteristically, it exhibits strong ovicidal and larvicidal activity, with some effect on nymphs, while generally having a limited impact on adult mites.
- **Mode of Action:** Etoxazole acts as a contact acaricide with limited systemic movement within the plant. Its primary mode of action involves the inhibition of chitin biosynthesis in mites. Chitin, a vital component of the exoskeleton in insects and mites, especially during molting, is targeted by etoxazole. By disrupting chitin synthesis, the acaricide primarily affects the eggs and immature stages (larvae and nymphs) of mites, preventing their successful development and molting, ultimately leading to their demise.
- Although its main target is the immature stages, etoxazole can also exert a sterilizing effect on adult female mites, thereby reducing their reproductive capacity.
- Etoxazole is known to be stable to hydrolysis under a wide range of neutral and basic pH conditions, but it can degrade under acidic conditions.

Label claim (DPPQ & S, Faridabad)

Etoxazole 10.00 % SC

Crop	Pests	Dose / ha		Waiting period (Days)
		Quantity of formulation (ml)	Volume of water (Lit.)	
Brinjal	Red spider mite	400	400-500	05
Tea	Red spider mite	400	400	05
Apple	European, Red spider mite	400	1000	43

Trade Names

Sr. No.	Trade Name	Company
1	Borneo	Sumitomo Chemical India Ltd.

FENAZAQUIN

Available formulations in India : 1. **Fenazaquin 10.00 % EC**
2. **Fenazaquin 18.30 % SC**

Classification on the basis of

Target organism : Acaricide, Insecticide
Mode of entry : Stomach and Contact poison
Mode of action : Energy metabolism
Chemical group : Quinazoline

IRAC classification

Main Group and Primary Site of Action : 21 Mitochondrial complex I electron transport inhibitors
Sub-group or exemplifying Active Ingredient : 21A METI acaricides and insecticides

Category of toxicity : Highly toxic
Toxicity label colour : Yellow

- Fenazaquin is a quinazoline-based acaricide and insecticide effective against various mites and certain insects, including whiteflies, in a wide array of crops.
- **Mode of action :** Its mode of action involves the disruption of insect mitochondrial function. More precisely, fenazaquin inhibits mitochondrial complex I electron transport, thus blocking electron transfer within the electron transport chain of the target pest's mitochondria.
- This interference leads to a rapid knockdown effect, and its contact and residual activity allows for control of all mite developmental stages (eggs, larvae, nymphs, adults).
- Fenazaquin degrades more rapidly in acidic solutions. Fenazaquin is more stable at a neutral to slightly alkaline pH. Degradation is slower at pH 7 and pH 9 compared to acidic conditions.

Label claim (DPPQ & S, Faridabad)

Fenazaquin 10.00% EC

Crop	Pests	Dose / ha		Waiting period (Days)
		Quantity of formulation (ml)	Volume of water (Lit.)	
Chilli	Yellow mite	1250	400-600	10
Okra	Red spider mite	1250	500	07
Brinjal	Red spider mite	1250	500	07
Tomato	Two spotted mite	1250	500	07
Tea	Red spider mite, Pink Mite, Purple mite	1000	400-600	07
	Scarlet mite	1250	400-600	07

Apple	Red spider mite, Two spotted mite	400	1000	30

Trade Names

Sr. No.	Trade Name	Company
1	Magister	DuPont Crop Protection
2	Magic	Global Crop Care

Fenazaquin 18.30% SC

Crop	Pests	Dose / ha		Waiting period (Days)
		Quantity of formulation (ml)	Volume of water (Lit.)	
Brinjal	Red spider mite	625	400-500	10
Apple	European red mite	200	1000	30
Chilli	Yellow mite	625	500	10
Okra	Red spider mite	625	500	07
Tomato	Two spotted spider mite	625	500	07
Tea	Red spider mite	400-500	400	07

Trade Names

Sr. No.	Trade Name	Company
1	Planar	Global Crop Care

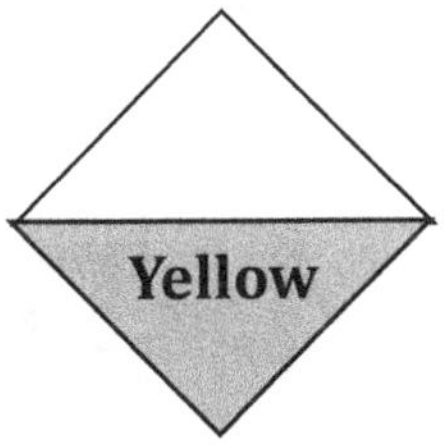

FENPROPATHRIN

Available formulations in India	:	**Fenpropathrin 30.00% EC**
Classification on the basis of		
Target organism	:	Insecticide
Mode of entry	:	Stomach and Contact poison
Mode of action	:	Nerve poison
Chemical group	:	Pyrethroid
IRAC classification		
Main Group and Primary Site of Action	:	3 Sodium channel modulators
Sub-group or exemplifying Active Ingredient	:	3A Pyrethroid
Category of toxicity	:	Highly toxic
Toxicity label colour	:	Yellow

Label claim (DPPQ & S, Faridabad)

Fenpropathrin 30.00% EC

Crop	Pests	Dose / ha		Waiting period (Days)
		Quantity of formulation (ml)	Volume of water (Lit.)	
Chilli	Thrips, Whitefly, Mite	250-340	750-1000	07
Brinjal	Whitefly, Shoot and fruit borer, Mite	250-340	750-1000	10
Okra	Whitefly, Shoot and fruit borer, Mite	250-340	750-1000	07
Tea	Mite	165-200	400-500	07

Brief information, mode of action, and trade names for this acaricide are mentioned in Part II: Sole Insecticides.

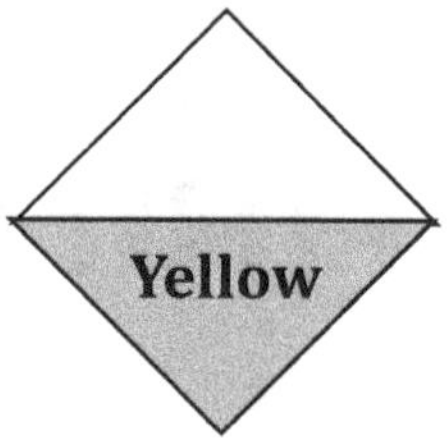

FENPYROXIMATE

Available formulations in India	:	**1. Fenpyroximate 05.00% EC** **2. Fenpyroximate 05.00 % SC**
Classification on the basis of		
Target organism	:	Insecticide, Acaricide
Mode of entry	:	Stomach and Contact poison
Mode of action	:	Energy metabolism
Chemical group	:	Pyrazole
IRAC classification		
Main Group and Primary Site of Action	:	21 Mitochondrial complex I electron transport inhibitors
Sub-group or exemplifying Active Ingredient	:	21A METI acaricides and insecticides
Category of toxicity	:	Highly toxic
Toxicity label colour	:	Yellow

- Fenpyroximate is a synthetic acaricide (miticide) used to control various types of mites in agriculture. While it primarily targets mites, it can also have some insecticidal activity against certain pests.
- **Mode of action :** It belongs to the METI acaricides and insecticides group (Mitochondrial Electron Transport Inhibitors) and specifically targets Mitochondrial complex I electron transport. Fenpyroximate disrupts the energy generation process in the mitochondria of the target pest's cells. It inhibits mitochondrial complex I electron transport, specifically blocking ubiquinone reduction, which is similar to the action of rotenone (another insecticide/acaricide). This disruption leads to a rapid cessation of all biological activities, including feeding and reproduction, ultimately causing paralysis and death of the mites.
- It has a quick knockdown effect on larvae, nymphs, and adults.
- It also exhibits molting and oviposition inhibitory action on nymphs.
- Fenpyroximate is a stable acaricide, particularly at neutral to slightly alkaline pH levels. While it is less stable in acidic environments.

Label claim (DPPQ & S, Faridabad)

Fenpyroximate 05.00% EC

Crop	Pests	Dose / ha		Waiting period (Days)
		Quantity of formulation (ml)	Volume of water (Lit.)	
Tea	Red spider mite, Pink mite, Purple mite	300-600	400-500	07
Chilli	Yellow mite	300-600	300-500	07
Cotton	Jassid, Mite	750	500	15

Coconut	Eriopyid mite (Root feeding)	10 ml / Lit.	As required	-
	Eriopyid mite (Spraying)	0.75-1.00 ml / Lit.	As required	-

Trade Names

Sr. No.	Trade Name	Company
1	Neon	Tata Rallis India Limited
2	Pyromite	Excel Crop Care Ltd.
3	Formite	Parijat Industries India Pvt. Ltd.
4	Mitigate	Isagro Asia Agrochemicals Pvt. Ltd.
5	Akio	Nichino India
6	Fenprox	Natural Crop Science
7	Akhanda	Koronation Bio-Science
8	K-Mit	Krishi Rasayan
9	Maximite	Global Crop Care
10	Signet	Preventis Chem India (P) Ltd.

Fenpyroximate 05.00% SC

Crop	Pests	Dose / ha		Waiting period (Days)
		Quantity of formulation (ml)	Volume of water (Lit.)	
Tea	Red spider mite, Pink mite, Purple mite	600-1200	400	07
Chilli	Yellow mite	300-600	500-750	03

Trade Names

Sr. No.	Trade Name	Company
1	Sedna	Tata Rallis India Limited

FLUFENOXURON

Available formulations in India : **Flufenoxuron 10.00 % DC**

Classification on the basis of

Target organism : Acaricide, Insecticide
Mode of entry : Contact and Stomach poison
Mode of action : Growth regulator
Chemical group : Benzoylureas

IRAC classification

Main Group and Primary Site of Action : 15 Inhibitors of chitin biosynthesis affecting CHS1
Sub-group or exemplifying Active Ingredient : Benzoylureas

Category of toxicity :
Toxicity label colour :

- Flufenoxuron is an important insecticide and acaricide belonging to the benzoylurea chemical class. It is primarily known for its activity as an insect growth regulator (IGR).
- **Mode of action** : Its mode of action is to inhibit chitin synthesis. Chitin is a crucial component of the exoskeleton of insects and mites. By disrupting the formation of new chitin, flufenoxuron interferes with the molting process. Immature stages (larvae and nymphs) cannot properly form new cuticles during molting, leading to their death at or during the next molt. It can also affect the viability of eggs laid by treated adults, leading to non-hatching eggs or the death of newly hatched larvae. In some cases, treated adult female mites may lay fewer eggs, and those eggs may be infertile.
- It primarily works through contact and stomach action, meaning pests must either ingest it or come into direct contact with the treated surface.
- Flufenoxuron is generally stable to hydrolysis in acidic and neutral conditions, but its stability decreases as the pH becomes more alkaline.

Label claim (DPPQ & S, Faridabad)

Flufenoxuron 10.00 % DC

Crop	Pests	Dose / ha		Waiting period (Days)
		Quantity of formulation (ml)	Volume of water (Lit.)	
Rose	Mite	500	500-1000	06

Trade Names

Sr. No.	Trade Name	Company
1	Flufenoxuron 10% DC	Hindustan Crop Science

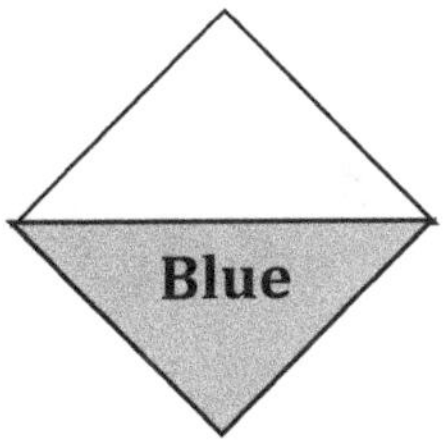

FLUFENZINE / DIFLOVIDAZIN

Available formulations in India : **Flufenzine 20.00 % SC**

Classification on the basis of

Target organism : Acaricide
Mode of entry : Stomach and Contact poison
Mode of action : Growth regulator
Chemical group : Tetrazine

IRAC classification

Main Group and Primary Site of Action : 10 Mite growth inhibitors affecting CHS1
Sub-group or exemplifying Active Ingredient : 10A Diflovidazin

Category of toxicity : Moderately toxic
Toxicity label colour : Blue

- Flufenzine, also known as diflovidazin, is a selective acaricide (mite killer) and has some insecticidal properties.
- **Mode of action** : Flufenzine has a unique mode of action. It appears to primarily inhibit the hatching of mite eggs. When feeding female mites consume the active ingredient, it sterilizes the developing eggs within their bodies. It also destroys all motile larval and nymphal forms of mites.
- It possesses translaminar properties, meaning it can move from the upper surface of a leaf to the lower surface, providing protection to mites feeding there.
- Flufenzine is stable in acidic to neutral solutions but degrades rapidly in alkaline conditions. This pH dependency is a crucial factor for its effective use as an acaricide.

Label claim (DPPQ & S, Faridabad)

Flufenzine 20.00 % SC

Crop	Pests	Dose / ha		Waiting period (Days)
		Quantity of formulation (ml)	Volume of water (Lit.)	
Brinjal	Mite	400-500	500-1000	05
Tea	Pink mite, Purple mite	400-500	500-1000	07
	Red spider mite	500-600	500-1000	07

Trade Names

Sr. No.	Trade Name	Company
1	Aquazin	Aquarius Agro Chemicals

Green

HEXYTHIAZOX

Available formulations in India : **Hexythiazox 05.45 % EC**

Classification on the basis of

Target organism : Acaricide

Mode of entry : Contact and Stomach poison

Mode of action : Growth regulator

Chemical group : Thiazolidinecarboxamide

IRAC classification

Main Group and Primary Site of Action : 10 Mite growth inhibitors affecting CHS1

Sub-group or exemplifying Active Ingredient : 10A Hexythiazox

Category of toxicity : Slightly toxic

Toxicity label colour : Green

- Hexythiazox is a widely used acaricide (miticide) specifically designed to control mites, particularly during their immature stages.
- **Mode of action** : It functions as a mite growth regulator (MGR) by inhibiting chitin synthesis, a process vital for the development of the arthropod exoskeleton. By disrupting chitin formation, hexythiazox prevents molting in immature mite stages (eggs, larvae, and nymphs), ultimately leading to their death. While it doesn't directly kill adult mites, it significantly reduces their egg-laying capacity and the viability of their eggs, which helps to gradually reduce the overall mite population.
- Hexythiazox acts primarily through contact and stomach action. Although it effectively penetrates the plant epidermis, it does not exhibit significant systemic movement within the plant. Its efficacy across the egg, larval, and nymph stages provides comprehensive control by breaking the mite life cycle.
- For optimal stability and effectiveness, it is generally recommended to mix hexythiazox in water with a neutral to slightly acidic pH (5.0-7.0).

Label claim (DPPQ & S, Faridabad)

Hexythiazox 05.45 % EC

Crop	Pests	Dose / ha		Waiting period (Days)
		Quantity of formulation (g)	Volume of water (Lit.)	
Tea	Scarlet mite, Red spider mite	300-500	400	05
Chilli	Yellow mite	300-500	625	03
Apple	European red mite	0.04%	10 lit/tree	15
Grapes	Red spider mite	500	1000	05
Rose	Red spider mite	400-500	500	05
Brinjal	Red spider mite	500	500	07
Okra	Red spider mite	500	500	07

Trade Names

Sr. No.	Trade Name	Company
1	Mitolin	Crystal Crop Protection Ltd.
2	K-Aradite	Godrej Agrovet Ltd.
3	Maiden, Xmite	Biostadt India Ltd.
4	Endurer	Coromandel International Ltd.
5	Macario	Isagro Asia Agrochemicals Pvt. Ltd.
6	Cubax	Best Agrolife Ltd.
7	Synred	Synwood Agro Ltd.
8	End	Global Crop Care
9	K Myte	Green India Care
10	Hyzox	Preventis Chem India (P) Ltd.

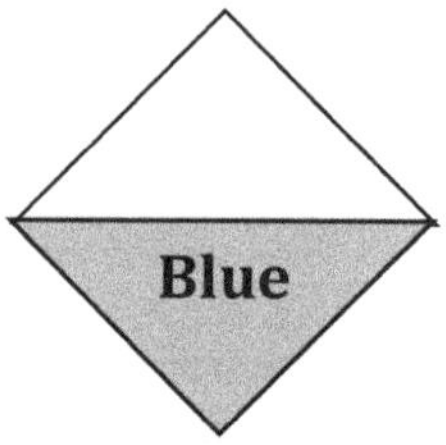

ISOCYCLOSERAM

Available formulations in India : **Isocycloseram 09.20% (10.00%) DC**

Classification on the basis of

Target organism : Insecticide, Acaricide

Mode of entry : Stomach and Contact poison

Mode of action : Nerve poison

Chemical group : Isoxazoline

IRAC classification

Main Group and Primary Site of Action : Group 30. GABA-gated channel allosteric modulators

Sub-group or exemplifying Active Ingredient : Isoxazolines

Category of toxicity : Slightly toxic

Toxicity label colour : Blue

Label claim (DPPQ & S, Faridabad)

Isocycloseram 09.20% (10.00%) DC

Crop	**Pests**	**Dose / ha**		**Waiting period (Days)**
		Quantity of formulation (ml)	**Volume of water (Lit.)**	
Brinjal	Jassid, Red spider mite	200	500	05
	Shoot and fruit borer	600	500	05
Chilli	Yellow mite	200	500	07
	Thrips, Fruit borer	600	500	07

Brief information, mode of action, and trade names for this acaricide are mentioned in Part II: Sole Insecticides.

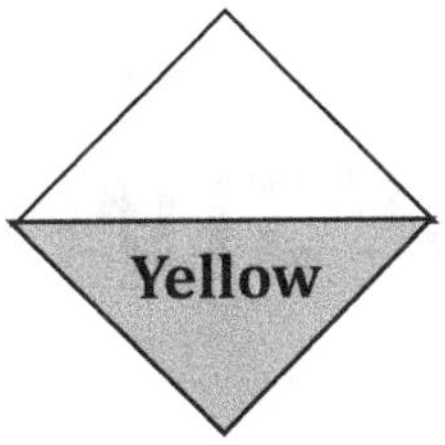

LAMBDA-CYHALOTHRIN

Available formulations in India : **Lambda-cyhalothrin 05.00 % EC**

Classification on the basis of

Target organism : Insecticide
Mode of entry : Stomach and Contact poison
Mode of action : Nerve poison
Chemical group : Pyrethroid

IRAC classification

Main Group and Primary Site of Action : 3 Sodium channel modulators
Sub-group or exemplifying Active Ingredient : 3A Pyrethroid

Category of toxicity : Highly toxic
Toxicity label colour : Yellow

Label claim (DPPQ & S, Faridabad)

Lambda-cyhalothrion 05.00 % EC

Crop	Pests	Dose / ha		Waiting period (Days)
		Quantity of formulation (ml)	Volume of water (Lit.)	
Chilli	Thrips, Mite, Pod borer	300	400-600	05

<u>Brief information, mode of action, and trade names for this acaricide are mentioned in Part II: Sole Insecticides.</u>

MILBEMECTIN

Available formulations in India	:	**Milbemectin 01.00 % EC**
Classification on the basis of		
Target organism	:	Insecticide, Acaricide
Mode of entry	:	Stomach and Contact poison
Mode of action	:	Nerve and muscle poison
Chemical group	:	Milbemycins
IRAC classification		
Main Group and Primary Site of Action	:	6 Glutamate-gated chloride channel (GluCl) allosteric modulators
Sub-group or exemplifying Active Ingredient	:	Milbemycins
Category of toxicity	:	
Toxicity label colour	:	

- Milbemectin is an important insecticide and acaricide (miticide) widely used in agriculture. It is a mixture of two closely related macrocyclic lactone compounds: milbemycin A3 and milbemycin A4, typically in a ratio of about 30% A3 and 70% A4.
- Milbemectin is derived from natural products produced by the soil bacterium Streptomyces hygroscopicus subsp. aureolacrimosus. This microbial origin gives it a favorable profile, making it suitable for certain organic farming practices in some regions.
- **Mode of action :** Milbemectin acts on the nervous system of the target pests. It works as an allosteric modulator that primarily targets glutamate-gated chloride channels in the neurons and myocytes (muscle cells) of invertebrates. By opening these channels, it leads to an influx of chloride ions, causing hyperpolarization of the nerve and muscle cells. This disrupts signal transmission, leading to paralysis and ultimately death of the pest. It is also believed to interact with insect GABA (gamma-aminobutyric acid) receptors.

Label claim (DPPQ & S, Faridabad)

Milbemectin 01.00 % EC

Crop	Pests	Dose / ha		Waiting period (Days)
		Quantity of formulation (ml)	Volume of water (Lit.)	
Rose	Two spotted spider mite	450	1000	05
Chilli	Yellow, White mite	325	500	07

Trade Names

Sr. No.	Trade Name	Company
1	Milbeknock	Mitsui Chemicals Crop & Life Solutions

PHENTHOATE

Available formulations in India : **Phenthoate 02.00% DP**

Classification on the basis of

Target organism : Insecticide
Mode of entry : Stomach and Contact poison
Mode of action : Nerve poison
Chemical group : Organophosphates

IRAC classification

Main Group and Primary Site of Action : 1 Acetylcholinesterase (AChE) inhibitors
Sub-group or exemplifying Active Ingredient : 1B Organophosphates

Category of toxicity : Highly toxic
Toxicity label colour : Yellow

Label claim (DPPQ & S, Faridabad)

Phenthoate 02.00 % DP

Crop	Pests	Dose / ha		Waiting period (Days)
		Quantity of formulation (Kg)	Volume of water (Lit.)	
Sorghum	Red spider mite, Pink mite, Purple mite, Scarlet mite	20	--	After 90% Emergence of earhead

<u>Brief information, mode of action, and trade names for this acaricide are mentioned in Part II: Sole Insecticides.</u>

Yellow

PHOSALONE

Available formulations in India : **Phosalone 35.00 % EC**

Classification on the basis of

Target organism : Insecticide, Acaricide
Mode of entry : Stomach and Contact poison
Mode of action : Nerve poison
Chemical group : Organophosphates

IRAC classification

Main Group and Primary Site of Action : 1 Acetylcholinesterase (AChE) inhibitors
Sub-group or exemplifying Active Ingredient : 1B Organophosphates

Category of toxicity : Highly toxic
Toxicity label colour : Yellow

Label claim (DPPQ & S, Faridabad)

Phosalone 35.00 % EC

Crop	Pests	Dose / ha		Waiting period (Days)
		Quantity of formulation (ml)	Volume of water (Lit.)	
Jute	Red spider mite	1000	500-1000	--
Tea	Aphid, Pink mite, Purple mite	1028	500-1000	--

Brief information, mode of action, and trade names for this acaricide are mentioned in Part II: Sole Insecticides.

Yellow

PROPARGITE

Available formulations in India : **Propargite 57.00% EC**

Classification on the basis of

Target organism : Aacricide
Mode of entry : Stomach and Contact poison
Mode of action : Energy metabolism
Chemical group : Sulfite ester

IRAC classification

Main Group and Primary Site of Action : 12 Inhibitors of mitochondrial ATP synthase
Sub-group or exemplifying Active Ingredient : 12C Propargite

Category of toxicity : Highly toxic
Toxicity label colour : Yellow

- Propargite is a highly effective miticide (acaricide) widely used in agriculture to control a variety of mite species.
- Belonging to the sulfite ester group, it works through contact and fumigant action, providing a rapid and powerful knockdown of mites at all life stages.
- This powerful pesticide acts in three ways: through direct contact, residual contact, and vapor action within dense crop canopies.
- It targets the mites' metabolism and respiration by inhibiting mitochondrial ATP synthase, which leads to paralysis and death.
- An added benefit is its effectiveness against mites that have developed resistance to other types of miticides.
- Propargite is most stable in the pH range of 5 to 7. Its effectiveness diminishes significantly in alkaline conditions (pH greater than 7), as it breaks down much more quickly.

Label claim (DPPQ & S, Faridabad)

Propargite 57.00% EC

Crop	Pests	Dose / ha		Waiting period (Days)
		Quantity of formulation (ml)	Volume of water (Lit.)	
Tea	Red spider mite, Pink mite, Purple mite, Scarlet mite	750-1250	400	07
Chilli	Mite	1500	500-625	07
Apple	European mite, Rwo spotted mite	5-10 ml / tree	10 lit / tree	09
Brinjal	Two spotted spider mite	1000	400	06

Trade Names

Sr. No.	Trade Name	Company
1	Simbaa	PI Industries
2	Mastamite	UPL Ltd.
3	Mainichi	IFFCO-MC Crop Science Private Ltd.
4	Propamite	Swal Corporation Ltd.
5	Omite	Dhanuka Agritech Ltd.
6	Mite Naashak	Anu Products Ltd.
7	Ocean	Crop Chemicals India Ltd.
8	Mitesik	Sikko Industries Ltd.
9	Willo-Mite	Willowood Crop Science Pvt. Ltd.
10	G-Pro	Essential Biosciences
11	Picador	HPM Chemicals and Fertilizers Ltd.
12	Progite	R K Chemicals
13	Nomite	Tropical Agrosystem India (P) Ltd.
14	Zero Mite	JU Agri Sciences Pvt. Ltd.
15	Oriana	RACCOLTO Agritech India Pvt. Ltd.

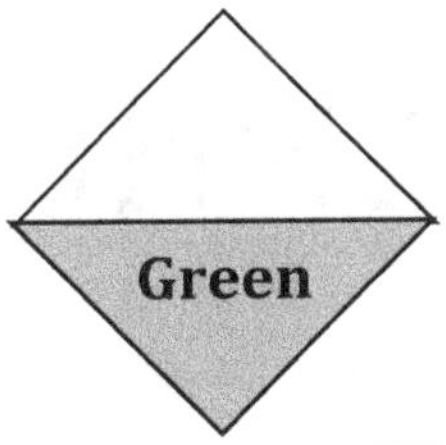

PYRIDABEN

Available formulations in India : **Pyridaben 20.00% WP**

Classification on the basis of

Target organism : Insecticide, Acaricide

Mode of entry : Stomach and Contact poison

Mode of action : Energy metabolism

Chemical group : Pyridazinone

IRAC classification

Main Group and Primary Site of Action : 21 Mitochondrial complex I electron transport inhibitors

Sub-group or exemplifying Active Ingredient : 21A METI acaricides and insecticides

Category of toxicity : Slightly toxic

Toxicity label colour : Green

Label claim (DPPQ & S, Faridabad)

Pyridaben 20.00% WP

Crop	Pests	Dose / ha		Waiting period (Days)
		Quantity of formulation (g)	Volume of water (Lit.)	
Tea	Red spider mite	500	500	07
Chilli	Yellow mite	375-500	500	05
Brinjal	Red spider mite	500	500	03
Apple	European red mite	5 ml	10 lit / tree	05

Brief information, mode of action, and trade names for this acaricide are mentioned in Part II: Sole Insecticides.

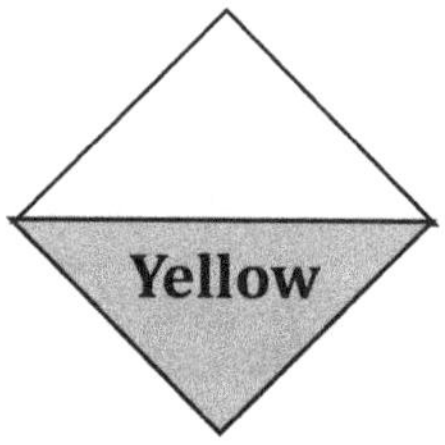

QUINALPHOS

Available formulations in India	:	**Quinalphos 25.00% EC**
Classification on the basis of		
Target organism	:	Insecticide
Mode of entry	:	Stomach and Contact poison
Mode of action	:	Nerve poison
Chemical group	:	Organophosphates
IRAC classification		
Main Group and Primary Site of Action	:	1 Acetylcholinesterase (AChE) inhibitors
Sub-group or exemplifying Active Ingredient	:	1B Organophosphates
Category of toxicity	:	Highly toxic
Toxicity label colour	:	Yellow

Label claim (DPPQ & S, Faridabad)

Quinalphos 25.00 % EC

Crop	Pests	Dose / ha		Waiting period (Days)
		Quantity of formulation (ml)	Volume of water (Lit.)	
Wheat	Aphid	1000	500-1000	--
	Earhead caterpillar, Mite	1600	500-1000	--
Chilli	Aphid	1000	500-1000	--
	Mite	1500	500-1000	--

Brief information, mode of action, and trade names for this acaricide are mentioned in Part II: Sole Insecticides.

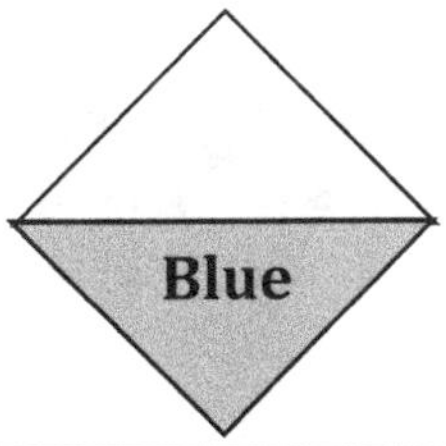

SPIROMESIFEN

Available formulations in India : **Spiromesifen 22.90% SC**

Classification on the basis of

Target organism : Insecticide, Acaricide

Mode of entry : Stomach and Contact poison

Mode of action : Lipid synthesis and growth regulator

Chemical group : Tetronic and Tetramic acid derivatives

IRAC classification

Main Group and Primary Site of Action : 23 Inhibitors of acetyl CoA carboxylase

Sub-group or exemplifying Active Ingredient : Tetronic and Tetramic acid derivatives

Category of toxicity : Moderately toxic

Toxicity label colour : Blue

Label claim (DPPQ & S, Faridabad)

Spiromesifen 22.90% SC

Crop	Pests	Dose / ha		Waiting period (Days)
		Quantity of formulation (ml)	Volume of water (Lit.)	
Brinjal	Red spider mite	400	500	05
Cotton	Whitefly, Mite	600	500	10
Apple	European red mite, Red spider mite	300	1000	30
Chilli	Yellow mite	400	500-750	07
Tea	Red spider mite	400	400	07
Okra	Red spider mite	400-500	500	03
Tomato	Whitefly, Mite	625	500	03

Brief information, mode of action, and trade names for this acaricide are mentioned in Part II: Sole Insecticides.

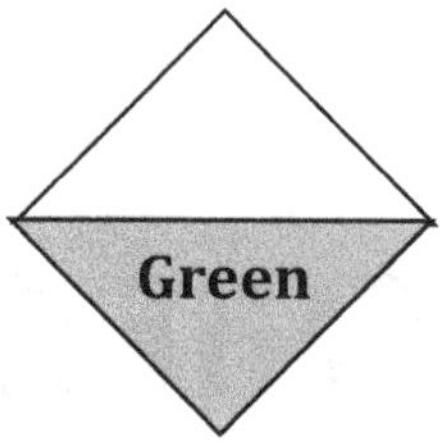

SPIROTETRAMAT

Available formulations in India : **Spirotetramat 15.31% OD**

Classification on the basis of

Target organism	:	Insecticide, Acaricide
Mode of entry	:	Systemic, Stomach and Contact poison
Mode of action	:	Lipid synthesis and growth regulator
Chemical group	:	Keto-enol

IRAC classification

Main Group and Primary Site of Action	:	23 Inhibitors of acetyl CoA carboxylase
Sub-group or exemplifying Active Ingredient	:	Keto-enol

Category of toxicity : Slightly toxic

Toxicity label colour : Green

Label claim (DPPQ & S, Faridabad)

Spirotetramat 15.31% OD

Crop	Pests	Dose / ha		Waiting period (Days)
		Quantity of formulation (ml)	Volume of water (Lit.)	
Okra	Aphid, Whitefly, Mite	600	500	03
Grapes	Mealybug, Mite	700	500-1000	60
Citrus	Psylla, Mite	300	500	30

Brief information, mode of action, and trade names for this acaricide are mentioned in Part II: Sole Insecticides.

TEBUFENPYRAD

Available formulations in India : **Tebufenpyrad 20.00% WP**

Classification on the basis of

Target organism : Acaricide, Insecticide
Mode of entry : Stomach and Contact poison
Mode of action : Energy metabolism
Chemical group : Pyrazole

IRAC classification

Main Group and Primary Site of Action : 21 Mitochondrial complex I electron transport inhibitors
Sub-group or exemplifying Active Ingredient : 21A METI acaricides anf insecticides

Category of toxicity :
Toxicity label colour :

- Tebufenpyrad is a widely used insecticide and acaricide, primarily for controlling mites and certain insects.
- Tebufenpyrad belongs to the pyrazole class of insecticides. It is also classified as a Mitochondrial Electron Transport Inhibitor (METI).
- Mode of action : Tebufenpyrad's primary mode of action is as a mitochondrial complex I inhibitor. It disrupts the electron transport chain within the mitochondria of pests. This inhibition prevents the production of ATP, the main energy source for cells, leading to a lack of energy, cell death, and ultimately, the death of the pest. This mechanism is similar to that of rotenone, another well-known mitochondrial toxicant.
- It has both contact and stomach action, meaning it works when the pest touches the treated surface or ingests the chemical. It is non-systemic.
- Tebufenpyrad is stable to hydrolysis across a wide range of pH values, specifically from pH 4 to pH 9.

Label claim (DPPQ & S, Faridabad)

Tebufenpyrad 20.00% WP

Crop	Pests	Dose / ha		Waiting period (Days)
		Quantity of formulation (g)	Volume of water (Lit.)	
Tea	Red spider mite	375-500	500	07
Tomato	Red spider mite	375-500	500	20

Trade Names

Sr. No.	Trade Name	Company
1	Tebufenpyrad 20.00% WP	PI Industries Ltd.

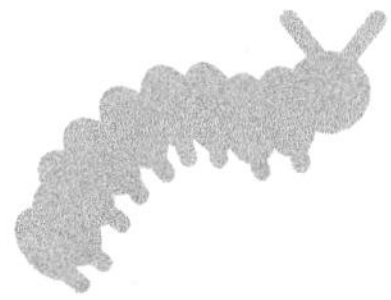

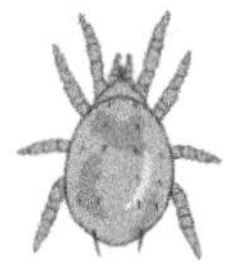

Insecticide + Acaricide Combinations

Buprofezin + Acephate

Available formulations in India : **Buprofezin 20.00% + Acephate 50.00% WP**

Classification on the basis of		Buprofezin	Acephate
Target organism	:	Insecticide	Insecticide
Mode of entry	:	Contact and Stomach poison	Systemic, Contact and Stomach poison
Mode of action	:	Insect Growth Regulator (IGR)	Nerve poison
Chemical group	:	Thiadiazine	Organophosphate
Category of toxicity : Moderately toxic **Toxicity label colour** : Blue			

Label claim (DPPQ & S, Faridabad)

Buprofezin 20.00% + Acephate 50.00% WP

Crop	Pests	Dose / ha		Waiting period (Days)
		Quantity of formulation (g)	Volume of water (Lit.)	
Chilli	Thrips, Aphid, Yellow mite	1250	500	07

Brief information, mode of action, and trade names for this combination are mentioned in Part III: Insecticide Combinations.

Chlorantraniliprole + Abamectin

Available formulations in India : **Chlorantraniliprole 04.30% + Abamectin 01.70% SC**

Classification on the basis of		Chlorantraniliprole	Abamectin
Target organism	:	Insecticide	Acaricide, Insecticide
Mode of entry	:	Systemic, Contact and Stomach poison	Contact and Stomach poison
Mode of action	:	Muscle poison	Nerve poison
Chemical group	:	Anthranilic diamide	Avermectin (Macrocyclic Lactone)
Category of toxicity : Moderately toxic **Toxicity label colour** : Blue			

Label claim (DPPQ & S, Faridabad)

Chlorantraniliprole 04.30% + Abamectin 01.70% SC

Crop	Pests	Dose / ha		Waiting period (Days)
		Quantity of formulation (ml)	Volume of water (Lit.)	
Chilli	Thrips, Mites, Fruit borer	625	500	05
Tomato	Fruit borer, Leaf miner, Mites	500	500	05

<u>Brief information, mode of action, and trade names for this combination are mentioned in Part III: Insecticide Combinations.</u>

Cyantraniliprole + Diafenthiuron

Available formulations in India : **Cyantraniliprole 08.00% + Diafenthiuron 40.00% SC**

Classification on the basis of		Cyantraniliprole	Diafenthiuron
Target organism	:	Insecticide	Insecticide, Acaricide
Mode of entry	:	Systemic, Contact and Stomach poison	Contact and Stomach poison
Mode of action	:	Muscle poison	Energy metabolism
Chemical group	:	Anthranilic Diamide	Thiourea
Category of toxicity : Moderately toxic **Toxicity label colour** : Blue			

Label claim (DPPQ & S, Faridabad)

Cyantraniliprole 08.00% + Diafenthiuron 40.00% SC

Crop	Pests	Dose / ha		Waiting period (Days)
		Quantity of formulation (ml)	Volume of water (Lit.)	
Chilli	Thrips, Mites, Whitefly, Fruit borer	625	500	05

Brief information, mode of action, and trade names for this combination are mentioned in Part III: Insecticide Combinations.

Diafenthiuron + Acetamiprid

Available formulations in India : **Diafenthiuron 40.10% + Acetamiprid 03.90% WP**

Classification on the basis of		Diafenthiuron	Acetamiprid
Target organism	:	Insecticide, Acaricide	Insecticide
Mode of entry	:	Contact and Stomach poison	Systemic, Contact and Stomach poison
Mode of action	:	Energy metabolism	Nerve poison
Chemical group	:	Thiourea	Neonicotinoid
Category of toxicity : Moderately toxic **Toxicity label colour** : Blue			

Label claim (DPPQ & S, Faridabad)

Diafenthiuron 40.10% + Acetamiprid 03.90% WP

Crop	Pests	Dose / ha		Waiting period (Days)
		Quantity of formulation (g)	Volume of water (Lit.)	
Cotton	Thrips, Whitefly, Mites	500-600	500	21
Chilli	Thrips, Whitefly, Mites	500-600	500	03

Brief information, mode of action, and trade names for this combination are mentioned in Part III: Insecticide Combinations.

Diafenthiuron + Dinotefuran

Available formulations in India : **Diafenthiuron 48.00% + Dinotefuran 08.00% WG**

Classification on the basis of		Diafenthiuron	Dinotefuran
Target organism	:	Insecticide, Acaricide	Insecticide
Mode of entry	:	Contact and Stomach poison	Systemic, Contact and Stomach poison
Mode of action	:	Energy metabolism	Nerve poison
Chemical group	:	Thiourea	Neonicotinoid
Category of toxicity : **Toxicity label colour :**			

Label claim (DPPQ & S, Faridabad)

Diafenthiuron 48.00% + Dinotefuran 08.00% WG

Crop	Pests	Dose / ha		Waiting period (Days)
		Quantity of formulation (g)	Volume of water (Lit.)	
Cotton	Thrips, Jassid, Whitefly, Mites	625	500	43
Brinjal	Thrips, Jassid, Whitefly, Mites	625	500	03

Brief information, mode of action, and trade names for this combination are mentioned in Part III: Insecticide Combinations.

Diafenthiuron + Pyriproxyfen

Available formulations in India :

1. **Diafenthiuron 30.00% + Pyriproxyfen 08.00% SC**
2. **Pyriproxyfen 05.00% + Diafenthiuron 25.00% SE**
3. **Pyriproxifen 08.00% + Diafenthiuron 30.00% SE**

Classification on the basis of		Diafenthiuron	Pyriproxyfen
Target organism	:	Insecticide, Acaricide	Insecticide
Mode of entry	:	Contact and Stomach poison	Contact and Stomach poison
Mode of action	:	Energy metabolism	Insect Growth Regulator (IGR)
Chemical group	:	Thiourea	Pyridine ether
Category of toxicity : Moderately toxic **Toxicity label colour** : Blue			

Label claim (DPPQ & S, Faridabad)

Diafenthiuron 30.00% + Pyriproxyfen 08.00% SC

Crop	Pests	Dose / ha		Waiting period (Days)
		Quantity of formulation (ml)	Volume of water (Lit.)	
Chilli	Mites, Whitefly	1000	500	03
Cotton	Whitefly	1000	500	35

Pyriproxyfen 05.00% + Diafenthiuron 25.00% SE

Crop	Pests	Dose / ha		Waiting period (Days)
		Quantity of formulation (ml)	Volume of water (Lit.)	
Chilli	Whitefly, Thrips, Yellow mite	750-1000	500	10

Pyriproxifen 08.00% + Diafenthiuron 30.00% SE

Crop	Pests	Dose / ha		Waiting period (Days)
		Quantity of formulation (ml)	Volume of water (Lit.)	
Chilli	Mites, Whitefly	1000	500	03

Brief information, mode of action, and trade names for this combination are mentioned in Part III: Insecticide Combinations.

Emamectin benzoate + Chlorantraniliprole + Deltamethrin

Available formulations in India : **Emamectin benzoate 01.90% + Chlorantraniliprole 05.00% + Deltamethrin 01.00% EC**

Classification on the basis of		Emamectin benzoate	Chlorantraniliprole	Deltamethrin
Target organism	:	Insecticide, Acaricide	Insecticide	Insecticide
Mode of entry	:	Systemic, Contact and Stomach poison	Systemic, Contact and Stomach poison	Contact and Stomach poison
Mode of action	:	Nerve poison	Muscle poison	Nerve poison
Chemical group	:	Avermectin (Macrocyclic lactones)	Anthranilic diamide	Synthetic pyrethroid
Category of toxicity :				
Toxicity label colour :				

Label claim (DPPQ & S, Faridabad)
Emamectin benzoate 01.90% + Chlorantraniliprole 05.00% + Deltamethrin 01.00% EC

Crop	Pests	Dose / ha		Waiting period (Days)
		Quantity of formulation (ml)	Volume of water (Lit.)	
Chilli	Thrips, Mites, Fruit borers (*Helicoverpa armigera, Spodoptera litura*)	375	500	28

<u>Brief information, mode of action, and trade names for this combination are mentioned in Part III: Insecticide Combinations.</u>

Emamectin benzoate + Diafenthiuron

Available formulations in India : **Emamectin benzoate 01.10% + Diafenthiuron 30.00% SC**

Classification on the basis of		**Emamectin benzoate**	**Diafenthiuron**
Target organism	:	Insecticide, Acaricide	Insecticide, Acaricide
Mode of entry	:	Systemic, Contact and Stomach poison	Contact and Stomach poison
Mode of action	:	Nerve poison	Energy metabolism
Chemical group	:	Avermectin (Macrocyclic lactones)	Thiourea
Category of toxicity : Highly toxic **Toxicity label colour** : Yellow			

Label claim (DPPQ & S, Faridabad)

Emamectin benzoate 01.10% + Diafenthiuron 30.00% SC

Crop	Pests	Dose / ha		Waiting period (Days)
		Quantity of formulation (ml)	Volume of water (Lit.)	
Brinjal	Thrips, Aphid, Whitefly, red spider mite, Shoot and fruit borer	1000	500	08

<u>Brief information, mode of action, and trade names for this combination are mentioned in Part III: Insecticide Combinations.</u>

Emamectin benzoate + Lambda-cyhalothrin

Available formulations in India : **Emamectin benzoate 03.50% + Lambda-cyhalothrin 05.00% WP**

Classification on the basis of		Emamectin benzoate	Lambda-cyhalothrin
Target organism	:	Insecticide, Acaricide	Insecticide
Mode of entry	:	Systemic, Contact and Stomach poison	Contact and Stomach poison
Mode of action	:	Nerve poison	Nerve poison
Chemical group	:	Avermectin (Macrocyclic lactones)	Synthetic pyrethroid
Category of toxicity : **Toxicity label colour** :			

Label claim (DPPQ & S, Faridabad)

Emamectin benzoate 03.50% + Lambda-cyhalothrin 05.00% WP

Crop	Pests	Dose / ha		Waiting period (Days)
		Quantity of formulation (g)	Volume of water (Lit.)	
Chilli	Fruit borer, Thrips, Mites	250	500-700	07

Brief information, mode of action, and trade names for this combination are mentioned in Part III: Insecticide Combinations.

Emamectin benzoate + Lufenuron

Available formulations in India :

1. **Emamectin benzoate 05.00% + Lufenuron 40.00% WG**
2. **Lufenuron 04.00% + Emamectin benzoate 01.50% EC**

Classification on the basis of		**Emamectin benzoate**	**Lufenuron**
Target organism	:	Insecticide, Acaricide	Insecticide
Mode of entry	:	Systemic, Contact and Stomach poison	Contact and Stomach poison
Mode of action	:	Nerve poison	Insect Growth Regulator (IGR)
Chemical group	:	Avermectin (Macrocyclic lactones)	Benzoylurea
Category of toxicity : Moderately toxic **Toxicity label colour** : Blue			

Label claim (DPPQ & S, Faridabad)

Emamectin benzoate 05.00% + Lufenuron 40.00% WG

Crop	**Pests**	**Dose / ha**		**Waiting period (Days)**
		Quantity of formulation (g)	**Volume of water (Lit.)**	
Chilli	Fruit borer, Tobacco caterpillar, Thrips, Yellow mite	60	500	03

Brief information, mode of action, and trade names for this combination are mentioned in Part III: Insecticide Combinations.

Emamectin benzoate + Profenofos

Available formulations in India : **Emamectin benzoate 01.50% + Profenofos 35.00% WDG**

Classification on the basis of		Emamectin benzoate	Profenofos
Target organism	:	Insecticide, Acaricide	Insecticide, Acaricide
Mode of entry	:	Systemic, Contact and Stomach poison	Contact and Stomach poison
Mode of action	:	Nerve poison	Nerve poison
Chemical group	:	Avermectin (Macrocyclic lactones)	Organophosphate
Category of toxicity : Moderately toxic **Toxicity label colour** : Blue			

Label claim (DPPQ & S, Faridabad)

Emamectin benzoate 01.50% + Profenofos 35.00% WDG

Crop	Pests	Dose / ha		Waiting period (Days)
		Quantity of formulation (ml)	Volume of water (Lit.)	
Chilli	Thrips, Fruit borer, Yellow mite	700	500	07

Brief information, mode of action, and trade names for this combination are mentioned in Part III: Insecticide Combinations.

Ethion + Cypermethrin

Available formulations in India : **Ethion 40.00% + Cypermethrin 05.00% EC**

Classification on the basis of		Ethion	Cypermethrin
Target organism	:	Insecticide, Acaricide	Insecticide
Mode of entry	:	Contact and Stomach poison	Contact and Stomach poison
Mode of action	:	Nerve poison	Nerve poison
Chemical group	:	Organophosphate	Synthetic pyrethroid
Category of toxicity : Highly toxic **Toxicity label colour** : Yellow			

Label claim (DPPQ & S, Faridabad)

Ethion 40.00% + Cypermethrin 05.00% EC

Crop	Pests	Dose / ha		Waiting period (Days)
		Quantity of formulation (ml)	Volume of water (Lit.)	
Chilli	Mites, Thrips	1500	500	10

Brief information, mode of action, and trade names for this combination are mentioned in Part III: Insecticide Combinations.

Etoxazole + Abamectin

Available formulations in India : **Etoxazole 06.00% + Abamectin 01.50% SC**

Classification on the basis of		Etoxazole	Abamectin
Target organism	:	Acaricide	Acaricide, Insecticide
Mode of entry	:	Contact and Stomach poison	Contact and Stomach poison
Mode of action	:	Insect Growth Regulator (IGR)	Nerve poison
Chemical group	:	Diphenyloxazoline	Avermectin (Macrocyclic lactone)
Category of toxicity : Moderately toxic **Toxicity label colour** : Blue			

- Etoxazole is an insect growth regulator affecting mite development, specifically inhibiting molting in immature stages (larvae, nymphs) by blocking chitin biosynthesis necessary for cuticle formation. It is highly selective for mites, disrupting their life cycle and preventing completion to adulthood (ovicidal and larvicidal action).
- Abamectin, a macrocyclic lactone derived from the soil bacterium Streptomyces avermitilis, acts on the insect nervous system by activating glutamate-gated chloride channels, which results in increased chloride ion influx, hyperpolarization, paralysis, and death of mites and some insects. It offers rapid contact and stomach action on mites, providing adulticidal activity that complements Etoxazole's life cycle disruption.
- The combination provides a quick knockdown effect from Abamectin on motile stages (adults, nymphs) and long-lasting residual control from Etoxazole on the immature stages (eggs, larvae, nymphs), leading to more thorough and sustained mite control.

Label claim (DPPQ & S, Faridabad)

Etoxazole 06.00% + Abamectin 01.50% SC

Crop	Pests	Dose / ha		Waiting period (Days)
		Quantity of formulation (ml)	Volume of water (Lit.)	
Tea	Red spider mite, Purple mite, Scarlet mite	450	500	03
Brinjal	Red spider mite	450	500	05

Trade Names

Sr. No.	Trade Name	Company
1	Portion	Sumitomo Chemical India Ltd.

Fenazaquin + Bifenthrin

Available formulations in India : **Fenazaquin 10.00% + Bifenthrin 04.00% EC**

Classification on the basis of		**Fenazaquin**	**Bifenthrin**
Target organism	:	Insecticide, Acaricide	Insecticide
Mode of entry	:	Contact and Stomach poison	Contact and Stomach poison
Mode of action	:	Mitochondrial complex I electron transport inhibitor	Nerve poison
Chemical group	:	Quinazoline	Synthetic pyrethroid
Category of toxicity : **Toxicity label colour :**			

Label claim (DPPQ & S, Faridabad)

Fenazaquin 10.00% + Bifenthrin 04.00% EC

Crop	**Pests**	**Dose / ha**		**Waiting period (Days)**
		Quantity of formulation (ml)	**Volume of water (Lit.)**	
Tomato	Red spider mite, Whitefly	1250	500	05
Tea	Ted spider mite, Mosquito bug	1000	400	07
Cotton	Whitefly, Mite	1250	500	74

Brief information, mode of action, and trade names for this combination are mentioned in Part III: Insecticide Combinations.

Fipronil + Propargite

Available formulations in India : **Fipronil 02.50% + Propargite 35.00% SE**

Classification on the basis of		**Fipronil**	**Propargite**
Target organism	:	Insecticide	Acaricide
Mode of entry	:	Contact and Stomach poison	Contact poison
Mode of action	:	Nerve poison	Affect energy metabolism
Chemical group	:	Phenylpyrazole	Sulfite ester
Category of toxicity : **Toxicity label colour :**			

Label claim (DPPQ & S, Faridabad)

Fipronil 02.50% + Propargite 35.00% SE

Crop	**Pests**	**Dose / ha**		**Waiting period (Days)**
		Quantity of formulation (ml)	**Volume of water (Lit.)**	
Chilli	Thrips, Aphid, Mites, Fruit borer	2000	500	20

Brief information, mode of action, and trade names for this combination are mentioned in Part III: Insecticide Combinations.

Fipronil + Hexythiazox

Available formulations in India : **Fipronil 07.00% + Hexythiazox 02.00% SC**

Classification on the basis of		Fipronil	Hexythiazox
Target organism	:	Insecticide	Acaricide
Mode of entry	:	Contact and Stomach poison	Contact and Stomach poison
Mode of action	:	Nerve poison	Insect Growth Regulator (IGR)
Chemical group	:	Phenylpyrazole	Thiazolidinone
Category of toxicity : Moderately toxic **Toxicity label colour** : Blue			

Label claim (DPPQ & S, Faridabad)

Fipronil 07.00% + Hexythiazox 02.00% SC

Crop	Pests	Dose / ha		Waiting period (Days)
		Quantity of formulation (ml)	Volume of water (Lit.)	
Chilli	Mites, Thrips	1000	500	07

Brief information, mode of action, and trade names for this combination are mentioned in Part III: Insecticide Combinations.

Fipronil + Tolfenpyrad + Abamectin

Available formulations in India : **Fipronil 07.00% + Tolfenpyrad 15.00% + Abamectin 01.25% SC**

Classification on the basis of		Fipronil	Tolfenpyrad	Abamectin
Target organism	:	Insecticide	Insecticide	Acaricide, Insecticide
Mode of entry	:	Contact and Stomach poison	Stomach and Contact poison	Stomach and Contact poison
Mode of action	:	Nerve poison	Energy metabolism	Nerve poison
Chemical group	:	Phenylpyrazole	Pyrazole	Avermectin
Category of toxicity : Highly toxic **Toxicity label colour** : Yellow				

Label claim (DPPQ & S, Faridabad)
Fipronil 07.00% + Tolfenpyrad 15.00% + Abamectin 01.25% SC

Crop	Pests	Dose / ha		Waiting period (Days)
		Quantity of formulation (ml)	Volume of water (Lit.)	
Okra	Aphid, Jassid, Thrips, Whitefly, Mites, Fruit borer	750	500	03
Chilli	Aphid, Thrips, Mites, Fruit borer	750	500	10

Brief information, mode of action, and trade names for this combination are mentioned in Part III: Insecticide Combinations.

Fluxametamide + Pyridaben

Available formulations in India : **Fluxametamide 03.80% + Pyridaben 09.50% SC**

Classification on the basis of		Fluxametamide	Pyridaben
Target organism	:	Insecticide	Insecticide, Acaricide
Mode of entry	:	Contact and Stomach poison	Contact and Stomach poison
Mode of action	:	Nerve poison	Energy metabolism
Chemical group	:	Isoxazoline	Pyridazinone
Category of toxicity : **Toxicity label colour** :			

Label claim (DPPQ & S, Faridabad)

Fluxametamide 03.80% + Pyridaben 09.50% SC

Crop	Pests	Dose / ha		Waiting period (Days)
		Quantity of formulation (ml)	Volume of water (Lit.)	
Brinjal	Leafhopper, Thrips, Whitefly, Shoot and fruit borer, Red spider mite	500	375	05
Chilli	Thrips, Fruit borer, Tobacco caterpillar, Yellow mite	1000	500	05

Brief information, mode of action, and trade names for this combination are mentioned in Part III: Insecticide Combinations.

Hexythiazox + Diafenthiuron

Available formulations in India : Hexythiazox 03.5% + Diafenthiuron 42.00% WDG

Classification on the basis of		Hexythiazox	Diafenthiuron
Target organism	:	Acaricide	Insecticide, Acaricide
Mode of entry	:	Contact and Stomach poison	Contact and Stomach poison
Mode of action	:	Mite Growth Regulator	Energy metabolism
Chemical group	:	Thiazolidinone	Thiourea
Category of toxicity : Moderately toxic **Toxicity label colour** : Blue			

Label claim (DPPQ & S, Faridabad)

Hexythiazox 03.5% + Diafenthiuron 42.00% WDG

Crop	Pests	Dose / ha		Waiting period (Days)
		Quantity of formulation (g)	Volume of water (Lit.)	
Chilli	Mites, Thrips, Jassid, Aphid, Whitefly	650	500	07

Brief information, mode of action, and trade names for this combination are mentioned in Part III: Insecticide Combinations.

Lufenuron + Chlorantraniliprole + Pyriproxyfen

Available formulations in India : **Lufenuron 20.00% + Chlorantraniliprole 10.00% + Pyriproxyfen 05.00% WG**

Classification on the basis of		**Lufenuron**	**Chlorantraniliprole**	**Pyriproxyfen**
Target organism	:	Insecticide	Insecticide	Insecticide
Mode of entry	:	Contact and Stomach poison	Systemic, Contact and Stomach poison	Contact and Stomach poison
Mode of action	:	Insect Growth Regulator	Muscle poison	Insect Growth Regulator (IGR)
Chemical group	:	Benzoylurea	Anthranilic diamide	Pyridine ether
Category of toxicity : **Toxicity label colour** :				

Label claim (DPPQ & S, Faridabad)

Lufenuron 20.00% + Chlorantraniliprole 10.00% + Pyriproxyfen 05.00% WG

Crop	**Pests**	**Dose / ha**		**Waiting period (Days)**
		Quantity of formulation (g)	**Volume of water (Lit.)**	
Chilli	Thrips, Mites, Fruit borers (*Helicoverpa armigera, Spodoptera litura*)	112.50	500	19

Brief information, mode of action, and trade names for this combination are mentioned in Part III: Insecticide Combinations.

Methoxyfenozide + Emamectin benzoate

Available formulations in India : **Methoxyfenozide 18.00% + Emamectin benzoate 01.80% SC**

Classification on the basis of		Methoxyfenozide	Emamectin benzoate
Target organism	:	Insecticide	Insecticide, Acaricide
Mode of entry	:	Stomach and Contact poison	Systemic, Contact and Stomach poison
Mode of action	:	Insect Growth regulator (IGR)	Nerve poison
Chemical group	:	Diacylhydrazines	Avermectin (Macrocyclic lactones)
Category of toxicity : **Toxicity label colour :**			

Label claim (DPPQ & S, Faridabad)

Methoxyfenozide 18.00% + Emamectin benzoate 01.80% SC

Crop	Pests	Dose / ha		Waiting period (Days)
		Quantity of formulation (ml)	Volume of water (Lit.)	
Chilli	Fruit borer, Thrips, Yellow mite	500	500	03

Brief information, mode of action, and trade names for this combination are mentioned in Part III: Insecticide Combinations.

Phenthoate + Emamectin benzoate

Available formulations in India : **Phenthoate 35.00% + Emamectin benzoate 03.50% EC**

Classification on the basis of		Phenthoate	Emamectin benzoate
Target organism	:	Insecticide	Insecticide, Acaricide
Mode of entry	:	Contact and Stomach poison	Systemic, Contact and Stomach poison
Mode of action	:	Nerve poison	Nerve poison
Chemical group	:	Organophosphate	Avermectin (Macrocyclic lactones)
Category of toxicity : **Toxicity label colour :**			

Label claim (DPPQ & S, Faridabad)
Phenthoate 35.00% + Emamectin benzoate 03.50% EC

Crop	Pests	Dose / ha		Waiting period (Days)
		Quantity of formulation (ml)	Volume of water (Lit.)	
Chilli	Mites, Thrips, Fruit borer	750	500	05

Brief information, mode of action, and trade names for this combination are mentioned in Part III: Insecticide Combinations.

Profenofos + Fenpyroximate

Available formulations in India : **Profenofos 40.00% + Fenpyroximate 02.50% EC**

Classification on the basis of		Profenofos	Fenpyroximate
Target organism	:	Insecticide	Insecticide, Acaricide
Mode of entry	:	Contact and Stomach poison	Stomach and Contact poison
Mode of action	:	Nerve poison	Energy metabolism
Chemical group	:	Organophosphate	Pyrazole
Category of toxicity : Highly toxic **Toxicity label colour** : Yellow			

Label claim (DPPQ & S, Faridabad)

Profenofos 40.00% + Fenpyroximate 02.50% EC

Crop	Pests	Dose / ha		Waiting period (Days)
		Quantity of formulation (ml)	Volume of water (Lit.)	
Chilli	Thrips, Mites, Fruit borer	1000	500	07
Brinjal	Whitefly, Mites, Shoot and fruit borer	1000	500	07

<u>Brief information, mode of action, and trade names for this combination are mentioned in Part III: Insecticide Combinations.</u>

Propargite + Bifenthrin

Available formulations in India : **Propargite 50.00% + Bifenthrin 05.00% SE**

Classification on the basis of		Propargite	Bifenthrin
Target organism	:	Acaricide	Insecticide
Mode of entry	:	Stomach and Contact poison	Contact and Stomach poison
Mode of action	:	Energy metabolism	Nerve poison
Chemical group	:	Sulfite ester	Synthetic pyrethroid
Category of toxicity : Moderately toxic **Toxicity label colour** : Blue			

Label claim (DPPQ & S, Faridabad)

Propargite 50.00% + Bifenthrin 05.00% SE

Crop	Pests	Dose / ha		Waiting period (Days)
		Quantity of formulation (ml)	Volume of water (Lit.)	
Okra	Mite, Whitefly, Jassid	1100-1150	500	05
Tomato	Mite, Whitefly, Jassid	1100-1150	500	05

Brief information, mode of action, and trade names for this combination are mentioned in Part III: Insecticide Combinations.

Propargite + Hexythiazox

Available formulations in India : **Propargite 42.00% + Hexythiazox 02.00% EC**

Classification on the basis of		Propargite	Hexythiazox
Target organism	:	Acaricide	Acaricide
Mode of entry	:	Stomach and Contact poison	Contact and Stomach poison
Mode of action	:	Energy metabolism	Mite Growth Regulator
Chemical group	:	Sulfite ester	Thiazolidinone
Category of toxicity : Moderately toxic **Toxicity label colour** : Blue			

- The combination of Propargite and Hexythiazox forms a powerful, dual-action miticide (acaricide) widely used in agriculture to control a broad spectrum of phytophagous mites. The primary benefit of this combination is its comprehensive action against all mite life stages and its utility in resistance management by having two distinct modes of action.
- Propargite, a sulfite ester, is known as a true miticide that provides rapid contact and vapor action, offering quick knockdown of the active life stages, such as adult mites and nymphs, and often exhibits ovicidal (egg-killing) activity.
- Conversely, Hexythiazox, belonging to the thiazolidinone group and classified as an Insect Growth Regulator (IGR), primarily acts as a contact and stomach poison with strong ovicidal, larvicidal, and nymphicidal properties. It is particularly effective against mite eggs and immature stages and exhibits translaminar activity, meaning it can penetrate the leaf tissue to reach and control hidden mites on the underside of leaves.
- The dual mechanism—Propargite providing immediate relief and Hexythiazox offering long-lasting residual control, especially on the eggs and early instars—ensures complete life cycle control and helps delay the onset of resistance in mite populations, which is a key principle in Integrated Pest Management (IPM) strategies.

Label claim (DPPQ & S, Faridabad)

Propargite 42.00% + Hexythiazox 02.00% EC

Crop	Pests	Dose / ha		Waiting period (Days)
		Quantity of formulation (ml)	Volume of water (Lit.)	
Tea	Red spider mite	1250	400-500	07
Chilli	Yellow mite	1250	500	07
Apple	European red mite	10 ml / 10 lit. water	10-15 lit/tree	15
Brinjal	Mite	1250	500	07

Trade Names

Sr. No.	Trade Name	Company
1	Ceasemite	Indofil Industries Ltd.
2	Clearmite	Essential Biosciences
3	Rigel	Synovita Crop science Pvt. Ltd.
4	Stopamite	Katyayani Organics

Pyriproxyfen + Bifenthrin

Available formulations in India : **Pyriproxyfen 10.00% + Bifenthrin 10.00% EC**

Classification on the basis of		Pyriproxyfen	Bifenthrin
Target organism	:	Insecticide	Insecticide
Mode of entry	:	Contact and Stomach poison	Contact and Stomach poison
Mode of action	:	Insect Growth Regulator (IGR)	Nerve poison
Chemical group	:	Pyridine ether	Synthetic pyrethroid
Category of toxicity : Moderately toxic **Toxicity label colour** : Blue			

Label claim (DPPQ & S, Faridabad)

Pyriproxyfen 10.00% + Bifenthrin 10.00% EC

Crop	Pests	Dose / ha		Waiting period (Days)
		Quantity of formulation (ml)	Volume of water (Lit.)	
Chilli	Yellow mite, Thrips, Whitefly, Fruit borer	1250	500	15

<u>Brief information, mode of action, and trade names for this combination are mentioned in Part III: Insecticide Combinations.</u>

Spiromesifen + Hexythiazox + Abamectin

Available formulations in India : **Spiromesifen 15.00% + Hexythiazox 02.20% + Abamectin 01.15% SC**

Classification on the basis of		Spiromesifen	Hexythiazox	Abamectin
Target organism	:	Insecticide, Acaricide	Acaricide	Acaricide, Insecticide
Mode of entry	:	Stomach and Contact poison	Contact and Stomach poison	Stomach and Contact poison
Mode of action	:	Lipid synthesis and growth regulator	Growth regulator	Nerve poison
Chemical group	:	Tetronic and Tetramic acid derivatives	Thiazolidinecarboxamide	Avermectin
Category of toxicity : Moderately toxic **Toxicity label colour** : Blue				

Label claim (DPPQ & S, Faridabad)
Spiromesifen 15.00% + Hexythiazox 02.20% + Abamectin 01.15% SC

Crop	Pests	Dose / ha		Waiting period (Days)
		Quantity of formulation (ml)	Volume of water (Lit.)	
Chilli	Yellow mite, Black thrips	625	500	16

Brief information, mode of action, and trade names for this combination are mentioned in Part III: Insecticide Combinations.

Spirotetramat + Imidacloprid

Available formulations in India : **Spirotetramat 11.01% + Imidacloprid 11.01% SC**

Classification on the basis of		Spirotetramat	Imidacloprid
Target organism	:	Insecticide, Acaricide	Insecticide
Mode of entry	:	Systemic, Contact poison	Systemic, Contact and Stomach poison
Mode of action	:	Nerve poison	Nerve poison
Chemical group	:	Tetronic / Tetramic acid derivatives (Ketoenol)	Neonicotinoid
Category of toxicity : Highly toxic **Toxicity label colour** : Yellow			

Label claim (DPPQ & S, Faridabad)

Spirotetramat 11.01% + Imidacloprid 11.01% SC

Crop	Pests	Dose / ha		Waiting period (Days)
		Quantity of formulation (ml)	Volume of water (Lit.)	
Okra	Red spider mite	500	500	03
Brinjal	Whitefly, Red spider mite	500	500	05
Grapes	Thrips, Mites, Mealybug	750	1000	60

Brief information, mode of action, and trade names for this combination are mentioned in Part III: Insecticide Combinations.

Tetraniliprole + Spirotetramat

Available formulations in India : **Tetraniliprole 10.81% + Spirotetramat 21.62% SC**

Classification on the basis of		Tetraniliprole	Spirotetramat
Target organism	:	Insecticide	Insecticide, Acaricide
Mode of entry	:	Systemic, Stomach poison	Systemic, Contact poison
Mode of action	:	Nerve poison	Nerve poison
Chemical group	:	Anthranilic diamide	Tetronic / Tetramic acid derivatives (Ketoenol)
Category of toxicity : **Toxicity label colour :**			

Label claim (DPPQ & S, Faridabad)
Tetraniliprole 10.81% + Spirotetramat 21.62% SC

Crop	Pests	Dose / ha		Waiting period (Days)
		Quantity of formulation (ml)	Volume of water (Lit.)	
Okra	Shoot and fruit borer, Aphid, Whitefly, Mite	375	500	03
Cabbage	Diamond back moth, Tobacco caterpillar, Aphid	375	500	07
Chilli	Fruit borer, Thrips, Whitefly, Mite	375	500	05
Brinjal	Shoot and fruit borer, Whitefly, Mite	375	500	05
Tomato	Fruit borer, Leaf miner, Thrips, Whitefly	375	500	05

Brief information, mode of action, and trade names for this combination are mentioned in Part III: Insecticide Combinations.

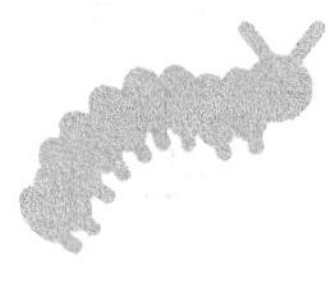

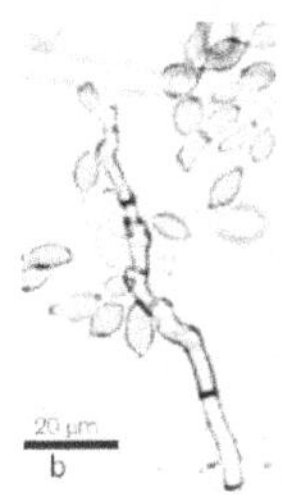

PART V

Insecticide + Fungicide Combinations

Azoxystrobin + Fipronil

Available formulations in India : **Azoxystrobin 10.00% + Fipronil 05.00% SC**

Classification on the basis of		Azoxystrobin	Fipronil
Target organism	:	Fungicide	Insecticide
Mode of entry	:	Systemic	Stomach and Contact poison
Mode of action	:	Mitochondrial respiration inhibitor	Nerve poison
Chemical group	:	Strobilurin	Phenylpyrazole
Category of toxicity : Moderately toxic **Toxicity label colour** : Blue			

- The pre-mix of Azoxystrobin + Fipronil is an agricultural product designed to offer a single solution for controlling both fungal diseases and insect pests in one application, utilizing two distinct modes of action for comprehensive crop protection.
- **Azoxystrobin (Fungicide Component)**
- Azoxystrobin is a broad-spectrum, systemic fungicide belonging to the strobilurin chemical class (FRAC Group 11). It is derived from naturally occurring antifungal compounds. Its primary role is to control a wide range of fungal diseases, such as rusts, mildews, and blights, on various crops. Azoxystrobin exhibits preventive, curative, and residual activity. It is systemic, meaning it is absorbed by the plant (through leaves, stems, or roots) and moves internally, offering protection to new growth and hard-to-reach tissues.
- The Mode of Action for Azoxystrobin is as a Quinone outside Inhibitor (QoI). It specifically targets the fungal pathogen's respiratory system by binding to the Qo site of the cytochrome bc1 complex (Complex III) in the mitochondria. This action blocks electron transfer, which immediately halts the process of ATP (energy) synthesis in the fungal cell. Without energy, key life processes like spore germination and mycelial growth are completely disrupted, leading to the death of the fungus and effective disease control.
- **Fipronil (Insecticide Component)**
- Fipronil is a broad-spectrum insecticide belonging to the phenylpyrazole chemical class (IRAC Group 2B). It is highly effective against a variety of insect pests, including stem borers, leaf folders, plant hoppers, and thrips, especially in crops. Fipronil is considered to have primarily contact and stomach (ingestion) modes of action but it also exhibits some systemic activity. It provides quick knockdown and a long residual effect as it can be absorbed and translocated within the plant.
- The Mode of Action for Fipronil is to disrupt the insect's central nervous system. It acts as a GABA-gated chloride channel blocker and a Glutamate-gated chloride (GluCl) channel blocker. By binding to the GABAA receptor in the insect's nerves, Fipronil prevents chloride ions from entering the nerve cell. This results in hyperexcitation of the

neurons and muscles, leading to uncontrolled nerve activity, paralysis, and ultimately, the death of the insect pest.

Label claim (DPPQ & S, Faridabad)

Azoxystrobin 10.00% + Fipronil 5.00% SC

Crop	Pests	Dose / ha		Waiting period (Days)
		Quantity of formulation (ml)	Volume of water (Lit.)	
Chilli	**Insect pests :** Thrips, Fruit borer **Diseases :** Powdery mildew, Anthracnose / Fruit rot	1000	500	05
Rice	**Insect pests :** Yellow stem borer, Leaf folder, Brown plant hopper **Diseases :** Sheath blight, Leaf blast	1250	500	53

Trade Names

Sr. No.	Trade Name	Company
1	Tipsy	Gharda Chemicals Ltd.

Azoxystrobin + Tebuconazole + Thiamethoxam

Available formulations in India : **Azoxystrobin 01.30% + Tebuconazole 0.22% + Thiamethoxam 25.90% FS**

Classification on the basis of		Azoxystrobin	Tebuconazole	Thiamethoxam
Target organism	:	Fungicide	Fungicide	Insecticide
Mode of entry	:	Systemic	Systemic	Systemic, Stomach and Contact poison
Mode of action	:	Mitochondrial respiration inhibitor	Interferes cell membrane synthesis	Nerve poison
Chemical group	:	Strobilurin	Triazole	Neonicotinoid
Category of toxicity :				
oxicity label colour :				

The combination of Azoxystrobin, Tebuconazole, and Thiamethoxam brings together a powerful mix of both fungicides and an insecticide to provide broad-spectrum protection for crops. This triple-action product is designed to simultaneously control a wide range of fungal diseases and problematic insect pests.

Azoxystrobin (Fungicide) : Azoxystrobin is a broad-spectrum systemic fungicide belonging to the Strobilurin chemical group (also known as a Quinone outside Inhibitor, or QoI). Its primary mode of action is to inhibit mitochondrial respiration in fungal pathogens by blocking the electron transport chain. This effectively stops the fungus from producing energy, leading to the inhibition of spore germination, mycelial growth, and overall fungal development. Azoxystrobin has a systemic and translaminar movement, meaning it is absorbed by the plant and moves within the tissues (acropetally, up the plant) and across the leaf surface, providing both preventive and some curative/eradicant action. It is highly valued for promoting greener foliage (a phyto-tonic effect) and enhancing crop quality and yield.

Tebuconazole (Fungicide) : Tebuconazole is a systemic fungicide from the Triazole chemical group (a Demethylation Inhibitor, or DMI). Its mode of action is to inhibit the biosynthesis of ergosterol, a vital component of fungal cell membranes. By disrupting ergosterol production, it compromises the structural integrity of the fungal cell, ultimately causing the fungus to die. Tebuconazole is known for its curative and eradicant properties, making it highly effective against established fungal infections. When combined with Azoxystrobin, the two fungicides provide a dual mode of action (dual-site action: respiration and cell membrane) which is crucial for enhanced efficacy, managing fungicide resistance, and offering both immediate and long-lasting protection against a broad spectrum of fungal diseases like rusts, blights, and mildews.

Thiamethoxam (Insecticide) : Thiamethoxam is a powerful systemic insecticide belonging to the Neonicotinoid class. It acts as a neurotoxin by interfering with the nicotinic acetylcholine receptors in the insect's central nervous system, leading to paralysis and death. Due to its systemic nature, Thiamethoxam is readily absorbed by the plant and translocated throughout the vascular system, protecting all parts of the plant, including new growth, from feeding pests. It is particularly effective against sucking and chewing insects like aphids, whiteflies, thrips, and leafhoppers. Its inclusion in a combination with the two fungicides transforms the product into a comprehensive protection tool, particularly useful for seed treatment or early-stage crop spray applications to guard young plants against both early-season diseases and insect infestations simultaneously.

Label claim (DPPQ & S, Faridabad)

Azoxystrobin 01.30% + Tebuconazole 0.22% + Thiamethoxam 25.90% FS

Crop	Pests	Dose / Kg Seed		Waiting period (Days)
		Quantity of formulation (ml)	Volume of water (Lit.)	
Okra	**Insect pests :** Aphid, Jassid **Diseases** : Damping off (*Rhizoctina spp. and Pythium spp.*)	6	--	05

Trade Names

Sr. No.	Trade Name	Company
1		

Azoxystrobin + Thiophanate Methyl + Thiamethoxam

Available formulations in India : **Azoxystrobin 2.5% + Thiophanate Methyl 11.25% + Thiamethoxam 25% FS**

Classification on the basis of		Azoxystrobin	Thiophanate Methyl	Thiamethoxam
Target organism	:	Fungicide	Fungicide	Insecticide
Mode of entry	:	Systemic	Systemic	Systemic, Stomach and Contact poison
Mode of action	:	Mitochondrial respiration inhibitor	Interferes cell division	Nerve poison
Chemical group	:	Strobilurin	Triazole	Neonicotinoid
Category of toxicity : Moderately toxic **Toxicity label colour** : Blue				

- The formulation Azoxystrobin 2.5% + Thiophanate Methyl 11.25% + Thiamethoxam 25% FS is a potent combination product used primarily for seed treatment in agriculture.
- It is an innovative, dual-action product, combining two fungicides (Azoxystrobin and Thiophanate Methyl) and one insecticide (Thiamethoxam) to provide comprehensive early-season protection to crops like soybeans and pulses.
- Azoxystrobin (a strobilurin fungicide) works by inhibiting fungal respiration, providing broad-spectrum control against various fungal pathogens.
- Thiophanate Methyl (a systemic fungicide) interferes with fungal cell division, offering both preventive and curative action against soil-borne and seed-borne diseases such as Fusarium and Rhizoctonia root rots.
- Thiamethoxam (a neonicotinoid insecticide) is a systemic neurotoxin that controls early-stage sucking pests (like shoot fly) and other insects (like white grub and termites), ensuring the seedling can establish well.
- This triple-mix is absorbed by the plant and translocated throughout the tissues, providing systemic, long-lasting protection right from germination. This enhances crop establishment, promotes healthy root development, and ultimately contributes to improved plant vigor and higher yields.

Label claim (DPPQ & S, Faridabad)

Azoxystrobin 2.5% + Thiophanate Methyl 11.25% + Thiamethoxam 25% FS

Crop	Pests	Dose / Kg Seed		Waiting period (Days)
		Quantity of formulation (ml)	Volume of water (Lit.)	
Soybean	**Insect pests :** Stem fly, White grub, Termite	10	--	--

	Diseases : *Fusarium* root rot, *Phytophthora* root rot, *Rhizoctonia* seedling blight, *Pythium* seedling blight			

Trade Names

Sr. No.	Trade Name	Company
1	Warden	Best Agrolife Ltd.
2	Electron	UPL Ltd.
3	Cascade	Swal Corporation Ltd.
4	Multi-Star	Jai Shree Ram Agro Pvt. Ltd.

Carboxin + Thiram + Imidacloprid

Available formulations in India : **Carboxin 22.50% + Thiram 22.50% + Imidacloprid 18.00% WS**

Classification on the basis of		**Carboxin**	**Thiram**	**Imidacloprid**
Target organism	:	Fungicide	Fungicide	Insecticide
Mode of entry	:	Systemic	Contact	Systemic, Contact and Stomach poison
Mode of action	:	SDHI Inhibitor	Multi-site inhibitor	Nerve poison
Chemical group	:	Anilide	Dithiocarbamate	Neonicotinoid
Category of toxicity : Moderately toxic **Toxicity label colour** : Blue				

- This combination product (Carboxin + Thiram + Imidacloprid) is a highly effective triple-action seed treatment designed to give the emerging seedling complete protection against both fungi and insect pests right from the moment of sowing.
- Carboxin is a systemic fungicide of the SDHI (Succinate Dehydrogenase Inhibitor) class, disrupting mitochondrial function in fungal cells by inhibiting complex II of the respiratory electron transport chain. It is widely used as a seed treatment fungicide to control diseases like smuts and bunts in cereals and other crops.
- Thiram is a non-specific multi-site contact fungicide that inhibits various fungal enzymes, thereby preventing spore germination and mycelial growth. It provides protective fungicidal action mostly on seed surfaces and foliage.
- Imidacloprid is a systemic insecticide acting on the nicotinic acetylcholine receptors of the insect central nervous system. It provides control against many sucking pests like aphids, whiteflies, leafhoppers, and thrips. It is effective via seed treatment, soil treatment, or foliar application.

Label claim (DPPQ & S, Faridabad)

Carboxin 22.50% + Thiram 22.50% + Imidacloprid 18.00% WS

Crop	**Pests**	**Dose / Kg seed**		**Waiting period (Days)**
		Quantity of formulation (g)	**Volume of water (Lit.)**	
Wheat	**Insect pests :** Aphid, Termite **Diseases :** Loose smut, False smut	3.4	--	--

Trade Names

Sr. No.	**Trade Name**	**Company**
1	Chatbot	Chemley Agritech Pvt. Ltd.

Chlorantraniliprole + Tebuconazole + Clothianidin

Available formulations in India : **Chlorantraniliprole 0.64% + Tebuconazole 02.00% + Clothianidin 01.92% GR**

Classification on the basis of		**Chlorantraniliprole**	**Tebuconazole**	**Clothianidin**
Target organism	:	Insecticide	Fungicide	Insecticide
Mode of entry	:	Systemic, Contact and Stomach poison	Systemic	Systemic, Contact and Stomach poison
Mode of action	:	Muscle poison	Interferes cell membrane synthesis	Nerve poison
Chemical group	:	Anthranilic diamide	Triazole	Neonicotinoid
Category of toxicity : Moderately toxic **Toxicity label colour** : Blue				

- This formulation, containing Chlorantraniliprole 0.64% + Tebuconazole 2.00% + Clothianidin 1.92% in a Granular (GR) form, is an advanced, combination pesticide designed to provide broad-spectrum protection against both major insect pests and fungal diseases in a single application.
- The combination of three different active ingredients offers comprehensive crop defense during critical growth stages, leading to improved plant establishment, health, and ultimately, enhanced grain yield.
- The product features a powerful pairing of two systemic insecticides and one systemic fungicide.
- Chlorantraniliprole is a potent insecticide belonging to the diamide class (Group 28), which acts by activating ryanodine receptors in insect muscles, leading to paralysis and death. It provides excellent control against Lepidopteran pests.
- The second insecticide, Clothianidin (a Neonicotinoid, Group 4A), is systemic and works by disrupting the insect's central nervous system, offering control over sucking and chewing pests. Together, the two insecticides deliver immediate and long-lasting protection against a wide range of insects.
- Complementing the insect control is the fungicide component, Tebuconazole (a Triazole, Group 3), which is a systemic ergosterol biosynthesis inhibitor. Tebuconazole is highly effective against significant fungal diseases.
- The granular (GR) formulation is specifically engineered for easy broadcast application in flooded fields, ensuring the active ingredients are slowly released and absorbed by the plant roots, providing uniform, residual, and rainfast action without the need for traditional foliar spraying.

Label claim (DPPQ & S, Faridabad)

Chlorantraniliprole 0.64% + Tebuconazole 2.00% + Clothianidin 1.92% GR

Crop	Pests	Dose / ha		Waiting period (Days)
		Quantity of formulation (Kg)	Volume of water	
Rice	**Insect pests :** Yellow stem borer **Diseases :** Sheath blight	6.25	--	46

Trade Names

Sr. No.	Trade Name	Company
1	Eugloria	Chemley Agritech Pvt. Ltd.

Flubendiamide + Hexaconazole

Available formulations in India : **Flubendiamide 03.50% + Hexaconazole 05.00% WG**

Classification on the basis of		Flubendiamide	Hexaconazole
Target organism	:	Insecticide	Fungicide
Mode of entry	:	Contact and Stomach poison	Systemic
Mode of action	:	Muscle poison	Interferes cell membrane synthesis
Chemical group	:	Diamide	Triazole
Category of toxicity : Moderately toxic **Toxicity label colour** : Blue			

- Flubendiamide 3.50% + Hexaconazole 5.00% WG formulation is a highly effective, dual-action product combining an insecticide and a fungicide, designed for comprehensive crop protection.
- Flubendiamide targets Ryanodine Receptors (RyR) in insect muscles, leading to rapid cessation of feeding and pest death, primarily effective against Lepidopteran larvae.
- Hexaconazole is a systemic triazole fungicide with protective and eradicative action. It inhibits fungal growth by interfering with cell membrane synthesis.

Label claim (DPPQ & S, Faridabad)

Flubendiamide 03.50% + Hexaconazole 05.00% WG

Crop	Pests	Dose / ha		Waiting period (Days)
		Quantity of formulation (g)	Volume of water (Lit.)	
Rice	**Insect pests :** Stem borer, Leaf folder **Diseases :** Sheath blight, Sheath rot	1000	500	30
Groundnut	**Insect pests :** Tobacco caterpillar **Diseases :** Tikka leaf spot, Rust	1500	500	31
Chilli	**Insect pests :** Tobacco caterpillar, Fruit borer **Diseases :** Leaf spot, Powdery mildew, Anthracnose	1500	500	10
Tomato	**Insect pests :** Tobacco caterpillar, fruit borer **Diseases :** Early blight, Leaf spot	1250	500	05

Trade Names

Sr. No.	Trade Name	Company
1	Origin	Tata Rakkies India Ltd., Global Crop Care
2	Flexaking	Mankind Agritech
3	Dhruv	Prisha Chemicals

Flubendiamide + Kresoxim-Methyl

Available formulations in India : **Flubendiamide 07.50% + Kresoxim-Methyl 37.50% SC**

Classification on the basis of		Flubendiamide	Kresoxim-Methyl
Target organism	:	Insecticide	Fungicide
Mode of entry	:	Contact and Stomach poison	Meso-systemic with strong translaminar and vapor-phase activity
Mode of action	:	Muscle poison	Inhibits mitochondrial respiration
Chemical group	:	Diamide	Strobilurin
Category of toxicity : Moderately toxic **Toxicity label colour** : Blue			

- Flubendiamide 7.50% + Kresoxim-Methyl 37.50% SC is a pesticide combination formulation used in agriculture.
- Flubendiamide is an insecticide acting as a ryanodine receptor modulator, effective primarily against leaf-eating and stem-boring pests. It targets pests like stem borer and leaf folder in crops such as rice and fruit borer in tomato.
- Kresoxim-Methyl is a broad-spectrum strobilurin fungicide with protective, curative, and eradicative actions. It inhibits spore germination and provides control against fungal diseases like powdery mildew. It also has good residual activity and extends the duration of disease control.
- This combination leverages Flubendiamide's insecticidal activity and Kresoxim-Methyl's fungicidal spectrum to provide broad crop protection.

Label claim (DPPQ & S, Faridabad)

Flubendiamide 07.50% + Kresoxim-Methyl 37.50% SC

Crop	Pests	Dose / ha		Waiting period (Days)
		Quantity of formulation (ml)	Volume of water (Lit.)	
Rice	**Insect pests :** Stem borer, Leaf folder **Diseases :** Leaf and Neck blight, Sheath blight	667	500	30
Tomato	**Insect pests :** Tobacco caterpillar **Diseases :** Early and Late blight	667	500	07

Trade Names

Sr. No.	Trade Name	Company
1	Taakat	Tata Rallies India Ltd.

Imidacloprid + Hexaconazole

Available formulations in India : **Imidacloprid 18.50% + Hexaconazole 01.50% FS**

Classification on the basis of		Imidacloprid	Hexaconazole
Target organism	:	Insecticide	Fungicide
Mode of entry	:	System, Contact and Stomach poison	Systemic
Mode of action	:	Nerve poison	Interferes cell membrane synthesis
Chemical group	:	Neonicotinoid	Triazole
Category of toxicity : Moderately toxic **Toxicity label colour :** Blue			

- This combination offers dual protection (insect + disease) in a single application, protecting the seed during storage and the seedling during the initial, most vulnerable stages of growth.
- Imidacloprid is a systemic insecticide that moves from the treated seed throughout the growing plant (acropetally, via the xylem). It works by acting on the central nervous system of insects, causing irreversible blockage of post-synaptic nicotinoid acetylcholine receptors, leading to paralysis and death. It provides protection against sucking and soil pests from the time of germination.
- Hexaconazole is also a systemic fungicide belonging to the triazole group. It inhibits the biosynthesis of ergosterol, which is a key component of fungal cell membranes. This disrupts the fungal growth and reproduction, offering protection against various seed and soil-borne diseases.

Label claim (DPPQ & S, Faridabad)

Imidacloprid 18.50% + Hexaconazole 01.50% FS

Crop	Pests	Dose / Kg seed		Waiting period (Days)
		Quantity of formulation (ml)	Volume of water (Lit.)	
Groundnut	**Insect pests :** Termite, Thrips, Jassid, Root grub **Diseases :** Collar rot, Stem rot, Tikka leaf spot, Rust	2.0	--	--
Wheat	**Insect pests :** Termite, Aphid **Diseases :** Smut, Rust	2.0	--	--

Trade Names

Sr. No.	Trade Name	Company
1	ImidaX	Agrostar
2	Beej Guard	Volkschem Crop Science Pvt. Ltd.
3	Cambio	RACCOLTO Agritech India Pvt. Ltd.
4	Avatar	Ryzochem Crop Science
5	Touch-Up	Gujarat Pesticides
6	Neonix	Tata Rallies India Ltd.
7	Haxida	Plazma Crop Science

Isoprothiolane + Fipronil

Available formulations in India : **Isoprothiolane 28.00% + Fipronil 05.00% EC**

Classification on the basis of		Isoprothiolane	Fipronil
Target organism	:	Fungicide	Insecticide
Mode of entry	:	Systemic	Contact and Stomach poison
Mode of action	:	Disrupt cell membrane	Nerve poison
Chemical group	:	Dithiolane	Phenylpyrazole
Category of toxicity : Moderately toxic **Toxicity label colour** : Blue			

- The combination provides a synergistic effect, promoting better crop establishment and overall plant health (often called a phytotonic effect) by managing early-stage stress from both insects and disease.
- Isoprothiolane is a systemic fungicide with both protective and curative action. It prevents the formation of infection pegs and the elongation of infecting hyphae, thereby controlling the spread of the fungal infection (specifically Magnaporthe oryzae which causes Rice Blast).
- Fipronil is a broad-spectrum insecticide that works by contact and stomach action. It interferes with the insect's central nervous system by blocking GABA-gated chloride channels, leading to hyperexcitation, paralysis, and death of the target insect. It provides good residual control, protecting the crop for a longer duration.

Label claim (DPPQ & S, Faridabad)

Isoprothiolane 28.00% + Fipronil 05.00% EC

Crop	Pests	Dose / ha		Waiting period (Days)
		Quantity of formulation (ml)	Volume of water (Lit.)	
Rice	**Insect pests :** Stem borer, Brown plant hopper, Green leaf hopper, Whorl maggot **Diseases :** Leaf blast, Neck blast	1000	500	58

Trade Names

Sr. No.	Trade Name	Company
1	Ingus	Global Crop Care
2	Isonil	Agrostar
3	Aadat	Parijat Industries (India) Ltd.
4	Fiprokind	Mankind Agritech

Lambda-cyhalothrin + Clothianidin + Carbendazim

Available formulations in India : **Lambda-cyhalothrin 05.00% + Clothianidin 05.00% + Carbendazim 37.00% SC**

Classification on the basis of		Lambda-cyhalothrin	Clothianidin	Carbendazim
Target organism	:	Insecticide	Insecticide	Fungicide
Mode of entry	:	Stomach and Contact poison	Systemic, Contact and Stomach poison	Systemic
Mode of action	:	Nerve poison	Nerve poison	Inhibits fungal cell division
Chemical group	:	Synthetic pyrethroid	Neonicotinoid	Benzimidazole
Category of toxicity :				
Toxicity label colour :				

- This product, combining Lambda-cyhalothrin, Clothianidin, and Carbendazim, is a triple-mix formulation designed to offer broad-spectrum protection against both insect pests and fungal diseases.
- Lambda-cyhalothrin is a synthetic pyrethroid insecticide that acts by disrupting the nervous system of insects, providing contact and stomach poisoning effects to control pests like aphids, leafhoppers, and other sap-feeders.
- Clothianidin is a neonicotinoid insecticide that works systemically by targeting nicotinic acetylcholine receptors in insect nervous systems, effective against sucking insects such as aphids and leafhoppers.
- Carbendazim is a systemic fungicide from the benzimidazole class that inhibits fungal cell division, controlling a wide range of fungal diseases including rust, powdery mildew, and leaf spots.
- The systemic activity of Clothianidin and Carbendazim provides protection from within the plant as it grows, while the contact action of Lambda-cyhalothrin gives a quick initial kill.

Label claim (DPPQ & S, Faridabad)

Lambda-cyhalothrin 05.00% + Clothianidin 05.00% + Carbendazim 37.00% SC

Crop	Pests	Dose / ha		Waiting period (Days)
		Quantity of formulation (ml)	Volume of water (Lit.)	
Rice	**Insect pests :** Thrips **Diseases :** Cercospora leaf spot	500-625	375	03

Trade Names

Sr. No.	Trade Name	Company
1		

Pymetrozine + Tebuconazole

Available formulations in India : **Pymetrozine 30.00% + Tebuconazole 37.00% WG**

Classification on the basis of		Pymetrozine	Tebuconazole
Target organism	:	Insecticide	Fungicide
Mode of entry	:	Systemic, Stomach poison	Systemic
Mode of action	:	Nerve poison	Interferes cell membrane synthesis
Chemical group	:	Pyridine azomethine	Triazole
Category of toxicity : Moderately toxic **Toxicity label colour** : Blue			

- This combination is engineered to manage the most economically damaging complex of sucking pests and fungal diseases.
- Pymetrozine has a unique mode of action. It quickly stops the feeding activity of sucking insects (like BPH) by interfering with the nervous regulation of their feeding behavior. It is absorbed by the foliage and moves throughout the plant, providing residual protection, even on the underside of leaves or new growth. Insects stop feeding within hours of application and eventually die from starvation, which significantly reduces the transmission of viral diseases.
- Tebuconazole is a broad-spectrum triazole fungicide that works by inhibiting the biosynthesis of ergosterol (a vital component of fungal cell membranes). It is known for its excellent protective, curative, and eradicative properties, making it highly effective against fungal pathogens like rice blast.

Label claim (DPPQ & S, Faridabad)

Pymetrozine 30.00% + Tebuconazole 37.00% WG

Crop	Pests	Dose / ha		Waiting period (Days)
		Quantity of formulation (g)	Volume of water (Lit.)	
Rice	**Insect pests** : Brown plant hopper **Diseases** : Sheath blight, Brown spot, Blast, Glume discoloration (Dirty panicle)	500	500	33

Trade Names

Sr. No.	Trade Name	Company
1	Runway	GSP Crop Science Pvt. Ltd.

Pymetrozine + Dinotefuran + Pyraclostrobin

Available formulations in India : **Pymetrozine 30.00% + Dinotefuran 10.00% + Pyraclostrobin 20.00% WG**

Classification on the basis of		Pymetrozine	Dinotefuran	Pyraclostrobin
Target organism	:	Insecticide	Insecticide	Fungicide
Mode of entry	:	Systemic, Stomach poison	Systemic, Contact and Stomach poison	Contact
Mode of action	:	Nerve poison	Nerve poison	Inhibits mitochondrial respiration
Chemical group	:	Pyridine azomethine	Neonicotinoid	Strobilurin
Category of toxicity : Moderately toxic **Toxicity label colour** : Blue				

- This combination provides comprehensive, dual-action control of significant insect pests and blast disease in rice.
- Pymetrozine provides a rapid anti-feeding effect, quickly stopping hoppers from damaging the crop and spreading diseases like Tungro. It paralyzes the mouthparts of sucking pests (like BPH), causing them to stop feeding immediately and die from starvation.
- Dinotefuran offers strong systemic and contact action, providing long-lasting kill and controlling pests that may be resistant to other chemical classes. It acts as a neurotoxin by disrupting the insect's central nervous system, leading to rapid paralysis and death.
- Pyraclostrobin is a highly effective, broad-spectrum fungicide that controls rice blast and offers a phytotonic effect (improving plant health, increasing chlorophyll content, and promoting better yields). It inhibits mitochondrial respiration in fungal cells, effectively shutting down their energy source and preventing spore germination and growth. It has translaminar (moves through the leaf tissue) and locally systemic action.

Label claim (DPPQ & S, Faridabad)

Pymetrozine 30.00% + Dinotefuran 10.00% + Pyraclostrobin 20.00% WG

Crop	Pests	Dose / ha		Waiting period (Days)
		Quantity of formulation (g)	Volume of water (Lit.)	
Rice	**Insect pests :** Brown plant hopper, White backed plant hopper **Diseases :** Blast	375	500	22

Trade Names

Sr. No.	Trade Name	Company
1	Defender	Best Agrolife Ltd.

Pymetrozine + Thiamethoxam + Hexaconazole

Available formulations in India : **Pymetrozine 25.00% + Thiamethoxam 17.50% + Hexaconazole 12.50% WG**

Classification on the basis of		Pymetrozine	Thiamethoxam	Hexaconazole
Target organism	:	Insecticide	Insecticide	Fungicide
Mode of entry	:	Systemic, Stomach poison	Systemic, Contact and Stomach poison	Systemic
Mode of action	:	Nerve poison	Nerve poison	Interferes cell membrane synthesis
Chemical group	:	Pyridine azomethine	Neonicotinoid	Triazole
Category of toxicity :				
Toxicity label colour :				

- Pymetrozine disrupts the insect's nervous system, causing sucking pests to stop feeding immediately (within hours). They die from starvation. Pymetrozine offers a unique, fast-acting stop to feeding, instantly protecting the crop from immediate damage and virus transmission.
- Thiamethoxam binds to nicotinic acetylcholine receptors in the insect's central nervous system, causing paralysis and quick death. It provides long-lasting residual and systemic control. Thiamethoxam offers a reliable, broad-spectrum kill and is fully systemic, protecting new growth and insects hiding in parts of the plant not directly hit by the spray.
- Hexaconazole inhibits the biosynthesis of ergosterol, a vital component of fungal cell membranes, leading to cell collapse and death. It provides protective and curative action. Hexaconazole provides strong systemic action against major fungal diseases, especially the complex of sheath blight and leaf blast, which often occur concurrently with hopper infestations.
- The systemic properties of all three components (Thiamethoxam, Pymetrozine, and Hexaconazole) allow for uniform distribution throughout the plant, controlling pests and diseases present on the lower parts of the plant and new foliage.

Label claim (DPPQ & S, Faridabad)

Pymetrozine 25.00% + Thiamethoxam 17.50% + Hexaconazole 12.50% WG

Crop	Pests	Dose / ha		Waiting period (Days)
		Quantity of formulation (g)	Volume of water (Lit.)	
Rice	**Insect pests :** Brown plant hopper, White backed plant hopper, Green leaf hopper, Yellow stem borer, Leaf folder **Diseases :** Sheath blight, Leaf blast	400	500	19

Trade Names

Sr. No.	Trade Name	Company
1	Respect	Agrostar

Pyraclostrobin + Fipronil

Available formulations in India : **Pyraclostrobin 10.00% + Fipronil 05.00% SC**

Classification on the basis of		Pyraclostrobin	Fipronil
Target organism	:	Fungicide	Insecticide
Mode of entry	:	Contact	Stomach and Contact poison
Mode of action	:	Inhibits mitochondrial respiration	Nerve poison
Chemical group	:	Strobilurin	Phenylpyrazole
Category of toxicity : Moderately toxic **Toxicity label colour** : Blue			

- This product offers comprehensive protection due to the complementary mechanisms of its two active ingredients.
- Pyraclostrobin has translaminar (moves through the leaf tissue) and locally systemic action. It inhibits the mitochondrial respiration (electron transport chain) in the fungal cells, specifically at Complex III. By blocking this process, it prevents the fungus from generating the energy (ATP) needed for spore germination and mycelial growth, leading to death. It provides preventive and curative action. It is known to have a phytotonic effect, improving overall plant health, greening, and stress tolerance, which contributes to higher yields.
- Fipronil has contact, stomach, and some systemic action. It interferes with the insect's central nervous system by blocking the GABA-gated chloride channels. This prevents the passage of chloride ions, leading to hyperexcitation, paralysis, and death of the insect. It provides effective control against both sucking pests (like hoppers and thrips) and chewing pests (like stem borer).

Label claim (DPPQ & S, Faridabad)

Pyraclostrobin 10.00% + Fipronil 05.00% SC

Crop	Pests	Dose / ha		Waiting period (Days)
		Quantity of formulation (ml)	Volume of water (Lit.)	
Rice	**Insect pests :** Brown plant hopper, Yellow stem borer, Leaf folder **Diseases :** Sheath blight, Leaf blast	1000	500	53
Chilli	**Insect pests :** Thrips, Fruit borer **Diseases :** Powdery mildew, Anthracnose / Fruit rot	1000	500	07

Trade Names

Sr. No.	Trade Name	Company
1	Joker	Gharda Chemicals Ltd.

Pyraclostrobin + Thiram + Clothianidin

Available formulations in India : **Pyraclostrobin 03.50% + Thiram 15.00% + Clothianidin 22.50% FS**

Classification on the basis of		Pyraclostrobin	Thiram	Clothianidin
Target organism	:	Fungicide	Fungicide	Insecticide
Mode of entry	:	Contact	Contact	Systemic, Contact and Stomach poison
Mode of action	:	Inhibits mitochondrial respiration	Multi-site inhibitor	Nerve poison
Chemical group	:	Strobilurin	Dithiocarbamate	Neonicotinoid
Category of toxicity :				
Toxicity label colour :				

- This is a sophisticated Triple-Mix Seed Treatment product, designed to provide comprehensive protection to seeds and emerging seedlings against both fungal pathogens and insect pests.
- Pyraclostrobin has translaminar/ locally systemic action. It is absorbed by the emerging roots and seedling tissue. It acts as a QoI (Quinone outside Inhibitor), blocking mitochondrial respiration in the fungal cells. This prevents energy production. It offers protection against fungi inside the seed (internal seed-borne) and provides a phytotonic effect (enhanced root development and early plant vigor).
- Thiram is contact/ protective fungicide. It remains largely on the seed surface and is toxic to fungal spores in the immediate vicinity of the seed. It inhibits spore germination and fungal growth by interfering with key fungal enzyme systems. It provides primary protection against seed-borne (on the seed coat) and soil-borne fungi that cause rot and damping-off before germination.
- Clothianidin is fully systemic. It is absorbed by the seed and moves up into the developing seedling via the xylem. It acts as a neurotoxin, selectively interfering with the insect's central nervous system (nicotinic acetylcholine receptors). It provides long-lasting control against early-season sucking pests and soil-dwelling pests like grubs and termites.

Label claim (DPPQ & S, Faridabad)

Pyraclostrobin 03.50% + Thiram 15.00% + Clothianidin 22.50% FS

Crop	Pests	Dose / Kg seed		Waiting period (Days)
		Quantity of formulation (ml)	Volume of water (Lit.)	
Groundnut	**Insect pests :** Aphid, Jassid, Termite, White grub **Diseases :** Collar rot, Stem rot	7.00	--	--

Trade Names

Sr. No.	Trade Name	Company
1	Peack	Parijat Industries (India) Ltd.
2	Petra 3D	Agrostar
3	PCT-410	GSP Crop Science Pvt. Ltd.
4	SeedKind TCP	Mankind Agritech
5	Citrum	Sudarshan Farm Chemicals Pvt. Ltd.

Sedaxane + Azoxystrobin + Thiamethoxam

Available formulations in India : **Sedaxane 12.61% + Azoxystrobin 03.15% + Thiamethoxam 22.06% FS**

Classification on the basis of		Sedaxane	Azoxystrobin	Thiamethoxam
Target organism	:	Fungicide	Fungicide	Insecticide
Mode of entry	:	Systemic	Systemic	Systemic, Contact and Stomach poison
Mode of action	:	SDHI (Succinate Dehydrogenase Inhibitors)	Mitochondrial respiration inhibitor	Nerve poison
Chemical group	:	Pyrazole Carboxamide	Strobilurin	Neonicotinoid
Category of toxicity :				
Toxicity label colour :				

- The effectiveness comes from the synergistic action of the three active ingredients, each having distinct roles and movement within the plant.
- Sedaxane is systemic fungicide. It is unique in that it is highly effective in the soil and is taken up by the roots. It creates a halo of protection around the seed and developing root system. It belongs to the SDHI (Succinate Dehydrogenase Inhibitor) group. It blocks the Complex II in the fungal respiration process, halting energy production. It is crucial for controlling Rhizoctonia and other soil-borne diseases that attack the roots and crown. It also provides a rooting power benefit, promoting healthier and more robust root systems.
- Azoxystrobin is translaminar and systemic to protect the foliage. It acts as a QoI (Quinone outside Inhibitor), blocking Complex III in the fungal respiration chain, thus starving the fungus of energy. It provides control against early blast and brown spot and contributes to a phytotonic effect (enhanced photosynthesis, improved stress tolerance).
- Thiamethoxam is highly systemic and absorbed by the seed and moves rapidly up into the foliage (acropetally via the xylem) as the plant grows. It acts as a neurotoxin, disrupting the central nervous system of insects by affecting the nicotinic acetylcholine receptors. It provides long-lasting protection for the emerging seedling against sucking and early-season chewing pests.

Label claim (DPPQ & S, Faridabad)

Sedaxane 12.61% + Azoxystrobin 03.15% + Thiamethoxam 22.06% FS

Crop	Pests	Dose / Kg seed		Waiting period (Days)
		Quantity of formulation (ml)	Volume of water	
Rice	**Insect pests :** Green leafhopper, Thrips, Whorl maggot **Diseases :** Sheath blight, Leaf blast, Brown spot	3.00	8-10 ml/Kg seed for pre-soaked seed and 15-20 ml/Kg seed for dry seed	--

Trade Names

Sr. No.	Trade Name	Company
1	Vibrance Integral	Syngenta India Pvt. Ltd.

Sedaxane + Fludioxonil + Thiamethoxam

Available formulations in India : **Sedaxane 02.20% + Fludioxonil 02.20% + Thiamethoxam 22.80% FS**

Classification on the basis of		Sedaxane	Fludioxonil	Thiamethoxam
Target organism	:	Fungicide	Fungicide	Insecticide
Mode of entry	:	Systemic	Contact	Systemic, Contact and Stomach poison
Mode of action	:	SDHI (Succinate Dehydrogenase Inhibitors)	Inhibits the germination of conidia and mycelial growth	Nerve poison
Chemical group	:	Pyrazole Carboxamide	Phenylpyrrole	Neonicotinoid
Category of toxicity :				
Toxicity label colour :				

- The combination of Sedaxane, Fludioxonil, and Thiamethoxam is widely used as a seed treatment formulation that provides broad-spectrum protection in crops. This mix targets early season insects and a range of seed and soil-borne fungal pathogens.
- Sedaxane belongs to the SDHI (succinate dehydrogenase inhibitor, FRAC Group 7) class of fungicides. It is effective against seed decay, seedling blight, and damping-off caused by Rhizoctonia and other soil-borne fungi. Sedaxane is commonly used as a protective fungicide in seed treatments and acts by inhibiting fungal respiration at the mitochondrial level.
- Fludioxonil is a non-systemic phenylpyrrole fungicide with long residual action and a unique mode of action (FRAC Group 12). It disrupts the osmotic signal transduction in fungi, inhibiting the germination of conidia and mycelial growth. Fludioxonil is effective against a broad spectrum of seed and soil-borne fungal pathogens, particularly Fusarium, Rhizoctonia, and Botrytis, making it valuable for both seed treatment and post-harvest disease management.
- Thiamethoxam is a systemic seed treatment insecticide from the neonicotinoid class (IRAC Group 4A). It is taken up by the germinating seed, protecting the crop against a range of early season sucking and chewing insect pests through both contact and ingestion. Thiamethoxam offers protection against pests.

Label claim (DPPQ & S, Faridabad)

Sedaxane 02.20% + Fludioxonil 02.20% + Thiamethoxam 22.80% FS

Crop	Pests	Dose / Kg seed		Waiting period (Days)
		Quantity of formulation (ml)	Volume of water	
Cotton	**Insect pests :** Aphid, Jassid, Whitefly **Diseases :** Fungal wilt, Root rot	4.00	Seeds are treated with the required quantity of test chemical (slurry volume of 10 - 15 ml/kg of wheat seeds and 10 – 12 ml/kg cotton seeds). The treated seeds are dried in shade before sowing	--
Wheat	**Insect pests :** Aphid, Termite **Diseases :** Loose smut, Karnal bunt	1.00-2.00		--

Trade Names

Sr. No.	Trade Name	Company
1	Vibrance Maxx	Syngenta India Pvt. Ltd.

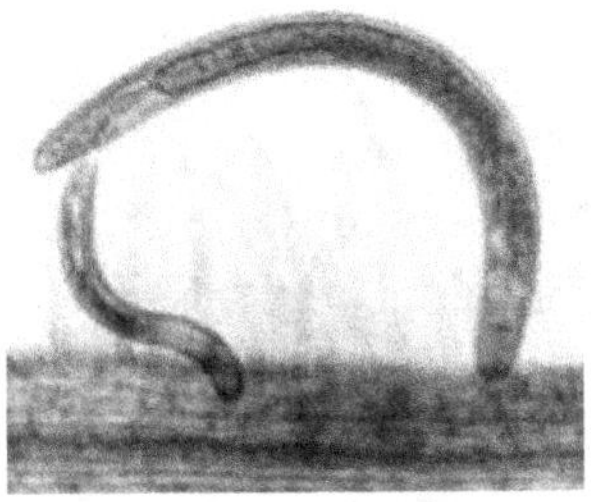

PART VI

Nematicides

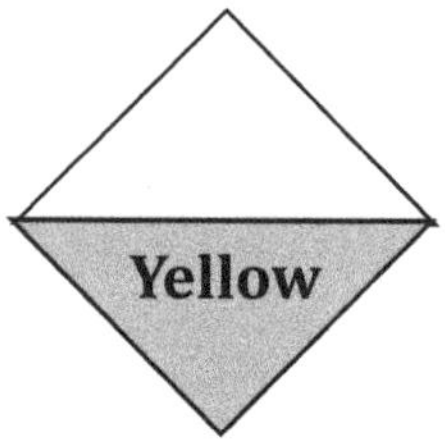

CARBOFURAN

Available formulations in India : **Carbofuran 03.00 % CG**

Classification on the basis of

Target organism : Insecticide, Nematicide, Acaricide

Mode of entry : Stomach and Contact poison, Systemic

Mode of action : Nerve poison

Chemical group : Carbamate

IRAC classification

Main Group and Primary Site of Action : 1 Acetylcholinesterase (AChE) inhibitors

Sub-group or exemplifying Active Ingredient : 1A Carbamate

Category of toxicity : Highly toxic

Toxicity label colour : Yellow

Label claim (DPPQ & S, Faridabad)

Carbofuran 03.00 % CG

Crop	Pests	Dose / ha		Waiting period (Days)
		Quantity of formulation (Kg)	Volume of water (Lit.)	
Wheat	Ear cockle nematode	10.00	--	--
	Cereal cyst nematode	66.60	--	--
Barley	Aphid, Cyst nematode	33.30	--	--
	Jassid	41.60	--	--
Rice	Brown plant hopper, Green leaf hopper, Gall midge, Stem borer, Hispa	25.00	--	--
	Nematode	50.00	--	--
Soybean	Root knot nematode	50.00	--	--
Jute	Nematodes	33.30	--	--
Brinjal	Root knot nematode, Reniform nematode	66.60	--	--
Cabbage	Nematode	50.00	--	--
Citrus	Nematode	12.00	--	--
	Leaf miner	50.00	--	--
Banana	Rhizome weevil	33 g / sucker	--	--
	Aphid	166g / sucker	--	--
	Nematode	50 g / Sucker	--	--

Blue

CYCLOBUTRIFLURAM

Available formulations in India : **Cyclobutrifluram 38.40 % SC (Import only)**

Classification on the basis of

Target organism : Nematicide, Fungicide
Mode of entry : Systemic, Contact poison
Mode of action : Succinate Dehydrogenase Inhibitor (SDHI)
Chemical group : Chiral phenyl-cyclobutyl-pyridineamide

IRAC classification

Main Group and Primary Site of Action : IRAC Group N-3 and FRAC Group 7
Sub-group or exemplifying Active Ingredient : -

Category of toxicity : Moderately tox
Toxicity label colour : Blue

- Cyclobutrifluram is a novel, high-potency active ingredient developed by Syngenta. It is a dual-purpose chemical that acts as both a nematicide (targeting parasitic worms) and a fungicide (targeting soil-borne diseases).
- **Mode of action** : Cyclobutrifluram is classified as a Succinate Dehydrogenase Inhibitor (SDHI). It inhibits Complex II of the mitochondrial electron transport chain. By blocking the enzyme succinate dehydrogenase, it disrupts cellular respiration and energy production (ATP) in the target pest. This leads to rapid paralysis and death of nematodes and stops the growth of fungal pathogens.
- The chemical is designed to be highly selective, providing long-lasting residual control in the rhizosphere (root zone).
- It is effective against all major economically significant plant-parasitic nematodes, including Root-knot, Cyst, Lesion, Stubby-root, and Reniform nematodes.
- It is also particularly effective against *Fusarium* species (e.g., Fusarium wilt and head blight) and other soil-borne pathogens.

Label claim (DPPQ & S, Faridabad)

Cyclobutrifluram 38.40 % GR

Crop	Pests	Dose / ha		Waiting period (Days)
		Quantity of formulation	Volume of water (Lit.)	
Banana	Burrowing nematode	500 ml/ha	1.2 lit./plant	--
Okra	Root knot nematode	375 ml/ha	1.2 lit./plant	--
Cucumber	Root knot nematode	375 ml/ha	50-100 ml/plant	--
Tomato	Root knot nematode	375 ml/ha	50-100 ml/plant	--

Trade Names

Sr. No.	Trade Name	Company
1	Vaniva	Syngenta India Private Limited

DAZOMET

Available formulations in India : **Dazomet**

Classification on the basis of

Target organism : Insecticide, Nematicide, Herbicide, Fungicide
Mode of entry : Fumigant
Mode of action : Nerve poison
Chemical group : Methyl isothiocyanate generator

IRAC classification

Main Group and Primary Site of Action : 8 Miscellaneous non-specific (multi-site) inhibitors
Sub-group or exemplifying Active Ingredient : 8F Methyl isothiocyanate generators

Category of toxicity :
Toxicity label colour :

- Dazomet is a broad-spectrum soil fumigant (methyl isothiocyanate generator) used as a herbicide, nematicide, fungicide, and insecticide, primarily for pre-plant soil sterilization in nurseries, golf courses, turf, and potting soils as an alternative to methyl bromide.
- **Mechanism of action** : Dazomet itself is an inactive solid, typically formulated as granules. Its effectiveness comes from its decomposition process in the soil. Dazomet is applied to bare, moist soil and thoroughly incorporated (mixed) into the planting bed. In the presence of soil moisture, Dazomet undergoes a chemical breakdown. The decomposition releases a volatile, highly toxic gas called Methyl Isothiocyanate (MITC). The MITC gas diffuses through the soil air spaces, sterilizing the soil by killing harmful organisms.
- Broadcast or furrow-applied at 750-1050 g of 98% granules per 100 m^2 (sand/clay soils), then watered in, covered with plastic sheeting (24-48 hours), and aerated for 7-14 days before planting.

Label claim (DPPQ & S, Faridabad)

Dazomet

Crop	Pests	Dose / ha		Waiting period (Days)
		Quantity of formulation (Kg)	Volume of water (Lit.)	
Tobacco Nursery	Root-knot nematode, Stunt nematode, Reni-form nematode	30-40	--	--
Tomato Nursery	Root-knot nematode	30-40	--	--
Floriculture (Carnation and Gerbera)	Root-knot nematode	30-40	--	--

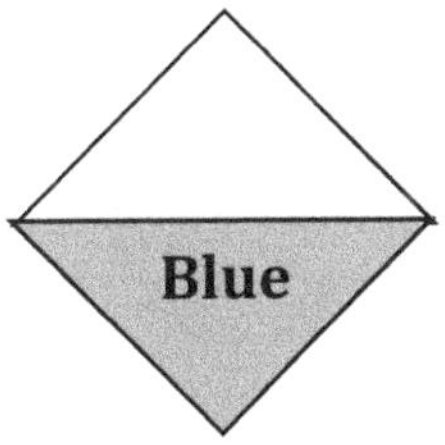

FLUENSULFONE

Available formulations in India : **Fluensulfone 02.00 % GR**

Classification on the basis of

Target organism : Nematicide
Mode of entry : Systemic, Contact poison
Mode of action : Inhibit physiological processes
Chemical group : Fluoroalkenyl sulfones

IRAC classification

Main Group and Primary Site of Action : Compounds with unknown or uncertain MoA
Sub-group or exemplifying Active Ingredient : -

Category of toxicity : Moderately tox
Toxicity label colour : Blue

- Fluensulfone is a modern nematicide used in agriculture to combat plant-parasitic nematodes. It is a member of a unique chemical class known as fluoroalkenyl sulfones, and it offers a distinct mode of action compared to older nematicides like organophosphates and carbamates.
- Fluensulfone is a pre-plant nematicide that effectively controls a wide range of plant-parasitic nematodes, including *Meloidogyne* spp. (root-knot nematodes). It is a "true nematicide," meaning it kills the nematodes rather than just paralyzing them.
- **Mode of action :** Its mode of action is distinct from other nematicides. While the exact mechanism is not fully understood, research suggests it has "pleiotropic" (multiple) effects. It inhibits various physiological processes in nematodes, including development, egg-laying and egg-hatching, feeding and locomotion, disruption of lipid homeostasis, which impacts the nematode's ability to utilize energy.
- Fluensulfone is typically applied to the soil before planting. Common application methods include broadcast or banded spraying with mechanical incorporation, drip irrigation.
- Foliar application has also been studied, but its systemic activity can be crop-specific and may pose a risk of phytotoxicity at high rates.
- Fluensulfone is stable to hydrolysis across a wide range of pH levels, including acidic, neutral, and basic conditions.

Label claim (DPPQ & S, Faridabad)

Fluensulfone 02.00 % GR

Crop	Pests	Dose / ha		Waiting period (Days)
		Quantity of formulation	Volume of water (Lit.)	
Tomato	Root knot nematode	1.0 g/plant and 22.2-25.6 Kg/ha	--	07

Cucumber	Root knot nematode	1.0 g/plant and 8-10 Kg/ha	--	55
Okra	Root knot nematode	1.0 g/plant and 40-45.5 Kg/ha	--	07
Pomegranate	Root knot nematode	10.0 g/dripper and 40 g/plant	--	91

Trade Names

Sr. No.	Trade Name	Company
1	Nimitz	ADAMA India Private Limited

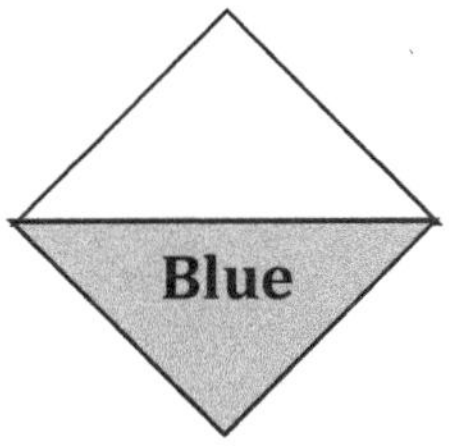

FLUOPYRAM

Available formulations in India : **Fluopyram 34.48 % SC**

Classification on the basis of

Target organism : Fungicide, Nematicide
Mode of entry : Systemic
Mode of action : Succinate dehydrogenase inhibitors
Chemical group : Pyridinyl ethylbenzamides

FRAC classification

Target Site of Action : Complex II; succinate-dehydrogenase
Group Name : SDHI (Succinate dehydrogenase inhibitors)

Category of toxicity : Moderately toxic
Toxicity label colour : Blue

- Fluopyram is a significant agricultural chemical with broad-spectrum activity, primarily used as a fungicide and nematicide.
- **Mode of action :** It is an SDHI fungicide, meaning it inhibits succinate dehydrogenase (Complex II) within the fungal mitochondrial respiratory chain. This action disrupts electron transport and energy production in the fungus, leading to its death. It inhibits spore germination, germ tube elongation, mycelial growth, and sporulation.
- The exact nematicidal mode of action is not fully known. However, its effectiveness is linked to its ability to disrupt key physiological processes in nematodes.
- Within plants, fluopyram shows translaminar activity (movement across the leaf) and some movement within the xylem (systemic movement), offering good protection.
- Fluopyram can be applied in several ways, depending on the target pest and crop such as foliar spraying, soil application and seed treatment.
- Fluopyram is stable to hydrolysis across a wide range of pH levels, including acidic, neutral, and basic conditions.

Label claim (DPPQ & S, Faridabad)

Fluopyram 34.48 % SC

Crop	Pests	Dose / ha		Waiting period (Days)
		Quantity of formulation (ml)	Volume of water (Lit.)	
Tomato	Root knot nematode	625 ml (2 applications) or 1250 ml (Single application)	1000	05
Cucumber	Root knot nematode	625 ml (2 applications) or 1250 ml (Single application)	1000	05

Trade Names

Sr. No.	Trade Name	Company
1	Velum Prime	Bayer Crop Science Ltd.

PART VII

Rodenticide

Red

ALUMINIUM PHOSPHIDE

Available formulations in India :

1. **Aluminium phosphide 56.00% (3g Tablet, 10g Pouch)**
2. **Aluminium phosphide 06.00% Tablet**

Classification on the basis of

Target organism : Insecticide, Rodenticide and Acaricide

Mode of entry : Respiratory poison

Mode of action : Nerve poison

Chemical group : Phosphides

IRAC classification

Main Group and Primary Site of Action : 24 Mitochondrial complex IV electron transport inhibitors

Sub-group or exemplifying Active Ingredient : 24A Phosphides

Category of toxicity : Extremely toxic

Toxicity label colour : Red

Label claim (DPPQ & S, Faridabad)

Aluminium phosphide 56.00% (3g Tablet, 10g Pouch)

Commodity	Target pests	Dose
Rodents Burrows	Rodents	01 Tablet/ Burrow

Aluminium phosphide 06.00% Tablet

Commodity	Target pests	Dose
Crop and Non-crop area	Field rodents	01 Tablet of 12g / Burrow

Red

BRODIFACOUM

Available formulations in India : **Brodifacoum 00.005 % BB**

Classification on the basis of
Target organism : Rodenticide
Category of toxicity : Extremely toxic
Toxicity label colour : Red

- Brodifacoum is one of the most powerful and dangerous chemicals in the rodenticide class. It is often referred to as a "super-warfarin" due to its extreme potency and long-lasting effects.
- Primarily it is used for controlling rats and mice, especially those resistant to older anticoagulants like warfarin.
- It acts as a single feed poison at very low bait concentrations (typically 0.005% a.i.), but has very high persistence and secondary poisoning risk, so it is usually classed as a restricted or professional use product in many countries.
- Second-Generation Anticoagulant Rodenticide (SGAR): It is highly effective and often marketed as a single-feeding rodenticide, meaning a lethal dose can be consumed in just one feeding session.
- **High Toxicity**: Brodifacoum is significantly more potent than other SGARs like bromadiolone, and hundreds of times more potent than first-generation anticoagulants like warfarin.
- **Mechanism of action** : Like all anticoagulant rodenticides, brodifacoum is a Vitamin K antagonist that disrupts the blood clotting process. It strongly and persistently inhibits the enzyme Vitamin K epoxide reductase in the liver. This enzyme is essential for recycling inactive Vitamin K back into its active form. Without active Vitamin K, the body cannot synthesize critical blood-clotting factors (specifically factors II, VII, IX, and X). The depletion of clotting factors leads to a progressive failure of the blood's ability to clot, resulting in fatal internal haemorrhage (bleeding).
- **Antidote**: The primary and only effective antidote for brodifacoum poisoning is high-dose, prolonged administration of Vitamin K1 (Phytomenadione).

Label claim (DPPQ & S, Faridabad)

Brodifacoum 00.005 % BB

Pest	Dose rate	Manner of application / Use pattern
Field rats/Bandicoot rats (*Bandoicota bengalensis; B. indica*) Indian house rat / Black, Indian house rat /black rat/ roof rat (*Rattus rattus; R. meltade*),	One bait of 0.005% (a block of 20 g each) per baiting station as a single feed	In and around premises (Residential, commercial, institutional, industrial public service premises, cold storage, Godowns, ware house, municipal locations, grain mandis, crop store

House mouse/ Field mouse (*Mus musculus*)		rooms, burrow baiting, livestock rearing facilities, damp premises such as sewer etc.)

Trade Names

Sr. No.	Trade Name	Company
1	Klerat WB	Syngenta India Pvt. Ltd.

Red

BROMADIOLONE

Available formulations in India : **1. Bromadiolone 00.25 % CB**
2. Bromadiolone 00.005 % RB

Classification on the basis of
Target organism : Rodenticide
Category of toxicity : Extremely toxic
Toxicity label colour : Red

- Bromadiolone is a second-generation anticoagulant rodenticide (super-warfarin) used mainly for control of rats and mice in and around buildings, sewers and other structures. It is highly potent, effective even against rodents resistant to first-generation anticoagulants like warfarin, and is typically formulated at 0.005% a.i. in cereal or wax baits.
- **Mechanism of action** : Bromadiolone is a Vitamin K antagonist. It is absorbed into the body (mammals and birds). It works by inhibiting the enzyme Vitamin K epoxide reductase in the liver. This enzyme is necessary to recycle Vitamin K, which is required to produce essential blood-clotting factors (specifically, Factors II, VII, IX, and X). By blocking this recycling, the body runs out of active clotting factors, leading to a failure in the blood-clotting mechanism. This results in uncontrolled internal bleeding (haemorrhage) and eventual death.
- Bromadiolone is extremely toxic to rats, mice, and non-target animals, including pets (dogs and cats) and humans.
- **Antidote**: The specific antidote for bromadiolone poisoning in both animals and humans is Vitamin K1. Treatment usually involves repeated doses until the patient's blood clotting ability returns to normal.
- **Professional use:** Due to the high risks to children and wildlife, bromadiolone products are often restricted and intended only for sale to licensed professionals in many regions.
- **Safety measures:** Baits are typically placed in secured, tamper-proof bait stations to prevent access by non-target animals and children. They are often dyed (blue-green or red) to aid in identification if ingested.

Label claim (DPPQ & S, Faridabad)

Bromadiolone 00.25 % CB

Crop	Pests	Dose (%)
Rice	Field rat, Large Bandicota, Indian house rat, Indian field mouse	0.005
Wheat Gram	Field rat, Indian house rat	0.005
Groundnut, Sugarcane	Field rat, Large Bandicota	0.005
Coconut, Bamboo	Indian house rat	0.005
Residential premises	Field rat, Large Bandicota	0.005
Poultry Farm	Indian house rat, House mouse	0.005

Trade Names

Sr. No.	Trade Name	Company
1	Kalrat	Kalyani Industries Ltd.
2	Ratcon CB	A. Saj Agricare Pvt. Ltd.
3	Roban CB	Pawan Pest Control Services

Bromadiolone 00.005 % RB

Crop	Pests	Dose (%)
Rice	Field rat, Large Bandicota, Indian house rat, Indian field mouse	0.005
Wheat Gram	Field rat, Indian house rat	0.005
Groundnut, Sugarcane	Field rat, Large Bandicota	0.005
Coconut, Bamboo	Indian house rat	0.005
Residential premises	Field rat, Large Bandicota	0.005
Poultry Farm	Indian house rat, House mouse	0.005

Trade Names

Sr. No.	Trade Name	Company
1	Racumin Sure	Bayer Crop Science Ltd.
2	Kalrat RB	Kalyani Industries Ltd.
3	Dr Pest	Saya Sales Pvt. Ltd.
4	Arbuda Run-Rat	Top One Pest Control Services
5	Roban RB	Pawan Pest Control Services
6	Ratonil	Acme Organic Pvt. Ltd.
7	Herenba Rat Kill	Herenba Industries Ltd.
8	Ratban	Sandhya Organic Chemicals Ltd.
9	Mooshkill	Fytocare Chemicals Pvt. Ltd.

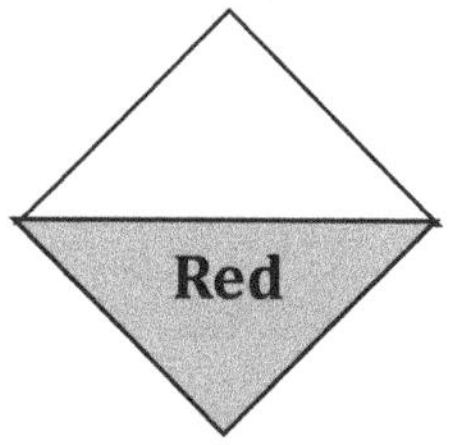

COUMATETRALYL

Available formulations in India : **1. Coumatetralyl 00.75 % Gel,**
2. Coumatetralyl 00.0375 % Bait

Classification on the basis of
Target organism : Rodenticide
Category of toxicity : Moderately toxic
Toxicity label colour : Yellow

- Coumatetralyl is a first-generation anticoagulant rodenticide of the 4-hydroxycoumarin type, introduced in the 1950s as an alternative to warfarin for controlling rats and mice, particularly in commensal settings like farms, stores and buildings.
- First-generation anticoagulants like Coumatetralyl generally require multiple feedings over several days for a lethal dose to accumulate, unlike the more potent second-generation anticoagulants (SGARs) which are often effective with a single dose.
- **Mechanism of action** : Coumatetralyl, like all anticoagulant rodenticides, works by interfering with the body's essential blood clotting process. The chemical acts as a Vitamin K antagonist (VKA). It specifically blocks the enzyme Vitamin K epoxide reductase (VKOR) in the liver. This enzyme is responsible for recycling inactive Vitamin K epoxide back into active Vitamin K1. Active Vitamin K1 is a crucial cofactor required for the synthesis of functional blood clotting factors (specifically Factors II, VII, IX, and X). By preventing the recycling of Vitamin K, the body rapidly depletes its supply of active clotting factors. Once the existing factors are used up (typically taking 3 to 7 days), the rodent's blood loses its ability to clot, leading to internal and external bleeding (haemorrhage) and eventually, death from haemorrhagic shock or severe anaemia. This delayed action is one of the advantages of anticoagulants, as the rodent does not immediately associate the bait consumption with the sickness, thereby preventing bait shyness.
- **Non-target risk** : Like all anticoagulant rodenticides, it poses a risk to non-target animals (pets, livestock, wildlife) who may consume the bait directly (primary poisoning) or by eating poisoned rodents (secondary poisoning).
- **Antidote** : The specific antidote for Coumatetralyl poisoning in mammals is Vitamin K1 (Phytomenadione). Treatment must be administered promptly and often continued for several weeks.
- **Resistance** : Coumatetralyl, along with Warfarin, was one of the first anticoagulants to face resistance in rodent populations, which led to the development of the more potent second-generation compounds.

Label claim (DPPQ & S, Faridabad)

Coumatetralyl 00.75% Gel

Place	Pest	Dose
Indoor or Outdoor	Rats (*Rattus rattus, Rattus norvegicus, Bandicota bengalensis, Bandicota indica, Tetra indica, Meriones hurrianae*)	2.5 per spot
Indoor	Mice	2.5 per spot

Trade Names

Sr. No.	Trade Name	Company
1	Rodent	Global Crop Care

Coumatetralyl 00.0375% Bait

Place	Pest	Dose
Indoor or Outdoor	Rats (*Rattus rattus, Rattus norvegicus, Bandicota bengalensis, Bandicota indica, Tetra indica, Meriones hurrianae*)	2.5 per spot
Indoor	Mice	2.5 per spot

Trade Names

Sr. No.	Trade Name	Company
1	Racumin	Bayer Crop Science

Red

ZINC PHOSPHIDE

Available formulations in India : **Zinc Phosphide 80.00% Powder**

Classification on the basis of
Target organism : Rodenticide
Category of toxicity : Extremely toxic
Toxicity label colour : Red

- Zinc phosphide is an inorganic acute rodenticide used to control commensal and field rodents such as rats, mice, voles, ground squirrels, prairie dogs, gophers, nutria and jackrabbits in agricultural, non-crop and structural settings.
- It is formulated at 0.75–2% a.i. in baits like wheat, oats, pellets or concentrates mixed with attractive foods, featuring a strong garlic-like odour from phosphine that attracts rodents but repels many non-targets.
- **Mechanism of toxicity** : The compound's toxicity is due to its reaction inside the body of the target animal. The bait is eaten. Upon contact with the hydrochloric acid in the stomach, the zinc phosphide reacts via a process called hydrolysis to release the highly toxic gas, phosphine. The phosphine gas is rapidly absorbed into the bloodstream. Its primary mechanism of action is the disruption of cellular energy production, mainly by inhibiting cytochrome c oxidase in Complex IV of the mitochondrial respiratory chain. This leads to cellular hypoxia, cell death, and rapid multi-organ dysfunction (especially affecting the heart, lungs, liver, and kidneys).
- **Fatal risk** : Zinc phosphide is highly toxic to humans and other non-target animals if ingested or inhaled as dust. It has a high mortality rate, particularly when ingested with suicidal intent, and there is no specific antidote.

Label claim (DPPQ & S, Faridabad)

Crop	Pests	Dosage	Technical
For rodent control in field and residential premises (to be used under the supervision of trained personal)	*Rattus rattus, Bandicota bengalensis, Rattus meltade, Tatera indica, Meriones hurrianae, Mus platythrix, Mus musculus, Rattus norvegicus, Mus booduga, Suncus caeruleus*	1.5-2.5 % active ingredient in bait	Mix 10 g of Zinc phosphide with 10g of edible oil and then mix with 380g of food material. Keep 10g of poisoned bait at each point.

Trade Names

Sr. No.	Trade Name	Company
1	Ratol	UPL Ltd.
2	Zeephos	Anu Products Ltd.
3	RodentNil	Global Crop Care
4	Ratox 80%	Royal EDB Tube Mfg. Co.

PART VIII

Molluscicide

METALDEHYDE

Available formulations in India : **Metaldehyde 5 % DP**

- Metaldehyde is a molluscicide used mainly to control snails and slugs in agricultural and horticultural crops, applied as baits or granules in soil or on the crop row.
- **Active group:** Organic molluscicide (cyclic tetramer of acetaldehyde)
- Main target pests: Terrestrial slugs and snails in vegetables, fruits, ornamentals, field crops, plantations, and home gardens.
- **Mode of action on molluscs :** It acts mainly by ingestion, also by contact, disrupting mucus production and epithelial cells in the foot and body wall. It hydrolyses to acetaldehyde in the snail/slug leads to excessive mucus loss, dehydration, reduced mobility and feeding, and death over hours to a couple of days.

Application

- Typically applied as spot or banded bait on soil surface, in furrows or around plants, avoiding direct placement on edible plant parts; ready-to-use pellets are spread at low rates per hectare because a.i. content is low.
- Baits remain active for several days (often up to about 7–10 days depending on weather), after which re-application may be needed if pest pressure continues

Precautions

- Use only where slugs/snails are a proven economic problem; integrate with cultural measures such as hand collection, habitat management, barriers and trapping.
- Place bait in small, well-distributed spots or bands rather than broadcasting heavily, to reduce non-target exposure and cost while still giving sufficient encounter points for molluscs.
- Keep bait away from livestock, pets, poultry and fish ponds; store and transport as a hazardous pesticide, following label and disposal instructions, and consider alternative molluscicides or non-chemical methods in highly sensitive areas.
- Use gloves, avoid eating/drinking/smoking during application, and store in a cool, dry, locked place; avoid contamination of ponds, fish tanks and animal feed.
- Metaldehyde is a toxic substance, with poisoning most frequently reported in dogs and cats due to the palatability of the baits. It can also be toxic to humans and other wildlife.

Label claim (DPPQ & S, Faridabad)

Crop	Pests	Use
Citrus, Rubber, Rice, Tea, Vegetables	Snails, Slugs, Giant African snail	Available in ready to use dry pellets

Trade Names

Sr. No.	Trade Name	Company
1	Snailkill	PI Industries

Dose

- **Small areas:** About 50–80 g bait per 100 sq ft, placed in the evening near crawling paths, at plant base and between crop rows in infested patches.
- **Field scale:** Approximate field recommendation is about 15–25 kg product per acre, depending on pest pressure and label guidance.

PART IX

Biopesticides

AZADIRACHTIN

Available formulations in India	:	1. **Azadirachtin 0.15% EC (1500 ppm)** Neem Seed Kernel Based 2. **Azadirachtin 0.30% EC (3000 ppm)** Neem Seed Kernel Based 3. **Azadirachtin 01.00% EC (10000 ppm)** Neem Based 4. **Azadirachtin 0.03% EC (300 ppm)** Neem Oil Based 5. **Azadirachtin 0.03% WSP (300 ppm)** Neem Oil Based 6. **Azadirachtin 05.00% w/w (50000 ppm)** Neem Extract Concentrates

Classification on the basis of

Target organism	:	Insecticide
Mode of entry	:	Stomach and Contact poison
Mode of action	:	Insect Growth Regulator (IGR), Antifeedant, Repellent
Chemical group	:	Botanical insecticide
IRAC classification		
Main Group and Primary Site of Action	:	Compounds of unknown or uncertain MoA
Sub-group or exemplifying Active Ingredient	:	Azadirachtin
Category of toxicity	:	Slightly toxic
Toxicity label colour	:	Green

Azadirachtin is one of the most important naturally occurring compounds in the field of pest management, especially in sustainable and organic farming.

It is the main active ingredient extracted from the seeds of the Neem tree (Azadirachta indica). Azadirachtin is considered a broad-spectrum, multi-action biopesticide, meaning it affects a wide variety of insect species through several different ways.

It is generally considered safe for humans, animals, and, crucially, for most beneficial insects like honeybees, ladybugs, and predatory mites, when applied correctly.

Modes of Action

Azadirachtin disrupts insect life cycles by several mechanisms:

- **Antifeedant effect:** It deters insects from feeding, often by interfering with their gustatory (taste) receptors.
- **Growth regulation:** It antagonizes ecdysteroids and juvenile hormones, suppressing molting, pupation, and adult emergence, leading to malformed or non-viable adults.
- **Reproductive disruption:** It can cause sterility, reduce mating success, and interfere with yolk protein synthesis, thereby limiting pest population growth.
- **Repellent and Oviposition deterrent:** Azadirachtin repels adults from treated surfaces and deters egg-laying.

Azadirachtin efficacy is highly dependent on water pH. It achieves optimal stability and potency in mildly acidic conditions (pH 4 to 6.5), which extends its half-life. Conversely, its effectiveness is rapidly diminished by alkaline hydrolysis when the water pH exceeds 7, with degradation accelerating significantly at pH 8 and above. To maximize performance, it is practically recommended to adjust the spray water pH to between 5.5 and 6.5 to prevent premature breakdown in the tank.

Label claim (DPPQ & S, Faridabad)

Azadirachtin 0.15% EC (1500 ppm) Neem Seed Kernel Based

Crop	Pests	Dose / ha		Waiting period (Days)
		Quantity of formulation (ml)	Volume of water (Lit.)	
Cotton	Whitefly, Bollworms	2500-5000	500-1000	05
Rice	Thrips, Brown planthopper, Stem borer, Leaf folder	1500-2500	500	05

Trade Names

Sr. No.	Trade Name	Company
1	Dynamite	Auxin Crop Care
2	Neemdhan	Anu Products Ltd.
3	Dr. Neem 1500	Katyayani Organics
4	All in One	Essential Biosciences
5	Nimola	Synwood Agro Ltd.
6	Astha Killer 15	Lila Agrotech

Azadirachtin 0.30% EC (3000 ppm) Neem Seed Kernel Based

Crop	Pests	Dose / ha		Waiting period (Days)
		Quantity of formulation (ml)	Volume of water (Lit.)	
Cotton	American bollworm	4000	1000	05

Trade Names

Sr. No.	Trade Name	Company
1	K-Nine	Koronation Bio-Science

Azadirachtin 0.03% EC (300 ppm) Neem Oil Based

Crop	Pests	Dose / ha		Waiting period (Days)
		Quantity of formulation (ml)	Volume of water (Lit.)	
Cotton	American bollworm, Aphid	2500-5000	500	05
Rice	Leaf roller, Stem borer, Brown planthopper	2000	1000	05

Trade Names

Sr. No.	Trade Name	Company
1	Mint 3000	Herenba Industries Ltd.
2	Dragon Killer	Linuxcrop Chemicals Pvt. Ltd.
3	Azamax	Coromandel International

4	Niimpa 3000	Titan Agritech Ltd.
5	Dr. Anand Neem	Anand Agro Care
6	Multiplex Multineem	Multiplex Bio-tech Pvt. Ltd.
7	Neembin	Katra Fertilizers and Chemicals Pvt. Ltd.

Azadirachtin 0.03% WSP (300 ppm) Neem Oil Based

Crop	Pests	Dose / ha		Waiting period (Days)
		Quantity of formulation (g)	Volume of water (Lit.)	
Gram	Pod borer	2500-5000	500-1000	07
Tur	Pod borers	2500-5000	500-1000	07
Cotton	Aphid, Jassid, Whitefly, Bollworms	2500-5000	500-1000	07
Okra	Fruit borer, Whitefly, Leafhopper	2500-5000	500-1000	07
Brinjal	Shoot and fruit borer, Beetles	2500-5000	500-1000	07
Cabbage	Aphid, Diamondback moth, Cabbage worm	2500-5000	500-1000	07
Jute	Semilooper, Hairy caterpillar	2500-5000	500-1000	07

Trade Names

Sr. No.	Trade Name	Company
1	Neemaxin	Auxin Crop Care
2	Azasik	Sikko Industries Ltd.
3	Neem Pro	Natural Crop Science
4	Dr Neem 300	Katyayani Organics
5	Virgin Neem Oil	Vikalp

Azadirachtin 01.00% EC (10000 ppm) Neem Based

Crop	Pests	Dose / ha		Waiting period (Days)
		Quantity of formulation (ml)	Volume of water (Lit.)	
Tea	Thrips	400-500	450	01
	Red spider mite	400-500	600	01
Tomato	Fruit borer	1000-1500	500	03
Brinjal	Shoot and fruit borer	1000-1500	500	03

Trade Names

Sr. No.	Trade Name	Company
1	Simba	Koronation Bio-Science
2	Neemaxin Gold	Auxin Crop Care
3	Mint 10000	Herenba Industries Ltd.
4	Neemsik	Sikko Industries Ltd.
5	Neem Pro 10K	Natural Crop Science

6	Niimpa	Titan Agritech Ltd.
7	Bio Neem	Zeal Biologicals
8	Neemol	Greenpeace Agro Industries
9	Dr. Anand Neem	Anand Agro Care
10	VBL-Nema	Vidarbha Biotech Lab
11	Astha Killer	Lila Agritech

Azadirachtin 05.00% EC (50000 ppm) Neem Extract Concentrates

Crop	Pests	Dose / ha		Waiting period (Days)
		Quantity of formulation (ml)	Volume of water (Lit.)	
Tea	American Caterpillar, Pink mite, Red spider mite, Thrips	200	400	05
Tobacco	Tobacco caterpillar, Aphid	200	400	05
Rice	Brown planthopper, Leaf folder, Stem borer	200	400	05
Cotton	Whitefly, Leafhopper, Aphid, American bollworm	375	750	05
Cauliflower	Tobacco caterpillar, Diamondback moth, Aphid	200	400	05
Okra	Leafhopper, Whitefly, Aphid, Pod borer	200	400	05
Tomato	Aphid, Whitefly, Fruit borer	200	400	05

Trade Names

Sr. No.	Trade Name	Company
1	Dr. Anand Neem	Anand Agro Care, Nashik
2	Dr. Neem Prime	Katyayani Organics

BACILLUS THURINGIENSIS

Available formulations in India :

1. ***Bacillus thuringiensis*** **var. *galleriae* 1593 M** Serotype H 59 5b, 1.3% flowable concentrate, Potency 1500 IU/mg
2. ***Bacillus thuringiensis*** **var. *kurstaki***
3. ***Bacillus thuringiensis*** **var. *kurstaki*,** Serotype H-39, 3B, Strain Z-52
4. ***Bacillus thuringiensis*** **serovar *kurstaki*** (3a, 3b, 3c) 5.0% WP Potency 55000 SU (*Spodoptera* unit based) (5 x 10^7 spore/mg)
5. ***Bacillus thuringiensis*** **var. *kurstaki* 0.5% WP** Serotype 3a, 3b, 3c, Strain DOR Bt-1, Potency 9000 IU/mg min. U/s 9(3b)
6. ***Bacillus thuringiensis*** var. ***kurstaki*** **0.5% WP** Serotype 3a, 3b, 3c, Strain DOR Bt-1 NAIMCC-B-01118, Potency 13329 IU/mg min. U/s 9(3b)
7. ***Bacillus thuringiensis*** **var. *kurstaki* 0.5% WP** Serotype 3a, 3b, 3c, Strain DOR Bt-1, Potency 9000 IU/mg min. U/s 9(3b)
8. ***Bacillus thuringiensis*** **var. *kurstaki* 0.5% WP** Serotype 3a, 3b, 3c, Strain DOR Bt-1, Potency 16000 IU/mg min.
9. ***Bacillus thuringiensis*** **var. *kurstaki* 2.5% AS** (Spicbio-BTK AS)
10. ***Bacillus thuringiensis*** **var. *krustaki*,** Serotype H-3a, 3b, Strain Z-52 Potency : 3000 IU/mg min on Gypsy moth, 32000 IU/mg min on *Trichoplusia ni*, 50000 IU/mg min on *Helicoverpa armigera* and 55000 IU/mg min on *Spodoptera exiqua*
11. ***Bacillus thuringiensis*** **var. *kurstaki*** Strain HD-1, serotype 3a, 3b, 3.5% ES for Import &repack. Potency 17600 IU/mg
12. ***Bacillus thuringiensis*** **var. *kurstaki*** Serotype 3a, 3b, SA II WG Potency: 53000 SU/mg, 32000 IU/mg
13. ***Bacillus thuringiensis*** **var. *kurstaki* 10% WSL** (CFU: 2 x 10^9 /gm min.) Strain- NBAIR-BtG4, Accession no- JN120763, JN120765, Potency- 14245 IU/ml min.

Classification on the basis of

Target organism : Insecticide

Mode of entry : Stomach poison

Mode of action : Disruptors of insect midgut membranes

Chemical group : Microbial insecticide

IRAC classification

Main Group and Primary Site of Action : 11 Microbial disruptors of insect midgut membranes

Sub-group or exemplifying Active Ingredient : 11A *Bacillus thuringiensis* and the insecticidal proteins they produce

Category of toxicity : Slightly toxic

Toxicity label colour : Green

Bacillus thuringiensis (Bt) is the single most important microbial insecticide used worldwide in pest management. It's a naturally occurring, soil-dwelling bacterium that provides a safe and highly targeted alternative to chemical pesticides, widely used in both conventional and organic farming, as well as in public health programs.

Mode of action:

The effectiveness of *Bacillus thuringiensis* stems from its production of insecticidal proteins, primarily Cry proteins (Crystal proteins or delta-endotoxins), during the sporulation phase of its life cycle.

- **Ingestion:** The insect must ingest the Bt spores and the protein crystals sprayed onto foliage. Bt is a stomach poison, not a contact poison.
- **Activation:** The crystals are inert until they reach the highly alkaline environment (pH 9 to 11) of the target insect's midgut. This high pH and specific digestive enzymes (proteases) dissolve the insoluble crystal, releasing and activating the protoxin into a smaller, lethal toxin.
- **Binding and Pore Formation:** The active toxin then binds to specific receptor proteins on the midgut epithelial cell membrane. This binding is highly specific, which is why Bt only affects certain insect groups. The toxin inserts itself into the cell membrane, creating pores or holes.
- **Death:** The pore formation disrupts the cell's osmotic balance and causes the cells to swell and lyse (rupture). This leads to gut paralysis, the insect stops feeding within hours, and eventually dies from starvation or septicemia (bacteria entering the body cavity) within a few days.

Bt Strain (Subspecies)	Primary Target Pests	Common Application
kurstaki (Btk)	Lepidoptera (Caterpillars/Moths): Cabbage looper, tomato hornworm, gypsy moth, European corn borer.	Spray applied to vegetables, fruits, and ornamentals.
israelensis (Bti)	Diptera (Flies): Mosquito larvae, black fly larvae, fungus gnats.	Applied to standing water (briquets, granules) for public health/aquatic pest control.
tenebrionis (Btt)	Coleoptera (Beetles): Colorado potato beetle, elm leaf beetle.	Spray applied to potato and tree crops.
aizawai (Bta)	Lepidoptera: Wax moth and certain armyworms, often used in stored product protection.	Stored grain pest control and specialty crops.

For optimal results and stability, most experts recommend an ideal spray water pH range for Bt products of 4.0 to 8.0 (neutral to slightly acidic).

Label claim (DPPQ & S, Faridabad)

***Bacillus thuringiensis* var. *galleriae* 1593 M** Serotype H 59 5b, 1.3% flowable concentrate, Potency 1500 IU/mg

Crop	Pests	Dose / ha		Waiting period (Days)
		Quantity of formulation (g)	Volume of water (Lit.)	
Cauliflower, Cabbage	Diamondback moth	600-1000	500	--
Tomato	Fruit borer	1000-1500	500	--
Okra	Shoot and fruit borer	1000-1500	500	--
Chilli	Tobacco caterpillar	1500-2000	1000	--
Cotton	American bollworm	2000-2500	1000	--
Rice	Leaf folder	1000-3000	1000	--

Bacillus thuringiensis* var. *kurstaki

Crop	Pests	Dose / ha		Waiting period (Days)
		Quantity of formulation (g)	Volume of water (Lit.)	
Cotton	Bollworm	750-1000	750-1000	--

***Bacillus thuringiensis* var. *kurstaki*,** Serotype H-39, 3B, Strain Z-52

Crop	Pests	Dose / ha		Waiting period (Days)
		Quantity of formulation (g)	Volume of water (Lit.)	
Cotton	Bollworms, Tobacco caterpillar	750-1000	500-750	--
Rice	Stem borer, Leaf folder	1500	500-750	--
Gram	Pod borer	750	500-750	--
Pigeonpea	Pod borer	750	500-750	--
Soybean	Tobacco caterpillar, Gram pod borer, Bihar hairy caterpillar, Semilooper, Leaf miner	750	500-750	--
Tobacco	Tobacco caterpillar, Gram pod borer	1500-2000	500-750	--
Castor	Hairy caterpillar, Castor semilooper	1000	500-750	--
Teak	Defoliator, Skeletonizer	250-500	500-750	--

Bacillus thuringiensis* serovar *kurstaki (3a, 3b, 3c) 5.0% WP Potency 55000 SU (*Spodoptera* unit based) (5 x 10^7 spores/mg)

Crop	Pests	Dose / ha		Waiting period (Days)
		Quantity of formulation (g)	Volume of water (Lit.)	
Cotton	American bollworm	500-1000	500-1000	--
	Spotted bollworm	750-1000	500-1000	--
Pigeonpea	Pod borer	1000-1250	500-1000	--
Cabbage	Diamondback moth	500-1000	500-1000	--

Bacillus thuringiensis **var. *kurstaki* 0.5% WP** Serotype 3a, 3b, 3c, Strain DOR Bt-1, Potency 9000 IU/mg min. U/s 9(3b)

Crop	Pests	Dose / ha		Waiting period (Days)
		Quantity of formulation (g)	Volume of water (Lit.)	
Castor	Castor semilooper	250	250-300	--

Bacillus thuringiensis var. ***kurstaki*** **0.5% WP** Serotype 3a, 3b, 3c, Strain DOR Bt-1 NAIMCC-B-01118, Potency 13329 IU/mg min. U/s 9(3b)

Crop	Pests	Dose / ha		Waiting period (Days)
		Quantity of formulation (g)	Volume of water (Lit.)	
Pigeonpea	Pod borer	1000-1250	1000	--

Bacillus thuringiensis **var. *kurstaki* 0.5% WP** Serotype 3a, 3b, 3c, Strain DOR Bt-1, Potency 9000 IU/mg min. U/s 9(3b)

Crop	Pests	Dose / ha		Waiting period (Days)
		Quantity of formulation (g)	Volume of water (Lit.)	
Castor	Castor semilooper	250-375	250	--

Bacillus thuringiensis **var. *kurstaki* 0.5% WP** Serotype 3a, 3b, 3c, Strain DOR Bt-1, Potency 16000 IU/mg min.

Crop	Pests	Dose / ha		Waiting period (Days)
		Quantity of formulation (ml)	Volume of water (Lit.)	
Chickpea	Pod borer	1000-1500	500	--

Bacillus thuringiensis **var. *kurstaki* 2.5% AS** (Spicbio-BTK AS)

Crop	Pests	Dose / ha		Waiting period (Days)
		Quantity of formulation (ml)	Volume of water (Lit.)	
Chickpea	Pod borer	1000-1500	500	--

Bacillus thuringiensis **var. *krustaki*,** Serotype H-3a, 3b, Strain Z-52 Potency : 3000 IU/mg min on Gypsy moth, 32000 IU/mg min on *Trichoplusia ni*, 50000 IU/mg min on *Helicoverpa armigera* and 55000 IU/mg min on *Spodoptera exiqua*

Crop	Pests	Dose / ha		Waiting period (Days)
		Quantity of formulation (g)	Volume of water (Lit.)	
Cotton	Bollworms, Tobacco caterpillar	750-1000	500-750	--
Rice	Stem borer, Leaf folder	1500	500-750	--
Chickpea	Pod borer	750	500-750	--
Pigeonpea	Pod borer	750	500-750	--

Soybean	Tobacco caterpillar, Gram pod borer, Bihar hairy caterpillar, Semilooper, Leaf miner	750	500-750	--
Tobacco	Tobacco caterpillar, Gram pod borer	1500-2000	500-750	--
Teak	Defoliator, Skeletonizer	0.25-0.50%	As required	--

Bacillus thuringiensis* var. *kurstaki Strain HD-1, serotype 3a, 3b, 3.5% ES for Import &repack. Potency 17600 IU/mg

Crop	Pests	Dose / ha		Waiting period (Days)
		Quantity of formulation (ml)	Volume of water (Lit.)	
Cotton	Bollworms	750-1000	750-1000	--
Tea	Leaf roller	1000	750-1000	--

Bacillus thuringiensis* var. *kurstaki Serotype 3a, 3b, SA II **WG** Potency: 53000 SU/mg, 32000 IU/mg

Crop	Pests	Dose / ha		Waiting period (Days)
		Quantity of formulation (g)	Volume of water (Lit.)	
Cabbage, Cauliflower	Diamondback moth	500	500-700	--

***Bacillus thuringiensis* var. *kurstaki* 10% WSL** (CFU: 2 x 10^9 /gm min.) Strain- NBAIR-BtG4, Accession no- JN120763, JN120765, Potency- 14245 IU/ml min.

Crop	Pests	Dose / ha		Waiting period (Days)
		Quantity of formulation (Lit.)	Volume of water (Lit.)	
Pigeonpea	Pod borer	10	200	12 hrs.

Trade Names

Sr. No.	Trade Name	Company
1	Dipel	Sumitomo Chemicals India Ltd.
2	Samhar	Katyayani Organics
3	Delfin WG	Bharat Certis AgriScience Ltd.
4	EBS Bacillus thuringiensis	Essential Biosciences
5	Utkarsh BT	Utkarsh Agrichem
6	Mahastra	IPL Biologicals
7	Ekotrump BT	Mankind Agritech
8	Astha BT	Lila Agrotech
9	VBt 0.5% WP	Varsha Bioscience
10	Minchu Plus	Multiplex Bio-Tech Pvt. Ltd.
11	Bio Orthocide	K. N. Bio Sciences
12	Sun Bio Bt	Star Bio Science

BEAUVERIA BASSIANA

Available formulations in India :

1. ***Beauveria bassiana*** **1.15% WP**
2. ***Beauveria bassiana*** **1.15% WP**
3. ***Beauveria bassiana*** **1.15% WP** (1 x 10^8 /gm min) Strain BB-ICAR-RJP, Accession No – MCC 1022
4. ***Beauveria bassiana*** **1.15% WP** (Strain : BB – 5372, own R & D Isolate)
5. ***Beauveria bassiana*** **1.15% WP** (1 x 10^8 /gm min) Strain ICAR, Research Complex, Umiam, Meghalaya, Accession No – NAIMCC-F-03045
6. ***Beauveria bassiana*** **1.15% WP** (1 x 10^8 /gm min) Accession No – NAIMCC-F-03045, Strain No. NBAIM, MAU
7. ***Beauveria bassiana*** **1.15% WP** (1 x 10^8 /spores/ml) Strain BCRL, Accession No – BCRL Bbpx-6892
8. ***Beauveria bassiana*** **1.0% WP,** Strain No: NBRI – 9947 (1 x 10^8 CFU/gm Min.)
9. ***Beauveria bassiana*** **1.0% WP** (1 x 10^9 CFU/gm min), Strain No. IPL/BB/MI/01
10. ***Beauveria bassiana*** **1.0% WP** (1 x 10^8 CFU/gm min), Strain No. SVBPU/CSP/Bb-10, Accession No. ITCC 7520
11. ***Beauveria bassiana*** **5.0% WP,** (1 x 10^8 CFU/gm min) Strain IARI, Accession No. ITCC-7353
12. ***Beauveria bassiana*** **5.0% SC,** Strain: NBAII, Bangalore, Accession No. ITCC-7102, (Strain Isolated by Project Directorate of Bio-logical control, Bangalore)
13. ***Beauveria bassiana*** **5.0% AS** Strain: BB-AAU-RJP Accession No. MCC – 1024
14. ***Beauveria bassiana*** **1.15% WP** (1 x 10^8 /gm min) Accession No – NAIMCC-F-03048
15. ***Beauveria bassiana*** **10.00% SC**
16. ***Beauveria bassiana*** **1.5% Liquid Formulation** (CFU count 1 x 10^8) Accession No. MTCC-5171
17. ***Beauveria bassiana*** **1.15% WP** (1 x 10^6 CFU /gm min)

Classification on the basis of

Target organism : Insecticide
Mode of entry : Contact poison
Mode of action :
Chemical group : Microbial insecticide

IRAC classification

Main Group and Primary Site of Action : UNF Fungal agents of unknown or uncertain MoA
Sub-group or exemplifying Active Ingredient : --

Category of toxicity : Slightly toxic
Toxicity label colour : Green

Beauveria bassiana is one of the most widely used and effective entomopathogenic fungi in pest management. It acts as a mycoinsecticide, infecting and killing insects through direct contact, making it a cornerstone of organic and integrated pest management (IPM) programs worldwide.

Mode of action:

Unlike bacterial insecticides (like *Bt*) which must be ingested, *B. bassiana* is a contact pathogen. It causes a fatal condition known as White Muscardine Disease.

- **Attachment**: The microscopic fungal spores (conidia) land and adhere to the insect's outer shell (cuticle).
- **Penetration:** Under favorable conditions (moderate temperature and high humidity), the spores germinate, and the resulting hyphae produce specialized enzymes (chitinases, proteases, lipases) that dissolve and bore through the insect's cuticle.
- **Invasion:** Once inside the body cavity (haemocoel), the fungus multiplies rapidly, producing toxins (beauvericin, bassianolide) that suppress the insect's immune system and disrupt vital physiological processes. The insect stops feeding and becomes lethargic.
- **Death and Sporulation:** The insect dies, and the fungus grows back out of the cadaver, covering it in a distinctive layer of white, powdery mold. This dead, mold-covered insect serves as an inoculum to spread new infective spores to other pests.

B. bassiana is a broad-spectrum biocontrol agent, effective against pests in most insect orders. The performance of *B. bassiana* in the field is highly dependent on environmental factors because it is a living organism:

- **Humidity**: High humidity is critical for spore germination and cuticle penetration.15 Efficacy is often highest in greenhouses or during high-dew periods.
- **Temperature:** Optimal killing occurs at moderate temperatures, typically 18-290C
- **Light:** The spores are sensitive to UV radiation (sunlight), which can rapidly inactivate them. Applications are best done in the late afternoon, evening, or on cloudy days to maximize spore viability.
- **Compatibility:** It is incompatible with most synthetic fungicides, which will kill the fungus. It is generally safe to use with most chemical and biological insecticides.

Label claim (DPPQ & S, Faridabad)

Beauveria bassiana **1.15% WP**

Crop	Pests	Dose / ha		Waiting period (Days)
		Quantity of formulation (g)	Volume of water (Lit.)	
Cotton	Bollworms	400	750-1000	--

Beauveria bassiana **1.15% WP**

Crop	Pests	Dose / ha		Waiting period (Days)
		Quantity of formulation (g)	Volume of water (Lit.)	
Cotton	Bollworm	2000	400	--
Rice	Leaf folder	2500	750-850	--

***Beauveria bassiana* 1.15% WP** (1 x 10^8 /gm min) Strain BB-ICAR-RJP, Accession No – MCC 1022

Crop	Pests	Dose / ha		Waiting period (Days)
		Quantity of formulation (g)	Volume of water (Lit.)	
Rice	Leaf folder	2500	750-850	--

***Beauveria bassiana* 1.15% WP** (Strain : BB – 5372, own R & D Isolate)

Cop	Pests	Dose / ha		Waiting period (Days)
		Quantity of formulation (g)	Volume of water (Lit.)	
Rice	Leaf folder	2500	600-750	--

***Beauveria bassiana* 1.15% WP** (1 x 10^8 /gm min) Strain ICAR, Research Complex, Umiam, Meghalaya, Accession No – NAIMCC-F-03045

Crop	Pests	Dose / ha		Waiting period (Days)
		Quantity of formulation (g)	Volume of water (Lit.)	
Rice	Leaf folder	2500	750-850	--

***Beauveria bassiana* 1.15% WP** (1 x 10^8 /gm min) Accession No – NAIMCC-F-03045, Strain No. NBAIM, MAU

Crop	Pests	Dose / ha		Waiting period (Days)
		Quantity of formulation (g)	Volume of water (Lit.)	
Rice	Leaf folder	2500	750	--

***Beauveria bassiana* 1.15% WP** (1 x 10^8 /spores/ml) Strain BCRL, Accession No – BCRL Bbpx-6892

Crop	Pests	Dose / ha		Waiting period (Days)
		Quantity of formulation (g)	Volume of water (Lit.)	
Cabbage	Diamondback moth	1000-1500	500-750	--

***Beauveria bassiana* 1.0% WP,** Strain No: NBRI – 9947 (1 x 10^8 CFU/gm Min.)

Crop	Pests	Dose / ha		Waiting period (Days)
		Quantity of formulation (g)	Volume of water (Lit.)	
Chickpea	Pod borer	3000	500	--

***Beauveria bassiana* 1.0% WP** (1 x 10^9 CFU/gm min), Strain No. IPL/BB/MI/01

Crop	Pests	Dose / ha		Waiting period (Days)
		Quantity of formulation (g)	Volume of water (Lit.)	
Okra	Fruit borer, Shoot and fruit borer	3750-5000	400-500	--

Beauveria bassiana **1.0% WP** (1 x 10^8 CFU/gm min), Strain No. SVBPU/CSP/Bb-10, Accession No. ITCC 7520

Crop	Pests	Dose / ha		Waiting period (Days)
		Quantity of formulation (g)	Volume of water (Lit.)	
Chickpea	Pod borer	3000	500	--

Beauveria bassiana **5.0% WP,** (1 x 10^8 CFU/gm min) Strain IARI, Accession No. ITCC-7353

Crop	Pests	Dose / ha		Waiting period (Days)
		Quantity of formulation (g)	Volume of water (Lit.)	
Cabbage	Diamondback moth	2000	500	--

Beauveria bassiana **5.0% SC,** Strain: NBAII, Bangalore, Accession No. ITCC-7102, (Strain Isolated by Project Directorate of Bio-logical control, Bangalore)

Crop	Pests	Dose / ha		Waiting period (Days)
		Quantity of formulation (ml)	Volume of water (Lit.)	
Tomato	Fruit borer	500	500	--

Beauveria bassiana **5.0% AS** Strain: BB-AAU-RJP Accession No. MCC – 1024

Crop	Pests	Dose / ha		Waiting period (Days)
		Quantity of formulation (ml)	Volume of water (Lit.)	
Tomato	Fruit borer	500	500	--

Beauveria bassiana **1.15% WP** (1 x 10^8 /gm min) Accession No – NAIMCC-F-03048

Crop	Pests	Dose / ha		Waiting period (Days)
		Quantity of formulation (g)	Volume of water (Lit.)	
Chickpea	Pod borer	2500	500	--

Beauveria bassiana **10.00% SC**

Crop	Pests	Dose / ha		Waiting period (Days)
		Quantity of formulation (ml)	Volume of water (Lit.)	
Cabbage	Diamondback moth	1000-1500	500-750	--

Beauveria bassiana **1.5% Liquid Formulation** (CFU count 1 x 10^8) Accession No. MTCC-5171

Crop	Pests	Dose / ha		Waiting period (Days)
		Quantity of formulation (ml)	Volume of water (Lit.)	
Tomato	Fruit borer	2000	500	--

***Beauveria bassiana* 1.15% WP** (1 x 10^6 CFU /gm min)

Crop	Pests	Dose / ha		Waiting period (Days)
		Quantity of formulation (g)	Volume of water (Lit.)	
Cotton	Bollworm	2000	400	--

Trade Names

Sr. No.	Trade Name	Company
1	*Beauveria bassiana,* Bhasam	Katyayani Organics
2	*EBS Beauveria bassiana*	Essential Biosciences
3	Beveroz-P	Utkarsh Agrochem
4	Daman, Daman L	IPL Biologicals
5	Beveria	Hedoag India Pvt. Ltd.
6	Baba	Multiplex Bio-Tech Pvt. Ltd.
7	Mycota	Bharat Agro Chemicals and Fertilizers
8	Dr. Batco's 5G Brave	Anand Agro care
9	Wonder	Vidarbha Biotech Ltd.

METARHIZIUM ANISOPLIAE

Available formulations in India :

1. ***Metarhizium anisopliae* 1% WP** (1 x 10^8 CFU/gm min) Strain No. IPL/KC/44, Accession No. ITCC-6895
2. ***Metarhizium anisopliae* 1.15% WP** (1 x 10^8 CFU/gm min) Accession No. MTCC – 5173
3. ***Metarhizium anisopliae* 1.0% WP** (1 x 10^8 CFU/gm min) Strain No. IPL/KC/44 (Own R & D Isolate), Accession No. 6895
4. ***Metarhizium anisopliae* 1.15% WP** (1 x 10^8 CFU/gm min) Strain No. AAI, Allahabad, Accession No. NAIMCC-F-03037
5. ***Metarhizium anisopliae* 1.5% Liquid Formulation** (Accession No. MTCC-5173, T Stanes Ma-1 Strain)
6. ***Metarhizium anisopliae* 10% GR** (CFU count 1 X 10^8 /gm. min.) Strain BCRL– Me, Accession No. ITCC 6911

Classification on the basis of

Target organism	:	Insecticide
Mode of entry	:	Contact poison
Mode of action	:	
Chemical group	:	Microbial insecticide
IRAC classification		
Main Group and Primary Site of Action	:	UNF Fungal agents of unknown or uncertain MoA
Sub-group or exemplifying Active Ingredient	:	--
Category of toxicity	:	Slightly toxic
Toxicity label colour	:	Green

Metarhizium anisopliae is another major globally used entomopathogenic fungus with a long history in biological pest control. Like Beauveria bassiana, it is a contact biopesticide and is particularly renowned for controlling soil-dwelling pests.

It is often referred to as the "Green Muscardine" fungus due to the distinctive green color of its spores on a dead insect.

Mode of action:

The infection mechanism of Metarhizium anisopliae is the same as other mycoinsecticides—it is a contact pathogen that does not need to be ingested.

- **Adhesion:** The spores (conidia) attach to the insect's outer shell (cuticle), aided by hydrophobic interactions and specific surface proteins.
- **Germination and Penetration:** Under suitable conditions (moisture and temperature), the spore germinates, forming a germ tube and a specialized penetration structure called an appressorium. The fungus then produces enzymes (proteases, chitinases) that break down the cuticle, allowing the fungus to bore directly into the insect's body cavity (haemocoel).

- **Invasion and Toxins:** Inside the host, the fungus multiplies and produces toxins, notably the cyclic peptides called destruxins, which suppress the insect's immune system and quickly disrupt its physiology. The insect becomes paralyzed, stops feeding, and dies.
- **Sporulation:** After the host dies, the fungus grows out from the carcass and covers it in a dense, powdery layer of green spores, which are then ready to infect new hosts.

M. anisopliae is a generalist pathogen with a very wide host range, infecting over 200 species across many insect orders. It is particularly effective against pests that spend part of their life cycle in the soil.

The efficacy of *Metarhizium anisopliae*, an entomopathogenic fungus used in pest management, is strongly influenced by environmental factors such as temperature, relative humidity, ultraviolet (UV) radiation, and soil conditions.

- Optimal temperature: 25-30/33°C; reduced activity above 35°C or below 15°C.
- High humidity (>85%) needed for spore germination and infection.
- UV radiation rapidly degrades spores; protection from sunlight improves control.
- Soil moisture and microbial environment modulate fungal persistence.

M. anisopliae is an essential tool in sustainable pest management, providing long-term control, especially for soil-dwelling pests, and its safety to non-target organisms makes it highly valuable for IPM programs.

Label claim (DPPQ & S, Faridabad)

***Metarhizium anisopliae* 1% WP** (1 x 10^8 CFU/gm min) Strain No. IPL/KC/44, Accession No. ITCC-6895

Crop	Pests	Dose / ha		Waiting period (Days)
		Quantity of formulation (g)	Volume of water (Lit.)	
Brinjal	Shoot and fruit borer	2500-5000	500-750	--

***Metarhizium anisopliae* 1.15% WP** (1 x 10^8 CFU/gm min) Accession No. MTCC – 5173

Crop	Pests	Dose / ha		Waiting period (Days)
		Quantity of formulation (g)	Volume of water (Lit.)	
Rice	Brown planthopper	2500	500	--

***Metarhizium anisopliae* 1.0% WP** (1 x 10^8 CFU/gm min) Strain No. IPL/KC/44 (Own R & D Isolate), Accession No. 6895

Crop	Pests	Dose / ha		Waiting period (Days)
		Quantity of formulation (g)	Volume of water (Lit.)	
Brinjal	Shoot and fruit borer	2500-5000	500-750	--

***Metarhizium anisopliae* 1.15% WP** (1 x 10^8 CFU/gm min) Strain No. AAI, Allahabad, Accession No. NAIMCC-F-03037

Crop	Pests	Dose / ha		Waiting period (Days)
		Quantity of formulation (g)	Volume of water (Lit.)	
Chickpea	Pod borer	2500	500	--

***Metarhizium anisopliae*1.5% Liquid Formulation** (Accession No. MTCC-5173, T Stanes Ma-1 Strain)

Crop	Pests	Dose / ha		Waiting period (Days)
		Quantity of formulation (Lit.)	Volume of water (Lit.)	
Paddy	Brown plant hopper	2.0 Lit.	500	--

***Metarhizium anisopliae* 10% GR** (CFU count 1 X 10^8 /gm. min.) Strain BCRL– Me, Accession No. ITCC 6911

Crop	Pests	Dose / ha / application		Waiting period (Days)
		Quantity of formulation	Method of application	
Potato	White grub	60 Kg	Mix *Metarhizium anisopliae* with FYM @ 1 : 10 and apply at the root base by broadcasting method @ 6 gm/m^2 (=60 kg/ha) along the furrows at the time of sowing and next after one month at the time of earthing up.	--

Trade Names

Sr. No.	Trade Name	Company
1	EBS *Metarhizium anisopliae*	Essential Biosciences
2	Metacare	Gujrat Green Crop care
3	Dr. Batco's Meta	Anand Agro Care
4	*Metarhizium anisopliae*	Katyayani Organics
5	Metarhoz-P	Utkarsh Agrochem
6	*Metarhizium anisopliae*	Manidharma Biotech Pvt. Ltd.
7	Pacer	Agrilife (India) Pvt. Ltd.
8	*Metarhizium anisopliae*	Varsha Bioscience and Technology India Pvt. Ltd.
9	Meta Power	K. N. Bio Sciences India Pvt. Ltd.
10	Kalichakra	IPL Biologicals
11	Hedo Meta	Hedoag India Pvt. Ltd.

LECANICILLIUM LECANII (VERTICILLIUM LECANII)

Available formulations in India	:	1. ***Verticillium lecanii* 1.15% WP,** (1 x 10^8 CFU/gm min) Strain – AS MEGH-VL Accession No – MCC-1028 2. ***Verticillium lecanii* 1.50% Liquid Formulation,** (1 x 10^8 CFU/ml. min.) Strain – T Stanes VI-1, Accession No – MTCC-5172 3. ***Verticillium lecanii* 3.0% AS,** (strain: Accession No. MCC-1127, Strain No. MPKV / Biocontrol/ RVN/ VL-01 4. ***Verticillium lecanii* 5.0% SC,** (Strain: Accession No. NFCCI – 2638 5. ***Verticillium lecanii* 5.0% SC,** (1 x 10^8 CFU/gm Min.) Strain – Own Red Isolate, Strain No. VI-17874, MTCC No.5716

Classification on the basis of

Target organism	:	Insecticide
Mode of entry	:	Contact poison
Mode of action	:	
Chemical group	:	Microbial insecticide
IRAC classification		
Main Group and Primary Site of Action	:	UNF Fungal agents of unknown or uncertain MoA
Sub-group or exemplifying Active Ingredient	:	--
Category of toxicity	:	Slightly toxic
Toxicity label colour	:	Green

Verticillium, specifically *Verticillium lecanii*, is widely used as a biological control agent in pest management. It is an entomopathogenic fungus effective against a broad range of soft-bodied and sucking insect pests such as aphids, mealybugs, whiteflies, thrips, red spider mites, scales, leafhoppers, and termites.

Mode of action:

- **Infection by contact:** The fungal spores come into contact with the outer layer (cuticle) of the target insect.
- **Penetration:** The spores germinate and produce hyphae, which penetrate the insect's cuticle, often aided by secreted enzymes like chitinases.
- **Colonization and Death:** Once inside, the fungus proliferates, consumes the insect's internal tissues, and may produce insecticidal toxins (like bassianolide), leading to the insect's death, usually within a few days.
- **"White Halo" fungus:** After the insect dies, the fungus typically grows out of the body and produces white, cottony mycelial growth and spores, which can then infect other pests.

For maximum efficacy and pest control potential, the fungus requires moderate temperatures and high humidity. Limiting exposure to direct UV light is also crucial, as UV radiation can significantly reduce the fungus's biological activity.

V. lecanii is therefore ideally suited for applications in controlled environments such as greenhouses, naturally humid climates, or during cooler, high-moisture times of the day (like dawn or dusk) in field settings.

Label claim (DPPQ & S, Faridabad)

***Verticillium lecanii* 1.15% WP,** (1 x 10^8 CFU/gm min) Strain – AS MEGH-VL Accession No – MCC-1028

Crop	Pests	Dose / ha		Waiting period (Days)
		Quantity of formulation (g)	Volume of water (Lit.)	
Cotton	Whitefly	2500	500	--
Citrus	Mealybug	2500	550	--

***Verticillium lecanii* 1.50% Liquid Formulation,** (1 x 10^8 CFU/ml. min.) Strain – T Stanes VI-1, Accession No – MTCC-5172

Crop	Pests	Dose / ha		Waiting period (Days)
		Quantity of formulation (ml)	Volume of water (Lit.)	
Tomato	Whitefly	2000	500	--

***Verticillium lecanii* 3.0% AS,** (strain: Accession No. MCC-1127, Strain No. MPKV / Biocontrol/ RVN/ VL-01

Crop	Pests	Dose / ha		Waiting period (Days)
		Quantity of formulation (ml)	Volume of water (Lit.)	
Onion	Thrips	2000-2500	500	--

***Verticillium lecanii* 5.0% SC,** (Strain: Accession No. NFCCI – 2638

Crop	Pests	Dose / ha		Waiting period (Days)
		Quantity of formulation (ml)	Volume of water (Lit.)	
Cabbage	Diamondback moth	500	500	--

***Verticillium lecanii* 5.0% SC,** (1 x 10^8 CFU/gm Min.) Strain – Own Red Isolate, Strain No. VI-17874, MTCC No.5716

Crop	Pests	Dose / ha		Waiting period (Days)
		Quantity of formulation (ml)	Volume of water (Lit.)	
Rice	White backed planthopper	3125	600	--

Trade Names

Sr. No.	Trade Name	Company
1	Verticoz-P	Utkarsh Agrochem
2	Verticillium lecanii	Katyayani Organics
3	EBS Verticillium lecanii	Essential Biosciences
4	Verti	Hedoag India Pvt. Ltd.
5	Dr. Batco's 5G Vertigo	Anand Agro Care
6	Wertisphere	Bharat Agro Chemicals and Fertilizers

PSEUDOMONAS FLUORESCENS

Available formulations in India : ***Pseudomonas fluorescens*** **1.0% WP** (Strain No. IIHR-PF-2, Accession No. ITCC- B0034)

Classification on the basis of

Target organism	:	Nematicide
Mode of entry	:	Contact poison
Mode of action	:	
Chemical group	:	Microbial insecticide

IRAC classification

Main Group and Primary Site of Action	:	
Sub-group or exemplifying Active Ingredient	:	--

Category of toxicity : Slightly toxic
Toxicity label colour : Green

Pseudomonas fluorescens is a member of the Plant Growth-Promoting Rhizobacteria (PGPR) group and is an excellent alternative to chemical nematicides due to its eco-friendly nature and dual action of pest control and plant growth promotion.

It is an effective biological control agent used as a nematicide primarily against root-knot nematodes (*Meloidogyne* spp.) and other plant-parasitic nematodes.

Mode of action

P. fluorescens suppresses plant-parasitic nematodes, particularly the destructive Root-Knot Nematodes (Meloidogyne spp.), through multiple mechanisms:

- **Direct Toxicity (Antibiosis):** The bacteria produce various secondary metabolites and antibiotics that are toxic to nematodes, reducing their motility, hatching, and survival. Key compounds include:
  - 2,4-Diacetylphloroglucinol (DAPG): This potent antibiotic has been shown to directly reduce nematode penetration and egg production.
  - Hydrogen Cyanide (HCN): Acts as a volatile toxin and repellent to nematodes at high concentrations.
  - Phenazines and Pyrrolnitrin: Other broad-spectrum antimicrobial compounds that contribute to nematode suppression.
- **Lytic Enzymes:** Some strains produce enzymes like chitinases which can degrade the eggs or the cuticle (outer layer) of the nematodes.
- **Induced Systemic Resistance (ISR):** *P. fluorescens* colonizes the root surface (rhizosphere) and triggers the plant's natural defense mechanisms. This physiological change strengthens the host plant's ability to resist nematode invasion and reproduction.

- **Competition:** As a successful colonizer, the bacteria rapidly consume the root exudates (nutrients) that the nematodes would otherwise use to find and invade the roots, effectively reducing the nematode's chance of infection.

For maximal nematicidal and other biocontrol effects, *Pseudomonas fluorescens* requires moderate temperatures, sufficient moisture, favorable soil nutrient conditions, and protection from harsh environmental stresses like UV and desiccation. These factors must be considered in field applications to optimize pest management outcomes.

Label claim (DPPQ & S, Faridabad)

***Pseudomonas fluorescens* 1.0% WP** (Strain No. IIHR-PF-2, Accession No. ITCC- B0034)

Crop	Pests	Application
Tomato, Brinjal, Carrot, Okra	Root-knot nematode	Treat the seed with *Pseudomonas fluorescens* 1.0% WP @ 20 gm/kg of seeds & treat the nursery beds with the *Pseudomonas fluorescens* 1.0% WP @ 50 gm/sq.m and apply *Pseudomonas fluorescens* 1.0% WP @ 5 kg/ha enriched FYM @ 5 tons/ha to the soil before transplanting.

Trade Names

Sr. No.	Trade Name	Company
1	Sudoz-P	Utkarsh Agrochem
2	Pseudomonas	Pioneer Agro Industry
3	Striker	Katyayani Organics
4	Multiplex Sparsha	Multiplex Bio-tech Pvt. Ltd.
5	Sudo	Hedoag India Pvt. Ltd.
6	Phasal Rakshak	IPL Biologicals
7	Averse	Bharat Agro Chemicals and Fertilizers
8	Fluro 4K	Anand Agro Care Nashik
9	Bio-Disease Control	BloomBuddy
10	Pseudo-Pep	Titan Agritech Ltd.

TRICHODERMA HARZIANUM

Available formulations in India	:	1. ***Trichoderma harzianum* 1.0% WP** (Strain No. IIHR-TH-2 Accessions No. ITCC 6888) 2. ***Trichoderma harzianum* 1.5% WP** (Strain No. IIHR-TV-5 Accessions No. ITCC 6889)

Classification on the basis of

Target organism	:	Nematicide
Mode of entry	:	Contact poison
Mode of action	:	
Chemical group	:	Microbial insecticide
IRAC classification		
Main Group and Primary Site of Action	:	UNF Fungal agents of unknown or uncertain MoA
Sub-group or exemplifying Active Ingredient	:	--
Category of toxicity	:	Slightly toxic
Toxicity label colour	:	Green

The fungus *Trichoderma harzianum* is an extremely important and commonly used biocontrol agent in agriculture, with proven efficacy against plant-parasitic nematodes like Root-Knot Nematodes (Meloidogyne spp.).

While *Trichoderma* is often primarily known for its control of fungal plant diseases (mycoparasitism), its nematicidal effect is significant and achieved through a combination of mechanisms.

Nematicidal Mechanisms of *T. harzianum*

T. harzianum uses several different strategies to suppress nematode populations and protect plants:

- **Direct Toxicity (Antibiosis):** *T. harzianum* produces various secondary metabolites (such as viridin, gliotoxin, and trichodermin) that are directly toxic or repellent to nematodes. These compounds can interfere with nematode mobility, egg hatching, and survival.
- **Lytic Enzyme Secretion:** The fungus secretes enzymes that can degrade the structural components of the nematode and its eggs. Key lytic enzymes include:
    - Chitinases: Degrade the chitin present in the eggshell of nematodes (like Meloidogyne spp.), preventing hatching.
    - Proteases, Lipases, and Glucanases: These enzymes help break down the outer cuticle and internal body components of the nematodes.
- **Induced Systemic Resistance (ISR):** *T. harzianum* colonizes the plant roots and triggers the plant's natural defense system, making the plant physiologically stronger and more resistant to nematode invasion. This is often achieved by modulating phytohormones like jasmonic acid and salicylic acid, which ramp up the plant's

production of defense-related enzymes (e.g., peroxidase, chitinase) and secondary metabolites. This is the most critical and consistent mechanism.

- **Competition for Space and Nutrients:** The fungus is a fast-growing, highly competitive organism in the soil (rhizosphere). It out-competes nematodes and other pathogens for essential nutrients and root colonization sites, denying the nematodes a place to establish and feed.
- **Synergism:** Crucially, *T. harzianum* often works synergistically with other beneficial microbes, such as *Pseudomonas fluorescens,* enhancing the production of nematicidal compounds (like DAPG) by the bacteria and leading to a much greater reduction in nematode populations than either organism alone.

The nematicidal efficacy of *Trichoderma harzianum* is maximized under moderate temperatures (25-30°C), optimal moisture, suitable soil pH (4-7), and protection from environmental stresses. These conditions support fungal growth, enzymatic activity against nematodes, and plant defense induction for effective nematode management.

Label claim (DPPQ & S, Faridabad)

***Trichoderma harzianum* 1.0% WP** (Strain No. IIHR-TH-2 Accessions No. ITCC 6888)

Crop	Pests	Application
Tomato, Brinjal	Root-knot nematode	Treat the seeds with *Trichoderma harzianum* 1.0% WP @ 20 gm/kg of seeds & nursery beds with the *Trichoderma harzianum* 1.0% WP @ 50 gm/sq.m and also apply *Trichoderma harzianum* 1.0% WP @ 5 kg/ha enriched FYM @ 5 tons/ha to the soil before transplanting.
Carrot, Okra	Root-knot nematode	Treat the seeds with *Trichoderma harzianum* 1.0% WP @ 20 gm/kg of seeds and apply *Trichoderma harzianum* 1.0% WP (@ 5 kg/ha enriched FYM @ 5 tons/ha to the soil before sowing.
Gerbera, Tuberose	Root-knot nematode	Apply the Nemastin @ 50 gm/sq.m at the time of planting
Banana	Root-knot nematode	Apply 2 Kg Nemastin enriched FYM @ 2 Kg/pant at the time of planting and at an interval of 3 months after planting for a period of one year
Acid lime	Citrus nematode	Apply 2 Kg Nemastin enriched FYM @ 2 Kg/pant at the time of planting and at an interval of 3 months after planting for a period of one year
Papaya	Root-knot and Reniform nematode	Apply 2 Kg Nemastin enriched FYM @ 2 Kg/pant at the time of planting and at an interval of 3 months after planting for a period of one year

Trade Names

Sr. No.	Trade Name	Company
1		

***Trichoderma harzianum* 1.5% WP** (Strain No. IIHR-TV-5 Accessions No. ITCC 6889)

Crop	Pests	Application
Tomato, Brinjal, Carrot, Okra	Root-knot nematode	Treat the seed with *Trichoderma harzianum* 1.5% WP @ 20 gm/kg of seeds & treat the nursery beds with the *Trichoderma harzianum* 1.5% WP @ 50 gm/sq.m and also apply *Trichoderma harzianum* 1.5% WP @ 5 kg/ha enriched FYM @ 5 tons/ha to the soil before transplanting.

Trade Names

Sr. No.	Trade Name	Company
1	Biowrap	T. Stanes and Company Ltd.
2	Hattrick	Katyayani Organics
3	EBS *Trichoderma harzianum*	Essential Biosciences
4	Sangathi Plus	R K Chemicals
5	Bioharz	IPL Biologicals
6	Dr. Batco's 5G Herz	Anand Agro Care

TRICHODERMA VIRIDE

Available formulations in India	:	***Trichoderma viride* 1.5% WP** (Strain No. IIHR-TV-5 Accessions No. ITCC 6889)
Classification on the basis of		
Target organism	:	Nematicide
Mode of entry	:	Contact poison
Mode of action	:	
Chemical group	:	Microbial insecticide
IRAC classification		
Main Group and Primary Site of Action	:	UNF Fungal agents of unknown or uncertain MoA
Sub-group or exemplifying Active Ingredient	:	--
Category of toxicity	:	Slightly toxic
Toxicity label colour	:	Green

Trichoderma viride is increasingly recognized as an effective biocontrol agent for managing plant-parasitic nematodes, especially root-knot nematodes (*Meloidogyne* spp.), through direct and indirect mechanisms.

Mechanisms of Nematicidal Action

T. viride employs multiple mechanisms to suppress nematode populations and protect plants:

- **Direct Parasitism (Mycoparasitism):** The fungus can attack the nematode eggs and juveniles (larvae). It produces lytic enzymes such as chitinases, proteases, lipases, and glucanases. These enzymes break down the nematode's eggshell (which contains chitin) and the cuticle of the larvae, leading to their death.
- **Antibiosis and Toxin Production:** *T. viride* produces inhibitory metabolites and compounds like gliotoxin and viridin. These metabolites can directly affect the nematode's mobility, reproduction, and survival.
- **Competition:** The fungus is highly competitive and rapidly colonizes the soil and plant root zone (rhizosphere), effectively competing with nematodes and other pathogens for space and nutrients.
- **Induced Systemic Resistance (ISR):** *T. viride* interacts with the plant's root system, which can trigger the plant's natural defense mechanisms. This induced resistance makes the host plant more tolerant or resistant to nematode attack, often by modulating phytohormones like jasmonic acid and ethylene.
- **Plant Growth Promotion:** While not directly nematicidal, *T. viride* often promotes plant growth, resulting in stronger, healthier roots that are better able to withstand or compensate for nematode damage.

For maximum nematicidal efficacy, the application of *T. viride* must be timed and managed to coincide with optimal soil moisture and temperature (25-30^{0}C) conditions to ensure the fungus actively grows, sporulates, and produces its toxic metabolites and lytic enzymes.

Trichoderma viride acts as an effective, environmentally safe nematicide by direct antagonism of nematodes and plant growth promotion, making it a valuable addition to integrated nematode management programs.

Label claim (DPPQ & S, Faridabad)

***Trichoderma viride* 1.5% WP** (Strain No. IIHR-TV-5 Accessions No. ITCC 6889)

Crop	Pests	Application
Tomato, Brinjal, Carrot, Okra	Root-knot nematode	Treat the seeds with *Trichoderma viride* l.5% WP @ 20 gm/kg of seeds & nursery beds with the *Trichoderma viride* l.5 % WP @ 50 gm/sq.m. and also apply *Trichoderma viride* 1.5% WP (@ 5 kg/ha enriched FYM @ 5 tons/ha to the soil before transplanting.

Trade Names

Sr. No.	Trade Name	Company
1	Trichoz-P	Utkarsh Agrochem
2	Rom	Romvijay Bio Tech Pvt. Ltd.
3	Sanjeevni	IPL Biologicals
4	Fighter	Navratan Bio Science Pvt. Ltd.
5	Tricho	Hedoag India Pvt. Ltd.
6	Tyson	Katyayani Organics
7	EBS Trichoderma viride	Essential Biosciences
8	Nisarga	Multiplex Bio-tech Pvt. Ltd.
9	Dr. Batco's Dermus 4K	Anand Agro Care
10	Trichoderma viride	BloomBuddy
11	Trident	Bharat Agro Chemicals and Fertilizers
12	Tricho-Pep V	Titan Agritech Ltd.
13	Trichoderma viride	Pioneer Agro Industry
14	Biocure F	T. Stanes and Company Ltd.
15	Fasitry	Vidarbha Biotech Lab

VERTICILLIUM CHLAMYDOSPORIUM

Available formulations in India	:	***Verticillium chlamydosporium* 1.0% WP,** (2 x 10^6 CFU/g min) Strain – IIHR-VC-3 Accession No–ITCC-6898
Classification on the basis of		
Target organism	:	Nematicide
Mode of entry	:	Contact poison
Mode of action	:	
Chemical group	:	Microbial insecticide
IRAC classification		
Main Group and Primary Site of Action	:	UNF Fungal agents of unknown or uncertain MoA
Sub-group or exemplifying Active Ingredient	:	--
Category of toxicity	:	Slightly toxic
Toxicity label colour	:	Green

Verticillium chlamydosporium (now *Pochonia chlamydosporia*) is one of the most promising fungal agents for the sustainable management of sedentary plant-parasitic nematodes, particularly Root-Knot Nematodes (*Meloidogyne* spp.) and Cyst Nematodes (*Globodera* and *Heterodera* spp.).

Primary Mode of Action: Egg Parasitism

The main mechanism by which *V. chlamydosporium* controls nematodes is through egg parasitism (oviparasitism).

- **Rhizosphere Colonization:** The fungus, applied as chlamydospores (robust resting spores) or conidia, first colonizes the rhizosphere (the area immediately surrounding the plant roots).
- **Attraction to Egg Masses:** The fungus is specifically attracted to the egg masses and sedentary female nematodes on the root surface, where it thrives.
- **Infection and Penetration:** The fungal hyphae (filaments) adhere to and penetrate the nematode eggshell, which is composed of chitin. This penetration is facilitated by the secretion of lytic enzymes, primarily serine proteases and chitinases.
- **Internal Feeding:** Once inside, the fungus feeds on the contents of the egg, preventing the juvenile nematode (J2) from developing or hatching.
- **Long-Term Control:** The fungus forms thick-walled resting structures called chlamydospores within the parasitized egg, which allows it to survive in the soil for long periods and build up a suppressive population over successive crop cycles.

Secondary Mechanisms

- **Toxin Production:** The fungus produces nematicidal metabolites, such as phomalactone, which can have a toxic effect on the nematode, reducing its mobility and ability to invade the root.
- **Root Colonization:** *V. chlamydosporium* can colonize the root surface (rhizoplane) and, in some cases, may exhibit endophytic behavior (growing inside the root without causing harm). This colonization is key for its survival and its ability to encounter the nematode egg masses on the root surface.
- **Plant Growth Promotion (Indirect):** By significantly reducing nematode damage and improving root health, it indirectly promotes overall plant growth and yield.

The successful control with *V. chlamydosporia* relies on creating a soil environment that maximizes its growth (aided by organic matter and optimal temperaturepH) while maintaining conditions where the nematode eggs are exposed and susceptible to fungal penetration.

Verticillium chlamydosporium is a promising, environmentally safe fungal nematicide that controls root-knot and other nematodes through egg parasitism and enzyme action, making it suitable for sustainable nematode management in various crops.

Label claim (DPPQ & S, Faridabad)

***Verticillium chlamydosporium* 1.0% WP,** (2 x 10^6 CFU/gm min) Strain – IIHR-VC-3 Accession No – ITCC-6898

Crop	Pests	Application
Tomato, Brinjal, Carrot, Okra	Root-knot nematode	Treat the seeds with *Verticillium chlamydosporium* 1.0% WP @ 20 gm/kg of seeds & nursery beds with the *Verticillium chlamydosporium* 1.0% WP @ 50 gm/sq.m and also apply *Verticillium chlamydosporium* 1.0% WP @ 5 kg/ha enriched FYM @ 5 tons/ha to the soil before transplanting.

Trade Names

Sr. No.	Trade Name	Company
1	Nematode Plus	Katyayani Organics
2	Nematofree+	IPL Biologicals
3	Dr. Batco's 5G Nemos	Anand Agro Care
4	Nematoz-P	Utkarsh Agrochem
5	Verti-Guard	Ajay Bio-Tech (India) Ltd.

PAECILOMYCES LILACINUS

Available formulations in India :

1. ***Paecilomyces lilacinus* 01.15% WP** (Accession No. MTCC No. 5175, T-Stanes PI-1 Strain
2. ***Paecilomyces lilacinus* 1.15% WP** (CFU 1 x 10^8 /gm. Min) (T Stanes–PI-1 Strain Accession No. MTCC 5175)
3. ***Paecilomyces lilacinus* 01.50% LF** (CFU count 1 x 10^8/ml min.) (Accession No. MTCC No. 5175, T-Stanes PI-1 Strain)

Classification on the basis of

Target organism	:	Nematicide
Mode of entry	:	Contact poison
Mode of action	:	
Chemical group	:	Microbial insecticide
IRAC classification		
Main Group and Primary Site of Action	:	UNF Fungal agents of unknown or uncertain MoA
Sub-group or exemplifying Active Ingredient	:	--
Category of toxicity	:	Slightly toxic
Toxicity label colour	:	Green

Paecilomyces lilacinus is a widely used biological nematicide fungus effective against various plant-parasitic nematodes, especially root-knot nematodes (*Meloidogyne* spp.). It works predominantly by parasitizing nematode eggs and juveniles, significantly reducing nematode populations and root damage.

Nematicidal Mechanism of Action

P. lilacinus employs a multi-faceted approach to control nematodes, combining direct physical attack with the secretion of powerful toxic compounds:

- **Egg Parasitism (Oviparasitism) - The Primary Action:** The fungus is an opportunistic egg parasite. Its spores germinate in the soil and grow towards nematode egg masses (especially those produced by Root-Knot and Cyst Nematodes). It adheres to the eggshell and secretes lytic enzymes, primarily serine proteases and chitinases, which degrade the chitin layer of the eggshell. The fungal hyphae then penetrate the egg, destroying the developing juvenile embryo inside, thereby preventing the next generation of nematodes from hatching.
- **Juvenile and Adult Female Infection:** While its primary target is the egg, P. lilacinus can also infect and penetrate the cuticles of juveniles (J2) and sedentary adult females (through natural openings like the anus or vulva). Once inside, the fungus consumes the internal contents of the nematode, leading to death.
- **Toxin Production (Antibiosis):** The fungus produces various nematicidal toxins and metabolites, such as leucinostatins (also called paecilotoxins), which have neurotoxic effects. These toxins can cause paralysis and ultimately death in susceptible nematode species.

- **Plant Growth Promotion (Indirect Effect):** *P. lilacinus* is also known to enhance root development, nutrient uptake, and overall plant biomass. By reducing nematode stress and potentially producing phytohormones (like gibberellins and IAA), it helps the plant grow healthier and tolerate minor nematode damage.

Its tolerance for a wider range of conditions, especially temperature, makes *P. lilacinum* one of the most reliable fungal bio-nematicides for field application.

Paecilomyces lilacinus is a potent, eco-friendly biological nematicide effective mainly through egg parasitism, enzymatic degradation, and toxin production, providing sustainable nematode control with long-lasting effects in diverse cropping systems.

Label claim (DPPQ & S, Faridabad)

***Paecilomyces lilacinus* 01.15% WP** (Accession No. MTCC No. 5175, T-Stanes PI-1 Strain

Crop	Pests	Dose / ha		Waiting period (Days)
		Quantity of formulation (g)	Volume	
Brinjal	Root-knot nematode	3000	500 Kg Organic manure / Organic fertilizer	--

***Paecilomyces lilacinus* 1.15% WP** (CFU 1 x 10^8 /gm. Min) (T Stanes–Pl-1 Strain Accession No. MTCC 5175)

Crop	Pests	Dose / ha		Waiting period (Days)
		Quantity of formulation (Kg)	Volume	
Chilli	Root-knot nematode	5-6	--	--
Cucumber	Root-knot nematode	5-6	--	--

***Paecilomyces lilacinus* 01.50% LF** (CFU count 1 x 10^8/ml min.) (Accession No. MTCC No. 5175, T-Stanes PI-1 Strain)

Crop	Pests	Dose / ha		Waiting period (Days)
		Quantity of formulation (ml)	Volume of water (Lit.)	
Tomato	Root-knot nematode	6000	500	--

Trade Names

Sr. No.	Trade Name	Company
1	Nema Cure	Katyayani Organics
2	Agri Nematode	Amruth Farming
3	Dr.Batco's Nemos	Anand Agro Care
4	EBS *Paecilomyces lilacinus*	Essential Biosciences

NUCLEAR POLYHEDROSIS VIRUS (NPV) OF HELICOVERPA ARMIGERA

Available formulations in India :

1. **Nuclear Polyhedrosis Virus of *Helicoverpa armigera* 0.43% AS** (1 x 10^9 POB/ml)
2. **Nuclear Polyhedrosis Virus of *Helicoverpa armigera* 2.0% AS,** Strain No. GBS/HNPV -01 (1 x 10^9 POB/ml Min.)
3. **Nuclear Polyhedrosis Virus of *Helicoverpa armigera* 2.0% AS,** Strain No. NBRI-8821 (1 x 10^9 POB/ml Min.)
4. **Nuclear Polyhedrosis Virus of *Helicoverpa armigera* 2.0% AS,** Strain No. IBH-17268 (1 x 10^9 POB/ml Min.)
5. **Nuclear Polyhedrosis Virus of *Helicoverpa armigera* 2.0% AS,** Strain No. BIL/HV-9 POB (1 x 10^9 POB/ml Min.)
6. **Nuclear Polyhedrosis Virus of *Helicoverpa armigera* 2.0% AS,** Strain No. IBL-17268
7. **Nuclear Polyhedrosis Virus of *Helicoverpa armigera* 0.43% AS,** Strain No. BIL/HV-9 (1 x 10^9 POB/ml Min.)
8. **NPV of *Helicoverpa armigera* 0.5% AS,** (1 x 10^9 POB/ml Min.)
9. **NPV of *Helicoverpa armigera* 2.0% AS,** (1 x 10^9 POBs count / ml min) Biological Insecticide

Classification on the basis of

Target organism : Insecticide
Mode of entry : Stomach poison
Mode of action :
Chemical group : Microbial insecticide

IRAC classification

Main Group and Primary Site of Action :
Sub-group or exemplifying Active Ingredient : --

Category of toxicity : Slightly toxic
Toxicity label colour : Green

The Nuclear Polyhedrosis Virus (NPV) of *Helicoverpa armigera* (abbreviated as *Ha*NPV) is a highly effective and host-specific microbial biopesticide used worldwide to control the *Helicoverpa armigera* pest, commonly known as the gram pod borer or cotton bollworm. It is a crucial component of Integrated Pest Management (IPM) strategies due to its environmental safety and non-harm to beneficial insects.

Mechanism and Effect

The virus works by causing a fatal disease (virosis) in the larval stage of the pest:

- **Ingestion :** *H. armigera* larvae must ingest the virus particles, known as Polyhedral Occlusion Bodies (POBs) or Polyhedral Inclusion Bodies (PIBs), which are sprayed onto the crop foliage.
- **Infection:** Once inside the alkaline gut of the caterpillar, the POBs dissolve, releasing the infectious viral particles. These particles infect the cells lining the gut.

- **Replication:** The virus then moves to and replicates massively within the host's nuclei (hence "Nucleo-polyhedrosis"), primarily in the fat body, skin, and blood cells.
- **Death and Spread:** The infected larva stops feeding within a day or two and becomes sluggish. The larva's body swells, turns dark, and eventually ruptures, releasing billions of new POBs onto the crop surface and soil, which then spread the infection to other larvae and help establish an epizootic (natural outbreak) in the pest population.
- **Timing and Efficacy:** The HaNPV is most effective against early larval instars (1st and 2nd). Older larvae (3rd instar and above) are much more resistant, requiring significantly higher doses or showing less mortality.

For *Ha*NPV to be effective, farmers must primarily manage the negative impacts of UV radiation and ensure the spray reaches the pest at its most susceptible stage (young larvae), often requiring the use of adjuvants and appropriate timing (evening application).

Helicoverpa armigera NPV is a potent, host-specific biocontrol agent effective principally on young larvae and suitable for integrated pest management, with efficacy enhanced by proper formulation and integrated use with other control agents.

Label claim (DPPQ & S, Faridabad)

Nuclear Polyhedrosis Virus of *Helicoverpa armigera* 0.43% AS (1 x 10^9 POB/ml)

Crop	Pests	Dose / ha		Waiting period (Days)
		Quantity of formulation (ml)	Volume	
Cotton	American bollworm	2700	400-600	--
Tomato	Fruit borer	1500	400-600	--

Nuclear Polyhedrosis Virus of *Helicoverpa armigera* 2.0% AS, Strain No. GBS/HNPV -01 (1 x 10^9 POB/ml Min.)

Crop	Pests	Dose / ha		Waiting period (Days)
		Quantity of formulation (ml)	Volume	
Pigeonpea	Pod borer	250-500	500-750	--
Chickpea	Pod borer	250-500	500-750	--

Nuclear Polyhedrosis Virus of *Helicoverpa armigera* 2.0% AS, Strain No. NBRI-8821 (1 x 10^9 POB/ml Min.)

Crop	Pests	Dose / ha		Waiting period (Days)
		Quantity of formulation (ml)	Volume	
Pigeonpea	Pod borer	500	500	--

Nuclear Polyhedrosis Virus of *Helicoverpa armigera* 2.0% AS, Strain No. IBH-17268 (1 x 10^9 POB/ml Min.)

Crop	Pests	Dose / ha		Waiting period (Days)
		Quantity of formulation (ml)	Volume	
Pigeinpea	Pod borer	250-500	500-750	--
Chickpea	Pod borer	250-500	500-750	--

Nuclear Polyhedrosis Virus of *Helicoverpa armigera* 2.0% AS, Strain No. BIL/HV-9 POB (1 x 10^9 POB/ml Min.)

Crop	Pests	Dose / ha		Waiting period (Days)
		Quantity of formulation (ml)	Volume	
Pigeinpea	Pod borer	250-500	500-750	--
Chickpea	Pod borer	250-500	500-750	--
Tomato	Fruit borer	250-500	500-750	--

Nuclear Polyhedrosis Virus of *Helicoverpa armigera* 2.0% AS, Strain No. IBL-17268

Crop	Pests	Dose / ha		Waiting period (Days)
		Quantity of formulation (ml)	Volume	
Pigeinpea	Pod borer	250-500	500-750	--
Chickpea	Pod borer	500-1000	500-750	--

Nuclear Polyhedrosis Virus of *Helicoverpa armigera* 0.43% AS, Strain No. BIL/HV-9 (1 x 10^9 POB/ml Min.)

Crop	Pests	Dose / ha		Waiting period (Days)
		Quantity of formulation (ml)	Volume	
Cotton	American bollworm	2700	400-600	--
Tomato	Fruit borer	1500	400-600	--

NPV of *Helicoverpa armigera* 0.5% AS, (1 x 10^9 POB/ml Min.)

Crop	Pests	Dose / ha		Waiting period (Days)
		Quantity of formulation (ml)	Volume	
Chickpea	Pod borer	250	500	--

NPV of *Helicoverpa armigera* 2.0% AS, (1 x 10^9 POBs count / ml min) Biological Insecticide

Crop	Pests	Dose / ha		Waiting period (Days)
		Quantity of formulation (ml)	Volume	
Chickpea	Pod borer	250	600	--

Trade Names

Sr. No.	Trade Name	Company
1	Bio Kill-H	Ganesh Biocontrol System
2	Sun Bio HANPV	Sonkul
3	Utkranti	Vidarbha Biotech Lab

NUCLEAR POLYHEDROSIS VIRUS (NPV) OF SPODOPTERA LITURA

Available formulations in India	:	**Nuclear Polyhedrosis Virus of *Spodoptera litura* 0.5% AS,** (1 x 10^9 POB/ml Min.)

Classification on the basis of

Target organism	:	Insecticide
Mode of entry	:	Stomach poison
Mode of action	:	
Chemical group	:	Microbial insecticide
IRAC classification		
Main Group and Primary Site of Action	:	UNF Fungal agents of unknown or uncertain MoA
Sub-group or exemplifying Active Ingredient	:	--
Category of toxicity	:	Slightly toxic
Toxicity label colour	:	Green

This virus (designated as SlNPV or SpltNPV) is a type of baculovirus and is used as a highly effective, host-specific bio-pesticide or bio-control agent in Integrated Pest Management (IPM) programs.

Host-Specific: SlNPV is generally specific to *Spodoptera litura* and its close relatives, making it safe for beneficial insects (like pollinators and natural predators), humans, animals, and the environment.

Mechanism and Symptoms of Infection

- **Infection Process:** Larvae ingest the Polyhedral Occlusion Bodies (POBs)—the protective, crystalline form of the virus—when feeding on treated foliage.
- **Action:** Once inside the alkaline gut of the larva, the POBs dissolve, releasing the virus particles. The virus multiplies rapidly inside the host cells, leading to a systemic infection.
- **Symptoms:** Infected larvae become sluggish, flaccid, and show a characteristic change in color. They often climb to the tops of plants before they die, hanging upside down (a symptom known as the "tree top disease"). The body wall eventually ruptures, releasing millions of new POBs onto the crop to infect other larvae.
- **Time to Kill:** The virus is slower-acting than chemical pesticides, typically taking 3 to 10 days for the larvae to die, depending on the viral dose and environmental conditions.

It is most effective when applied against early larval instars (younger larvae) before they cause significant damage.

Since the virus is sensitive to UV radiation/sunlight, it is best sprayed in the late evening or during periods of low sunlight to maximize its effectiveness and persistence on the leaf surface.

Label claim (DPPQ & S, Faridabad)

Nuclear Polyhedrosis Virus of *Spodoptera litura* 0.5% AS, (1 x 10^9 POB/ml Min.)

Crop	Pests	Dose / ha		Waiting period (Days)
		Quantity of formulation (ml)	Volume	
Tobacco	Tobacco caterpillar	1500	400-600	--

Trade Names

Sr. No.	Trade Name	Company
1	Bio Kill-S	Ganesh Biocontrol System

PB Rope

The PB Rope (often sold as PB-Rope L or PB Knot) is an innovative and highly effective tool for managing the Pink Bollworm (*Pectinophora gossypiella*), a major pest of cotton. It is based on the principle of Mating Disruption, an environmentally friendly component of Integrated Pest Management (IPM).

Mechanism: Mating Disruption

The PB Rope is essentially a specialized, slow-release dispenser containing the synthetic version of the female pink bollworm's sex pheromone, called Gossyplure (ZZ/ZE-7,11-Hexadecadiene-1-yl-Acetate).

- **Over-Saturating the Air:** When applied in the cotton field, the ropes release a continuous, high concentration of Gossyplure into the air. Each rope releases a significantly higher dose of pheromone (around 158-160 mg per rope) than a single female moth (e.g., one commercial PB Knot may release 50 times the pheromone of a single female).
- **Confusing the Males:** This dense cloud of synthetic pheromone confuses the male moths (P. gossypiella). The males are unable to follow the subtle pheromone trail of actual female moths to locate them for mating.
- **Preventing Reproduction:** Since successful mating is severely disrupted or prevented entirely, the female moths lay unfertilized eggs. This dramatically reduces the next generation of larvae that cause boll damage, leading to a major suppression of the pest population over the season.

Label claim (DPPQ & S, Faridabad)

Crop	Pest	Dosage			PHI (Days)
		a.i. (mg/dispenser)	Pheromone dispensers	Recommended area of treatment (ha)	
Cotton	Pink bollworm	>140	9875	25	NA

Trade Names

Sr. No.	Trade Name	Company
1	PB Knot	PI Industries Ltd.

Timing of Application : It must be applied early, immediately before or at the start of moth emergence, typically at the pin square stage of the cotton plant (around 35-45 days after sowing).

Placement : Ropes are tied or twisted securely around the main stem or stakes near the bottom of the plant, avoiding contact with the soil. The ropes are applied uniformly across the field, often over larger areas (like 50 acres) for effectiveness.

Duration : A single application can provide effective disruption for approximately 60 to 120 days (2-4 months), depending on the specific product and environmental conditions. Some management strategies recommend a second application to cover the entire season.

Part X

Pests of Crops

Fibre Crops

Cotton

Aphid (*Aphis gossypii*)
Both nymphs and adults suck sap from the tender shoots and undersurface of leaves. Heavy infestation causes leaves to curl downwards and stunts plant growth. Excretes honeydew, which leads to the growth of sooty mold (black coating), hindering photosynthesis and contaminating lint.

Jassid (*Amrasca biguttula*)
Nymphs and adults suck sap from the undersurface of leaves and inject toxic saliva. This causes the leaves to turn pale yellow, with the margins curling downwards and turning red (a symptom called "hopperburn"). Severe damage leads to retarded growth.

Thrips (*Thrips tabaci*)
Nymphs and adults scrape the leaf surface (epidermis) and suck the oozing sap, causing a silvery or glazed shine on the undersurface. Leaves become wrinkled, distorted, and curl upwards. Severe damage stunts growth and causes the shedding of squares (flower buds).

Whitefly (*Bemisia tabaci*)
Nymphs and adults suck sap, primarily from the undersurface of the leaves. Causes irregular yellowing of leaves (chlorosis) and premature defoliation. It excretes honeydew, leading to sooty mold growth and, critically, is a vector for the devastating Cotton Leaf Curl Virus (CLCV) disease.

Mealybug (*Phenacoccus solenopsis*)
Colonies suck sap from all plant parts, including the stem, leaves, and bolls. Infestation causes stunting, distortion of shoots, curling of leaves, and development of a waxy, cottony mass. It also excretes honeydew, promoting sooty mold.

Spotted bollworm (*Earias vittella*)
Larvae first bore into the terminal shoot during the vegetative stage, causing it to dry up and droop. Later, they bore into squares (buds) and bolls, leading to shedding of fruiting parts and making bolls rot due to fungal infection. Bore holes are often covered with frass (excreta).

American bollworm (*Helicoverpa armigera*)
The larva is a voracious feeder that attacks squares, flowers, and bolls. It is known for its habit of thrusting its head alone inside the boll to feed while leaving the rest of its body outside. It creates large, regular, circular bore holes and can damage multiple bolls.

Pink bollworm (*Pectinophora gossypiella*)
Larvae bore into flower buds, causing "rosetted flowers" (petals failing to open normally). Their main damage is inside the boll, where they feed on the developing seeds and lint. They often join two adjoining seeds with silk, creating "double seeds". Damage is internal, leading to poor quality, stained lint.

Tobacco leaf eating caterpillar (*Spodoptera litura*)

Larvae are polyphagous leaf feeders. Young larvae initially skeletonize the leaves, while older larvae devour the entire leaf, causing severe defoliation. They can also attack bolls and squares.

Cotton leaf roller (*Sylepta derogata*)

The larva spins silken threads and rolls the leaf margins into a trumpet or funnel shape, feeding from within the roll. This reduces the effective photosynthetic area and can cause heavy defoliation in severe infestations.

Cotton green semilooper (*Anomis flava*)

Larvae are typically leaf feeders that move with a characteristic looping motion. They feed on leaves, often starting from the margin, and can cause significant defoliation, particularly to the younger foliage.

Grey weevil (*Myllocerus spp.*)

The adult weevils feed on the leaf margins, creating characteristic semi-circular notches or U-shaped cuts. The grubs (larvae) live in the soil and feed on the roots, which can sometimes cause wilting and weakened plants.

Red cotton bug (*Dysdercus cingulatus*)

Adults and nymphs suck the sap from the developing seeds inside open bolls. Their feeding reduces the viability and oil content of the seeds, significantly lowering the market value of the cotton fiber.

Dusky cotton bug (*Oxycaranus hyalinipennis*)

Adults and nymphs are primarily found on mature or bursting bolls. They suck the sap from the developing and ripe seeds, which causes the lint to be stained black or discoloured and the seeds to become shrunken and non-viable.

Sunhemp and Mesta

Sunhemp hairy caterpillar *Utethesia pulchella*

Young larvae scrape the leaf tissue, causing skeletonization. Older larvae are voracious feeders that consume the entire leaf, leading to severe defoliation of the plant. Larvae bore into developing pods (seed capsules) and feed on the seeds, which significantly reduces seed yield.

Sugarcane

Early shoot borer (*Chilo infuscatellus*)

The larvae bore into the base of young shoots (2-5 months old), destroying the growing point and causing the central leaves to dry up, forming a 'dead heart' (which is easily pulled out). This results in profuse tillering (bushy growth) of side shoots, but ultimately reduces the number of millable canes.

Internode borer (*Chilo saccarifagus indicus*)
The larvae bore into and feed within the internodes of the mature cane, creating tunnels. This leads to the formation of bore holes, often covered with frass, and weakens the cane, making it prone to breakage. Feeding also reduces the sucrose content and juice purity.
Top shoot borer (*Scirpophaga nivella*)
The larvae bore into the top tender portion of the main shoot and tunnel downwards. This kills the growing point and the central whorl of leaves, forming a 'dead heart' (which cannot be easily pulled out). Later, the top portions of the cane become bunchy (bunchy top or side-shoot formation) due to the sprouting of auxiliary buds.
Whitefly (*Aleurolobus barodensis*)
Nymphs and adults suck the sap from the undersurface of the leaves. This causes the leaves to turn yellow (chlorosis) and eventually dry up. They excrete honeydew, which favors the growth of sooty mold (black fungal growth), interfering with photosynthesis and reducing cane yield and quality.
Pyrilla (Leaf hopper) (*Pyrilla perpusilla*)
Nymphs and adults suck the sap from the undersurface of the leaves. This causes the leaves to turn pale-yellow and wilt. Like whiteflies, they excrete a large amount of honeydew, leading to the growth of sooty mold which severely impairs photosynthesis.
White wooly aphid (*Ceratovacuna lanigera, C. graminum*)
Colonies of this aphid suck sap from the undersurface of the leaves. They are characterized by a white, cottony, waxy coating. Heavy infestation leads to yellowing, drying, and premature death of leaves, significantly weakening the plant and reducing cane weight and sugar recovery.
Mealy bug (*Saccharicoccus sacchari*)
These soft, oval insects, covered in a waxy powder, suck sap primarily from the sheath area of the stalk, often hidden by the leaf sheath. Heavy infestation leads to stunted growth, shriveling of internodes, and a decrease in juice quality. They may also excrete honeydew.
Scale insect (*Melanaspis glomerata*)
The insects are sedentary and covered by a hard, brownish-black scale. They suck sap from the internodes under the leaf sheaths. Heavy infestation makes the cane shriveled, dry, and fibrous, reducing cane yield and sugar content drastically. The cane may look charred.
Termite (*Odontotermes obesus*)
Termites attack the setts (seed pieces) after planting, leading to poor germination, or they bore into the base of young shoots (similar to early shoot borer) causing a dead heart. They also feed on the roots and underground parts of mature canes, leading to wilting and drying up of the plant.
White grub (Root grub) (*Holotrichia serrata*)
The C-shaped grubs (larvae) feed voraciously on the roots and underground parts of the cane. This cuts off the nutrient and water supply, causing the plants to wilt, yellow, and die in patches. Affected canes are often easily pulled out as the roots are damaged.

Cereal Crops

Rice

Paddy stem borer (*Scirpophaga incertulas*)
Larvae bore into the tiller stem. In the vegetative stage, this causes a 'dead heart' (central shoot dries up, easily pulled out). In the reproductive stage, the panicle emerges white and empty, known as a 'white ear head'.
Green leaf hopper (*Nephotettix spp.*)
Nymphs and adults suck plant sap, leading to yellowing and reduced vigor. They are the crucial vectors for the devastating Rice Tungro Virus (RTV), which causes severe stunting and yellow-orange discoloration of leaves.
Brown plant hopper (*Nilaparvata lugens*)
Nymphs and adults suck sap vigorously from the base of the tillers. High populations cause plants to dry up rapidly in circular patches, a condition called 'Hopper Burn'. Also transmits grassy stunt virus.
White backed plant hopper (*Sogatella furcifera*)
Nymphs and adults suck sap, primarily from the leaf sheaths and lower stems. Heavy infestation causes the plants to turn yellow and eventually dry, similar to hopper burn.
Paddy leaf roller (*Cnaphalocrocis medinalis*)
Larvae fold the leaf blade lengthwise and feed inside by scraping the chlorophyll. This leaves white, transparent streaks and severely reduces the photosynthetic area.
Paddy gall fly (*Orseolia oryzae*)
The larva bores into the growing point, which is then transformed into a hollow, non-productive, tube-like structure known as a 'silver shoot' or 'onion shoot'.
Paddy blue beetle (*Leptispa pygmaea*)
Adults and grubs feed on the leaf's chlorophyll by scraping the tissue, resulting in white, parallel streaky patches on the leaves.
Rice hispa (*Dicladispa armigera*)
Adults scrape the upper leaf surface, creating white, parallel streaks. Grubs (larvae) are miners that tunnel inside the leaf, causing irregular, blotchy white patches that lead to the drying of the leaf tip.
Swarming caterpillar (*Spodoptera mauritia*)
Larvae are voracious leaf feeders that move in large groups, rapidly cutting the seedlings/leaves just above the water level, causing massive defoliation in nurseries and fields.
Rice caseworm (*Nymphula depunctalis*)
Larvae cut the leaves to form a tubular case and feed on the green tissue, creating ladder-like markings and white patches on the remaining leaves.

Gundhi bug (*Leptocorisa acuta*)

Nymphs and adults suck the milky juice from the developing grains (dough stage). This results in the affected grains being chaffy, shriveled, and discolored, significantly reducing both yield and quality.

Paddy grass hopper (*Hieroglyphus banian*)

Nymphs and adults are mainly leaf feeders, cutting the leaf blades and sometimes the young earheads. Large numbers can cause significant defoliation.

Sorghum (Jowar)

Jowar shoot fly (*Atherigona soccata*)

The larva enters the base of the plant and feeds on the growing point of young seedlings (within 2-4 weeks after germination). This causes the central shoot to dry up and die, forming a characteristic 'dead heart' (easily pulled out), leading to compensatory tillering.

Jowar stem borer (*Chilo partellus*)

Larvae bore into the stem of older plants. Similar to the shoot fly, they cause 'dead hearts' in the early stage. In later stages, feeding leads to shot-hole symptoms on the unfurled leaves and the development of tunneling inside the stem, weakening the plant and causing lodging.

Jowar earhead midge (midge fly) (*Contarinia sorghicola*)

The larvae (maggots) feed by sucking the sap from the developing ovary of the individual flower inside the glumes before the grain is formed. This prevents grain formation, resulting in the earhead having a characteristic blighted appearance with many empty or chaffy spikelets.

Earhead caterpillar (*Helicoverpa armigera*)

Larvae feed on the developing grains within the earhead. They cause direct loss by consuming the grain and also contaminate the earhead with their large amounts of excreta (frass), promoting secondary fungal infections. They are voracious and polyphagous feeders.

Webworm (*Cryptoblabes angustipennella*)

Larvae feed on the developing grains and spin silken threads to form a web that holds the grains, flowers, and frass together, damaging the grain and making the earhead difficult to thresh.

Aphid (*Rhopalosiphum maidis*)

Colonies suck sap from the leaves, especially the funnel leaves and earheads. Heavy infestation causes yellowing, distortion of leaves, and stunting. They excrete sticky honeydew, which favors the growth of sooty mold, interfering with photosynthesis.

Delphacid (*Peregrinus maidis*)

Nymphs and adults suck the sap from the leaves and stem, causing yellowing and weakening. They are known to transmit Maize Mosaic Virus and Maize Stripe Virus, which also affect sorghum.

Surface grasshopper (*Chrotogonus spp.*)

Both nymphs and adults are polyphagous leaf feeders. They chew the leaves from the edges, causing ragged margins or entire defoliation, particularly on young seedlings.

Maize

Shoot fly (*Atherigona soccata*)
The larva enters the base of the plant and feeds on the growing point of young seedlings (within 2-4 weeks after germination). This results in the death and drying of the central shoot, forming a 'dead heart' (easily pulled out). This often leads to excessive, unproductive tillering.
Stem borer (*Chilo partellus*)
Larvae bore into the stem of the plant. In the early stage, they cause 'dead hearts'. Later, feeding on the developing tissues within the whorl results in characteristic 'shot-hole' symptoms on the leaves when they unfurl. Extensive tunneling in the stem weakens the plant, leading to lodging and reduced nutrient flow.
Fall armyworm (*Spodoptera frugiperda*)
Larvae are highly damaging, attacking all stages, but are most destructive when feeding deep inside the leaf whorl of young plants. Their feeding causes large, irregular ragged holes and deposits large amounts of wet, granular frass (excreta) in the whorl. In older plants, they attack the tassels, kernels, and cobs.
Armyworm (*Mythimna separata*)
Larvae feed gregariously, initially skeletonizing the leaves and then consuming the entire leaf blade, causing severe defoliation. They are known for their habit of marching from field to field in large groups when food is scarce.
Cob earworm (*Helicoverpa armigera*)
The larva attacks the tassel and then moves to feed primarily on the developing silks and kernels inside the ear (cob). The feeding damage provides entry points for molds and fungi, causing grain rot and contamination. It is often found feeding on the tip of the cob.

Pearl millet (Bajra)

Shoot fly (*Atherigona soccata*)
The larva enters the base of young seedlings and feeds on the growing point (within 2-4 weeks after germination). This results in the death and drying of the central shoot, forming a 'dead heart' (easily pulled out). This damage causes stunted growth and leads to the production of unproductive tillers.
Blister beetle (*Zonabris pustulata*)
Adult beetles are highly destructive to the earhead (panicle) during the flowering and grain-setting stage. They feed voraciously on the pollen, stigma, and developing kernels, resulting in poor grain setting and a significant loss of yield. The body fluid of the crushed beetle can cause blisters on human skin.

Wheat

Wheat stem borer (*Sesamia inferens*)
The larva bores into the stem of the wheat plant. In the vegetative stage, this causes the central shoot to dry up, forming a 'dead heart' (similar to rice/sorghum dead heart). In the booting/reproductive stage, it causes the earhead to emerge as a white, chaffy ear (similar to the white earhead in rice) because the nutrient supply is cut off.

Aphid (*Rhopalosiphum maidis*)
Colonies of nymphs and adults suck sap from the tender leaves, stem, and developing earheads. Heavy infestation causes yellowing, reduced vigor, and stunting. They excrete honeydew, which leads to the growth of sooty mold, fouling the leaves and grains, thus reducing both yield and quality.

Termite (*Odontotermes obesus*)
Termites attack the wheat crop at two main stages: they consume the sown seeds, leading to poor and patchy germination; and they feed on the roots and base of the stem of older plants. This causes the plants to dry up and die, resulting in a thin stand and reduced yield.

Minor millets

Shoot fly (*Atherigona soccata*)
The larva attacks young seedlings (within 2-4 weeks of germination). It enters the base of the plant and destroys the growing point, causing the central shoot to dry up and die, forming a 'dead heart' (easily pulled out). This leads to compensatory but often unproductive tillering.

Wheat / Ragi / Pink stem borer (*Sesamia inferens*)
The larva bores into the stem of the plant. In the vegetative stage, this results in the formation of a 'dead heart'. In the reproductive stage, it cuts off the nutrient supply, causing the earhead (panicle) to emerge as a white, chaffy ear that contains no grain.

Spotted stem borer (*Chilo partellus*)
Larvae bore into the stem, causing 'dead hearts' in younger plants. In older plants, feeding on the unfurled leaves creates characteristic 'shot-hole' damage. Extensive tunneling within the stem weakens the plant, leading to lodging and reduced grain filling.

White stem borer (*Saluria inficita*)
The larvae bore into the stem and feed internally, causing damage that is primarily visible as stunted growth and the formation of 'dead hearts'. It is a common stem borer in millets like Finger Millet (Ragi).

Leaf / Plum / Ragi aphid (*Hysteroneura setariae*)
Colonies suck sap from the leaves, leaf sheaths, and developing earheads. Heavy feeding causes yellowing (chlorosis), leaf distortion, and reduced plant vigor. They excrete sticky honeydew, which promotes the growth of black sooty mold, hindering photosynthesis.

Ragi root aphid (*Tetraneura nigriabdominalis*)
These aphids colonize the roots underground and suck the sap. This root feeding inhibits the uptake of water and nutrients, leading to a visible yellowing, wilting, and stunting of the entire plant. Severe infestation can cause plants to die prematurely.

Pulse Crops

Pigeonpea (Tur / Red gram)

Pod borer (*Helicoverpa armigera*)
The larva is a highly destructive feeder that attacks flowers, buds, and primarily, the pods. It creates a circular bore hole in the pod and thrusts only its head inside to feed on the developing seeds, leaving the rest of its body outside. This causes direct seed loss and secondary fungal infection (rot).
Tur plume moth (*Exelastis atomosa*)
The minute larva bores into the tender pods and feeds on the seeds inside. Damage is visible as a small pin-prick hole on the pod surface. It also feeds on buds and flowers, often webbing them together before boring, causing them to shed.
Tur pod fly (*Melanagromyza obtusa*)
The adult fly lays eggs on the pods, and the developing maggot (larva) tunnels into the seed. The damage is internal, as the maggot feeds inside the seed. The pod appears normal from the outside, but the seeds are destroyed, leaving behind a brown pupal case.
Spotted pod borer (*Maruca testulalis*)
Larvae bore into the buds and flowers and web them together with silken threads. They then bore into the pods and feed on the seeds. The characteristic sign is the formation of a silken web that binds flowers, buds, and pods together with frass (excreta).
Leaf webber (*Grapholita critica*)
Larvae web together the terminal leaves and tender shoots using silken threads and feed inside the protective structure. This causes the leaves to dry and the growing point to be damaged, leading to the drying and shedding of terminal shoots, which affects plant development.
Mite (*Aceria cajani*)
This is a tiny eriophyid mite that sucks sap from the tender growth points and leaves. Its feeding causes characteristic damage known as "sterility mosaic disease" symptoms (though it is a vector for the virus), or direct symptoms like leaf curling, malformation, and stunting of the terminal shoots.

Chickpea (Gram, Bengal gram)

Pod borer (*Helicoverpa armigera*)
The larva is the most destructive pest. It attacks the tender foliage in early stages, but its main damage is to the buds, flowers, and developing pods. It creates a circular bore hole in the pod

and feeds by thrusting only its head inside, leaving the rest of the body exposed. This causes direct seed loss and often leads to secondary fungal infections (pod rot).

Aphid (*Aphis craccivora*)

Colonies of nymphs and adults suck sap from the tender leaves, stems, and flowering parts. Heavy feeding causes the plants to become weak, stunted, and yellow (chlorotic). They excrete sticky honeydew, which promotes the growth of black sooty mold, interfering with photosynthesis. They can also act as vectors for virus diseases.

Cutworm (*Agrotis ipsilon*)

The dark-colored larvae are primarily active at night. They live in the soil during the day and come out at night to cut the young seedlings (or occasionally older plants) at the ground level (just above the soil surface), causing the plant to fall over. They can cause severe damage in patchy areas of the field shortly after germination.

Green gram (Mung) and Black gram (Udid)

Aphid (*Aphis craccivora*)

Nymphs and adults suck sap from tender leaves, stems, and flowering parts. Heavy feeding causes yellowing (chlorosis), stunting, and reduced vigor. They excrete sticky honeydew, leading to the growth of black sooty mold. Most importantly, they are key vectors for the devastating Mungbean Yellow Mosaic Virus (MYMV).

Pod borer (*Helicoverpa armigera*)

The larva is a serious pest, feeding on flowers, buds, and pods. It creates a bore hole in the pod and thrusts its head inside to feed on the developing seeds, leaving the rest of its body exposed. This causes direct seed loss and often leads to pod rot due due to secondary fungal infections.

Leaf eating caterpillar (*Agrius convolvuli*)

The large larva is an occasional pest that causes severe defoliation by consuming the entire leaf lamina. It can quickly strip the plants of their leaves, especially during the vegetative stage, which severely impacts the plant's ability to photosynthesize.

Semilooper (*Anomis flava*)

The larva is a leaf feeder that moves with a characteristic looping motion. It feeds on the leaves, causing defoliation. While polyphagous, in pulses like Mung/Udid, it can cause significant feeding damage during periods of high population.

Pea

Gram pod borer (*Helicoverpa armigera*)

The caterpillar (larva) is the most destructive stage. It feeds on the tender foliage and flowers, but its major damage is to the pods. It bores into the developing pea pods and feeds on the seeds inside. The caterpillar often keeps its head and thorax inside the pod and the rest of its body outside. The entry hole allows for secondary infection, causing the seeds to be damaged, rotten, and unfit for consumption.

Stem fly (*Ophiomyia phaseoli*)
The maggot (larva) bores into the stem of the young plant, tunneling downwards towards the root. This internal feeding disrupts the flow of water and nutrients. Infested plants show stunting, yellowing, and wilting, often leading to the death of the seedling (particularly at the unifoliate and primary leaf stages).
Pea leaf miner (*Chromatomyia horticola*)
The larvae (maggots) tunnel and feed within the leaf tissue between the epidermal layers. This feeding creates characteristic long, whitish, serpentine (winding) mines on the leaves. Heavy mining reduces the photosynthetic area, leading to reduced plant vigor and affecting the pod setting and final yield.
Pea aphid (*Acyrthosiphum pisum*)
These are sap-sucking pests that form dense colonies on tender shoots, leaves, flowers, and developing pods. Their feeding causes the leaves to curl, crinkle, and turn yellow. Heavy infestation leads to stunting, poor flower set, and shriveled, underdeveloped pods. They also excrete honeydew and are known vectors for several pea viruses.

Oilseed Crops

Soybean

Stem fly (*Melanagromyza sojae*)
The larva (maggot) bores into the stem and tunnels down the pith. This causes the central shoot of the young plant to dry up, forming a 'dead heart'. In older plants, the tunneling interferes with the transportation of nutrients and water, causing stunting and reduced pod formation.
Girdle beetle (*Obereopsis brevis*)
The adult female makes two parallel girdles (rings) on the stem or petiole and lays an egg in between. The part above the girdle dries up and droops. The larva then bores into the stem and tunnels downwards, causing the plant to wilt and die.
Leaf miner (*Aproaerema modicella*)
Young larvae tunnel (mine) into the leaf tissue, feeding on the mesophyll. This appears as short, brownish blotches or blister-like mines on the leaflets. Older larvae emerge from the mines, web the leaflets together, and continue feeding within the folds. This causes the leaves to roll, shrivel, dry up, and eventually drop off.
Tobacco leaf eating caterpillar (*Spodoptera litura*)
The larvae are voracious polyphagous feeders. Young larvae initially skeletonize the leaves, while older larvae consume the entire leaf lamina, leading to severe defoliation. They are known to move in masses, quickly destroying foliage.
Semilooper (*Gesonia gema*)
The larva feeds on the leaves, causing defoliation. It is characterized by its looping movement (measuring worm action) when crawling. Heavy infestation during the flowering and pod formation stage can significantly reduce yield.

Gram pod borer (*Helicoverpa armigera*)
Larvae feed on the tender foliage in early stages, but their major damage is to the pods. They bore into the pod and feed on the developing seeds inside. The entry holes allow moisture and pathogens to enter, often leading to pod rot.
Whitefly (*Bemisia tabaci*)
Nymphs and adults suck sap from the undersurface of the leaves, causing yellowing and weakening. Crucially, it is the vector for the devastating Soybean Mosaic Virus (SMV) and can also transmit Mungbean Yellow Mosaic Virus (MYMV), causing severe yellowing and stunting.

Groundnut

Leaf miner (*Aproaerema modicella*)
Young larvae tunnel (mine) into the leaf tissue, feeding on the mesophyll. This appears as short, brownish blotches or blister-like mines on the leaflets. Older larvae emerge from the mines, web the leaflets together, and continue feeding within the folds. This causes the leaves to roll, shrivel, dry up, and eventually drop off.
Red hairy caterpillar (*Amsacta spp.*)
The hairy larvae are gregarious (feed in groups) and highly polyphagous leaf feeders. They cause severe defoliation, often completely stripping the plants bare, especially in young crops. They march from field to field, causing widespread devastation.
Bihar hairy caterpillar (*Spilosoma obliqua*)
Similar to the red hairy caterpillar, the larvae feed gregariously, skeletonizing the leaves initially and later consuming the entire lamina, causing severe defoliation. They are a major threat, particularly during the vegetative stage.
Tobacco leaf eating caterpillar (*Spodoptera litura*)
The larvae are nocturnal and feed voraciously on the leaves, causing severe defoliation by eating large, irregular holes in the leaf lamina. They can also attack buds and flowers.
Aphid (*Aphis craccivora*)
Nymphs and adults suck sap from tender parts, stems, and growing points. Heavy feeding causes stunting and yellowing. Most importantly, it is the primary vector for the severe Peanut Mottle Virus (PMV) and Peanut Clump Virus (PCV), leading to widespread disease transmission.
Thrips (*Scirtothrips dorsalis, Thrips palmi*)
Nymphs and adults scrape the leaf surface (epidermis) and suck the sap. Their feeding causes the leaves to become crinkled, wrinkled, and bronzed (scab-like appearance), severely affecting growth and flower formation. They can also transmit Groundnut Bud Necrosis Virus (GBNV).
Pod sucking bug (*Elasmolomus sordidus*)
Adults and nymphs pierce the developing pods and suck the juice from the seeds inside. This causes the kernels to become shriveled, poor in quality, and reduced in oil content. The damage is often concentrated in pods lying near the soil surface.

White grub (*Holotrichia serrata*)
The C-shaped grubs (larvae) live in the soil and feed voraciously on the roots and developing pods (pegs and nuts). This damages the pod, prevents proper pod development, and causes the plants to wilt, yellow, and die in patches. Affected plants are easily pulled out due to root damage.

Sunflower

Capitulum borer / Gram pod borer (*Helicoverpa armigera*)
The larva feeds on the flower buds, florets, and developing seeds within the capitulum (flower head). It bores into the head and feeds on the seeds, causing direct yield loss and creating entry points for secondary fungal and bacterial infections, leading to head rot.
Hairy caterpillar (*Spilosoma obliqua*)
The larvae are voracious leaf feeders. They initially feed gregariously, skeletonizing the leaves, and later spread out to consume the entire leaf lamina, leading to severe defoliation of the plant, which severely impacts seed filling.
Jassid (*Empoasca devastans*)
Nymphs and adults suck sap from the undersurface of the leaves and inject toxic saliva. This causes the leaves to curl downwards and turn yellow, especially along the margins (a symptom known as "hopperburn"). Severe damage leads to the drying and shedding of leaves.
Thrips (*Scirtothrips dorsalis, Thrips palmi*)
Nymphs and adults scrape the leaf surface (epidermis) and suck the sap. This causes the leaves to appear crinkled, wrinkled, and scarred or silvered, especially on tender parts, leading to distorted growth.
Whitefly (*Bemisia tabaci*)
Nymphs and adults suck sap, leading to yellowing and reduced vigor. They also excrete honeydew, which promotes the growth of black sooty mold, hindering photosynthesis.
Stem borer (*Nupserha bicolor*)
The adult female makes girdles (rings) on the stem, and the larva bores into the stem and tunnels down the pith. This weakens the stem, leading to wilting, lodging, and breakage of the plant, especially near the damaged area.

Safflower

Aphid (*Uroleucon compositae*)
Colonies of nymphs and adults suck sap from tender shoots, stem, and especially the developing capitula (flower heads). Heavy infestation causes severe stunting of the plant, shriveling of the flower head, and prevents proper seed development, leading to significant yield loss. The aphid is often found covering the entire stem and head.

Capitulum borer / Gram pod borer (*Helicoverpa armigera*)
The larva feeds on the developing seeds inside the capitulum (flower head). It bores into the head and feeds internally, causing direct seed loss. The resulting damage creates entry points for moisture and pathogens, leading to head rot and reduced oil quality.
Gujhia weevil (*Tanymecus indicus*)
The adult weevils are particularly damaging to young seedlings. They feed on the tender leaves and cut the young plants at the ground level (collar region) during the day. This leads to poor plant stand and necessitates replanting. In older plants, they chew the leaves and cause irregular notching.

Mustard

Mustard aphid (*Lipaphis erysimi*)
Colonies of nymphs and adults suck the sap from tender leaves, stems, flower buds, and young pods (siliquae). Heavy infestation causes the plant to become stunted, leaves to curl and turn yellow, and the pods to shrivel and fail to develop seeds. They also excrete honeydew, leading to sooty mold growth.
Mustard sawfly (*Athalia proxima*)
The larva (false caterpillar) is the main damaging stage. It feeds voraciously on the leaves of young plants, consuming them from the margins inwards and leaving behind only the veins (skeletonization). They can cause severe defoliation, leading to seedling death or severe stunting.
Leaf webber (*Crocidolomia binotalis*)
The larvae feed on the leaves, shoots, and flower buds. They web the terminal leaves and inflorescence (flower cluster) together with silken threads, feeding protected inside the web. This causes the death of the growing point, leading to the drying and dropping of the terminal shoot and flowers.

Castor

Castor semilooper (*Achoea janata*)
The larva is the main damaging stage, known for its looping movement when crawling. It is a voracious feeder that causes severe defoliation, consuming the entire leaf lamina. When population density is high, it can quickly strip the plants, leading to huge yield losses.
Castor capsule borer (*Dichocrosis punctiferalis*)
The larva bores into the tender stem, shoots, and inflorescence, but its main damage is to the developing capsules (seed pods). It bores into the capsule and feeds on the seeds inside, often moving from one capsule to another. This causes direct seed loss and secondary capsule rot.

Tobacco leaf eating caterpillar (*Spodoptera litura*)

The larvae are highly polyphagous and cause severe defoliation by consuming the leaves, leaving behind only the major veins in severe cases. They are active at night and hide in the soil or debris during the day.

Castor jassid (*Empoasca distinguenda*)

Nymphs and adults suck the sap from the undersurface of the leaves and inject toxic saliva. This causes the leaf margins to curl downwards and turn yellow or reddish (a symptom known as "hopperburn"). Severe damage leads to the drying and premature shedding of leaves.

Sesamum

Til hawk moth (*Acherontia styx*)

The large larvae (caterpillars) are voracious, leaf-eating feeders. They consume the entire leaf lamina, leading to severe defoliation of the plant. They are typically spotted due to their size and characteristic anal horn.

Gall fly (*Asphondylia sesami*)

The adult fly lays eggs in the tender flower buds. The developing maggot (larva) feeds inside, causing the bud to swell and turn into a characteristic gall structure. This prevents the formation of normal flowers and capsules (seed pods), leading to significant loss of seed yield.

Leaf webber (*Antigastra catalaunalis*)

The larvae are the main damaging stage. They web together the leaves and terminal shoots with silk threads and feed inside the protective web. They also bore into the tender capsules. This feeding damages the growing point, causes stunting, and leads to the shedding of leaves, flowers, and capsules, severely impacting yield.

Linseed

Gall fly (*Dasineura lini*)

The adult fly lays eggs in the tender flower buds. The developing maggot (larva) feeds inside the bud, which prevents the bud from opening normally. Instead, the bud swells up abnormally to form a characteristic fleshy gall. This inhibits the formation of flowers and capsules (seed pods), leading to significant direct loss of seed yield.

Niger

Semilooper (*Thysanoplusia orichalcea*)

The larva is a leaf feeder characterized by its looping movement (measuring worm action) when crawling. It consumes the leaf tissue, causing defoliation. Heavy infestation during the flowering and seed-filling stages significantly impacts the plant's photosynthetic capacity and reduces seed yield.

Gram pod borer (*Helicoverpa armigera*)
The larva attacks the tender terminal shoots, flowers, and developing capitula (flower heads). It feeds on the florets and young seeds inside the head, causing direct seed loss. Damage to the growing point can also lead to stunting and reduced branching.

Vegetable Crops

Brinjal

Brinjal shoot & fruit borer (*Leucinodes orbonalis)*
Larvae bore into terminal shoots causing wilted/drooping shoots ('dead heart') and bore into fruits, making them unfit for consumption (holes plugged with frass/excreta).
Jassids/leaf hopper (*Amrasca biguttula biguttula)*
Nymphs and adults are sap-suckers on the lower leaf surface, causing yellowing of leaf margins, curling of leaves, and reduced plant vigor.
Aphids (*Aphis gossypii, Myzus persicae)*
Sap-suckers that cluster on tender shoots and leaf undersides, causing yellowing, stunted growth, and secreting honeydew which leads to the growth of sooty mold.
Whitefly (*Bemisia tabaci)*
Nymphs and adults are sap-suckers (mainly on the underside of leaves), causing yellowing, wilting, reduced vigor, and acting as a vector for viral diseases (e.g., Little Leaf, if present).
Hadda Beetle (*Epilachna vigintioctopunctata)*
Grubs and adults scrape the chlorophyll from the leaf surface (skeletonizing the leaves), leaving a net-like or lace-like appearance.
Brinjal leaf roller (*Eublemma olivacea)*
Larvae roll or fold the leaves and feed inside the fold by scraping the green matter, leading to skeletonization and reduced photosynthetic area.
Lacewing bug (*Urentius centis)*
Nymphs and adults suck sap from the leaves, resulting in yellowish patches (often near the midrib) and general loss of plant vitality.
Stem borer (*Euzophera perticella)*
Larvae bore into the stem (often near ground level or in the branching area), causing the top shoots to droop/wither, and leading to stunted growth in older plants.
Red Spider Mite (*Tetranychus telarius)*
Mites feed on the underside of leaves, causing stippling/bronzing of leaves and the formation of fine webbing, leading to premature leaf drop and shriveling.

Okra

Shoot & fruit borer (*Earias vitella / E. insulana)*
Larvae bore into terminal shoots during the vegetative stage, causing them to wither and droop ('dead heart'). Later, they bore into fruits, making them deformed and unfit for sale/consumption.

Fruit borer (*Helicoverpa armigera)*
Larger caterpillars bore circular holes into fruits and feed inside, often leaving part of their body outside. They also feed on flowers and developing buds.
Jassid / Leafhoppers (*Amrasca bigutulla bigutulla)*
Nymphs and adults are sap-suckers on the lower leaf surface. Their feeding injects toxic saliva, causing yellowing of leaf margins, downward curling, and a typical "hopper burn" symptom where leaves turn brown, brittle, and crumble.
Aphid (*Aphis gossypii)*
Sap-suckers that colonize tender shoots and leaf undersides. They cause curling, crinkling, and stunting of growth. They excrete honeydew, leading to the growth of sooty mold, which interferes with photosynthesis.
Whitefly (*Bemisia tabaci)*
Nymphs and adults are sap-suckers, causing yellowing and reduced plant vigor. Most critically, they are the primary vector for the devastating Okra Yellow Vein Mosaic Virus (OYVMV).
Leaf roller (*Haritalodes (=Sylepta) derogata)*
Larvae roll the leaves and feed from within the fold by scraping the chlorophyll (skeletonizing the leaf tissue), reducing the photosynthetic area and causing defoliation.
Leaf miner (*Liriomyza trifoli)*
Larvae tunnel within the leaf tissues, creating serpentine mines (irregular, winding white/light green trails). High infestation reduces photosynthesis and causes premature leaf drop.
Flea beetle (*Podagrica spp.)*
Adults chew small, round holes or shot-holes in the leaves, especially on young seedlings, which can severely defoliate and stunt the plant.
Red spider mite (*Tetranychus spp.)*
Mites suck the cell sap, causing fine white/yellow speckles or stippling on the leaves. In severe cases, they cause a bronzing effect and create fine webbing on the leaf underside and growing points, leading to shrivelling and leaf drop.

Tomato

Fruit borer (*Helicoverpa armigera)*
Larvae feed on leaves and flower buds, but primarily cause direct loss by boring large, circular holes into green and ripe fruits. They thrust only the front half of their body inside, making the fruit unfit for consumption and prone to secondary rot.
Serpentine leaf miner (*Liriomyza trifolii)*
Larvae tunnel within the leaf mesophyll, creating long, winding, whitish trails (serpentine mines). High infestation leads to severe defoliation, reduced photosynthesis, and weakened plants.
Tomato pinworm (*Tuta absoluta)*
Highly destructive larvae cause blotch mines in leaves, bore into stems and apical buds (destroying the growing point), and create pinholes in the fruit (often near the calyx), tunneling inside and causing fruit decay.

Tobacco leaf eating caterpillar (*Spodoptera litura)*
Larvae are voracious leaf feeders, causing irregular holes or complete defoliation (skeletonization) under heavy infestation. Older larvae sometimes feed on the fruit surface.
Aphids (*Aphis gossypii, Myzus persicae)*
Sap-suckers that cluster on young leaves and shoots, causing yellowing, curling, and stunting. They secrete sticky honeydew, leading to the growth of black sooty mold, which hinders photosynthesis.
Thrips (*Thrips tabaci, Caliothrips indicus)*
Sap-suckers that scrape the leaf surface and suck exuding sap, causing silvery or bronzed patches and stippling on leaves, buds, and fruits. They are critical as vectors for the Tomato Spotted Wilt Virus (TSWV).
Whitefly (*Bemisia tabaci)*
Nymphs and adults are sap-suckers on the lower leaf surface, causing yellowing and stunted growth. Most importantly, they are the vector for the highly damaging Tomato Leaf Curl Virus (ToLCV), which causes severe leaf curling and plant stunting.
Mites *Tetranychus spp.* (Spider Mite), *Aceria spp.* (Russet Mite)
Spider Mites cause stippling, bronzing, and yellowing of leaves, often accompanied by fine webbing. Russet Mites cause the epidermis of stems and fruits to turn brown/russet and leathery, leading to fruit cracking and shrivelling.

Chilli and Capsicum

Thrips (*Scirtothrips dorsalis)*
Sap-suckers on tender leaves and shoots. Feeding causes leaves to become crinkled and curl upwards (upward curling is a key symptom). Infestation leads to stunted growth and buds becoming brittle and dropping off.
Fruit borer (*Helicoverpa armigera)*
Larvae bore circular holes into developing fruits, eating the internal contents, and often leaving their head inside and body outside. This renders the fruits unmarketable and susceptible to rot.
Tobacco leaf eating caterpillar (*Spodoptera litura)*
Young larvae scrape the leaf surface (skeletonizing it). Older, larger larvae are voracious feeders that cause irregular holes or complete defoliation of leaves and may bore into or scrape the fruit surface.
Whitefly (*Bemisia tabaci)*
Sap-suckers on the underside of leaves, causing yellowing and weakening the plant. Their main significance is as the vector for the highly destructive Chilli Leaf Curl Virus (ChiLCV).
Aphids (*Aphis gossypii)*
Sap-suckers that cluster on tender parts, causing curling, crinkling, and stunting. They excrete honeydew, which promotes the growth of black sooty mold, inhibiting photosynthesis.
Mite (Broad/Yellow Mite) (*Polyphagotarsonemus latus)*
Mites suck sap from the lower surface of tender leaves and growing points. This causes leaves to become leathery, thickened, and curl downwards (downward curling is a key symptom, sometimes called 'rat tail' or 'Murda' disease). Plants become severely stunted.

Potato

Potato tuber moth (*Phthorimaea operculella)*
Larvae mine the leaves, stems, and petioles causing irregular galleries. Most significant damage is to the tubers (in the field or in storage), where larvae bore tunnels and galleries, rendering them unfit for consumption and prone to secondary rot.

Cutworm (*Agrotis ipsilon)*
Older larvae are nocturnal feeders that cut off young plants at the stem (at or near ground level). Later in the season, they feed on exposed or shallow tubers, creating gouged out cavities or holes.

Tobacco leaf eating caterpillar (*Spodoptera litura)*
Young larvae scrape the chlorophyll (skeletonizing the leaves) while older, gregarious larvae are voracious feeders that can cause severe to complete defoliation of the foliage.

Thrips (*Thrips sp.)*
Nymphs and adults scrape the leaf epidermis and suck the oozing sap, resulting in silvery or yellowish streaks/stippling on the leaves, leading to bronzing and sometimes stunted growth of the plant.

Jassid / Leafhopper (*Amrasca devastans)*
Sap-suckers on the underside of leaves. Their feeding injects toxic saliva, causing leaves to turn yellow, develop downward curling margins, and eventually show a characteristic bronzing symptom known as "hopper burn."

Mite (Broad Mite) (*Polyphagotarsonemus latus)*
Mites suck sap from tender leaves and growing points, causing leaves to become thickened, curled downwards, and have a bronzed or leathery appearance (sometimes called 'mite burn'). Infestation leads to stunted growth and weakened plants.

Sweet Potato

Sweet potato weevil (*Cylas formicarius)*
Larvae bore tunnels and galleries packed with frass (excrement) into the storage roots (tubers), stems, and vines. Tubers become spongy, dark, and bitter/inedible due to the plant producing terpenes in response to the damage. This is the most destructive pest.

Sweet potato leaf eating caterpillar / Sphinx caterpillar *Agrius (=Herse) convolvulii*
Larvae are voracious leaf feeders that cause large, irregular holes in the leaves. They often eat the entire leaf blade, leaving only the main veins and the leaf stalk (petiole). Severe infestation, especially on young plants, can lead to complete defoliation of the field, which reduces the plant's photosynthetic ability and results in reduced yield (smaller tubers).

Cruciferous Vegetables (Cabbage, Cauliflower, Brocolli)

Diamondback moth (*Plutella xylostella)*
Larvae (caterpillars) initially mine the leaves. Later, they feed by scraping the underside of the leaves, leaving the upper epidermis intact, which creates a characteristic "window-panning" effect . Older larvae bite small holes in the leaves and may bore into cauliflower curds or cabbage heads, contaminating the produce.

Aphid (*Brevicoryne brassicae)*
Sap-suckers that cluster heavily on the underside of leaves and tender shoots. They cause yellowing, curling, and distortion of leaves. They secrete honeydew, leading to the growth of sooty mold, which covers the foliage and hinders photosynthesis.
Painted bug (*Nezara virudulla, Bagrada cruciferarum)*
Nymphs and adults are sap-suckers. On seedlings, heavy feeding can cause wilting and withering of young plants. On older plants, feeding leaves white, papery patches or scars on the leaves and can spoil developing seed pods.
Cabbage butterfly (*Pieris brassicae)*
Larvae (caterpillars) are voracious feeders that create large, irregular, ragged holes in the leaves (defoliation). They can bore into the heads of cabbage or cauliflower, contaminating the produce with frass (excreta) and making it unmarketable.
Leaf eating caterpillar (*Spodoptera litura)*
Larvae are polyphagous leaf feeders. Young larvae skeletonize leaves by scraping the surface; older larvae chew large, irregular holes in the foliage, leading to severe defoliation under heavy infestation.
Head borer (*Hellula undalis)*
Larvae web the tender leaves near the growing point together and feed from within, often boring into the terminal buds or growing points of young plants, preventing head formation. This is particularly damaging to cabbage and cauliflower seedlings.
Mustard sawfly (*Athalia proxima)*
Larvae (grubs) are voracious feeders that cause numerous shot holes or irregular holes in the leaves, often feeding from the margins inward. Heavy infestation, especially at the seedling stage, can lead to complete defoliation and seedling death.
Flea beetle (*Phyllotreta cruciferae)*
Adults chew small, rounded holes or pits (resembling shot-holes) in the leaves. This damage is most severe on young seedlings, where heavy feeding can stunt growth or completely destroy the plant.

Cucurbitaceous Vegetables (Cucumber, Pumpkin, Bitter guard, Ridge guard)

Fruit Flies (*Bactrocera cucurbitae, Dacus ciliatus, B. dorsalis)*
Adult females pierce young, tender fruits (or flowers/stems) with their ovipositor to lay eggs inside. The resulting maggots feed on the fruit pulp, causing the fruit to rot, become distorted, or drop prematurely. The puncture site often shows oozing resinous fluid and secondary fungal/bacterial infection.
Aphid (*Aphis gossypii)*
Sap-suckers that cluster heavily on young leaves and shoots. They cause yellowing, stunting, and deformation of leaves. They secrete honeydew, which facilitates the growth of sooty mold, interfering with photosynthesis. They can also act as virus vectors.
Leaf miner (*Liriomyza trifolii)*
Larvae tunnel within the leaf tissue, creating serpentine (winding, irregular) white/light trails known as mines. Heavy infestation reduces the plant's photosynthetic capacity, leading to premature leaf drop and reduced yield.

Whitefly (*Bemisia tabaci)*
Nymphs and adults suck sap (mainly from the underside of leaves), leading to yellowing, mottling, and general plant weakening. They are also known vectors for viral diseases that can affect cucurbits.
Thrips (*Thrips palmi)*
Sap-suckers that scrape the leaf surface, causing silvery or bronzed patches and stippling. Heavy feeding on young growth can lead to deformation and stunting. They are also known vectors for certain viruses like Tomato Spotted Wilt Virus (TSWV).
Pumpkin beetle (*Raphidopalpa foveicollis)*
Adults are highly destructive, especially to seedlings, by voraciously feeding on cotyledons and young leaves, creating irregular holes and sometimes consuming the entire plant, necessitating re-sowing. Grubs (larvae) feed on roots, underground stems, and fruits touching the soil, causing the plant to wilt.
Blister beetle (*Mylabris pustulata, M. phalerata)*
Adults are highly destructive, appearing in swarms and voraciously feeding on flowers and flower buds. This feeding destroys the reproductive parts, leading to poor pollination and a failure of fruit formation (no pods/fruit set).

Colocasia

Leaf eating caterpillar / Sphinx caterpillar (*Agrius convolvuli)*
Larvae are large, voracious feeders that consume the entire leaf blade, leaving only the main veins and the petiole. They cause severe defoliation, which hampers photosynthesis and reduces the corm yield (tubers).
Tobacco leaf eating caterpillar (*Spodoptera litura)*
Young larvae feed gregariously by scraping the chlorophyll (skeletonizing the leaves). Older larvae spread out and create large, irregular holes or completely devour the foliage, leading to significant defoliation and poor plant growth.
Cotton aphid (*Aphis gossypii)*
Sap-suckers that colonize the underside of leaves and tender shoots. Their feeding causes yellowing, curling, and distortion of the leaves. They excrete sticky honeydew, which leads to the growth of black sooty mold, reducing photosynthetic efficiency.
Banana aphid (*Pentalonia nigronervosa)*
Sap-suckers primarily found in the lower region of the leaf along the midrib and on the petioles. The main damage is indirect: this aphid is the vector for the Dasheen mosaic virus (DMV), which causes mosaic symptoms and feathering patterns on the leaves.

Moringa / Drumstick

Leaf caterpillar or Leaf worm (*Noorda blitealis)*
Larvae feed on the tender leaves, often webbing them together to form a cluster or nest and feeding from within. Heavy infestation causes severe defoliation, which hampers tree growth and subsequent pod/flower production.
Moringa budworm (*Noorda moringae)*
Larvae bore into and feed on the flower buds, flowers, and young fruits/pods, causing them to drop prematurely and resulting in heavy yield losses of pods.

Bark eating caterpillar (*Indarbela tetraonis*)

Larvae bore a tunnel into the main stem or thick branches and live inside it. They feed externally on the bark, covering the feeding area with a conspicuous silken web/mesh mixed with their frass (excreta) and wood particles. This ring-barking weakens the branch or trunk and can cause it to break off.

Beans (French bean, Cowpea)

Bean aphid (*Aphis craccivora*)

Nymphs and adults suck sap from tender shoots, buds, and pods, causing leaf curling, wilting, and stunting. They excrete honeydew, leading to sooty mould. Most critically, they transmit viral diseases.

Whitefly (*Bemisia tabaci*)

Nymphs and adults suck sap from the underside of leaves, causing yellowing. They are major vectors for viral diseases (e.g., Yellow Mosaic Virus), and excrete honeydew, causing sooty mould.

Leafhopper (*Empoasca fabae*)

Nymphs and adults suck sap, often causing the leaf margins to curl, turn yellow, and dry up- a symptom known as 'hopper burn'- reducing the photosynthetic area.

Pod sucking bug (*Riptortus spp., Clavigralla gibbosa*)

Suck sap from developing pods and seeds. This feeding causes the seeds to become shrivelled, poor quality, and reduces pod filling. Feeding punctures may also lead to seed discoloration.

Green stink bug (*Nezara viridula*)

Suck sap from developing pods and seeds. This results in punctured pods and shrivelled, darkened, non-viable seeds, severely reducing the harvestable yield and quality.

Red spider mite (*Tetranychus telarius*)

Pierces leaf cells and sucks sap, causing fine white or yellowish stippling on the leaves. Severe infestation leads to bronzing, webbing, and premature leaf drop, reducing plant vigour.

French bean pod borer (*Adisura atkinsoni*)

The caterpillar bores into the pods and feeds on the developing seeds inside, often leaving frass (excreta). The damage results in mutilated, rotting seeds and unmarketable pods.

Plume moth (*Exelastis atomosa*)

The caterpillar bores into the pods and feeds on the seeds. It is often found with its hind end outside the pod. Damage causes pod deformity and destruction of the seeds.

Spotted pod borer (*Maruca vitrata*)

The caterpillar bores into flower buds, flowers, and pods, often webbing them together. Feeding on the reproductive structures causes flower and pod drop and direct damage to seeds inside the pods.

Pulse/pea pod borer (*Etiella zinckenella*)

The caterpillar bores into the pods and consumes the seeds. It often moves between seeds inside the same pod, destroying the marketable contents.

Blue butterfly (*Lampides boeticus, Euchrysops cnejus*)
The caterpillar bores into the pods and feeds on the inner contents, often leaving the feeding hole covered with frass. Damage results in damaged, rotting seeds.
Gram pod borer (*Helicoverpa armigera*)
The caterpillar bores into the pods to feed on the seeds, often leaving part of its body outside the pod. Causes heavy damage to seeds, leading to rotten, unmarketable pods.
Stem fly (*Ophiomyia phaseoli*)
The maggot (larva) bores into the main stem near the ground level, tunneling upwards. This causes swelling and cracking of the stem and leads to stunting, yellowing, and eventual death of the young seedling due to the disruption of vascular tissue.
Blister beetle (*Mylabris pustulata*)
Adult beetles feed gregariously on flowers and tender foliage. Their primary damage is the complete destruction of the flowers, leading to a direct failure of pod set.
Til hawk moth (*Acherontia styx*)
The large caterpillar is a voracious leaf feeder, capable of causing rapid and severe defoliation, especially on younger plants.
Bihar hairy caterpillar (*Spilosoma obliqua*)
The caterpillars are polyphagous and feed gregariously, skeletonizing the leaves initially, and later consuming the entire leaf tissue. They cause heavy defoliation.
Tobacco leaf eating caterpillar (*Spodoptera litura*)
The caterpillars are highly destructive, causing large, irregular holes in the leaves. In severe cases, they lead to complete defoliation of the plants.
Hairy caterpillar (*Amsacta albistriga*)
The caterpillars feed aggressively, often in large numbers, resulting in complete defoliation of the crop over wide areas.

Fruit Crops

Citrus

Lemon butterfly (*Papilio demoleus)*
Larvae (caterpillars) are voracious leaf feeders, especially on tender, new flushes and seedlings. They feed from the leaf margin inwards, causing severe defoliation .
Citrus psylla (*Diaphorina citri)*
Nymphs and adults are sap-suckers on young shoots, causing curling, twisting, and dieback of new growth. Most significantly, it is the vector for Citrus Greening disease (Huanglongbing), which is highly destructive.
Leaf miner (*Phyllocnistis citrella)*
Larvae tunnel just beneath the epidermis of young leaves and tender stems, creating characteristic silvery, winding, serpentine mines. This damages the tissue, leading to leaf distortion, curling, and provides entry points for canker disease.

Whitefly (*Dialeurodes citri)*
Nymphs and adults are sap-suckers on the underside of leaves, causing yellowing and weakening of the plant. They secrete large amounts of honeydew, leading to the growth of black sooty mold on leaves and fruits.
Blackfly (*Aleurocanthus woghlumi)*
Nymphs and adults are sap-suckers that infest the underside of leaves. They excrete sticky honeydew, resulting in sooty mold covering the plant parts, which inhibits photosynthesis and spoils fruit appearance.
Fruit sucking moth (*Eudocima spp., Achoea Janata)*
Adult moths are destructive: they pierce the ripe or nearly ripe fruit skin with their strong proboscis to suck the juice. The puncture marks serve as entry points for fungi and bacteria, causing the fruit to rot and drop prematurely.
Aphid (*Toxoptera aurantia)*
Sap-suckers that cluster on tender growth, causing leaf curl, distortion, and stunting of new shoots. They also secrete honeydew and are vectors for Citrus Tristeza Virus (CTV).
Citrus red scale (*Aonidiella aurantia)*
Nymphs and females settle on fruits, leaves, and twigs, sucking sap from the plant cells. Severe infestation causes yellow spots on leaves, premature leaf drop, dieback of twigs, and discoloration/pitting of fruits, reducing their market value.
Mealybug (*Planococcus citri)*
Sap-suckers that aggregate in large, waxy colonies on twigs, leaves, and fruit clusters. They cause stunting and leaf drop, and their honeydew promotes the growth of sooty mold, spoiling the fruit's appearance.
Thrips (*Scirtothrips dorsalis)*
Sap-suckers that feed on tender leaves, flowers, and young fruits. Feeding on young fruits causes a characteristic ring-like scar or a silvered/bronzed patch around the stem end, which reduces fruit quality.
Bark borer (*Indarbela quadrinotata)*
Larvae bore tunnels into the trunk or main branches and live inside. They feed on the bark externally, covering the feeding area with a webbing of silk, frass (excreta), and wood pieces. These feeding girdles and weakens the branches, potentially leading to breakage.
Fruit borer (*Deudoris (=Virachola) Isocrates)*
Larvae bore into the young fruit (often near the stem end) and feed internally on the pulp and seeds. The damaged fruit turns yellow, rots, and drops prematurely.
Red mite (*Tetranychus citri)*
Mites suck the sap from leaves), twigs, and fruit. Their feeding causes fine stippling (tiny spots) on the leaves and a silvering or bronzing effect on the fruit surface, reducing the peel's quality.

Mango

Mango hoppers (*Amritodus spp., Idioscopus spp.)*
Nymphs and adults are sap-suckers on tender parts, primarily new flushes and inflorescences. Heavy infestation causes flower and fruit drop. They secrete honeydew, leading to the growth of black sooty mold on leaves and flowers, interfering with photosynthesis and fruit setting.

Mango stem borer (*Batocera rufomaculata, B. rubus*)
Larvae bore into the main trunk and thick branches, feeding on the inner bark and wood, creating irregular tunnels. This weakens the tree, resulting in dieback of branches or the entire tree, often signaled by gummy exudation and frass (wood dust mixed with excreta) at the entrance hole.

Mango stone weevil (*Sternochetus mangiferae*)
Adult female lays eggs on developing fruit. Larvae bore through the fruit pulp and reach the stone, where they feed inside, damaging the cotyledons (embryo). While external damage is minimal, it contaminates the seed and is a major quarantine issue.

Oriental fruit fly (*Bactrocera dorsalis*)
Adult females pierce the ripening fruit skin with their ovipositor and lay eggs inside. The resulting maggots feed on the pulp, causing the fruit to rot, become distorted, and drop prematurely.

Mealybug (*Drosicha mangiferae*)
Nymphs and wingless females crawl up the trunk to suck sap from tender shoots and inflorescences. They cause stunting, withered flowers and fruits, and secrete large amounts of honeydew, leading to heavy sooty mold development.

Shoot borer (*Chlumetia transversa*)
Larvae bore into the tender terminal shoots, killing the growing point. This results in wilting and drying of the new shoot tips and causes the plant to produce multiple side branches (bushy appearance).

Flower / Inflorescence midge (*Procystiphora spp., Erosomyia indica, Dasineura spp.*)
Larvae infest flower buds, flowers, and tiny developing fruit. The midges cause malformation, gall formation, and premature shedding of buds, flowers, and fruit, leading to significant yield loss.

Leaf gall midge (*Procontarinia matteiana*)
Larvae feed within the leaf tissue, inducing the formation of characteristic small, raised, dome-shaped galls on both surfaces of the leaves, primarily on new flushes. Heavy infestation reduces photosynthetic area.

Slug caterpillar (*Parasa lepida*)
Larvae feed gregariously by scraping the green matter and eventually chewing large, irregular holes in the leaves, leading to defoliation. The caterpillar has stinging spines that can cause irritation.

Thrips (*Caliothrips spp., Rhipiphorothrips spp., Scirtothrips spp.*)
Sap-suckers on tender leaves and flowers. Feeding causes silvery patches or streaks on the leaves and can lead to the shedding of flowers and scarring on young fruit.

Mango leaf webber (*Orthaga exvinacea*)
Larvae feed on the leaves, webbing them together into a characteristic cluster or nest . They feed from within the web, causing defoliation and damage to inflorescences, especially during the post-monsoon period.

Grapevine

Flea beetle / Udadya beetle (*Scelodonta strigicollis*)
Adult beetles feed on the tender buds (swollen eyes), newly emerged leaves, and developing flowers/inflorescences, causing them to dry up. This leads to the failure of bud-break and subsequent yield loss. The larvae feed on the rootlets underground.

Thrips (*Rhipiphorothrips cruentatus)*
Sap-suckers on tender leaves and especially on young berries. Feeding causes scars and brownish patches on the developing berries (creating a "netting" or "scabbing" effect). Severe feeding on leaves causes silvery sheen and curling.
Stem girdler (*Sthenias grisator)*
Adult beetles girdle (ring-bark) the tender branches and shoots by making incisions/cuts around the stem before laying eggs. This causes the distal part of the branch/shoot to dry up and break off, leading to loss of bearing wood.
Mealybug (*Maconellicoccus hirsutus)*
Sap-suckers that colonize tender shoots, leaves, and grape bunches. They cause stunting and distortion of shoots and leaves. They excrete large amounts of sticky honeydew, which promotes the growth of black sooty mold on the bunches, severely reducing marketability.
Mite (Two-spotted spider mite) (*Tetranychus urticae)*
Mites suck the cell contents, mainly from the underside of leaves. This causes numerous fine yellow or white spots (stippling), which merge to give a bronzing or dusty appearance. Heavy infestation leads to premature leaf senescence (drying) and can affect fruit quality.

Guava

Fruit fly (*Bactrocera dorsalis, B. correcta*)
Females puncture the skin of the developing and ripening fruits to lay eggs. Upon hatching, the maggots feed inside the pulp, creating dark, rotting patches. This feeding leads to premature fruit drop and makes the fruit completely unfit for consumption.
Spiraling whitefly (*Aleurodicus dispersus*)
Both nymphs and adults suck sap from the leaves, leading to yellowing (chlorosis) and weakening of the plant. They excrete large amounts of honeydew, which promotes the growth of sooty mould (a black fungal layer). This mould interferes with photosynthesis, reducing plant vigor and fruit quality.
Bark eating caterpillar (*Indarbela tetraonis, I. quadrinotata*)
The caterpillars bore into the trunk and branches and feed on the bark tissues from under a protective web of silk and frass (excreta). This damage girdles the branches, interfering with the transport of water and nutrients, which can lead to dieback of affected branches and weakened overall tree structure.
Fruit Borers (*Conogethes punctiferalis, Deudorix isocrates, Rapala varuna*)
The larvae bore into the guava fruit. *C. punctiferalis* larvae often tunnel through the fruit, while the larvae of the butterfly species (*D. isocrates* and *R. varuna*) bore into the fruit from near the calyx or stem end. The internal feeding causes the fruit to rot and drop prematurely. The entry/exit holes also serve as points for secondary infection by pathogens.
Mealybug (*Ferrisia virgata*)
They are sap suckers that infest shoots, leaves, and fruits, often found in clusters at the growing points. Their feeding causes stunting, leaf deformation, and premature fruit drop. Like whiteflies, they excrete honeydew, leading to the development of sooty mould, which reduces fruit marketability and photosynthetic efficiency.
Green Scale (*Coccus viridis*)
These are sap-sucking insects that settle on the young stems, leaves, and fruits. Heavy infestation leads to chlorosis (yellowing) and wilting of leaves. They also produce honeydew,

causing a build-up of sooty mould, which hinders the tree's ability to photosynthesize and can stunt growth.

Banana

Root stock weevil (*Cosmopolites sordidus*)
The grubs (larvae) are the primary damaging stage. They bore extensive tunnels into the rhizome, the underground stem of the banana plant. This damage severely disrupts nutrient and water uptake, leading to reduced growth, yellowing of leaves, premature plant death, and increased susceptibility to wind lodging (the plant falling over).
Pseudostem borer / Stem weevil (*Odoiporus longocollis*)
The grubs (larvae) bore long, extensive tunnels through the pseudostem (the false stem formed by leaf sheaths). Initial symptoms include pin-sized holes and the exudation of gummy sap or 'jelly'. Extensive tunnelling weakens the pseudostem, leading to the snapping or toppling of the plant, especially during wind or when bearing a heavy bunch.
Fruit rust thrips (*Chaetanaphothrips signipennis*)
Nymphs and adults feed by rasping and sucking sap from the surface of the developing banana fruit fingers, especially where they touch. This causes a characteristic rusty-red or dark-brown discolouration on the peel, making the fruit skin rough and resulting in significant loss of market value. Severe feeding can cause longitudinal cracks on the fruit.
Aphid (*Pentalonia nigronervosa*)
Aphids are sap-sucking insects that form dense colonies, typically on the underside of leaves and near the growing points. Their main significance is as the vector for the highly destructive Banana Bunchy Top Virus (BBTV), which causes the leaves to become stunted, rigid, and 'bunched' at the crown, leading to a complete failure to produce a marketable bunch.
Tingid or Lace wing bug (*Stephanitis typica*)
Both nymphs and adults suck sap from the underside of the leaves. This feeding results in tiny white or yellowish spots (stippling) on the upper leaf surface and causes the leaves to turn pale, yellow, and dry up. Heavy infestation leads to premature leaf death, reducing the plant's photosynthetic capacity.
Banana skipper (*Erionota thrax*)
The caterpillars (larvae) feed on the leaves. They are famous for their habit of cutting leaf segments and rolling them up into a characteristic, protective tubular shelter within which they feed. This feeding reduces the leaf area, resulting in ragged, chewed leaves and loss of photosynthetic efficiency.
Leaf eating caterpillar (*Spodoptera litura*)
The larvae are voracious feeders that initially scrape the chlorophyll, but later feed by making large, irregular holes in the leaves. In severe cases, they can cause heavy defoliation, consuming large areas of the leaf lamina, which stresses the plant and negatively affects fruit development.

Papaya

Papaya mealybug (*Paracoccus marginatus*)
Both nymphs and adults are sap-sucking insects that feed on leaves, stems, and fruit. It causes leaf crinkling, yellowing, stunting, and deformation of the growing points (apical leaves). They

excrete copious honeydew, which leads to the growth of sooty mould (a black fungal layer) on all parts, severely reducing photosynthesis and making the fruit unmarketable.

Whitefly (*Bemisia tabaci*)

Nymphs and adults suck sap from the underside of leaves, causing them to turn yellow (chlorotic). The another damage is the excretion of honeydew, which encourages the development of sooty mould, leading to reduced plant vigour. Most critically, this pest is a vector for serious papaya viral diseases, such as Papaya Leaf Curl Virus (PaLCV).

Green peach aphid (*Myzus persicae*)

These sap-sucking insects cluster on new leaves, petioles, and young fruits. Direct feeding causes leaf curling, wilting, stunting of the plant, and sometimes premature fruit drop. Like other sucking pests, they excrete honeydew. Their main threat, however, is their role as a highly efficient vector for numerous plant viruses, which can lead to severe systemic diseases in the papaya plant.

Ash weevil (*Myllocerus spp.*)

The adult weevils cause damage by feeding on the leaves, creating characteristic notches or V-shaped cuts along the leaf margins. The grubs (larvae) are the most destructive stage, as they live and feed underground on the roots of the papaya plant. This root feeding causes young saplings to wilt and can lead to the death of severely infested plants due to impaired water and nutrient uptake.

Sapota

Chiku moth / Leaf Webber (*Nephopteryx eugraphella*)

The caterpillars web together young leaves, buds, and flowers with silken threads and feed from within the shelter. They scrape the chlorophyll content, causing the webbed leaves to dry up and hang from the shoots. They also bore into and damage flower buds and tender fruits, leading to the shedding of buds and flowers, which reduces fruiting.

Sapota seed borer (*Trymalitis margarias*)

The larvae bore into the semi-ripened fruit and feed almost exclusively on the endosperm of the seed (cotyledons). While the fruit may look normal externally at first, the internal damage results in fruit rot and deterioration of quality, making it unfit for consumption. The full-grown larva leaves an exit hole, allowing entry of fungi and ants.

Bud borer (*Anarsia achrasella*)

The larvae are primarily damaging, initially feeding by webbing and scraping the tissue of young leaves. Their major impact is on the flower buds, which they bore into and feed upon, causing the buds to wither and shed before they can set fruit, leading to significant yield loss.

Fruit fly (*Bactrocera dorsalis*)

The female fly punctures the skin of developing or ripening fruits to lay eggs. Upon hatching, the maggots bore into and feed on the pulp. This internal feeding causes the fruit to develop dark, rotten patches and ooze fluid. The fruit becomes soft, shrivelled, drops prematurely, and is completely unfit for human consumption.

Hairy caterpillar (*Metanastria hyrtaca*)

The caterpillars are voracious feeders that consume the leaves of the sapota tree. They cause heavy defoliation, especially in severe infestations, leading to loss of photosynthetic area, which stresses the tree and can affect overall vigour and productivity.

Stem borer (*Neoplocaederus ferrugineus*)

The grubs (larvae) bore into the trunk and main branches, particularly at the collar region (base of the trunk). Symptoms include gummosis (oozing of sap) and the extrusion of frass (sawdust-like excreta) from the boreholes. The tunnelling girdles the stem, obstructing nutrient and water flow, which results in the yellowing and shedding of leaves, drying of twigs, and the gradual death of the entire tree if the infestation is severe.

Pomegranate

Anar caterpillar / Fruit Borer (*Deudorix isocrates*)
The caterpillar (larva) is the most destructive stage. It bores into young and developing fruits, feeding on the pulp and seeds inside. The entry hole is often plugged with its excreta (frass) and is usually near the calyx. Internal feeding causes the fruit to rot and drop prematurely, leading to severe yield loss.
Fruit sucking moth (*Eudocima spp.*)
The adult moth is nocturnal and causes damage by piercing the skin of ripening fruits with its strong, barbed proboscis to suck the juice. The puncture marks are small, pin-like holes. These wounds allow secondary infections by fungi and bacteria, leading to the rotting of the fruit pulp and subsequent premature fruit drop, rendering the fruit unmarketable.
Thrips (*Rhipiphorothrips cruentatus, Scirtothrips dorsalis*)
Both nymphs and adults lacerate and suck the sap from tender parts, including leaves, buds, and young fruits. Feeding on young fruits causes scab formation (russeting or dark, rough patches) on the peel. Feeding on flowers leads to drying and shedding of flowers, severely affecting fruit setting and reducing the market value of the fruit.
Shot hole borer (*Euwallacea fornicates*)
The adult beetles bore into the main trunk, exposed roots, and basal part of the stem, creating numerous pin-sized or "shot" holes. This activity disrupts the vascular tissues, severely impairing the conduction of water and nutrients. This damage often leads to the wilting and drying of twigs, progressing to the death of the entire tree (a major component of Pomegranate Wilt Complex).
Bark eating caterpillar (*Indarbela spp.*)
The caterpillar bores into the main trunk or branches and remains inside the tunnel during the day. At night, it comes out to feed on the bark tissues under a protective web of silk and frass (excreta and wood dust). This girdles the stem, reducing the flow of sap and leading to the dieback of affected branches and weakened tree structure.
Mealybug (*Ferrisia virgata, Planococcus citri*)
Nymphs and adults are sap-sucking pests that cluster on young shoots, flowers, and fruits. Their feeding causes stunting, distortion of new growth, and wilting. They excrete copious honeydew, which facilitates the growth of sooty mould (a black fungal film) that covers the leaves and fruit, reducing photosynthesis and fruit quality.
Ash whitefly (*Siphorinus phillyreae*)
Nymphs and adults suck sap from the underside of leaves and tender shoots. Heavy infestation causes the leaves to yellow and curl. They also produce honeydew, leading to sooty mould development, which affects the plant's photosynthetic ability and reduces the aesthetic and market value of the fruit.
Aphid (*Aphis punicae*)
Nymphs and adults suck the sap from tender leaves, shoots, and flower buds. This feeding causes the leaves to curl and turn yellow (chlorotic). Like whiteflies and mealybugs, they

excrete honeydew, which promotes the growth of sooty mould, further hindering plant health and fruit development.

Apple

Codling moth (*Cydia pomonella*)
The caterpillar (larva) bores into the fruit, usually entering through the calyx (eye) or the side. It tunnels down to the core, feeding on the pulp and seeds. The tunnels are filled with frass (excreta). This damage causes the fruit to rot internally, leads to premature ripening and drop, and makes the fruit unmarketable.
San Jose scale (*Quadraspidiotus perniciosus*)
Nymphs and adults suck sap from the bark of twigs, branches, and fruit. Heavy infestation on branches causes reduced vigor, thinning foliage, cracked bark, and potential death of young limbs or entire trees. On fruit, feeding causes distinct reddish-purple spots or rings around the feeding site, reducing the fruit's cosmetic and market value.
Wooly aphid (*Eriosoma lanigerum*)
These aphids suck sap from the bark of branches, shoots, and the roots. Their feeding causes the formation of rough, corky galls or swellings on the roots and twigs. Above ground, they are visible as white, cottony masses in pruning scars and bark crevices. This damage stunts tree growth, reduces vigor, and the root damage makes the tree susceptible to other diseases.
Blossom thrips (*Thrips flavus*)
Nymphs and adults lacerate and suck sap from tender tissue, especially flower buds and blossoms. Heavy feeding causes the flowers to become brown, withered, and drop prematurely, resulting in poor fruit set and a significant reduction in yield.
Cottony cushion scale (*Icerya purchasi*)
Nymphs and adults suck sap from branches and stems. Heavy infestation leads to stunting and yellowing of foliage, and dieback of twigs. They excrete large amounts of honeydew, which promotes the growth of sooty mould (a black fungus), covering leaves and fruit, which interferes with photosynthesis.
Root borer (*Dorysthenes hugeli*)
The large, fleshy grubs (larvae) feed extensively on the roots of the apple tree. This internal feeding causes the tree to show symptoms of general decline, yellowing of foliage, loss of vigor, and sudden wilting. Severe root damage can lead to the death of the tree, especially in younger plants, as it cuts off nutrient and water supply.
Stem borer (*Apriona cinera, Aeolesthes sarta*)
The grubs (larvae) bore long, meandering tunnels into the trunk and main branches. Symptoms include frass (sawdust-like excreta) ejected from boreholes and accumulating at the base of the tree, and the oozing of sap (gummosis). The tunnelling weakens the structural integrity of the tree, leading to branch breakage and eventual tree death from girdling of the vascular system.
Shot hole borer / Bark beetle (*Scolytus nitidus, Euwallacea fornicatus*)
Adult beetles bore tiny, pinhole-sized "shot holes" into the bark of the trunk and branches, often attacking stressed or weakened trees. The female beetle excavates galleries under the bark where larvae develop. This tunnelling and the introduction of fungal spores (Ambrosia beetle) cause the wilting and drying of twigs and can contribute to the decline and death of the tree (Pomegranate/Apple Wilt Complex).

Spider mites (*Tetranychus urticae*)
These tiny mites pierce leaf cells and suck the contents, primarily on the underside of leaves. This feeding results in tiny yellow or white spots (stippling) on the upper leaf surface. In heavy infestations, the leaves turn bronzed or mottled, eventually drying up and dropping prematurely, which reduces the tree's photosynthetic capacity and fruit size.
European red mite (*Panonychus ulmi*)
Similar to spider mites, these mites suck sap from leaf cells. Heavy feeding causes the leaves to become bronzed or pale, reducing the ability of the leaves to produce food for the fruit. Serious infestations, particularly in early summer, can lead to reduced fruit bud formation for the following season and poor fruit sizing and color in the current year.

Ber

Ber fruit borer (*Meridarchis scyrodes*)
The caterpillar (larva) is the damaging stage. It bores into the developing fruit, feeding on the pulp and the soft seed inside. The entrance hole is sealed with frass (excreta). Internal feeding causes the fruit to rot, become deformed, and drop prematurely, leading to severe yield loss.
Ber fruit fly (*Carpomyia vesuviana*)
The maggots (larvae) are the most destructive stage. The adult female lays eggs beneath the fruit skin. Upon hatching, the maggots feed voraciously on the pulp, creating tunnels filled with excreta. This internal damage causes the fruit to rot, taste bitter, and drop prematurely. The infested fruits are often visibly deformed and arrested in growth.
Hairy caterpillar (*Thiacidas postica*)
The caterpillars feed on the leaves, especially in groups on the underside. They initially scrape the chlorophyll content, leaving the leaf looking skeletonized and whitish. In severe infestations, they can cause rapid and complete defoliation of the plant, which severely stresses the tree and reduces its ability to set fruit.
Lac insect (*Kerria lacca*)
The insect settles on the twigs and branches and is a sap-sucking scale insect. Its feeding, in large numbers, depletes the plant's sap (phloem), leading to weakened branches, reduced vigor, and poor fruit development. Although *K. lacca* is commercially grown for its resin (lac), its presence on fruit-bearing branches can be detrimental to fruit production.

Aonla (Indian gooseberry)

Bark eating caterpillar (*Indarbela tetraonis, I. quadrinotata*)
The caterpillar bores a tunnel into the trunk and branches for shelter. It is nocturnal, emerging at night to feed on the surrounding bark tissues under a protective, irregular web of silk, frass (excreta), and chewed wood particles. This feeding activity girdles the branches, obstructing the transport of nutrients and water, which can lead to dieback of affected limbs and a reduction in tree vigor and yield.
Apical twig gall maker (*Hypolamprus stylophora*)
The larva bores into the apical (terminal) portion of the shoot. Its feeding causes the infested twig to swell and deform into a characteristic, spindle-shaped gall (a localized swelling of the woody tissue). A single larva is found inside each gall, feeding on the succulent tissue. The

formation of these galls stunts the growth of the shoot, affects the tree's architecture, and hampers its flowering and fruiting potential.

Pomegranate butterfly / Fruit Borer (*Virachola isocrates*)

The caterpillar (larva) bores into the developing Aonla fruit, typically entering near the calyx or through minor cracks. It feeds internally on the seed and pulp. The entrance hole is often plugged with frass. This internal feeding causes the fruit to decay, rot, and ultimately drop prematurely, resulting in significant direct yield loss.

Fruit sucking moths (*Eudocima fullonica, E. materna, Achaea janata*)

The adult moth is nocturnal and causes damage by piercing the skin of ripening fruits with its barbed proboscis to suck the juice. This piercing creates small pin-hole wounds on the fruit surface. These wounds serve as entry points for secondary bacterial and fungal infections, leading to rotting of the fruit pulp, which then turns yellow and drops prematurely, rendering the fruit unfit for market.

Fig

Jassid (*Velu caricae*)

Both nymphs and adults are sap-sucking insects that feed on the underside of the leaves. This feeding results in tiny yellowish-white spots (stippling) on the upper leaf surface. Heavy and continuous feeding causes the leaves to curl, turn brown, and eventually dry up and drop prematurely. This loss of foliage reduces the plant's photosynthetic capacity, affecting overall vigor and fruit production.

Mite (*Eutetranychus hirsti*)

These tiny mites pierce the leaf cells and suck the sap, primarily on the upper surface of the leaves. Their feeding causes the affected areas to develop a dull, pale, or whitish appearance and eventually a brown or bronzed coloration. Heavy infestation leads to premature yellowing and leaf drop, significantly reducing the plant's health and fruit yield.

Custard apple

Mealy bug (*Maconellicoccus hirsutus, Ferrisia virgata*)

These are sap-sucking pests that form dense, cottony colonies on tender shoots, flower buds, and young fruits. Their feeding causes the leaves to curl, twist, and deform, and the shoots to be stunted (a condition known as "mealy bug wilt"). They excrete copious honeydew, which promotes the growth of sooty mould (a black fungal film). The mould covers the fruits, making them unmarketable, and reduces photosynthesis.

Fruit fly (*Bactrocera zonata*)

The female fly punctures the skin of developing and ripening fruits to lay eggs. Upon hatching, the maggots (larvae) feed internally on the fruit pulp. This feeding causes the tissues to rot, become soft, and darken. The fruits drop prematurely and are completely unfit for human consumption.

Scales (*Ceroplastes floridensis*)

These are sessile, sap-sucking pests that attach themselves to the twigs, leaves, and fruits. They suck phloem sap, leading to yellowing of leaves and stunting of new growth. Like mealy bugs, they excrete honeydew, resulting in the growth of sooty mould, which contaminates the fruit surface and reduces the plant's photosynthetic efficiency.

Fruit boring caterpillar (*Anonaepestis bengalella*)

The caterpillar (larva) is the most destructive stage. It bores into the developing fruit and feeds internally on the pulp and seeds. The presence of the larva and its frass (excreta) inside the fruit causes the fruit tissue to rot. The fruit often becomes misshapen or may drop prematurely, leading to direct yield loss.

Plantation and Spices Crops

Coconut

Rhinoceros beetle (*Oryctes rhinoceros*)

The adult beetle is the primary damaging stage. It bores into the tender unopened fronds (spindle) and cabbage (growing point) of the palm. This feeding results in characteristic triangular cuts or holes when the frond opens. Severe damage to the growing point can kill the palm. The feeding wounds also provide entry points for the highly destructive Red palm weevil.

Red palm weevil (*Rhynchophorus ferrugineus*)

The grub (larva) is the most dangerous stage. It bores and feeds internally within the soft tissues of the crown, stem (trunk), and leaf axils. Initial symptoms are often subtle, including the oozing of a thick brown fluid and the extrusion of chewed fibers from boreholes. Internal damage causes the crown to wilt and drop, leading to the death of the entire palm.

Black headed caterpillar (*Opisina arenosella*)

The caterpillar (larva) feeds exclusively on the green tissues of the lower surface of the leaflets. It constructs silken galleries intermixed with its frass (excreta) beneath which it feeds. This scraping causes the leaflets to turn brown and dry up, making them appear scorched. Severe infestation leads to complete drying of the fronds, drastically reducing the palm's photosynthetic area and nut yield.

Eriophyid mite (*Aceria guerreronis*)

These microscopic mites live and feed under the bracts of the tender developing nuts in the area normally covered by the perianth. Their feeding causes necrotic patches (dead tissue) that appear as brown or black triangular lesions at the base of the tender nuts. This damage leads to nut deformation, splitting, a reduction in nut size, and significant loss in copra and oil yield.

Rat (*Rattus rattus*)

The common house rat causes damage by gnawing holes into the young and mature nuts on the palm to feed on the tender kernel water and soft pulp. This typically occurs in the crown of the palm. The gnawing leaves distinct circular holes. The damaged nuts drop prematurely or decay, resulting in high nut loss.

Arecanut

Spindle Bug / Mirid bug (*Mircarvalhoia arecae*)
Both nymphs and adults suck sap from the tender, unopened spindle leaf and inner leaflets, typically inhabiting the innermost leaf axils. This feeding causes blackish-brown linear lesions on the emerging leaves. In severe cases, the lesions become necrotic patches that dry up and fall off, resulting in numerous shot holes and a twisted, stunted, and sickly appearance of the palm.
Inflorescence Caterpillar (*Tirathaba mundella*)
The caterpillar (larva) feeds on the tender flowers (female flowers) and rachillae (branches) of the inflorescence. It webs the rachillae together into a wet mass of silk and frass (excreta). This webbing prevents the spathe (flower sheath) from opening normally, delaying spathe opening and causing shedding of the flowers and young nuts, directly reducing yield.
Red palm weevil (*Rhynchophorus ferrugineus*)
The grub (larva) is an internal borer that tunnels through the soft tissues of the stem (trunk), particularly in younger palms, or through the crown. Infestation is often indicated by the oozing of a thick, brownish fluid and chewed fibers from boreholes, or a gnawing sound inside the trunk. Extensive internal feeding leads to the wilting and collapse of the central spindle (crown), causing the death of the palm.

Cashew

Tea mosquito bug (*Helopeltis antonii*)
Both nymphs and adults suck sap from tender shoots, inflorescences, and young nuts using their piercing mouthparts. They inject toxic saliva during feeding. This causes characteristic brown, necrotic, sunken patches on the tender parts. Severe infestation leads to the withering and drying of shoots (dieback), shedding of flowers, and failure of the nut to set, resulting in significant yield loss.
Cashew stem and root borer (*Plocaederus ferrugineus*)
The grub (larva) is the most destructive stage. It bores deep tunnels into the main trunk and roots, particularly near the collar region (base). The damage is often indicated by the oozing of gum mixed with frass (excreta) from boreholes and the piling of frass at the base of the tree. This boring activity girdles the trunk, disrupting the flow of water and nutrients, leading to yellowing of leaves, wilting, and eventually the death of the entire tree.
Thrips (*Rhipiphorothrips cruentatus, Selenothrips rubrocinctus*)
Nymphs and adults lacerate and suck the sap from tender shoots, leaves, and particularly the inflorescences (flower bunches). This feeding causes the flower buds and tender shoots to wither, turn brownish, and dry up. Damage to the flower panicles results in poor fruit set and significant crop loss.
Shoot and blossom webber (*Lamida moncusalis*)
The caterpillar (larva) is the damaging stage. It webs together tender shoots and developing inflorescences with silk, forming a protective shelter inside which it feeds on the soft tissue. This feeding destroys the flower buds and tender leaves, leading to the drying up and shedding of the webbed parts, thus reducing flowering and fruiting.
Apple and nut borer (*Thylocoptila panrosema*)
The larva (caterpillar) bores into the developing cashew apple and nut. It feeds internally, damaging the pulp of the apple and the kernel of the nut. The internal damage causes the apple

and nut to rot and drop prematurely. The feeding also makes the nuts unfit for processing and consumption.

Tea

Tea mosquito bug (*Helopeltis theivora, H. antonii*)
Both nymphs and adults suck sap from tender shoots, buds, and young leaves using their piercing mouthparts. They inject toxic saliva during feeding. This causes characteristic brown, necrotic, sunken spots on the attacked tissue. The tender shoots wither, turn black, and die back, resulting in a condition known as 'blight' and significant loss of harvestable two-leaves-and-a-bud material.
Tea Green leaf Hopper (*Empoasca flavescens*)
Nymphs and adults are sap-sucking pests that feed primarily on the veins and margins of tender leaves. This causes the leaf margin to curl downwards and turn brown or reddish-brown. The damage, often called 'hopper burn,' significantly reduces the quality and quantity of the flush, as the leaves fail to develop normally.
Mites (Red Spider, Pink, Purple, Scarlet, Yellow)
All listed mites are sap-sucking arachnids that feed on the surface cells of the tea leaves, primarily on the upper or lower side depending on the species. Their feeding causes the affected leaves to develop a bronzed, reddish-brown, or yellowish-mottled appearance.
Red Spider Mite (*Oligonychus coffeae*)
Feeds mainly on the upper surface of mature leaves, causing reddish-brown blotches and bronzing.
Pink and Purple Mites (*Acaphylla theae, Calacarus carinatus*)
Feeds mainly on the upper surface of mature leaves, causing a rusty or purple tinge and the leaves to lose their sheen.
Scarlet Mite (*Brevipalpus phoenicis*)
Feeds mainly on the lower surface of mature leaves, causing a reddish-brown discoloration and loss of vigor.
Yellow Mite (*Polyphagotasonemus latus*)
Feeds on the underside of the tenderest leaves and buds. Its toxic feeding causes the youngest leaves to become small, brittle, and curl downwards, leading to stunted, distorted growth (sometimes called 'crippled shoot') and failure to produce marketable flush.

Coffee

Coffee berry borer (*Hypothenemus hampei*)
The female adult beetle is the primary damaging stage. She bores a tiny, clean, circular hole, usually in the apical (end) region of the green or ripening coffee berry. She tunnels into the fruit to lay eggs inside the endosperm (the bean). Both the adults and the larvae (grubs) feed directly on the coffee bean, consuming the inner material. This internal feeding causes the bean to be pitted, discolored, and reduced in weight, severely reducing both the quality and quantity of the coffee yield. Infested berries often remain on the branch or dry up, providing shelter for the pest.

Turmeric and Ginger

Rhizome fly (*Mimegralla coeruleifrons*)
The maggot (larva) bores into and feeds on the rhizomes (the underground edible stems). This internal damage leads to rotting of the rhizome tissue. Above ground, the symptoms include yellowing of leaves and stunting of the plant, often resulting in the death of the pseudo-stem and poor final yield.
Shoot borer (*Conogethes punctiferalis*)
The caterpillar (larva) bores into the pseudo-stem (the rolled leaf sheaths) near the collar region. It feeds inside the stem, tunneling downwards. This internal damage causes the central spindle leaf to yellow and dry up (a symptom known as "dead heart"). Infestation is severe when the crop is 2-4 months old, leading to the death of the entire shoot and failure to produce rhizomes.
Rhizome scale (*Aspidiella hartii*)
The scale insects settle on the rhizomes during storage and in the field. They suck the sap from the rhizome surface. Heavy infestation in storage causes the rhizomes to shrivel, dry up, and lose viability (sprouting capacity) for the next planting season. In the field, it causes poor plant vigor and reduced yield.
Leaf roller / Skipper (*Udaspes folus*)
The caterpillar (larva) is a defoliator that rolls the leaves into a protective cigar-shaped cylinder using silken threads and feeds from within. This feeding activity reduces the effective photosynthetic leaf area. Heavy infestation causes the leaves to appear ragged and contributes to poor plant growth and reduced rhizome development.
Thrips (*Stenchaetothrips indicus*)
Nymphs and adults lacerate and suck sap from the leaves, especially on the underside of the foliage. This feeding causes the leaf margins to curl inwards and fold, leading to the formation of pale or white streaks along the veins. In severe cases, the leaves turn yellow and wither, severely hampering the plant's ability to photosynthesize.
White grub (*Holotrichia serrata*)
The grub (larva) is a soil pest that feeds on the roots and developing rhizomes. This feeding severs the root system, leading to the sudden wilting and death of entire clumps of the crop. Infested fields show patchy yellowing and plant death, resulting in significant yield reduction at harvest.

Onion and Garlic

Onion and garlic thrips (*Thrips tabaci*)
Nymphs and adults are sap-sucking insects that feed by rasping and puncturing the epidermal cells of the leaves. This causes characteristic silvery-white streaks, blotches, or spots on the leaves and scapes (flower stalks). In severe cases, the tips of the leaves wither, turn brown, and curl, severely reducing the plant's photosynthetic area and leading to small, underdeveloped bulbs.
Onion fly (*Delia antiqua*)
The maggot (larva) is the damaging stage. It bores into the base of the young seedlings and bulbs and feeds internally on the succulent tissue. Infested plants show wilting and yellowing of the central leaves, often followed by the death of the entire plant (dead heart symptom in young

seedlings). The feeding damage also allows for secondary decay by bacteria and fungi, causing the bulbs to rot.

Tobacco leaf eating caterpillar (*Spodoptera litura*)

The caterpillars are nocturnal defoliators that feed on the foliage. In young stages, they may scrape the leaf surface, but older larvae consume the entire leaf tissue, leaving only the main veins. This feeding can lead to severe defoliation, weakening the plants, and significantly impacting the growth and size of the harvested bulbs.

Cutworm (*Agrotis ipsilon*)

The caterpillar (larva) is a soil-dwelling pest that is active at night. It causes damage by cutting the young seedlings and transplants at the ground level (collar region). A single cutworm can destroy numerous plants in a single night. This results in gaps in the field stand and the death of young plants, leading to significant stand loss and necessitating costly replanting.

Curry leaf

Lemon butterfly (*Papilio demoleus*)

The caterpillar (larva) is the main damaging stage. It is a voracious defoliator that feeds on the tender leaves and shoots. The larva consumes the entire leaf lamina, often leaving behind only the main midrib and stalk. This leads to severe defoliation, which is critical for the curry leaf plant since the leaves are the economically valuable part of the crop.

Scale insect (*Unaspis citri*)

The nymphs and adult females are sedentary, sap-sucking pests that cluster on the stems, twigs, and leaves. They suck phloem sap, leading to yellowing (chlorosis) of the leaves, stunting of new growth, and dieback of twigs in heavy infestations. Severe sap depletion weakens the entire plant.

Psylla (*Diaphorina citri*)

Nymphs and adults are sap-sucking pests that prefer tender new flushes. Their feeding causes the new leaves to become twisted, curled, or deformed, and often fail to expand properly. They excrete honeydew, which promotes the growth of sooty mould (a black fungal film) on the leaves, reducing the plant's aesthetic value and interfering with photosynthesis. This pest is a major vector of the Citrus Greening Disease (Huanglongbing), which can also affect the related curry leaf plant.

Blackfly (*Aleurocanthus woglumi*)

Nymphs and adults settle on the underside of the leaves and suck sap. Heavy infestation causes the leaves to yellow and dry up. They excrete copious amounts of honeydew, which quickly leads to the heavy growth of sooty mould. The black mould covers the leaves, hinders photosynthesis, reduces the flavour/quality of the leaves, and makes them aesthetically undesirable for culinary use.

Black pepper

Pollu beetle / Flea beetle (*Lanka ramakrishnai*)

The grubs (larvae) and adult beetles both cause damage. The larvae feed inside the developing berries (peppercorns) after boring into them, resulting in the interior being completely eaten out. These damaged berries turn black and crumble when pressed, a condition known as "pollu" (meaning hollow). The adult beetles feed externally on tender leaves and berries, creating small holes. This damage leads to massive crop loss due to the hollowing of the developing peppercorns.

Mealy bug (*Planococcus sp.*)

These are sap-sucking pests that form colonies, often covered in a white, cottony, waxy coating, and settle on the tender shoots, spikes, and especially near the nodes. Their feeding leads to the yellowing of leaves, stunting of new growth, and wilting of the young shoots. They excrete honeydew, which promotes the growth of sooty mould (a black fungal film) on the leaves and stems, further affecting plant health and photosynthesis.

Coriander

Aphid (*Hyadaphis coriandri, Aphis gossypii*)

These are sap-sucking pests that form dense colonies on tender shoots, leaves, and developing umbels (flower heads). Their feeding causes the leaves to curl, crinkle, and turn yellow. Heavy infestation leads to stunted growth, shriveling of the flower heads, and severe reduction in seed setting. They also excrete honeydew, leading to the growth of sooty mould, which covers the plant.

Indigo caterpillar / Beet armyworm (*Spodoptera exigua*)

The caterpillar (larva) is a voracious defoliator. Young larvae initially scrape the leaf surface, but older larvae consume the entire leaf lamina. They attack the tender foliage, shoots, and developing seed heads. Heavy feeding results in severe defoliation, which stunts plant growth, reduces the photosynthetic area, and drastically impacts the final seed yield.

Mite (*Petrobia latens, Tetranychus telarius*)

These tiny mites pierce the leaf cells and suck the sap. Their feeding results in pale yellow or whitish spots (stippling) on the leaves. In severe infestations, the leaves may take on a bronzed or dusty appearance, become mottled, and dry up prematurely. This loss of functional leaf area negatively affects plant health, vigor, and overall seed development.

Cardamom

Cardamom thrips (*Sciothrips cardamomi*)

Both nymphs and adults lacerate and suck the sap primarily from the tender capsules (fruit) and, to a lesser extent, from flower buds and leaves. Feeding on the capsules causes the skin to develop scab-like, corky, brown encrustations. This damage leads to shriveling, splitting of the capsule skin, and the eventual dropping of the immature capsules, severely reducing both the quality and quantity of the final produce.

Rhizome borer (*Prodioctes haematicus*)

The grub (larva) bores into the rhizomes (underground stem) and feeds internally on the core tissue. This tunneling damages the growing points, leading to yellowing of the entire clump and stunting of new shoots. Severe damage causes the infested tillers to rot and die, eventually leading to the death of the entire plant clump.

Shoot/panicle borer (*Conogethes punctiferalis*)

The caterpillar (larva) bores into the tender pseudostem (shoot) or the panicle stalk (flower stalk). In the shoot, its feeding causes the central spindle leaf to wither and dry up ("dead heart"). In the panicle, it bores into the main axis, causing the entire flower/fruit bunch to dry up and drop prematurely, resulting in direct loss of the developing capsules.

Early capsule borer (*Jamides alecto*)

The caterpillar (larva) is a borer that attacks the very young, developing capsules. It bores into the fruit and feeds internally on the seeds and inner tissue. This internal feeding causes the capsule to wither and drop off when still very small, leading to early crop loss.

Cardamom root grub (*Basilepta fulvicorne*)

The grub (larva) is a soil-dwelling pest that feeds on the roots of the cardamom plant. This root feeding causes the plant to show signs of yellowing, wilting, and general decline. Severe damage to the root system affects the plant's ability to absorb water and nutrients, leading to reduced vigor, stunting, and poor capsule formation.

Floriculture

Rose, Gerbera, Carnation

Thrips (*Rhipiphorothrips spp.*)

Nymphs and adults lacerate and suck sap from tender tissues, especially flower buds and petals. This causes streaking, silvering, or bronzing on the petals, leading to distorted flowers that fail to open fully or develop normally. Feeding on young leaves causes scars and deformation. This damage severely reduces the market quality of cut flowers.

Whitefly (*Aleurocanthus sp.*)

Nymphs and adults suck sap from the underside of leaves, causing them to yellow and weaken. The main damage is through the excretion of honeydew, which promotes the growth of sooty mould (a black fungal film). The mould covers the leaves and buds, hindering photosynthesis and making the flowers and foliage unmarketable.

Bud borer (*Helicoverpa armigera*)

The caterpillar (larva) bores into the tender flower buds. It feeds on the internal developing floral parts. This damage causes the buds to fail to open or to open with mutilated, riddled petals. The presence of the larva or bore-hole damage makes the entire bloom unfit for sale.

Leaf miner (*Liriomyza trifoli*)

The larvae (maggots) tunnel and feed within the leaf tissue between the upper and lower epidermis. This feeding creates characteristic white, serpentine (winding) mines on the leaves. Heavy infestation reduces the photosynthetic area, causes the leaves to curl and drop prematurely, and decreases the overall vigor and aesthetic quality of the plant.

Mites (*Tetranychus spp.*)

These tiny mites pierce leaf cells and suck the sap, primarily on the underside of leaves. This feeding causes fine yellow or white spots (stippling) on the leaf surface. In severe cases, leaves turn bronzed or mottled and may be covered in fine webbing. This leads to premature leaf drop and a decline in plant health, reducing the size and quality of the flowers.

Stored pests

Rice weevil (*Sitophilus oryzae*)
The female beetle bores a small hole into the grain (e.g., rice, wheat, maize) where she lays an egg, then seals the hole. The grub (larva) feeds and develops entirely inside the grain. The adult emerges, leaving a characteristic round exit hole. This internal feeding causes significant weight loss and reduced viability of the grain.

Lesser grain borer (*Rhyzopertha dominica*)
The small, cylindrical adult beetles bore into the grain, particularly wheat and rice, creating circular holes. The larvae also feed inside the kernels. Both stages reduce the grain to flour and dust through their destructive feeding, leading to rapid deterioration of the grain mass.

Pulse beetle (*Callosobruchus chinensis*)
The female lays eggs on the surface of the pulses (legumes), either in the field or in storage. The grub (larva) bores into the seed and develops completely inside. The adult emerges, leaving a perfect circular hole. This internal feeding severely damages the seed viability and makes the pulse unfit for consumption.

Grain moth (*Sitotroga cerelella*)
The caterpillar (larva) bores into the grain (especially maize and paddy) and feeds internally, developing entirely within the kernel. The adult moth emerges, leaving a pinhole from which the larva pushed out a window. Damage often starts in the field and continues in storage, resulting in hollowed-out grains and contamination.

Khapra beetle (*Trogoderma granarium*)
The larva is the most destructive stage, feeding on a wide variety of grains and oilseeds. The larvae nibble only on the surface or embryo portion of the kernels, causing scratch marks. Their feeding is often accompanied by large amounts of shed skins and larval hairs, which severely contaminate the produce and can cause health issues.

Indian meal moth (*Plodia interpunctella*)
The caterpillar (larva) does not bore into individual kernels but feeds on the surface of the grain mass. The most noticeable damage is the copious amounts of silken webbing produced by the larvae, which binds the kernels together. This webbing and their frass (excreta) severely contaminate the grain, primarily damaging the surface layers of the stored commodity.

Rice moth (*Corcyra cephalonica*)
The caterpillar (larva) is a general feeder on grains, flours, and milled products. Similar to the Indian meal moth, the larvae produce thick mats of silk webbing over the surface of the grain or in the commodity, which binds the material together. This webbing and contamination with frass are the primary source of damage, especially in milled products and flour.

Rust red flour beetle (*Tribolium castaneum*)
Both adults and larvae feed on broken kernels, flour, and milled products, rather than whole, sound grains. Their presence causes the infested material to develop a foul, pungent odor and a grayish-red tint due to their secretions. This contamination makes the food unusable.

Saw toothed grain beetle (*Orzyaephilus surinamensis*)
Both adults and larvae are external feeders on processed foods, flour, and broken/damaged grains. They cannot infest sound whole grains. Their damage is mainly contamination of the food product with the insects, their shed skins, and excreta.
Long headed beetle (*Latheticus oryzae*)
Both adults and larvae feed primarily on flour, milled products, and broken rice. They are often found in grain dust and milled cereals. Their presence is an indicator of poor storage sanitation and their primary damage is contamination of the processed food material.

Non-insect pests

Plant parasitic nematode

Important nematodes
Root-knot nematode (*Meloidogyne* species)
Reniform nematode (*Rotylenchulus reniformis*)
Root lesion nematode (*Pratylenchus* species)
Cyst nematode (*Heterodera* and *Globodera* species)
Dagger nematode (*Xiphinema* species)
Citrus nematode (*Tylenchulus semipenetrans*)
Burrowing nematode (*Radopholus similis*)

Symptoms caused by plant parasitic nematodes
Symptoms produced by below ground feeders : Stunting, yellowing, wilting of above ground parts and in below ground parts, root gall, reduced root system, root lesions, roots of fleshy parts, excessive root branching is seen.
Symptoms produced by above ground feeders : Dead or devitalized buds, crinkled and distorted stems and foliage, seed galls, necrosis and discolouration, leaf lesions and leaf galls are seen.

Snail and Slug

Snails and slugs belong to invertebrate phylum Mollusca (class- Gastropoda). These are amphibious capable of living on land and waters. Snails differ from slugs in having a spirally coiled shell over their body.
Agriculturally important species of snails
Giant African snail- *Achatina fulica* : The snails live in humid condition. They are nocturnal and damages to vegetables, ornamental and plantation plants etc.
Slug- *Limax* Sp.

Rodents

Rodents include rate, mice and squirrels. Rats are some of the most troublesome and damaging rodents in India. They eat and contaminate food, damage structures and property, and transmit parasites and diseases to other animals and humans. Rats live and thrive in wide variety of climates and conditions and are often found in and around homes and other buildings, on farms and in gardens and open fields.

Nature of damage : Rodents are omnivorous and feed on grains, vegetables, fruits, meat and other products in the houses or in the field. Rodents are responsible for causing enormous losses to crops and stored grains. Besides feeding on these products, they destroy a substantial quantity by spillage and contamination with their dropping, urine, body hairs etc. They damage about 20 times the amount they actually consume, by their gnawing activity and by polluting the food grains. Rodents cause about 5-20% loss in food production alone.

References and Websites

References and Websites

Anirban Sil and Kaushik Pramanik. Recent promotion of agrochemical formulations for eco-friendly pest management. Agriallis, 2 (10) : 37-42.

A.S. Atwal and G.S. Dhaliwal. Agricultural Pests of South Asia and their Management

B. Vasantraj David and V. V. Rammurthy. Elements of Economic Entomology

B.P. Khare. Storage Entomology

Dhamo K. Butani. Insects and Fruits

Dhamo. K. Butani and M. G. Jotwani. Insects in Vegetables

Dr. K.D.Upadhyay, Dr. Kusum Dwived : A text book of Plant Nematology

Goi, (2021). Insecticides / Pesticides Registered under section 9(3) of the Insecticides Act, 1968 for use in the Country :(As on 01.03.2021)

GOI, (2020). Statistical Database | Directorate of Plant Protection, Quarantine & Storage | GOI. Retrieved from http://ppqs.gov.in/statistical-database

Janaki Rani A, Kavitha B, Abirami S, Narma-dha N, Murugan PP, Sumathi E, Bama SK, Manimekalai R. Pesticide in Indian agriculture: Consumption pattern and farmers perception. Plant Science Today. 2025; 12(2): 1-9

Manishekharan and Sudarrajan. Pest Management in Field Crops

P. Parvatha Reddy : Plant Nematology

Reddy, A.A.; Reddy, M.; Mathur, V. Pesticide Use, Regulation, and Policies in Indian Agriculture. Sustainability 2024, 16, 7839. https://doi.org/10.3390/su16177839

S. Pradhan : Insect pests of crops

R. Mounika, V. Saravanakumar, K. R. Karunakaran, and A. Suganthi. 2022. "Pesticide Consumption Trends in India". *Asian Journal of Agricultural Extension, Economics & Sociology* 40 (10):221–226.

Urvashi Kashyap, Shivani Garg, Pooja Arora. 2024. Pesticide pollution in India: Environmental and health risks, and policy challenges. Toxicology Reports, Volume 13, 2024, 101801, ISSN 2214-7500, https://doi.org/10.1016/j.toxrep.2024.101801.

V.B. Awasthi: Introduction of General and Applied Entomology.

Venu Gopal Rao. Insect Pest Management

Yawalkar Pesticides Limited. Pesticides in India : recent facts and figures

https://ppqs.gov.in/
https://www.ncbi.nlm.nih.gov/pmc/articles/PMC3162709/
https://www3.epa.gov/pesticides/chem_search/reg_actions/reregistration/fs_PC-066501_1-Dec-98.pdf
http://sitem.herts.ac.uk/aeru/ppdb/en/Reports/74.htm
Product – Shivalik (shivalikcropsciences.com)

https://sitem.herts.ac.uk/aeru/ppdb/en/Reports/3169.htm
https://sitem.herts.ac.uk/aeru/ppdb/en/Reports/1138.htm
http://npic.orst.edu/factsheets/chlorantraniliprole.html
https://www.sciencedirect.com/
https://www.researchgate.net/publication/375485765_Pesticides_and_Their_Formulations. Mukesh Kumar Mishra, Shailendra Kumar Mishra, Lakshmi Pandey and Arun Kumar. (2023). Pesticides and Their Formulations. 57-67.
https://www.canr.msu.edu/news/effect_of_water_ph_on_the_stability_of_pesticides
https://koronationbioscience.com/
https://farmmate.in/products/
https://www.adama.com
https://bestagrolife.com
https://www.crystalcropprotection.com
https://www.iffcomc.in
http://www.hil.gov.in
https://www.rallis.co.in
https://www.syngenta.co.in
https://www.piindustries.com
https://www.upl-ltd.com
https://www.swal.in
https://www.biostadt.com
https://crop-protection.basf.in
https://www.gharda.com
https://www.dhanuka.com
https://www.bayer.com
https://www.bighaat.com
https://coromandel.biz
https://ag.fmc.com
https://farmagritech.com
http://www.tropicalagro.in
https://www.anuproducts.in
https://www.insecticidesindia.com
https://www.indiamart.com
https://naclind.com
http://www.excelcropcare.com
https://www.croplifescience.com
https://parijatagrochemicals.com
http://sumichem.co.in
https://www.exportersindia.com
http://hemanigroup.com
https://chemleyagritech.com/product-type
https://www.heranba.co.in
https://sikkoindia.com/pesticide/
https://www.mordorintelligence.com/industry-reports/india-crop-protection-pesticides-market
https://www.imarcgroup.com/indian-pesticides-market
https://doi.org/10.9734/ajaees/2022/v40i1031064.
https://www.granthaalayahpublication.org/journals/granthaalayah/article/view/3930/3978
https://indianchemicalregulation.com/indian-pesticide/pesticide-regulation

https://ipca.org.in/rescources/pesticide-regulations/
https://chemindia.chemicals.gov.in/Publicationspdf/Statistics-at-a-Glance-2024.pdf
https://www.marketresearchfuture.com/reports/agricultural-pesticides-market-31453
https://www.marketgrowthreports.com/market-reports/pesticide-and-other-agricultural-chemicals-market-113363.
https://ijrap.net/admin/php/uploads/2119_pdf.pdf
https://foodsafety.institute/food-toxicology-public-health/classification-of-pesticides-and-uses/
https://www.who.int/publications-detail-redirect/9789240005662
https://niphm.gov.in/Recruitments/ASO-PMD.pdf
https://npic.orst.edu/ingred/ptype/insecticide.html
https://www.britannica.com/technology/insecticide
https://www.epa.gov/caddis/insecticides
https://irac-online.org/mode-of-action/classification-online/
https://npic.orst.edu/factsheets/formulations.html
https://www.researchgate.net/publication/375485765_Pesticides_and_Their_Formulations
https://ipft.gov.in/microemulsion-me-and-nanoemulsion-ne
https://en.wikipedia.org/wiki/Pesticide_formulation
https://extension.psu.edu/pesticide-formulation-demonstration
https://edis.ifas.ufl.edu/publication/PI231
https://patents.google.com/patent/WO2002082902A2/en
https://www.engebiotech.com/news/wdg-and-sg-formulation/
https://www.cipac.org/images/pdf/2014/andy%20fowles_CIPAC%20Formulation%20choice%20presentation%20for%20june%202014.pdf
https://theorchardmaster.in/understanding-different-pesticide-formulations-used-in-agriculture/
https://extension.psu.edu/pesticide-formulation-demonstration
https://patents.google.com/patent/JP2000512985A/en
https://www.sciencedirect.com/topics/agricultural-and-biological-sciences/pesticide-formulation
https://www.apvma.gov.au/about/communication-and-decision-making/section-6a-guidelines/formulation-types
https://www.crodaagriculture.com/en-gb/applications/water-dispersible-granule
https://patents.google.com/patent/US5846903A/en
https://www.sciencedirect.com/topics/agricultural-and-biological-sciences/wettable-powders
https://www.sciencedirect.com/topics/agricultural-and-biological-sciences/dust-formulations
https://www.phytojournal.com/archives/2019/vol8issue2/PartL/8-1-568-106.pdf
http://www.hort.cornell.edu/turf/a/BacktoBasics.pdf
https://www.crodaagriculture.com/en-gb/applications/dispersible-concentrate
https://www.extension.purdue.edu/extmedia/ppp/ppp-31.pdf
https://www.sciencedirect.com/topics/agricultural-and-biological-sciences/emulsifiable-concentrates
https://pmc.ncbi.nlm.nih.gov/articles/PMC6852281/
https://patents.google.com/patent/US20090143447A1/en
https://www.nature.com/articles/s41598-018-28626-0

https://www.sciencedirect.com/science/article/abs/pii/S0167732223017750
https://ipft.gov.in/capsule-suspension-cs
https://products.evonik.com/assets/42/21/BREAK_THRU_Article_Agropages_Seed_Treatment_EN_Asset_2214221.pdf
https://www.crodaagriculture.com/en-gb/applications/flowable-concentrate-for-seed-treatment
https://www.crodaagriculture.com/en-gb/applications/suspoemulsion
https://patents.google.com/patent/WO1994003056A1/en
https://ppqs.gov.in/en/divisions/integrated-pest-management/instruction-safe-use-pesticide
https://ipca.org.in/rescources/pesticide-regulations/
https://npic.orst.edu/health/safeuse.html
https://edis.ifas.ufl.edu/publication/PI061
https://www.epa.gov/pesticide-incidents/pesticide-safety-tips
https://agritech.tnau.ac.in/crop_protection/crop_prot_pesticides_safe%20use%20of%20pesticides.html

Abamectin

Arsic B., Petrovic S., Ilic B. S., Vrecl M., Trobec T., Sepcic K., Frangez R., Gkisic S. M. and Milicevic J. S. Inhibitory Potential of Boscalid and Abamectin Towards Acetylcholinesterase and Butyrylcholinesterase: Computational and In Vitro Studies. Int J Mol Sci. 2025 Mar 21;26(7):2865. doi: 10.3390/ijms26072865. PMID: 40243451; PMCID: PMC11988634.

Acequinocyl

https://www.tga.gov.au/resources/publication/scheduling-decisions-final/
https://pharmaffiliates.com/en/57960-19-7-acequinocyl-papst016465.html
https://sitem.herts.ac.uk/aeru/ppdb/en/Reports/10.
https://pubchem.ncbi.nlm.nih.gov/compound/Acequinocyl

Acetamiprid

https://echa.europa.eu/substance-information/-/substanceinfo/100.111.622
https://merckindex.rsc.org/monographs/m1318
Taillebois E, Cartereau A, Thany SH. Effect of Acetamiprid, a Neonicotinoid Insecticide, on Locomotor Activity of the American Cockroach. Insects. 2024 Jan 12;15(1):54. doi: 10.3390/insects15010054. PMID: 38249060; PMCID: PMC10816188.

Acynonapyr

https://www.smolecule.com/products/s823093
https://www.benchchem.com/product/b1384038
https://sitem.herts.ac.uk/aeru/ppdb/en/Reports/3240.
https://www.sciencedirect.com/science/article/pii/S0048357524003079
Jun Takahashi, Masahiro Kawaguchi, Koichi Hirata, Keiji Koizumi, Development of a novel acaricide, acynonapyr, Journal of Pesticide Science, 2025, Volume 50, Issue 3, Pages 102-104, Released on J-STAGE September 04, 2025, Advance online publication July 19, 2025, Online ISSN 1349-0923, Print ISSN 1348-589X, https://doi.org/10.1584/jpestics.J25-04, https://www.jstage.jst.go.jp/article/jpestics/50/3/50_J25-04/_article/-char/en
Hamamoto I, Kawaguchi M, Nakamura T, Yano M, Koizumi K, Takahashi J. Discovery of a novel acaricide, acynonapyr. J Pestic Sci. 2023 Nov 20;48(4):202-210. doi: 10.1584/jpestics.D23-028. PMID: 38090213; PMCID: PMC10710939.

Afidopyropen

Edery I. A stretch from the periphery helps brain clocks feel the daily heat. Neuron. 2009 Oct 29;64(2):157-60. doi: 10.1016/j.neuron.2009.10.013. PMID: 19874783.

Shafi MS, Iqbal N, Naqqash MN, Saeed S, Usman M, Abid AD, Shahzad MS, Riaz H, Bashir MA, Alajmi RA. Transgenerational effect of Afidopyropen on *Bemisia tabaci* Gennadius (Homoptera: Aleyrodidae). Sci Rep. 2023 Nov 15;13(1):19988. doi: 10.1038/s41598-023-46479-0. PMID: 37968272; PMCID: PMC10651898.

Wang R, Zhang Q, Zhou X, Zhang M, Yang Q, Su Q, Luo C. Characterization of Field-Evolved Resistance to Afidopyropen, a Novel Insecticidal Toxin Developed from Microbial Secondary Metabolites, in *Bemisia tabaci*. Toxins (Basel). 2022 Jul 1;14(7):453. doi: 10.3390/toxins14070453. PMID: 35878191; PMCID: PMC9320511.

Alphacypermethrin

https://www.fao.org/fileadmin/templates/agphome/documents/Pests_Pesticides/Specs/Alphacypermethrin09.pdf

Hartnik T, Sverdrup LE, Jensen J. Toxicity of the pesticide alpha-cypermethrin to four soil nontarget invertebrates and implications for risk assessment. Environ Toxicol Chem. 2008 Jun;27(6):1408-15. doi: 10.1897/07-385. Epub 2008 Jan 22. PMID: 18211124.

https://pubchem.ncbi.nlm.nih.gov/compound/Cypermethrin

https://www.heranba.co.in/technicals/alpha-cypermethrin/

https://www.mda.state.mn.us/sites/default/files/inline-files/nair-alphacyper.pdf

Aluminium phosphide

https://pubchem.ncbi.nlm.nih.gov/compound/aluminum%20phosphide

https://www.pomais.com/what-is-aluminum-phosphide-used-for/

https://www.sciencedirect.com/topics/agricultural-and-biological sciences /aluminium phosphide

https://www.intech.in/products/aluminum-phosphide/

Beta-cyfluthrin

https://www.cdpr.ca.gov/active-ingredient/beta-cyfluthrin

https://ipm.ucanr.edu/home-and-landscape/pesticide-active-ingredients-database/

https://sitem.herts.ac.uk/aeru/ppdb/en/Reports/74.

Benfuracarb

https://www.oat-agrio.co.jp/en/product/benfuracarb/

https://pubchem.ncbi.nlm.nih.gov/compound/Benfuracarb

Benzpyrimoxan

https://sitem.herts.ac.uk/aeru/ppdb/en/Reports/3262

https://pubchem.ncbi.nlm.nih.gov/compound/Benzpyrimoxan

https://hero.epa.gov/hero/index.cfm/reference/details/reference_id/9551751

Umetsu N, Shirai Y. Development of novel pesticides in the 21st century. J Pestic Sci. 2020 May 20;45(2):54-74. doi: 10.1584/jpestics. D20-201. PMID: 33132734; PMCID: PMC7581488.

Bifenazate

https://www.grosafe.co.nz/wp-content/uploads/2021/09/Grosafe-Organic-Root-Gro-SDS-20200920.pdf

P. Van Nieuwenhuyse, P. Demaeght, W. Dermauw, M. Khalighi, C.V. Stevens, B. Vanholme, L. Tirry, P. Lümmen, T. Van Leeuwen. 2012. On the mode of action of bifenazate: New evidence for a mitochondrial target site, Pesticide Biochemistry and Physiology, 104 (2), :88-95.

Bifenthrin

J.-M. Hougard, S. Duchon, M. Zaim, P. Guillet, Bifenthrin: A Useful Pyrethroid Insecticide for Treatment of Mosquito Nets, Journal of Medical Entomology, Volume 39, Issue 3, 1 May 2002, Pages 526–533, https://doi.org/10.1603/0022-2585-39.3.526

https://www.idph.state.il.us/envhealth/factsheets/pyrethroid.htm

https://npic.orst.edu/factsheets/bifgen.html
Buprofezin
https://pubchem.ncbi.nlm.nih.gov/compound/Buprofezin
https://sitem.herts.ac.uk/aeru/ppdb/en/Reports/100.html
https://edis.ifas.ufl.edu/publication/IN1379
https://www.aos.org/orchids/articles/insect-growth-regulators
Carbofuran
https://archive.epa.gov/pesticides/reregistration/web/html/carbofuran_ired_fs.html
https://pubchem.ncbi.nlm.nih.gov/compound/Carbofuran
Kempuraj D, Zhang E, Gupta S, Gupta RC, Sinha NR, Mohan RR. Carbofuran pesticide toxicity to the eye. Exp Eye Res. 2023 Feb;227:109355. doi: 10.1016/j.exer.2022.109355. Epub 2022 Dec 23. PMID: 36572166; PMCID: PMC9918712.
Carbosulfan
Lamb T, Selvarajah LR, Mohamed F, Jayamanne S, Gawarammana I, Mostafa A, Buckley NA, Roberts MS, Eddleston M. High lethality and minimal variation after acute self-poisoning with carbamate insecticides in Sri Lanka - implications for global suicide prevention. Clin Toxicol (Phila). 2016 Sep;54(8):624-31. doi: 10.1080/15563650.2016.1187735. Epub 2016 Jun 2. PMID: 27252029; PMCID: PMC4950420.
https://pubchem.ncbi.nlm.nih.gov/compound/Carbosulfan
https://ag.fmc.com/in/en/insecticides/marshal-insecticide
Cartap hydrochloride
Seog-Jong Lee, Motohiro Tomizawa, and John E. Casida. 2003. Nereistoxin and Cartap Neurotoxicity Attributable to Direct Block of the Insect Nicotinic Receptor/Channel. Journal of Agricultural and Food Chemistry, 51 (9), 2646-2652. DOI: 10.1021/jf021149s
https://pubchem.ncbi.nlm.nih.gov/compound/Cartap-hydrochloride
https://www.inchem.org/documents/jmpr/jmpmono/v076pr08.htm
https://www.dhanuka.com/insecticide/caldan-50-sp
https://juagrisciences.com/CARTAP-GR
Chlorantraniliprole
https://npic.orst.edu/factsheets/chlorantraniliprole.html
https://sitem.herts.ac.uk/aeru/ppdb/en/Reports/1138.htm
https://www.chemicalbook.com/ChemicalProductProperty_EN_CB01404014.htm
Chlorfenapyr
https://www3.epa.gov/pesticides/chem_search/reg_actions/registration/fs_PC-129093_01-Jan-01.pdf
https://ordspub.epa.gov/ords/pesticides/f?p=CHEMICALSEARCH:3:::NO:1,3,31,7,12,25:P3_XCHEMICAL_ID:1793
https://pubchem.ncbi.nlm.nih.gov/compound/Chlorfenapyr
Chlorfluazuron
https://pubchem.ncbi.nlm.nih.gov/compound/Chlorfluazuron
https://sitem.herts.ac.uk/aeru/ppdb/en/Reports/139.htm
https://www.chemicalbook.com/ChemicalProductProperty_EN_CB6749358.htm
Chlorpyrifos
https://npic.orst.edu/factsheets/chlorpgen.html
https://www.epa.gov/ingredients-used-pesticide-products/chlorpyrifos
https://indianseeds.in/pesticide-glossary/chlorpyrifos/
https://wwwn.cdc.gov/TSP/PHS/PHS.aspx?phsid=493&toxid=88
Chromafenozide
https://pubchem.ncbi.nlm.nih.gov/compound/Chromafenozide

https://sitem.herts.ac.uk/aeru/ppdb/en/Reports/1136.html
https://www.chemicalbook.in/product/chromafenozide-6412469
Mikio Yanagi, Yoshihisa Tsukamoto, Tetsuo Watanabe and Akiyoshi Kawagishi. 2006. Development of a novel Lepidopteran insect control agent, chromafenozide. Journal of Pesticide Science ,31(2):163-164. DOI: 10.1584/jpestics.31.163

Clothianidin

https://sumichem.co.in/products/spl/insecticides/dantotsu.php
https://sitem.herts.ac.uk/aeru/ppdb/en/Reports/171.htm
https://www3.epa.gov/pesticides/chem_search/cleared_reviews/csr_PC-044309_24-Nov-04_a.pdf
https://www.acs.org/molecule-of-the-week/archive/c/clothianidin.html

Cyantraniliprole

https://www.sigmaaldrich.com/IN/en/product/sial/32372
https://pubchem.ncbi.nlm.nih.gov/compound/Cyantraniliprole#section=InChIKey
https://www.greencastonline.com/products/mainspring-gnl-insecticide/ornamental
Tiwari S, Stelinski LL. Effects of cyantraniliprole, a novel anthranilic diamide insecticide, against Asian citrus psyllid under laboratory and field conditions. Pest Manag Sci. 2013 Sep;69(9):1066-72. doi: 10.1002/ps.3468. Epub 2013 Feb 1. PMID: 23371893.

Cyclaniliprole

https://sitem.herts.ac.uk/aeru/ppdb/en/Reports/2632.htm
https://ontosight.ai
https://cultree.in/products/coromandel-prachand-cyclaniliprole
Lu Z, Yue W, Ren W, Wang Y, Li Y. Determination of Cyclaniliprole in Fruits and Vegetables Using Disposable Pipette Extraction Cleanup and Ultrahigh-Performance Liquid Chromatography-Tandem Mass Spectrometry. Molecules. 2022 Sep 30;27(19):6464. doi: 10.3390/molecules27196464. PMID: 36235002; PMCID: PMC9570770.

Cyclobutrifluram

https://sitem.herts.ac.uk/aeru/ppdb/en/Reports/3205.htm
https://www.epa.gov/pesticides/epa-registers-new-pesticide-active-ingredient-cyclobutrifluram
https://pubchem.ncbi.nlm.nih.gov/compound/Cyclobutrifluram
Liu W, Shao H, Qi D, Huang X, Chen J, Zhou L, Guo K. The New Nematicide Cyclobutrifluram Targets the Mitochondrial Succinate Dehydrogenase Complex in *Bursaphelenchus xylophilus*. Int J Mol Sci. 2024 Jun 24;25(13):6914. doi: 10.3390/ijms25136914. PMID: 39000026; PMCID: PMC11241274.

Cyflumetofen

https://pubchem.ncbi.nlm.nih.gov/compound/Cyflumetofen#section=Structures
https://www.sigmaaldrich.com/IN/en/product/sial/32068

Cypermethrin

Shilpakar O, Karki B. Cypermethrin poisoning manifesting with prolonged bradycardia: A case report. Toxicol Rep. 2020 Dec 10;8:10-12. doi: 10.1016/j.toxrep.2020.12.005. PMID: 33364178; PMCID: PMC7750177.
https://pubchem.ncbi.nlm.nih.gov/compound/Cypermethrin
https://d3n8a8pro7vhmx.cloudfront.net/ncap/pages/26/attachments/original/1428423343/cypermethrin.pdf?1428423343

Deltamethrin

https://pubchem.ncbi.nlm.nih.gov/compound/deltamethrin
https://npic.orst.edu/factsheets/DeltaGen.html
https://indianseeds.in/pesticide-glossary/deltamethrin/

Diafenthiuron

Jeon Y, Kang G, Cho S, Kim TH. Diafenthiuron: 1-tert-butyl-3-(2,6-diisopropyl-4-phen-oxy-phen-yl)thio-urea. Acta Crystallogr Sect E Struct Rep Online. 2014 Jun 25;70(Pt 7):o807. doi: 10.1107/S1600536814014214. PMID: 25161588; PMCID: PMC4120544.

https://indianseeds.in/pesticide-glossary/diafenthiuron/

https://sitem.herts.ac.uk/aeru/ppdb/en/Reports/210.htm

https://pubchem.ncbi.nlm.nih.gov/compound/Diafenthiuron

Diflubenzuron

https://pubchem.ncbi.nlm.nih.gov/compound/diflubenzuron

https://www.chemicalbook.com/Article/What-is-Diflubenzuron-Uses-Environmental-toxicity.htm

https://www.who.int/docs/default-source/wash-documents/wash-chemicals/

https://openknowledge.fao.org/server/api/core/bitstreams/fca0ca4b-da27-40f2-bcd4-a743223b2d5d/content

Dimethoate

https://www.sciencedirect.com/topics/chemistry/dimethoate

https://www.selleckchem.com/datasheet/dimethoate-E741500-DataSheet.html

https://inchem.org/documents/jmpr/jmpmono/v84pr19.htm

Dimpropyridaz

Hodges, Desirée. (2024). Discovery, research and development of axalion® active insecticide: dimpropyridaz. Pest Management Science. 81. 2529-2534. 10.1002/ps.8385.

https://www.hpc-standards.com/shop/ReferenceMaterials/Pesticides/1611064848_Dimpropyridaz.htm

http://www.wppdb.com/dimpropyridaz_3292.htm

Dinotefuron

https://pubchem.ncbi.nlm.nih.gov/compound/Dinotefuran

https://www.mc-croplifesolutions.com/en/randd/active-ingredients/dinotefuran/

Ghosh, Amalendu & samanta, arunava & M., L.. (2014). Dinotefuran: A third generation neonicotinoid insecticide for management of rice brown planthopper. African Journal of Agricultural Research. 9. 750-754. 10.5897/AJAR2013.7054.

Emamectin benzoate

https://pubchem.ncbi.nlm.nih.gov/compound/Emamectin_benzoate

https://www.scimplify.com/emamectin-benzoate-technical-95

https://jj-jebagro.com/products/emamectin-benzoate

Ghasemi V, Salehinejad A, Ghadamyari M, Jack CJ, Sharifi M. Toxic evaluation of Proclaim Fit® on adult and larval worker honey bees. Ecotoxicology. 2022 Nov;31(9):1441-1449. doi: 10.1007/s10646-022-02601-y. Epub 2022 Oct 27. PMID: 36301371.

Ethion

https://archive.epa.gov/pesticides/reregistration/web/html/ethion_red.html

https://www.ebi.ac.uk/chebi/searchId.do?chebiId=CHEBI:38663

https://www.ncbi.nlm.nih.gov/books/NBK11143/

Ethofenprox

Hemingway J. Efficacy of etofenprox against insecticide susceptible and resistant mosquito strains containing characterized resistance mechanisms. Med Vet Entomol. 1995 Oct;9(4):423-6. doi: 10.1111/j.1365-2915. 1995. tb00017.x. PMID: 8541596.

Matsuo N. Discovery and development of pyrethroid insecticides. Proc Jpn Acad Ser B Phys Biol Sci. 2019;95(7):378-400. doi: 10.2183/pjab.95.027. PMID: 31406060; PMCID: PMC6766454.

https://chemicalwarehouse.com/blogs/active-ingredients/etofenprox

https://us.mitsuichemicals.com/service/product/etofenprox/index.htm

Ethylene dichloride + Carbon tetrachloride (3:1)

R. T. Cotton, R. C. Roark, Ethylene Dichloride-Carbon Tetrachloride Mixture; A New Non-Burnable, Non-Explosive Fumigant, Journal of Economic Entomology, Volume 20, Issue 4, 1 August 1927, Pages 636–639, https://doi.org/10.1093/jee/20.4.636

https://inchem.org/documents/jmpr/jmpmono/v65apr07.htm

https://era.ed.ac.uk/bitstream/handle/1842/5247/A1.P7%20Fumigation%20with%20the%20Liquid%20Fumigants%20Carbon%20Tetrachloride,%20Ethylene%20Dichloride%20and%20Ethylene%20Dibromide.pdf;jsessionid=190F93ABCB09986ECF966C8FBB5DB45E?sequence=1

https://www.jiomart.com/p/homeandkitchen/u-l-t-i-m-a-s-granol-100ml-insect-killer-for-grains-rice-pulses-pack-of-1/611616201

https://extoxnet.orst.edu/pips/edb.htm

Etoxazole

https://pubchem.ncbi.nlm.nih.gov/compound/Etoxazole

https://sitem.herts.ac.uk/aeru/ppdb/en/Reports/284.htm

Xingxing Guo, Jinhui Cheng, Peilin Liu, Ting Lin, Tongyang Wang, Kaiyang Feng, Ping Zhang, Lin He, and Zhifeng Xu. Enantioselectivity Mechanism of Etoxazole Enantiomers in *Tetranychus cinnabarinus*: Acaricidal Activity, Metabolic Difference, and Target Binding. Journal of Agricultural and Food Chemistry 2025 73 (1), 180-188. DOI: 10.1021/acs.jafc.4c07841

Fenazaquin

https://pubchem.ncbi.nlm.nih.gov/compound/86356

https://sitem.herts.ac.uk/aeru/ppdb/en/Reports/292.

Bahreini R, Nasr M, Docherty C, Muirhead S, de Herdt O, Feindel D. Miticidal activity of fenazaquin and fenpyroximate against Varroa destructor, an ectoparasite of Apis mellifera. Pest Manag Sci. 2022 Apr;78(4):1686-1697. doi: 10.1002/ps.6788. Epub 2022 Jan 23. PMID: 34994089; PMCID: PMC9303763.

Kurosu M, Begari E. Vitamin K2 in electron transport system: are enzymes involved in vitamin K2 biosynthesis promising drug targets? Molecules. 2010 Mar 10;15(3):1531-53. doi: 10.3390/molecules15031531. PMID: 20335999; PMCID: PMC6257245.

Fenmezoditiaz

https://agribegri.com/hi/products/basf-valexio-insecticide

https://www.tridge.com/news/basfs-latest-insecticidal-ingredient-fenmezo-rjdekf

Huang H, Dickhaut J, Weisel M, Mao L, Rankl N, Takeda H, Stam LF, Peacock QM, Höffken HW. Discovery and biological characterization of a novel mesoionic insecticide fenmezoditiaz. Pest Manag Sci. 2025 May;81(5):2535-2552. doi: 10.1002/ps.8108. Epub 2024 Apr 10. PMID: 38554053; PMCID: PMC11981979.

Fenobucarb

Zhu XY, Wu YY, Xia B, Dai MZ, Huang YF, Yang H, Li CQ, Li P. Fenobucarb-induced developmental neurotoxicity and mechanisms in zebrafish. Neurotoxicology. 2020 Jul;79:11-19. doi: 10.1016/j.neuro.2020.03.013. Epub 2020 Apr 2. PMID: 32247646.

https://sitem.herts.ac.uk/aeru/ppdb/en/Reports/1183.htm#:~:text=Transforming%20Agroecosystems%20Together-,Fenobucarb

https://esteam.com/wp-content/uploads/2019/04/TUFT-GUARD-II-CONCENTRATE

Fipronil

Simon-Delso N, Amaral-Rogers V, Belzunces LP, Bonmatin JM, Chagnon M, Downs C, Furlan L, Gibbons DW, Giorio C, Girolami V, Goulson D, Kreutzweiser DP, Krupke CH, Liess M, Long E, McField M, Mineau P, Mitchell EA, Morrissey CA, Noome DA, Pisa L, Settele J,

Stark JD, Tapparo A, Van Dyck H, Van Praagh J, Van der Sluijs JP, Whitehorn PR, Wiemers M. Systemic insecticides (neonicotinoids and fipronil): trends, uses, mode of action and metabolites. Environ Sci Pollut Res Int. 2015 Jan;22(1):5-34. doi: 10.1007/s11356-014-3470-y. Epub 2014 Sep 19. PMID: 25233913; PMCID: PMC4284386.

Flonicamid

https://www.pomais.com/product/flonicamid-50-wdg-insecticide/

https://agribegri.com/products/swal-panama-insecticide-flonicamid-50-wg.php

https://www.nbinno.com/article/insecticides/understanding-flonicamid-mechanism-use-and-safety-rh

Zhang T, Xu Y, Zhou X, Liang X, Bai Y, Sun F, Zhang W, Wang N, Pang X, Li Y. Dissipation Kinetics and Safety Evaluation of Flonicamid in Four Various Types of Crops. Molecules. 2022 Dec 6;27(23):8615. doi: 10.3390/molecules27238615. PMID: 36500708; PMCID: PMC9738400.

Flometoquin

https://www.benchchem.com/product/b1456449

https://www.tantuchemicals.com/pesticide-intermediates/intermediates-of-flometoquin/flometoquin-intermediates-cas-1204518-38-6.html

https://www.researchgate.net/publication/375688290_Discovery_of_flometoquin_a_novel_quinoline_insecticide

Flubendiamide

https://revistacultivar.com/index.php/plant-health/flubendiamide-flubendiamide

https://meghmani.com/product/flubendiamide-technical

https://www.researchgate.net/publication/242494613_Flubendiamide_the_first_insecticide_with_a_novel_mode_of_action_on_insect_ryanodine_receptors

https://www.kisanshop.in/product/ju-flubeng-flubendiamide-20-insecticide

Fluopyram

https://pubchem.ncbi.nlm.nih.gov/compound/Fluopyram

https://sitem.herts.ac.uk/aeru/ppdb/en/Reports/1362.htm

https://biologyinsights.com/fluopyram-mechanisms-chemistry-and-nematode-control/

Schleker, A.S.S., Rist, M., Matera, C. et al. Mode of action of fluopyram in plant-parasitic nematodes. Sci Rep 12, 11954 (2022). https://doi.org/10.1038/s41598-022-15782-7

Flupyradifurone

Nauen R, Jeschke P, Velten R, Beck ME, Ebbinghaus-Kintscher U, Thielert W, Wölfel K, Haas M, Kunz K, Raupach G. Flupyradifurone: a brief profile of a new butenolide insecticide. Pest Manag Sci. 2015 Jun;71(6):850-62. doi: 10.1002/ps.3932. Epub 2014 Nov 27. PMID: 25351824; PMCID: PMC4657471.

Taha, Maria & Cartereau, Alison & Taillebois, Emiliane & Thany, Steeve. (2024). Flupyradifurone activates DUM neuron nicotinic acetylcholine receptors and stimulates an increase in intracellular calcium through the ryanodine receptors. Pesticide Biochemistry and Physiology. 205. 106147. 10.1016/j.pestbp.2024.106147.

Issa KA, Wosula EN, Stephano F, Legg JP. Evaluation of the Efficacy of Flupyradifurone against *Bemisia tabaci* on Cassava in Tanzania. Insects. 2022 Oct 12;13(10):920. doi: 10.3390/insects13100920. PMID: 36292868; PMCID: PMC9604256.

https://chemicalwarehouse.com/blogs/active-ingredients/flupyradifurone

Fang, Nan & Lu, Zhongbin & Hou, Zhiguang & Zhang, Changpeng & Zhao, Xueping. (2022). Hydrolysis and photolysis of flupyradifurone in aqueous solution and natural water: Degradation kinetics and pathway. Chemosphere. 298. 134294. 10.1016/j.chemosphere.2022.134294.

Fluvalinate
https://pubchem.ncbi.nlm.nih.gov/compound/fluvalinate
https://law.resource.org/pub/in/bis/S06/is.13172.1991.pdf
https://www.sciencedirect.com/topics/agricultural-and-biological-sciences/fluvalinate

Fluxametamide
https://cultree.in/products/godrej-agrovet-gracia-fluxametamide
Asahi M, Kobayashi M, Kagami T, Nakahira K, Furukawa Y, Ozoe Y. Fluxametamide: A novel isoxazoline insecticide that acts via distinctive antagonism of insect ligand-gated chloride channels. Pestic Biochem Physiol. 2018 Oct;151 : 67-72. doi: 10.1016/j.pestbp.2018.02.002. Epub 2018 Feb 6. PMID: 30704715.
Gope, A.; Chakraborty, G.; Ghosh, S.M.; Sau, S.; Mondal, K.; Biswas, A.; Sarkar, S.; Sarkar, P.K.; Roy, D. Toxicity and Sublethal Effects of Fluxametamide on the Key Biological Parameters and Life History Traits of Diamondback Moth *Plutella xylostella* (L.). *Agronomy* **2022**, *12*, 1656. https://doi.org/10.3390/agronomy12071656

Fluensulfone
https://sitem.herts.ac.uk/aeru/ppdb/en/Reports/2895.
Giannakou, I. O., & Panopoulou, S. (2019). The use of fluensulfone for the control of root-knot nematodes in greenhouse cultivated crops: Efficacy and phytotoxicity effects. *Cogent Food & Agriculture*, *5*(1). https://doi.org/10.1080/23311932.2019.1643819
James Kearn, Elizabeth Ludlow, James Dillon, Vincent O'Connor, Lindy Holden-Dye. 2014. Fluensulfone is a nematicide with a mode of action distinct from anticholinesterases and macrocyclic lactones. Pesticide Biochemistry and Physiology, 109, : 44-57.
Khalil MS, Selim RE. Fluorinated nematicides: Novel classes in the way. J Plant Sci Phytopathol. 2021; 5: 014-016.
Morris KA, Langston DB, Davis RF, Noe JP, Dickson DW, Timper P. Efficacy of Various Application Methods of Fluensulfone for Managing Root-knot Nematodes in Vegetables. J Nematol. 2016 Jun;48(2):65-71. doi: 10.21307/jofnem-2017-010. PMID: 27418698; PMCID: PMC4930317.

Fluopyram
https://www.canada.ca/en/health-canada/services/consumer-product-safety/reports-publications/pesticides-pest-management/decisions-updates/evaluation-report/2014/fluopyram-erc2014-02.html
Yun MS, Choi H. Uptake of Fungicide Fluopyram from Soil by Scallions during Greenhouse Cultivation. Foods. 2023 May 15;12(10):1996. doi: 10.3390/foods12101996. PMID: 37238814; PMCID: PMC10217027.

Flupyrimin
https://www.chemicalbook.com/article/the-introduction-of-flupyrimin.
https://irac-online.org/active-ingredient/flupyrimin
Terajima T, Suzuki T, Horikoshi R, Doi S, Nakamura M, Kobayashi F, Durkin KA, Shimomura K, Nakamura S, Yamamoto K, Tomizawa M. Deciphering the Flupyrimin Binding Surface on the Insect Nicotinic Acetylcholine Receptor. J Agric Food Chem. 2021 Aug 25;69(33):9551-9556. doi: 10.1021/acs.jafc.1c03241. Epub 2021 Aug 10. PMID: 34374535.

Imidacloprid
https://www.acs.org/molecule-of-the-week/archive/i/imidacloprid.html
https://kunziteindia.com/imidacloprid.php
https://www.npic.orst.edu/factsheets/archive/imidacloprid.html
Chung KT, Chen LW, Tseng HW, Wu CH. Neonicotinoid Imidacloprid Affects the Social Behavior of Adult Zebrafish by Damaging Telencephalon Neurons through Oxidation

Stress, Inflammation, and Apoptosis. Life (Basel). 2023 Jun 20;13(6):1418. doi: 10.3390/life13061418. PMID: 37374200; PMCID: PMC10304127.

Zhu X, Hu CT, Erriah B, Vogt-Maranto L, Yang J, Yang Y, Qiu M, Fellah N, Tuckerman ME, Ward MD, Kahr B. Imidacloprid Crystal Polymorphs for Disease Vector Control and Pollinator Protection. J Am Chem Soc. 2021 Oct 20;143(41):17144-17152. doi: 10.1021/jacs.1c07610. Epub 2021 Oct 12. Erratum in: J Am Chem Soc. 2021 Dec 8;143(48):20525. doi: 10.1021/jacs.1c11855. PMID: 34634905.

Elizabeth P. Benton, Jerome F. Grant, Rebecca J. Nichols, R. Jesse Webster, John S. Schwartz, Joseph K. Bailey, Risk assessment of imidacloprid use in forest settings on the aquatic macroinvertebrate community, *Environmental Toxicology and Chemistry*, Volume 36, Issue 11, 1 November 2017, Pages 3108–3119, https://doi.org/10.1002/etc.3887

Isocycloserum

Willem G van Herk, Robert S Vernon, Ted Labun, Joshua Spies, Isocycloseram, a novel isoxazoline insecticide seed treatment for protection of wheat and barley and mortality of wireworms, *Limonius californicus* (Coleoptera: Elateridae), *Journal of Economic Entomology*, Volume 117, Issue 5, October 2024, Pages 1926–1937, https://doi.org/10.1093/jee/toae170

Zanetti R, Sanches JJ, Wenzel AVA, Haddi K, Ferreira H, Santos LV. Isocycloseram: A new active ingredient for leaf-cutting ants control. PLoS One. 2024 May 9;19(5):e0300187. doi: 10.1371/journal.pone.0300187. PMID: 38722866; PMCID: PMC11081378.

Zhou T, Wu W, Ma S, Chen J, Huang J, Qiao X. Effects of RDL GABA Receptor Point Mutants on Susceptibility to Meta-Diamide and Isoxazoline Insecticides in *Drosophila melanogaster*. Insects. 2024 May 6;15(5):334. doi: 10.3390/insects15050334. PMID: 38786890; PMCID: PMC11122182.

Cassayre, Jérôme & Smejkal, Tomas & Blythe, Judith & Hoegger, Patrik & Renold, Peter & Pitterna, Thomas & Prasanna, Chandika & Smits, Helmars & Godineau, Edouard & Luksch, Torsten & Berthon, Guillaume & Rawal, Girish & Patre, Rupesh & Lal, Mukul & Boussemghoune, Mohamed & Masala, Simonetta & Barreteau, Fabien & Flaeschel, Michael & Vogt, Juliane & El Qacemi, Myriem. (2021). The discovery of isocycloseram: A novel isoxazoline insecticide. 10.1016/B978-0-12-821035-2.00008-5.

https://www.syngenta.com.my/product/crop-protection/insecticide/simodis

https://pubchem.ncbi.nlm.nih.gov/compound/87323565

Lambda-cyhalothrin

Djouaka R, Soglo MF, Kusimo MO, Adéoti R, Talom A, Zeukeng F, Paraïso A, Afari-Sefa V, Saethre MG, Manyong V, Tamò M, Waage J, Lines J, Mahuku G. The Rapid Degradation of Lambda-Cyhalothrin Makes Treated Vegetables Relatively Safe for Consumption. Int J Environ Res Public Health. 2018 Jul 20;15(7):1536. doi: 10.3390/ijerph15071536. PMID: 30036953; PMCID: PMC6068899.

Choi JY, Chon K, Kim J, Vasamsetti BMK, Kim BS, Yoon CY, Hwang S, Park KH, Lee JH. Assessment of Lambda-Cyhalothrin and Spinetoram Toxicity and Their Effects on the Activities of Antioxidant Enzymes and Acetylcholinesterase in Honey Bee (*Apis mellifera*) Larvae. Insects. 2024 Aug 1;15(8):587. doi: 10.3390/insects15080587. PMID: 39194792; PMCID: PMC11354917.

https://ipm.ucanr.edu/home-and-landscape/pesticide-active-ingredients-database

https://www.rallis.com/international-business/Lambda-cyhalothrin

https://pubchem.ncbi.nlm.nih.gov/compound/lambda-Cyhalothrin

https://www.mda.state.mn.us/protecting/bmps/pollinators/beetoxicity

Lufenuron

Lenardon MD, Munro CA, Gow NA. Chitin synthesis and fungal pathogenesis. Curr Opin Microbiol. 2010 Aug;13(4):416-23. doi: 10.1016/j.mib.2010.05.002. Epub 2010 Jun 8. PMID: 20561815; PMCID: PMC2923753.

Rafaela Leão Soares P, Lucas Corrêa de Andrade A, Pinheiro Santos T, Caroline Barros Lucas da Silva S, Freitas da Silva J, Rodrigues Dos Santos A, Hugo Lima da Silva Souza E, Magliano da Cunha F, Wanderley Teixeira V, Sales Cadena MR, Bezerra de Sá F, Bezerra de Carvalho Júnior L, Gonçalves Cadena P. Acute and chronic toxicity of the benzoylurea pesticide, lufenuron, in the fish, *Colossoma macropomum*. Chemosphere. 2016 Oct;161:412-421. doi: 10.1016/j.chemosphere.2016.07.033. Epub 2016 Jul 21. PMID: 27448754.

Richard M. Davis, Erika Cleugh, Randall T. Smith, Curtis L. Fritz "Use of a chitin synthesis inhibitor to control fleas on wild rodents important in the maintenance of plague, *Yersinia pestis*, in California," Journal of Vector Ecology, 33(2), 278-284, (1 December 2008)

https://angelcrop.com/shop/lufenuron-ec

https://dailymed.nlm.nih.gov/dailymed/fda/

Malathion

https://www.health.ny.gov/publications/2740/

https://earthjustice.org/feature/organophosphate-pesticides-united-states/malathion

https://www.ncbi.nlm.nih.gov/books/NBK436793/

https://npic.orst.edu/factsheets/archive/malatech.html

https://synapse.patsnap.com/article/what-is-the-mechanism-of-malathion

Metaflumizone

Sun XX, Li HY, Jiang YJ, Zhang JX, Gu HL, Gao B, Ma JJ, Wang F, Zhou JC, Tian XR, Su J, Wang K. Resistance Risk Evaluated by Metaflumizone Selection and the Effects on Toxicities Over Other Insecticides in *Spodoptera exigua* (Lepidoptera: Noctuidae). J Econ Entomol. 2019 Sep 23;112(5):2354-2361. doi: 10.1093/jee/toz171. PMID: 31219572.

von Stein RT, Silver KS, Soderlund DM. Indoxacarb, Metaflumizone, and Other Sodium Channel Inhibitor Insecticides: Mechanism and Site of Action on Mammalian Voltage-Gated Sodium Channels. Pestic Biochem Physiol. 2013 Jul 1;106(3):101-112. doi: 10.1016/j.pestbp.2013.03.004. PMID: 24072940; PMCID: PMC3780446.

https://diypestcontrol.com/active-ingredients/metaflumizone

Methoxyfenozide

https://chemicalwarehouse.com/blogs/active-ingredients/methoxyfenozide

https://www.essencechem.com/product/methoxyfenozide.

https://sitem.herts.ac.uk/aeru/ppdb/en/Reports/461.

Carlson, Glenn & Dhadialla, Tarlochan & Hunter, Ricky & Jansson, Richard & Jany, Christine & Lidert, Zev & Slawecki, Richard. (2001). The chemical and biological properties of methoxyfenozide, a new insecticidal ecdysteroid agonist. Pest Management Science. 57. 115 - 119. 10.1002/1526-4998(200102)57:2<115::AID-PS245>3.0.CO;2-A.

Ahmed FS, Helmy YS, Helmy WS. Toxicity and biochemical impact of methoxyfenozide/spinetoram mixture on susceptible and methoxyfenozide-selected strains of Spodoptera littoralis (Lepidoptera: Noctuidae). Sci Rep. 2022 Apr 28;12(1):6974. doi: 10.1038/s41598-022-10812-w. PMID: 35484385; PMCID: PMC9050723.

Methyl bromide

https://www.unep.org/ozonaction/what-we-do/methyl-bromide

https://www.gov.uk/government/publications/methyl-bromide

https://www.cdc.gov/mmwr/preview/mmwrhtml/mm6027a4.htm

https://environment.govt.nz/acts-and-regulations/acts/ozone-layer-protection-act-1996/ozone-depleting-substances/
Barry KH, Koutros S, Lubin JH, Coble JB, Barone-Adesi F, Beane Freeman LE, Sandler DP, Hoppin JA, Ma X, Zheng T, Alavanja MC. Methyl bromide exposure and cancer risk in the Agricultural Health Study. Cancer Causes Control. 2012 Jun;23(6):807-18. doi: 10.1007/s10552-012-9949-2. Epub 2012 Apr 24. PMID: 22527160; PMCID: PMC3430844.

Milbemectin

https://www.mc-croplifesolutions.com/en/randd/active-ingredients/milbemectin/
https://www.bocsci.com/products/milbemycin-analogue-set-2402.html
https://www.toku-e.com/milbemectin/?setCurrencyId=19
https://sitem.herts.ac.uk/aeru/ppdb/en/Reports/518.htm
https://katyayanikrishidirect.com/products/katyayani-phenex-50-phenthoate-50-ec-insecticide
https://www.cnagrochem.com/news/phenthoate-insecticide.
https://sitem.herts.ac.uk/aeru/bpdb/Reports/472
https://www.croplife.org.au/resources/programs/resistance-management/insecticide-mode-of-action-classification-for-insecticides
https://irac-online.org/mode-of-action/classification-online/

Monocrotophos

Kaur, Ravneet & Goyal, Dinesh. (2019). Toxicity and degradation of the insecticide monocrotophos. Environmental Chemistry Letters. 17. 10.1007/s10311-019-00884-y.
https://www.inchem.org/documents/jmpr/jmpmono/v072pr22.

Nitenpyram

https://www.sinoagrochem.com.cn/Insecticide/Nitenpyram+Thiamethoxam-WG
https://www.chemicalbook.com/manufacturers-india/nitenpyram_2.htm
https://pubchem.ncbi.nlm.nih.gov/compound/Nitenpyram

Novaluron

https://sitem.herts.ac.uk/aeru/ppdb/en/Reports/487.html
https://www.sciencedirect.com/topics/agricultural-and-biological-sciences/novaluron
https://agribegri.com/products/ebs-nimon-novaluron-10-ec-insecticide.php
https://pubchem.ncbi.nlm.nih.gov/compound/Novaluron
http://www.globalsciencebooks.info/Online/GSBOnline/images/0706/PT_1(1)/PT_1(1)38-46o.pdf

Phenthoate

https://m.chemicalbook.com/ProductMSDSDetailCB9740624_EN.
https://www.inchem.org/documents/jmpr/jmpmono/v080pr31
https://www.canr.msu.edu/news/effect_of_water_ph_on_the_stability_of_pesticides

Phosalone

https://sitem.herts.ac.uk/aeru/ppdb/en/Reports/520.html
https://www.jindunchemical.com/h-nd-809.html
https://pubchem.ncbi.nlm.nih.gov/compound/Phosalone
https://www.inchem.org/documents/jmpr/jmpmono/v072pr25.htm

Propargite

https://sitem.herts.ac.uk/aeru/ppdb/en/Reports/547.
https://pubchem.ncbi.nlm.nih.gov/compound/Propargite
https://krushidukan.bharatagri.com/en/products/omite-insecticide
https://agribegri.com/products/acaricide-simbaa-propargite-57
https://www.hb-p.com/product/propargite

Pymetrozine
https://www.allpesticides.com/what-is-pymetrozine.html
https://www.awiner.com/product/pymetrozine-mechanism-of-action
https://sitem.herts.ac.uk/aeru/ppdb/en/Reports/562.
Tebufenpyrad
https://sitem.herts.ac.uk/aeru/ppdb/en/Reports/612.
https://www.pomais.com/product/tebufenpyrad-20-wp-insecticide/
https://en.wikipedia.org/wiki/Tebufenpyrad
Tetraniliprole
https://sitem.herts.ac.uk/aeru/ppdb/en/Reports/3190.
https://www.kisancenter.in/product/29777809/Bayer-Vayego-Insecticide--Tetraniliprole
Thiocyclam Hydrogen Oxalate
https://www.chemicalbook.com/ChemicalProductProperty_EN_CB4466039.
https://www.awiner.com/product/thiocyclam-hydrogen-oxalate-50-sp
https://sitem.herts.ac.uk/aeru/ppdb/en/Reports/753.htm
Triflumezopyrim
Cho SM, Lee HS, Park JS, Lee SJ, Shin HS, Chung YM, Choi HN, Jung YH, Oh JH, Yun SS. Determination of Residual Triflumezopyrim Insecticide in Agricultural Products through a Modified QuEChERS Method. Foods. 2021 Sep 3;10(9):2090. doi: 10.3390/foods10092090. PMID: 34574200; PMCID: PMC8472026.
Cordova, Daniel & Benner, Eric & Schroeder, Mark & Holyoke, Caleb & Zhang, Wenming & Pahutski, Thomas & Leighty, Robert & Vincent, Daniel & Hamm, Jason. (2016). Mode of action of triflumezopyrim: A novel mesoionic insecticide which inhibits the nicotinic acetylcholine receptor. Insect biochemistry and molecular biology. 74. 10.1016/j.ibmb.2016.04.008.
https://www.medchemexpress.com/triflumezopyrim.
https://www.glpbio.com/triflumezopyrim.
https://sitem.herts.ac.uk/aeru/ppdb/en/Reports/2668.
Acephate + Bifenthrin
https://raccolto.in/wp-content/uploads/2022/04/Acephate-50-Bifenthrin-10-WDG.pdf
https://universalagri.net/kilin-acephate-50-bifenthrin-10-wdg/
https://www.swal.in/insecticides/acenthrin
Acephate + Cypermethrin
https://patents.google.com/patent/US8142802B2/en
https://raccolto.in/wp-content/uploads/2022/04/ACEPHATE-45-CYPERMETHRIN-5-DF.pdf
https://perfectsolutions.best/insecticides-combinations/
Battu RS, Sahoo SK, Jyot G. Persistence of acephate and cypermethrin on cotton leaves, cottonseed, lint and soil. Bull Environ Contam Toxicol. 2009 Jan;82(1):124-8. doi: 10.1007/s00128-008-9541-3. Epub 2008 Sep 9. PMID: 18777148.
Acephate + Fenvalerate
https://www.rallis.com/international-business/Acephate
https://www.sciencedirect.com/topics/chemistry/acephate
https://patents.google.com/patent/US8142802B2/en
Acephate + Fipronil
https://patents.google.com/patent/CN104430503A/en
https://www.royalagrotech.com/agriculture-insecticides.html

https://angelcrop.com/shop/acephate-50-fipronil-5-wdg

Acephate + Imidacloprid

https://www.kisancenter.in/product/upl-lancergold-insecticide-acephate-imidacloprid-combination

https://agribegri.com/products/swal-starthene-power-acephate-50--imidacloprid-18-sp-insecticide.php

https://www.agrochemicalsindia.com/hi/acephate-50-imidacloprid-1-8-sp-9926652.html

https://agrostar.in/product/agrostar-agetate-gold-acephate-50-imidacloprid-18-sp-1kg/AGS-CP-1496?language=en&state=maharashtra

Acetamiprid 00.40%+Chlorpyriphos 20.00% EC

https://www.tradeindia.com/products/acetamiprid-0-4-chlorpyriphos-20-ec

https://peptechbio.com/products/acetamiprid-0-4-chlorpyrifos-20-w-w-ec

Benzpyrimoxam + Fipronil

https://www.chemicalbook.com/article/the-introduction-of-benzpyrimoxan

https://chemicalwarehouse.com/blogs/active-ingredients/fipronil

Beta-cyfluthrin + Imidacloprid

https://www.awiner.com/product/beta-cyfluthrin

https://www.katyayaniorganics.com/product/katyayani-insectora-chemical-insecticide-beta-cyfluthrin-8-49-imidacloprid-19-81

https://agribegri.com/products/buy-bayer-solomon-online--buy-beta-cyfluthrin--imidacloprid

https://cultree.in/products/bayer-solomon-beta-cyfluthrin-8-49-imidacloprid-19-81

https://www.farmersstop.com/products/solomon-beta-cyfluthrin-imidacloprid-300-od-bayer-india

Bifenthrin + Chlorpyriphos

https://www.globalcropcare.com/product/lansu-bifenthrin-3-chlorpyriphos-30-ec/

https://patents.google.com/patent/WO2017061959A1/en

https://www.pomais.com/product/chlorpyrifos-48ec/

https://analyticalsciencejournals.onlinelibrary.wiley.com/doi/10.1002/jat.4897

Bifenthrin + Clothianidin

https://analyticalsciencejournals.onlinelibrary.wiley.com/doi/10.1002/jat.4897

https://www.silversinebiotech.in/products/goldstar.html

https://www.pomais.com/product/chlorpyrifos-48ec/

Bifenthrin + Chlorantraniliprole

https://npic.orst.edu/factsheets/archive/biftech.html

https://www.fbn.com/community/blog/bifenthrin

https://www.pomais.com/chlorantraniliprole-vs-bifenthrin

Bifenthrin + Thiamethoxam

https://aceagrochem.com/pesticide/insecticide/mixture/Bifenthrin

https://patents.google.com/patent/CN102027986B/en

https://www.pomais.com/thiamethoxam-an-effective-solution-for-pest-control/

https://www.essencechem.com/product/bifenthrin-thiamethoxam

Buprofezin + Acephate

https://kisansewakendra.in/products/ebs-founder-buprofezin-15-acephate-35-wp

https://agribegri.com/products/buy-hpm-papro-buprofezin-15--acephate-35-wp-insecticide

https://peptechbio.com/products/buprofezin-15-acephate-35-wp/

https://parijatagrochemicals.com/product/bace-50/

https://highyieldsagro.com/product/adama-tapuz/
Buprofezin + Acetamiprid
https://www.natur-sim.com/showroom/acetamiprid-25-buprofezin-25-wp
https://patents.google.com/patent/CN101444220A/en
https://chemknock.com/view.do?no=194
https://a9162158112099b1.en.made-in-china.com/product/bmtUIjlrhXRF/China-Insecticide-Buprofezin-95-Tech-Acetamiprid
Buprofezin + Fipronil
https://www.awiner.com/product/buprofezin-mode-of-action/
https://angelcrop.com/shop/buprofezin-22-fipronil-3-sc
https://a9162158112099b1.en.made-in-china.com/product/bmtUIjlrhXRF/
https://katyayanikrishidirect.com/blogs/news/fipronil-insecticide-uses-and-dosage
https://www.naturalcropscience.com/pesticides.html
Cartap hydrochloride + Chlorantraniliprole
https://www.upl-ltd.com/in/crop-protection/insecticide/zevigo
https://www.swal.in/insecticides/convay
https://patents.google.com/patent/CN104585198B/en
Cartap hydrochloride + Buprofezin
https://agribegri.com/products/katyayani-cartazen-cartap-hydrochloride-50--buprofezin-10-wp-insecticide.
https://www.badikheti.com/insecticide/pdp/katyayani-cartazen-cartap-hydrochloride-50-buprofezin-10-wp-insecticide
https://www.defianceagrisciences.com/insecticides-1.html
https://perfectsolutions.best/insecticides-combinations/#:~:text=BUPROFEZIN%20
https://www.globalcropcare.com/product/meleaf-cartap-hydrochloride-50-buprofezin-10-wp/
Cartap hydrochloride + Emamectin benzoate
https://www.adama.com/india/en/crop-protection/insecticides/barroz
https://www.sigmaaldrich.com/IN/en/substance/cartaphydrochloride
https://highyieldsagro.com/product/adama-barroz/
Cartap hydrochloride + Fipronil
https://cultree.in/products/willowood-exotica-cartap-hydrochloride-4-fipronil-0-5-cg-insecticide
https://highyieldsagro.com/product-tag/cartap-hydrochloride-4fipronil-0-5-cg
Chlorantraniliprole + Abamectin
Suraphan N, Fan L, Liu B, Wu D. Co-delivery of chlorantraniliprole and avermectin with a polylactide microcapsule formulation. RSC Adv. 2020 Jul 3;10(43):25418-25425. doi: 10.1039/d0ra03825c. PMID: 35518616; PMCID: PMC9055352.
Kousika, J. & Krishnamoorthy, Sendha & Ragumoorthi, K.N.. (2016). Bioassays of chlorantraniliprole 4.3% + abamectin against 1.7% SC *Spodoptera litura* (F.) and *Tetranychus urticae* (Koch) on tomato. Indian Journal of Entomology. 78. 211. 10.5958/0974-8172.2016.00059.6.
https://www.awiner.com/chlorantraniliprole-abamectin
https://peptechbio.com/products/chlorantraniliprole-4-3-w-w-abamectin-1-7-w-w-sc
Chlorantraniliprole + Fipronil
Simon-Delso N, Amaral-Rogers V, Belzunces LP, Bonmatin JM, Chagnon M, Downs C, Furlan L, Gibbons DW, Giorio C, Girolami V, Goulson D, Kreutzweiser DP, Krupke CH, Liess M, Long E, McField M, Mineau P, Mitchell EA, Morrissey CA, Noome DA, Pisa L, Settele J, Stark JD, Tapparo A, Van Dyck H, Van Praagh J, Van der Sluijs JP, Whitehorn PR, Wiemers

M. Systemic insecticides (neonicotinoids and fipronil): trends, uses, mode of action and metabolites. Environ Sci Pollut Res Int. 2015 Jan;22(1):5-34. doi: 10.1007/s11356-014-3470-y. Epub 2014 Sep 19. PMID: 25233913; PMCID: PMC4284386.

Dhanapati, Gokul & Karjee, Mrinal & Sarkar, Prahlad. (2024). Bio-efficacy of Different Formulations of Chlorantraniliprole 0.53%+ Fipronil 0.8% GR Against Yellow Stem Borer of Rice. International Journal of Bio-resource and Stress Management. 15. 10.23910/1.2024.5069a.

https://agribegri.com/products/centran-sc-insecticide-rice-fipronil-chlorantraniliprole

https://peptechbio.com/products/fipronil-15-chlorantraniliprole-5-sc

https://chemrobotics.in/adama-launches-fmcs-patented-combination

Chlorantraniliprole + Fipronil + Emamectin benzoate

https://www.bestagrolife.com/product/fetagen

https://patents.google.com/patent/CN101554170A/en

Chlorantraniliprole + Lambda-cyhalothrin

https://synwoodagro.com/product/simigo-chlorantraniliprole-9-3-lambda-cyhalothrin-4-6-zc

https://agribegri.com/products/buy-gsp-helipower-chlorantraniliprole-93--lambda-cyhalothrin-46-insecticide-online

https://www.pomais.com/product/chlorantraniliprole-9-3-lambda-cyhalothrin-4-6-zc/

https://peptechbio.com/products/chlorantraniliprole-9-3lambda-cyhalothrin-4-6-zc-pesticide-manufacturer/

Chlorantraniliprole + Novaluron

https://chemrobotics.in/artificial-intelligence-based-market-intelligence-analysis-of-chlorantraniliprole-combination-approvals-for-indian-market-

https://npic.orst.edu/factsheets/chlorantraniliprole

https://www.allpesticides.com/chlorantraniliprole-mode-of-action.

Chlorantraniliprole + Novaluron + Emamectin benzoate

https://agribegri.com/products/best-agrolife-nemagen-chlorantraniliprole-novaluron-emametin-benzoate-insecticide--online.php

https://www.awiner.com/chlorantraniliprole-emamectin-benzoate/

https://peptechbio.com/products/novaluron-5-25-emamectin-benzoate-0-9-sc

https://www.spraykaro.com/product/best-agrolife-nemagen-chlorantraniliprole-4-5-novaluron-11-5-emametin-benzoate-1-5-se

Kumar S, Suby SB, Kumar N, Sekhar JC, Nebapure S, Mahapatro GK. Insecticide susceptibility vis-à-vis molecular variations in geographical populations of fall armyworm, *Spodoptera frugiperda* (J.E. smith) in India. 3 Biotech. 2022 Sep;12(9):241. doi: 10.1007/s13205-022-03303-2. Epub 2022 Aug 23. PMID: 36032511; PMCID: PMC9399317.

Chlorantraniliprole + Thiamethoxam

https://www.katyayaniorganics.com/product/katyayani-chakryaan-chlorantraniliprole- 8-8-thiamethoxam-17-5-w-w-sc-insecticide

https://peptechbio.com/products/chlorantraniliprole-8-8-thiamethoxam-17-5-sc/

https://agroshopy.com/index.php?route=product/product&product_id=379

https://www.pomais.com/product/thiamethoxam-20-chlorantraniliprole-10-sc/

Chlorpyrifos + Alphacypermethrin

https://kisansewakendra.in/products/ebs-metcyper-161-chlorpyriphos-16-alphacypermethrin-1-ec

https://www.indiamart.com/proddetail/chlorpyrifos-16-alphacypermethrin-1-e-c-insecticide

https://www.ncbi.nlm.nih.gov/books/NBK601883/

Chlorpyrifos + Cypermethrin

https://agribegri.com/products/super-d-insecticide.php
https://peptechbio.com/products/chlorpyriphos-50-cypermethrin-5-ec/
https://patents.google.com/patent/CN1183218A/en
https://katyayanikrishidirect.com/products/chlorpyriphos-50-cypermethrin-5-ec-docter-505-insecticide

Chlorpyrifos + Fipronil

https://titanagritech.com/product/chlorpyrifos-35-fipronil-3-5-ec/
https://peptechbio.com/products/chlorpyrifos-35-fipronil-3-5-w-w-ec/
https://www.globalcropcare.com/product/rustam-35-chlorpyrifos35-fipronil-3-5ec/
https://www.sciencedirect.com/science/article/abs/pii/S0048969720350361

Clothianidin + Pyriproxifen

https://sumichem.co.in/products/spl/insecticides/pyclome

Cypermethrin + Quinalphos

https://agribegri.com/products/upl-viraat-cypermethrin-3--quinalphos-20-ec.php
https://ebskrishibhandar.com/products/ebs-german-plus-cypermethrin-3-quinalphos-20-ec-insecticides-systemic-and-contact-action
https://angelcrop.com/shop/cypermethrin-3-quinalphos-20-ec
https://raccolto.in/wp-content/uploads/2022/04/Cypermethrin-3-Quinalphos-20-EC.pdf

Cyantraniliprole + Diafenthiuron

https://krushidukan.bharatagri.com/en/products/syngenta-plesiva-cyantraniliprole-diafenthiuron-insecticide
https://agrivruddhi.com/products/plesiva-insecticide
https://www.pomais.com/product/cyantraniliprole-10-sc/

Cyantraniliprole + Lufenuron

https://www.awiner.com/chlorantraniliprolelufenuron/
https://www.kisanshop.in/product/syngenta-minecto-xtra
https://patents.google.com/patent/CN103300037A/en
https://www.chemicalbook.com/article/pesticide-lufenuron

Cyantraniliprole + Pymetrozine

https://www.syngenta.co.zm/product/crop-protection/plesiva-star
https://www.pomais.com/product/cyantraniliprole-10-sc/

Cyantraniliprole + Thiamethoxam

https://www.syngenta.co.in/product/crop-protection/seedcare/fortenza-duo-cyantraniliprole-thiamethoxam
https://www.joshi-group.com/cyantraniliprole-19-8-thiamethoxam-19-8-fs-11950781.html
https://agrivruddhi.com/products/fortenza-duo-insecticide

Cypermethrin + Indoxacarb

https://www.indiamart.com/proddetail/intra-cypremethrin-10-indoxacarb-10-sc
https://www.engebiotech.com/hot-sale-cypermethrin-10%EC-indoxacarb-insecticide

Cypermethrin + Quinalphos

https://rskkindia.com/quinalphos-20-cypermethrin-3-ec
https://pubchem.ncbi.nlm.nih.gov/compound/Cypermethrin
https://cultree.in/products/upl-viraat-cypermethrin-3-quinalphos-20-ec-insecticide
https://kisansewakendra.in/hi/products/ebs-german-plus-cypermethrin-3-quinalphos-20-ec-insecticides

Deltamethrin + Buprofezin
https://www.indiamart.com/proddetail/deltamethrin-0-72-buprofezin-5-65-ec
https://skyncropcare.com/products/insecticide/buprothrin/
https://www.hpmindia.com/product-insecticides52
Diafenthiuron + Acetamiprid
https://www.perplexity.ai/search/acephate-50-00-bifenthrin-10-0-WYptQFpBRWCk6.nAV_BtnA
https://www.quickcompany.in/patents/a-synergistic-insecticidal-combination-of-diafenthiuron-and-acetamiprid-in-wettable-powder-form-and-a-process-of-prep
https://ppqs.gov.in/sites/default/files/diafenthiuron_40.1_acetamiprid_3.9_wp93finsecticides_india_limited_1.pdf
Diafenthiuron + Bifenthrin
https://highyieldsagro.com/product/adama-takaf
https://www.pomais.com/product/diafenthiuron-50g-l-bifenthrin-10g-l-sc/
https://angelcrop.com/shop/diafenthiuron-47-bifenthrin-sc
https://cultree.in/products/adama-takaf-diafenthiuron-47-bifenthrin-9-4-sc-insecticide
Diafenthiuron + Dinotefuran
https://krushidukan.bharatagri.com/en/products/pi-dorito-diafenthiuron-48-dinotefuran-8-wg-insecticide
https://tangyunbiotech.en.made-in-china.com/product/hnsReUKvEPcB/China-Diafenthiuron-35-Dinotefuran-5-Sc-Vegetable-Insecticide.
https://patents.google.com/patent/WO2018092033A1/en
Diafenthiuron 18% + Fipronil 8% + Dinotefuran 5% SC
https://tropicalagro.in/product/tag-fly-gold/
https://www.indiamart.com/proddetail/tag-fly-gold-diafenthiuron-18-fipronil-8-dinotefuran-5-sc-insecticide-2857769867748.html
Emamectin benzoate + Chlorantraniliprole + Deltamethrin
https://tropicalagro.in/product/tag-hikima/
https://www.awiner.com/chlorantraniliprole-emamectin-benzoate/
Diafenthiuron + Pyriproxyfen
https://cultree.in/products/coromandel-finio-diafenthiuron-25-pyriproxyfen-5-se-insecticide
https://parijatagrochemicals.com/product/xyfen-ultra-pyriproxifen-8-diafenthiuron-30-se
https://www.katyayaniorganics.com/product/pyriproxyfex/
https://mankindag.com/product/insecticide/diafenhuge-p/
https://agribegri.com/technical/pyriproxyfen-5--diafenthiuron-25-se.
Dimethoate + Cypermethrin
https://www.rallis.com/international-business/Dimethoate
https://www.rythukart.com/product/32504468/Kri-Kreate--Dimethoate-20----Cypermethrin-3--Ec-W-W---Krishi-Rasayan
Kariuki, D. K. (2005) "Synergistic Effects of & alpha;-Cypermethrin and Dimethoate against Insect Pests," East and Central African Journal of Pharmaceutical Sciences. African Journals Online (AJOL). doi: 10.4314/ECAJPS.V6I3.9703.
Okogwu OI, Elebe FA, Nwonumara GN. Combinations of cypermethrin and dimethoate alter behavior, hematology and histology of African Catfish, Clarias gariepinus. Environ Anal Health Toxicol. 2022 Dec;37(4):e2022028-0. doi: 10.5620/eaht.2022028. Epub 2022 Oct 11. PMID: 36916041; PMCID: PMC10014742.

Dinotefuran + Acephate
https://krushidukan.bharatagri.com/en/products/pi-industries-dinoace-dinotefuran-4-acephate-50-sg-insecticide
https://www.globalcropcare.com/product/dinotefuran-4-acephate-50-sg/
https://www.mc-croplifesolutions.com/en/randd/active-ingredients/dinotefuran/
Vilas K Patil, Jayant G Chandorkar, Yogesh B Wagh, Yogesh B More, Hemant P Narkhede, Determination of Acephate, Dinotefuran, and Emamectin Benzoate in a Pesticide Formulation: A Single-Laboratory Validation, *Journal of AOAC INTERNATIONAL*, Volume 106, Issue 3, May-June 2023, Pages 523–533, https://doi.org/10.1093/jaoacint/qsad007
Dinotefuran + Ethion
https://www.pomais.com/faq-items/what-is-dinotefuran-used-for/
https://sitem.herts.ac.uk/aeru/ppdb/en/Reports/1195.htm
https://www.rallis.com/international-business/Ethion
https://www.piindustries.com/what-we-do/crop-solutions/row-crops/insecticide-products/
Dinotefuran + Pymetrozine
https://www.katyayaniorganics.com/product/katyayani-bph-super-dinotefuran-15-pymetrozine-45-wg/
https://www.bestagrolife.com/product/axeman
Dinotefuran + Thiamethoxam + Pymetrozine
https://gramiti.com/products/ju-hemp
https://cropcaresolutions.com/dealer/ju-agri-science-private-limited/triston
https://www.global-agriculture.com/crop-protection/india-approves-27-new-crop-protection-products-under-section-93-at-464th-rc-meeting/
Emamectin benzoate + Diafenthiuron
https://longbin.en.made-in-china.com/product/gOltoaqjrXrS/China-Insecticide-Emamectin-Benzoate-1-5-Diafenthiuron-46-5-Sc
https://patents.google.com/patent/CN104522015A/en
https://sitem.herts.ac.uk/aeru/bpdb/Reports/1326.htm
Emamectin benzoate + Fipronil
https://katyayanikrishidirect.com/products/katyayani-sipahi-emamectin-benzonatye-1-5-fipronil-3-5-sc-insecticide
https://bacf.in/
https://kisansewakendra.in/products/ebs-kranti-emamectin-benzoate-1-5-fipronil-3-5-sc-insecticide
https://advancepesticides.com/products/licence/
Emamectin benzoate + Lambda-cyhalothrin
https://patents.google.com/patent/CN101406199A/en
https://molotuschem.en.made-in-china.com/product/vJxYMkhyCgVS/China-Emamectin-Benzoate-5-Lambda-Cyhalothrin-10-Wp-Mixture-Pesticides.html
Emamectin benzoate + Lambda-cyhalothrin + Bifenthrin
https://juagrisciences.com/blog/Crop/Ayaka-The-Patented-Innovation-Setting-New-Standards-in-Farming
https://gramiti.com/products/juayaka-emamectin-benzoate-1-25-lambda-cyhalothrin-1-5-bifenthrin-4-5-ec
https://agribegri.com/products/ju-ayaka-emamectin-benzoate-125--lambda-cyhalothrin-15--bifenthrin-45-ec-insecticide.
Emamectin benzoate + Lufenuron
https://www.awiner.com/emamectin-benzoate-and-lufenuron/

https://www.chemicalbook.com/article/pesticide-lufenuron-uses-mechanism-of-action-side-effects-and-toxicity.htm
https://www.pomais.com/emamectin-benzoate-5-sg-use-mode-of-action-and-effectiveness/
Emamectin benzoate + Permethrin
https://patents.google.com/patent/CN103931649A/en
https://krushidukan.bharatagri.com/en/products/coromandel-marvex-emamectin-benzoate-2-2-permethrin-15-3-ec-insecticide
Emamectin benzoate + Profenofos
https://krushidukan.bharatagri.com/en/products/sumitomo-karko-emamectin-benzoate-1-5-profenofos-35-wdg-insecticide
https://highyieldsagro.com/product/parijat-velektin/
Emamectin benzoate + Thiamethoxam
https://katyayanikrishidirect.com/products/emamectin-benzoate-3-thiamethoxam-12-wg-emathio-insecticide
https://kisansewakendra.in/products/ebs-emaan-plus-emamectin-benzoate-3-0-thiamethoxam-12-0-broad-spectrum-insecticide
https://www.awiner.com/emamectin-benzoate-and-thiamethoxam
https://www.pomais.com/product/emamectin-benzoate-25g-kg-thiamethoxam-125g-kg-wdg/
https://linuxcrop.com/product/emamectinbenzoate-thiamethoxam-wg
https://linuxcrop.com/product/emamectinbenzoate-thiamethoxam-wg
Ethion + Cypermethrin
https://www.rallis.com/international-business/Ethion
https://ebskrishibhandar.com/products/ebs-ethon-cyper-ethion-40-cypermethrin-5-ec
https://www.actagrochem.com/products/ethomite-ethion-40%25-%2B-cypermethrin-5%25-ec-
https://katyayanikrishidirect.com/products/ethiocyper-45-ethion-40-cypermethrin-5
Ethiprole + Imidacloprid
https://www.cropscience.bayer.in/en/protect-crops/insecticides-en/glamore
https://www.farmersstop.com/products/glamore-ethiprole-imidacloprid-80-wg-bayer-india
https://krushidukan.bharatagri.com/en/products/glamore-ethiprole-imidacloprid-40wg
Ethiprole + Pymetrozine
https://patents.google.com/patent/CN101669481A/en
https://www.cropscience.bayer.in/en/protect-crops/insecticides-en/curbix-pro
https://peptechbio.com/products/ethiprole-10-7-pymetrozine-40-wg
Etoxazole + Abamectin
https://www.awiner.com/etoxazole-6-abamectin-1-5-sc
https://aceagrochem.com/pesticide/insecticide/mixture/Abamectin%2030%20+%20 Etoxazole%20200%20SC.
https://sumichem.co.in/products/spl/insecticides/portion.php
Fenazaquin + Bifenthrin
https://saraswatiagrolifescience.com/our-products/sa-magnitude-fenazaquin-10-bifenthrin-4/
https://www.scimplify.com/fenazaquin-10-bifenthrin-4-ec

Fenobucarb + Buprofezin
https://raccolto.in/wp-content/uploads/2022/04/Fenobucarb-BPMC-20-Buprofezin-5-SE-for-indigenous-manufacture.pdf
https://patents.google.com/patent/CN102067846A/en
https://silversinebiotech.com/products/iffective-fenobucarb-bpmc-20-buprofezin-5-se.html
Fenobucarb + Buprofezin + Acephate
https://katyayanikrishidirect.com/products/katyayani-fenastra-fenobucarb-b-p-m-c-50-ec-insecticide
https://www.hpmindia.com/product-insecticide-papro.php
Fipronil + Acetamiprid
https://patents.google.com/patent/CN101292665A/en
https://parijatagrochemicals.com/product/supranil/
https://advancepesticides.com/products/Metabu/
https://www.badikheti.com/insecticide/pdp/katyayani-fantasy-plus-fipronil-4-acetamipirid-4-ww-sc-insecticides
https://www.hpmindia.com/product-insecticides44.php
https://agribegri.com/technical/fipronil-4--acetamiprid-4-sc.php
Fipronil + Buprofezin
https://patents.google.com/patent/CN105284837A/en
Fipronil + Chlorantraniliprole
https://agribegri.com/products/centran-sc-insecticide
https://peptechbio.com/products/fipronil-15-chlorantraniliprole-5-sc/
https://www.synovitacropscience.com/escort.html
Fipronil + Deltamethrin
https://essencechem.en.made-in-china.com/product/RnWYaDtSTGcL/China-Liquid-Insecticide-Deltamethrin-6-Fipronil-5-Sc.
https://www.gharda.com/product/branded-formulations-insecticides/
https://www.indiamart.com/proddetail/gharda-vajra-fipronil-deltamethrin-insecticides-2855775686155.html
Fipronil + Diafenthiuron
https://krushidukan.bharatagri.com/en/products/gharda-guru-fipronil-10-diafenthiuron-30-wg-insecticide
https://awiner-biotech.en.made-in-china.com/product/dFraSiUPHeTJ/China-Fipronil-10-Diafenthiuron-30-Wg-Insecticide-Organic-Thrip-Control.
https://krushidukan.bharatagri.com/en/products/dictator-fipronil-10-diafenthiuron-30-wg-insecticide
https://agribegri.com/products/gharda-guru-fipronil-10--diafenthiuron-30-wg-insecticide.php
Fipronil + Lambda-cyhalothrin
https://www.bigpesticides.com/news/lambda-cyhalothrin-vs-fipronil/
https://shbosman.en.made-in-china.com/product/KSBEQMAPheWL/China-Compound-Pesticides-Insecticides-Mixture-Lambda-Cyhalothrin-5-Fipronil-5-EC.
https://patents.google.com/patent/CN1947514A/en
Fipronil + Propargite
https://mankindag.com/product/insecticide/fiprogite/
https://cameochemicals.noaa.gov/chemical/4341
https://katyayanikrishidirect.com/blogs/news/fipronil-insecticide-uses-and-dosage

Fipronil + Imidacloprid
https://agribegri.com/products/buy-fimida-fipronil-40--imidacloprid-40-wg--online-agribegri.php
https://npic.orst.edu/factsheets/archive/fiptech.html
https://www.scimplify.com/fipronil-40-imidacloprid-40-wg
https://www.badikheti.com/insecticide/pdp/katyayani-nashak-fipronil-40-imidacloprid-40-wg-unique-insecticide/
Fipronil + Hexythiazox
https://agrivruddhi.com/products/mythri
https://www.bestagrolife.com/product/cubax-power
https://mankindag.com/product/insecticide/fiprokind-h-tm/
https://naclind.com/product/nagarjuna-mantle/
Fipronil + Tolfenpyrad + Abamectin
https://www.bestagrolife.com/product/best-man
https://www.indiamart.com/proddetail/500ml-fipronil-abamectin-tolfenpyrad-insecticide-2857649963188.html
https://www.linkedin.com/posts/sudarshanfarm_cropprotection-executiveinsecticide-sudarshanfarm-activity-7346863183019577344-VPt5
Fipronil + Flonicamid
https://agribegri.com/products/swal-oxalis-fipronil-15--flonicamid-15-wdg-insecticide.php
https://www.upl-ltd.com/in/crop-protection/insecticide/apache
https://vapiagro.com/products/agriculture-chemical/fipronil-15-flonicamid-15-wdg
https://www.awiner.com/product/flonicamid/
https://www.indiamart.com/proddetail/iconiq-flonicamid-insecticide-2857339858291.html
Fipronil + Thiamethoxam
https://krushidukan.bharatagri.com/en/products/katyayani-foxy-fipronil-4-thiamethoxam-4-sc-insecticide
https://peptechbio.com/products/thiamethoxam-0-9-fipronil-0-2-gr/
https://advancepesticides.com/products/Canoza/
Flubendiamide + Buprofezin
https://www.globalcropcare.com/product/rapid-flubendiamide-4-buprofenzin-20-sc
https://irac-online.org/active-ingredient/flubendiamide/
https://www.dhanuka.com/insecticide/apple
Flubendiamide + Deltamethrin
https://agribegri.com/products/fenos-quick-bayer.php
https://highyieldsagro.com/product/bayer-fenos-quick/
https://cultree.in/products/bayer-fenos-quick-flubendiamide-8-33-w-w-deltamethrin-5-56-w-w-sc-insecticide
https://www.cropscience.bayer.in/en/protect-crops/insecticides-en/fenos-quick
Flubendiamide + Thiacloprid
https://www.cropscience.bayer.in/en/protect-crops/insecticides-en/belt-expert
https://agribegri.com/products/bayer-belt-expert-flubendiamide-1992--thiacloprid-1992-ww-sc.php
https://farmsquare.ng/product/belt-expert-insecticide/
https://www.perplexity.ai/search/flubendiamide-deltamethrin-det-
Flonicamid + Diafenthiuron
https://patents.google.com/patent/CN103766363A/en

https://www.awiner.com/product/flonicamid/
https://chemsales.en.made-in-china.com/product/faKrUceubykq/China-Diafenthiuron-35-Flonicamid-5-SC-Diafenthiuron-Flonicamid-35-5-Pesticide-Mixture.html
https://orangeprotection.com.pk/products/bazigar-df-xtra

Fluxametamide + Bifenthrin

https://agrivruddhi.com/products/lanevo-inseticide
https://krushidukan.bharatagri.com/en/products/dhanuka-lanevo-bifenthrin-5-81-fluxametamide-5-81-ec-insecticide
https://highyieldsagro.com/product/dhanuka-lanevo/

Fluxametamide + Pyridaben

https://godrejagrovet.com/businesses/crop-protection/insecticides/rashinban
https://myagroshop.in/product/godrej-rashinban-400ml-fluxametamide-3-8-w-w-pyridaben-9-5-w-w-sc/
https://www.unitedagroindustries.in/pharmaceutical-chemical.html

Hexythiazox + Diafenthiuron

https://agrivruddhi.com/products/canister
https://krushidukan.bharatagri.com/en/products/coromandel-canister-hexythiazox-3-5-diafenthiuron-42-wdg-insecticide
https://agribegri.com/products/buy-coromandel-canister-hexythiazox-35--diafenthiuron-42-wdg-insecticide.php

Imidacloprid + Bifenthrin

https://www.rythukart.com/product/talextra-insecticide-fungicide
https://krepl.in/product/talextra-imidacloprid-20-bifenthrin-8-sl/
https://longbin.en.made-in-china.com/product/uwhfpWaEgBRy/China-Pesticide-Insecticide-Imidacloprid-3-Bifenthrin-2-Ec.html

Imidacloprid + Lambda-cyhalothrin

https://www.essencechem.com/product/lambda-cyhalothrin-imidacloprid.html
https://aceagrochem.com/pesticide/insecticide/mixture/Lambda-cyhalothrin%2050%20+%20Imidacloprid%20100%20SC.html
https://www.rallis.com/international-business/Lambda-cyhalothrin

Indoxacarb + Acetamiprid

https://agrosiaa.com/products/detail/gharada-kite-insecticide-50-ml-indoxacarb-14-acetamiprid-7-7-sc
https://www.hpmindia.com/product-insecticide-bajirao.php
https://agribegri.com/products/buy-gharda-chemical-kite-indoxacarb-145--acetamiprid-77-sc-online.php
https://www.katyayaniorganics.com/product/katyayani-ace-indox-indoxacarb-14-5-acetamiprid-7-7-sc-chemical-insecticide/

Indoxacarb + Fipronil

https://patents.google.com/patent/CN101473839A/en
https://agribegri.com/products/buy-gharda-black-indoxacarb-5--fipronil-5-sc-insecticide-online
https://agrivruddhi.com/products/black-insecticide
https://www.silversinebiotech.in/products/indoxa-prime

Indoxacarb + Thiamethoxam

https://www.allpesticides.com/thiamethoxam-vs-indoxacarb.html
https://agribegri.com/products/gharda-firrangi-indoxacarb-10--thiamethoxam-10-wg-insecticide.php
https://www.awiner.com/indoxacarb-thiamethoxam/

https://patents.google.com/patent/CN101790989A/en
Lufenuron + Chloantraniliprole + Pyriproxyfen
https://yougebiotech.com/lufenuron-insecticide
https://patents.google.com/patent/CN102960354A/en
Methoxyfenozide + Chlorantraniliprole
https://mankindag.com/product/insecticide/brizesta
https://agrivruddhi.com/products/liger-insecticide
https://agrostar.in/product/armda-methoxyfenozide-20-chlorantraniliprole-5-sc-1liter/AGS-CP-1759
Methoxyfenozide + Chlorpyrifos
https://patents.google.com/patent/CN103563980A/en
https://www.epa.gov/ingredients-used-pesticide-products/chlorpyrifos
Methoxyfenozide + Emamectin benzoate
https://krushidukan.bharatagri.com/en/products/gsp-fighter-methoxyfenozide-18-emamectin-1-8-sc-insecticide
https://agrivruddhi.com/products/fighter-insecticide
https://katyayanikrishidirect.com/blogs/news/emamectin-benzoate-5-sg
Nitenpyram + Fipronil
https://chicocrop.en.made-in-china.com/product/pyCxaDclhgWh/China-Pesticide-Nitenpyram-20-SL-Pest-Control.html
https://npic.orst.edu/factsheets/archive/fiptech.html
https://patents.google.com/patent/CN102396460B/en
Simon-Delso N, Amaral-Rogers V, Belzunces LP, Bonmatin JM, Chagnon M, Downs C, Furlan L, Gibbons DW, Giorio C, Girolami V, Goulson D, Kreutzweiser DP, Krupke CH, Liess M, Long E, McField M, Mineau P, Mitchell EA, Morrissey CA, Noome DA, Pisa L, Settele J, Stark JD, Tapparo A, Van Dyck H, Van Praagh J, Van der Sluijs JP, Whitehorn PR, Wiemers M. Systemic insecticides (neonicotinoids and fipronil): trends, uses, mode of action and metabolites. Environ Sci Pollut Res Int. 2015 Jan;22(1):5-34. doi: 10.1007/s11356-014-3470-y. Epub 2014 Sep 19. PMID: 25233913; PMCID: PMC4284386.
Novaluron + Emamectin benzoate
https://peptechbio.com/products/novaluron-5-25-emamectin-benzoate-0-9-sc/
https://krishibazaar.in/product/aakramak-insect-control
https://synwoodagro.com/product/novatin-novaluron-5-25-emamectin-benzoate-0-9-sc/
https://www.bestagrolife.com/product/mastery
Novaluron + Indoxacarb
https://highyieldsagro.com/product-tag/novaluron-5-25indoxacarb-4-5-sc/
https://agribegri.com/products/mahindra-wingman-novaluron-525--indoxacarb-450-sc-insecticide.php
https://pmc.ncbi.nlm.nih.gov/articles/PMC11724406/
Novaluron + Lambda-cyhalothrin
https://patents.google.com/patent/CN102224827A/en
https://pubmed.ncbi.nlm.nih.gov/33120258/
Phenthoate + Cypermethrin
https://patents.google.com/patent/WO2022034533A1/en
https://hellocrop.in/product/phendal-plus-phenthoate-45-cypermethrin-6%EC-coromandel-international/

Phenthoate + Emamectin benzoate
https://www.global-agriculture.com/crop-protection/india-approves-27-new-crop-protection-products-under-section-93-at-464th-rc-meeting/
https://www.bighaat.com/products/katyayani-phenex-50-phenthoate-50-ec-insecticides
Fanigliulo A, Sacchetti M. Emamectin benzoate: new insecticide against Helicoverpa armigera. Commun Agric Appl Biol Sci. 2008;73(3):651-3. PMID: 19226807.
Profenofos + Cypermethrin
https://www.bigpesticides.com/news/profenofos-cypermethrin-uses-crops-target-pests-application-tips/
https://krushidukan.bharatagri.com/en/products/pi-roket-profenofos-40-cypermethrin-4-ec-insecticide-1-1-combo
https://agribegri.com/products/proper-404-profenofos-cypermethrin.php
https://cultree.in/products/nagarjuna-nacl-profex-super-profenofos-40-cypermethrin-4-e-c
Profenofos + Fenpropathrin
https://patents.google.com/patent/CN104286005A/en
https://sumichem.co.in/products/spl/insecticides/meshi.php
https://www.bigpesticides.com/news/profenofos-cypermethrin-uses-crops-target-pests-application-tips/
https://www.sciencedirect.com/topics/chemistry/fenpropathrin
Profenofos + Fenpyroximate
https://cultree.in/products/excel-sumitomo-etna-profenophos-40-fenpyroximate-2-5-ec-insecticide
https://agribegri.com/products/buy-excel-sumitomo-etna-profenophos-40--fenpyroximate-25-ec.php
https://www.synovitacropscience.com/altair.html
https://krepl.in/product/galwan/
https://setu.atul.co.in/atul_setu/upload/5538/Insecticides/PDF/ORAX%20GOLD.pdf
Profenofos + Lambda-cyhalothrin
https://www.pomais.com/product/profenofos-300g-l-lambda-cyhalothrin-15g-l-ec/
https://engepesticides.en.made-in-china.com/product/VtorzJOcnbky/China-Insecticide-Profenofos-300g-L-Lambda-Cyhalothrin-15g-L-Ec-Pesticide-Combination.html
Propargite + Bifenthrin
https://agrosiaa.com/products/detail/indofil-dammu-insecticide-250-ml-propargite-50-bifenthrin-5-se
https://ag.fmc.com/in/en/insecticides/brigade-power-insecticide
https://agribegri.com/technical/propargite-50--bifenthrin-5-se.php
Propargite + Hexythiazox
https://agribegri.com/products/buy-indofil-ceasemite-propargite-42--hexythiazox-2-ec-insecticide-online.php
https://www.awiner.com/hexythiazox-propargite/
https://www.badikheti.com/insecticide/pdp/katyayani-stopamite-propargite-42-hexythiazox-2-ec-insecticide/zo3mhxi8
Pymetrozine + Fipronil + Dinotefuran
https://agribegri.com/products/buy-agriventure-dinopy-dinotefuron-15--pymetrozine-45-wg-insecticide-online.php
https://pubchem.ncbi.nlm.nih.gov/compound/Emamectin-Benzoate

https://lavaurchemical.en.made-in-china.com/product/lxpRvGujYsWY/China-Insecticide-Phenthoate-40-Me-50-Ec-60-Ec-High-Quality.html

Pymetrozine + Triflumezopyrim

https://patents.google.com/patent/WO2022180096A1/en
https://www.syngenta.co.in/product/crop-protection/insecticide/vestoria-pro
https://ag.fmc.com/in/en/insecticides/eltra-insecticide

Pyriproxyfen + Fenpropathrin

https://cultree.in/products/sumitomo-sumiprempt-pyriproxyfen-5-fenpropathrin-15-ec-insecticide
https://agribegri.com/products/buy-sumitomo-sumiprempt.php
https://www.globalcropcare.com/product/fenox-pyriproxyfen-5-fenpropathrin-15-ec/
https://sahibgroup.in/products/assets/images/Insecticides/debd2d64-58a2-437a-8770-4a81db079b6a-Rudram.png

Pyriproxyfen + Bifenthrin

https://parijatagrochemicals.com/product/leoniscomax/
https://www.katyayaniorganics.com/product/katyayani-fengo-10-pyriproxyfen-10-bifenthrin-10-ec-insecticide/
https://cropgenetechnology.com/shop/fly-gene-whitefly-control-insecticide-pyriproxyfen-10-bifenthrin-10-ec-for-cotton-crops/
https://www.gujaratpesticides.com/products/cease/

Pyriproxyfen + Dinotefuran + Diafenthiuron

https://www.bestagrolife.com/product/ronfen
https://agroshopy.in/product/sutathion-pyriproxyfen-8-dinotefuran-5-diafenthiuron-18-sc-250-ml/
https://preventis.in/product/insecticides/caliber-pyriproxyfen-8-dinotefuran-5-diafenthiuron-18-sc/
https://gramiti.com/products/jutridev-pyriproxyfen-8-diafenthiuron-18-dinotefuran-5-sc
https://patents.google.com/patent/WO2018092033A1/en

Spinetoram + Methoxyfenozide

https://epubs.icar.org.in/index.php/IJPP/article/view/106369
https://farme.in/product-view/engage
https://www.indiamart.com/proddetail/corteva-engage-spinetoram-2850247720933.html
https://www.apvma.gov.au/sites/default/files/publication/105531-trade_advice_notice_on_methoxyfenozide_and_spinetoram_in_the_product_intrepid_edge

Ahmed FS, Helmy YS, Helmy WS. Toxicity and biochemical impact of methoxyfenozide/spinetoram mixture on susceptible and methoxyfenozide-selected strains of Spodoptera littoralis (Lepidoptera: Noctuidae). Sci Rep. 2022 Apr 28;12(1):6974. doi: 10.1038/s41598-022-10812-w. PMID: 35484385; PMCID: PMC9050723.

Spiromesifen + Hexythiazox + Abamectin

https://pubchem.ncbi.nlm.nih.gov/compound/Hexythiazox#section=Use-Classification
https://sitem.herts.ac.uk/aeru/ppdb/en/Reports/598.html
https://patents.google.com/patent/WO2025003875A1/en
https://www.facebook.com/bestagrolifelimited/posts/-coming-soon-cubax-power-extra-a-never-before-triple-action-solution-is-about-to/1407031801434639/

Spirotetramat + Imidacloprid

https://www.cropscience.bayer.in/en/protect-crops/insecticides-en/movento-energy

https://cultree.in/products/bayer-movento-energy-spirotetramat-11-01-imidacloprid-11-01-w-w-sc-insecticide
https://www.cnagrochem.com/news/spirotetramat-imidacloprid-service.html

Spiropidon + Acetamiprid

https://www.syngenta.co.in/product/crop-protection/insecticide/elestal-neo
https://www.syngenta.com.my/product/crop-protection/elestal-54wg
https://www.sweechiang.com.my/showproducts/productid/4299136/cid/430755/spiropidion-300-ww-acetamiprid-240-ww-wg/
https://www.indianchemicalnews.com/general/syngenta-launches-new-insecticide-active-ingredient-for-crop-protection-6859
https://www.syngenta.com/media/media-releases/2020/syngenta-crop-protection-announces-launch-spiropidion-new-insecticide

Sulfoxaflor + Buprofezin

https://www.epa.gov/ingredients-used-pesticide-products/sulfoxaflor
https://pubs.acs.org/doi/10.1021/jf102765
https://www.uplcorp.com/lk/product-details/buprofezin-25-sc

Tetraniliprole + Fipronil

https://agribegri.com/products/bayer-bicota-tetraniliprole-04--fipronil-06-gr-insecticide.php
https://sitem.herts.ac.uk/aeru/ppdb/en/Reports/316.htm
https://xerces.org/systemic-insecticides/tetraniliprole

Tetraniliprole + Spirotetramat

https://www3.epa.gov/pesticides/chem_search/ppls/000264-01193-20210310.pdf
https://www3.epa.gov/pesticides/chem_search/ppls/000432-01591-20210310.pdf
https://www.lsuagcenter.com/profiles/aiverson/articles/page1548277319792
https://www.hbjrain.com/news/introduction-to-spirotetramat-insecticide-a-modern-approach-to-pest-control.html

Tetraniliprole + Thiacloprid

https://agribegri.com/products/shriram-chrone-tetraniliprole-1008--thiacloprid-3025-sc-insecticide
https://patents.google.com/patent/WO2020012312A1/en

Thiamethoxam + Lambda-cyhalothrin

https://peptechbio.com/products/thiamethoxam-12-6-lambda-cyhalothrin-9-5-zc/
https://agribegri.com/products/thiamethoxam-126---lambda-cyhalothrin-95--online--agrostore.php
https://www.chicocrop.com/products/optima-lambda-cyhalothrin-5thiamethoxam-10-15-sc-insecticide
https://agribegri.com/products/status-insecticide-thiamethoxam-12-6-lambda-cyhalothrin-9-5-zc.php
https://krushidukan.bharatagri.com/en/products/dhanuka-zapac-thiamethoxam-lambda-cyhalothrin-4-9-cs-insecticide-80-ml-1-1-combo

Thiamethoxam + Fipronil + Chlorantraniliprole

https://www.pomais.com/thiamethoxam-vs-fipronil-insecticide/
https://farmmate.in/products/chlorantraniliprole-8-8-thiamethoxam-17-5-sc-insecticide

Thiocyclam hydrogen oxalate + Clothianidin

https://agribegri.com/products/gsp-aurthor-thiocyclam-hydrogen-oxalate-30--clothianidin-12-gr-insecticide.php
https://www.benchchem.com/product/b058007

https://agrivruddhi.com/products/aurthor-insecticide
https://www.awiner.com/product/thiocyclam-hydrogen-oxalate-50-sp/

Tolfenpyrad + Bifenthrin

https://agrostar.in/product/treza-tolfenpyrad-15-bifenthrin-75-se-500-ml/AGS-CP
https://agrivruddhi.com/products/raavan-insecticide
https://www.entomoljournal.com/archives/2020/vol8issue5/PartF/8-4-502-283.pdf
https://www3.epa.gov/pesticides/chem_search/ppls/071711-00036-20140115.pdf

Tolfenpyrad + Emamectin benzoate

https://www.oasisagro.com/product/300.html
https://www.greenrivercn.com/product/452.html
https://www.chicocrop.com/products/smash-emamectin-benzoate-2tolfenpyrad-10-insecticide

Tolfenpyrad + Pyriproxyfen + Acetamiprid

https://www.facebook.com/bestagrolifelimited/posts/-introducing-trishanku-the-power-move-for-crop-protection-best-agrolife-ltd-brin/1400919582045861/
https://sitem.herts.ac.uk/aeru/ppdb/en/Reports/1687.htm
https://www.bestagrolife.com/products

Triflumezopyrim + Spinetoram

https://irac-online.org/active-ingredient/triflumezopyrim/
https://irac-online.org/active-ingredient/spinetoram
https://patents.google.com/patent/CN102396474A/en
https://krushidukan.bharatagri.com/en/products/corteva-delegate-spinetoram-11-7-sc-insecticide
https://agribegri.com/products/iil-sparcle-insecticide-triflumezopyrim-10-sc.php

Azoxystrobin + Fipronil

https://agribegri.com/products/buy-gharda-tipsy-azoxystrobin-10--fipronil-5-wiv-sc-online.php
https://krushidukan.bharatagri.com/en/products/gharda-tipsy-azoxystrobin-10-fipronil-5-sc-fungicide-insecticide
https://www.pomais.com/azoxystrobin-fungicide-mode-of-action/
https://katyayanikrishidirect.com/blogs/news/fipronil-insecticide-uses-and-dosage

Azoxystrobin + Tebuconazole + Thiamethoxam

https://agribegri.com/products/buy-agribegri-arostel-azoxystrobin-11--tebuconazole-183-sc-online--buy-broad-spectrum-fungicide.php
https://www.dhanuka.com/fungicide/spectrum
https://www.indiamart.com/proddetail/azoxystrobin-2-5-thiophanate-methyl-11-5-thiamethoxam-25-fs-2855981631873.html

Azoxystrobin + Thiophanate Methyl + Thiamethoxam

https://cultree.in/products/upl-electron-azoxystrobin-2-5-thiophanate-methyl-11-25-thiamethoxam-25-fs-insecticide-fungicide
https://www.jaishreeramagro.com/multi-star-azoxystrobin-2-5-thiophanate-methyl-11-25-thiamethoxam-25-fs--10814530.html

Flubendiamide + Hexaconazole

https://www.globalcropcare.com/product/origin-flubendiamide-3-5-hexaconazole-5-wg
https://hellocrop.in/product/origin-tata-rallies/

Flubendiamide + Kresoxim-Methyl

https://www.rallis.com/international-business/Kresoxim-methyl
https://pubchem.ncbi.nlm.nih.gov/compound/Flubendiamide

Carboxin + Thiram + Imidacloprid
https://www.essencechem.com/product/imidacloprid-carboxin-thiram.html
https://www.napanta.com/pesticide-products/carboxin-37-5-thiram-37-5-ds/dhanuka-vitavax-power/
https://essencechem.en.made-in-china.com/product/lKvJEXibljcN/China-Agrochemical-Pesticide-Imidacloprid-20-Thiram-6-Carboxin-6-Fs.html
Imidacloprid + Hexaconazole
https://www.gujaratpesticides.com/products/touch-up
https://agrostar.in/product/imidax-imidacloprid-185-hexaconazole-15-fs
https://www.rallis.com/international-business/Imidacloprid
https://raccolto.in/wp-content/uploads/2022/04/Imidacloprid-18.5-Hexaconazole-1.5-FS-for-indigenous-manufacture.pdf
Isoprothiolane + Fipronil
https://www.globalcropcare.com/product/isoprothiolane-28-fipronil-5-ec
https://agrostar.in/product/isonil-isoprothaline-28-fipronil-5-ec-50-ml/
https://parijatagrochemicals.com/product/aadat/
https://patents.google.com/patent/WO2019186575A1/en
Lambda-cyhalothrin + Clothianidin + Carbendazim
https://aceagrochem.com/pesticide/insecticide/mixture/Lambda-cyhalothrin%20130%20+%20Clothianidin%20120%20ZC.html
https://www.pomais.com/product/lambda-cyhalothrin-insecticide/
https://farmmate.in/products/carbendazim-12-mancozeb-63-wp-fungicide-star-plus
Pymetrozine + Tebuconazole
https://agribegri.com/products/buy-gsp-across-x-pymetrozine-50-insecticide-online--agribegri.php
https://www.sinoagrochem.com.cn/Insecticide/Pymetrozine-WG
https://ebskrishibhandar.com/products/ebs-insectastop-pymetrozine-50-wg
https://pubmed.ncbi.nlm.nih.gov/34748044/
https://www.indiamart.com/proddetail/gsp-runway-pymetrozine-tebuconazole-fungicide-2857573873373.html
Pymetrozine + Dinotefuran + Pyraclostrobin
https://www.bestagrolife.com/product/defender
https://www.indiamart.com/proddetail/defender-pymetrozine-30-dinotefuran-10-pyraclostrobin-20-wg-500-gm-2854768232655.html
https://www.chicocrop.com/products/padan-dinotefuran-10pymetrozine-40-50-wdg-insecticide/
Pymetrozine + Thiamethoxam + Hexaconazole
https://gramiti.com/products/ju-hemp
https://peptechbio.com/products/pymetrozine-25-0-thiamethoxam-17-5-hexaconazole-12-5-wg/
https://agrostar.in/product/respect-3-way-pymetrozine-25-thiamethoxam-175-hexaconazole-125-wdg-500-g/
https://patents.google.com/patent/CN113712036A/en
Pyraclostrobin + Fipronil
https://agribegri.com/products/buy-gharda-joker-fungicide--insecticide-online.php
https://pubmed.ncbi.nlm.nih.gov/29123242/
Pyraclostrobin + Thiram + Clothianidin
https://parijatagrochemicals.com/product/peack/
https://agrostar.in/product/petra-3d-pyraclostrobin-35-thiram-15-clothianidin-225-fs-

https://agribegri.com/products/buy-gsp-pct-410-pyraclostrobin-35--thiram-15--clothianidin-225-fs-fungicide-online.php

Sedaxane + Azoxystrobin + Thiamethoxam

https://www.syngenta-us.com/seed-treatment/vibrance

Sedaxane + Fludioxonil + Thiamethoxam

https://www.syngenta.co.in/sites/g/files/kgtney376/files/media/document/2025/02/10/0827p24-syn-vibrance_maxx-insert_leaflet_as_per_new_layout-in3092414-ctc.pdf
https://www.syngenta-us.com/seed-treatment/cruisermaxx-vibrance-pulses
https://www.sinoagrochem.com.cn/Fungicide/Fludioxonil+Metalaxyl-M-L-FS

Chlorantraniliprole + Tebuconazole + Clothianidin

https://www.krushivista.com/products/ad9b490b71/1477796000007444065
https://farmmate.in/products/chlorantraniliprole-18-5-sc-insecticide-tarzan

Dazomet

https://pubchem.ncbi.nlm.nih.gov/compound/Dazomet
https://www.chemicalbook.com/article/uses-and-toxicity-of-dazomet.htm
https://www.sciencedirect.com/topics/agricultural-and-biological-sciences/dazomet
https://cameochemicals.noaa.gov/chemical/18086
https://sitem.herts.ac.uk/aeru/ppdb/en/Reports/203.htm

Azadirachtin

https://patents.google.com/patent/US5001146A/en
https://indianjournals.com/article/prj-15-2-008
https://www.growertalks.com/Article/?articleid=17155
https://www.frontiersin.org/journals/agronomy/articles/10.3389/fagro.2021.676208/full
https://arccjournals.com/journal/agricultural-reviews/R-2692
https://www.isca.me/IJBS/Archive/v7/i12/8.ISCA-IRJBS-2018-057.pdf
https://www.sciencedirect.com/topics/agricultural-and-biological-sciences/azadirachtin
https://pharmacy.hebmu.edu.cn/trywhx/resources/43/20191229163918.pdf
https://pubmed.ncbi.nlm.nih.gov/19112026/

Bacillus thuringiensis

https://pubmed.ncbi.nlm.nih.gov/25264571/
https://bioprotectionportal.com/resources/bacillus-thuringiensis-biocontrol-agent/
https://en.wikipedia.org/wiki/Bacillus_thuringiensis
https://pmc.ncbi.nlm.nih.gov/articles/PMC11306962/
https://www.planetnatural.com/bacillus-thuringiensis/
https://www.sciencedirect.com/science/chapter/bookseries/abs/pii/S0065280623000176
https://pmc.ncbi.nlm.nih.gov/articles/PMC3918364/
https://agribegri.com/products/buy-sumitomo-dipel.php
https://www.varshabioscience.com/product/vbt-bacillus-thuringenisis-0-5wp/
https://mankindag.com/product/bio-pesticide/
https://lilaagrotech.com/product/astha-bt-bacillus-thuringiensis/
https://agriplexindia.com/products/multiplex-minchu-plus-bio-pesticide

Beauveria bassiana

https://www.researchgate.net/publication/370154888_Status_and_scope_of_entomopathogenic_fungus_Beauveria_bassiana_in_sustainable_pest_management_A_review
https://cals.cornell.edu/integrated-pest-management/outreach-education/fact-sheets/beauveria-bassiana

https://ipm.cahnr.uconn.edu/using-beauveria-bassiana-for-insect-management/
https://peptechbio.com/best-practices-for-using-beauveria-bassiana-in-crops/
https://www.mdpi.com/2075-4450/16/3/287
https://www.vidarbhabiotechlab.com/vbl-beauveria-bassiana-biopesticides.htm

Metarhizium anisopliae

https://cals.cornell.edu/integrated-pest-management/outreach-education/fact-sheets/metarhizium-spp-biocontrol-agent-factsheet
https://pmc.ncbi.nlm.nih.gov/articles/PMC5715920/
https://www.researchgate.net/publication/394147199_Metarhizium_anisopliae_Unlocking_green_solutions_for_sustainable_pest_management
https://pmc.ncbi.nlm.nih.gov/articles/PMC7560189/
https://www.nature.com/articles/s41598-022-11424-0

Leacanicillium lecanii (Verticillium lecanii)

https://bazodo.com/product/503-verticillium-lecanii-biopesticide-1kg-for-white-fly-pest
https://ebskrishibhandar.com/products/ebs-verticillium-lecanii-bio-pesticide
https://ipm.ucanr.edu/PMG/GARDEN/PLANTS/DISEASES/vertwilt.html
https://apsjournals.apsnet.org/doi/10.1094/PHYTO-03-17-0083-RVW
https://connectjournals.com/file_full_text/1143101H_1-14.pdf
https://www.sciencedirect.com/science/article/abs/pii/0167880985900763
https://onlinelibrary.wiley.com/doi/abs/10.1111/j.1439-0418.1992.tb01155.x
https://arccjournals.com/journal/legume-research-an-international-journal/LRF-706
https://www.sciencedirect.com/topics/agricultural-and-biological-sciences/lecanicillium-lecanii

Pseudomonas fluorescens

https://www.researchgate.net/publication/327829731_Pseudomonas_fluorescence_as_a_biocontrol_agent_of_plant_parasitic_nematodes
https://www.researchgate.net/publication/260184990_Soil_application_of_Pseudomonas_fluorescens_and_Trichoderma_harzianum_reduces_root-knot_nematode_Meloidogyne_incognita_on_tobacco
https://pmc.ncbi.nlm.nih.gov/articles/PMC3380499/
https://www.sciencedirect.com/science/article/abs/pii/S0038071703002694
http://www.globalsciencebooks.info/Online/GSBOnline/images/2011/PT_5(1)/PT_5(1)44-47o.pdf
https://ajs.journals.ekb.eg/article_83169_82b769ab62e7ffbc1cbe8207b79face2.pdf

Trichoderma harzianum

https://academic.oup.com/lambio/article-abstract/38/2/169/6703338
https://pubmed.ncbi.nlm.nih.gov/14746551/
https://ejaj.journals.ekb.eg/article_257669_659d39a836a22f1e2650c678a589d37a.pdf
https://www.frontiersin.org/journals/microbiology/articles/10.3389/fmicb.2023.1160551/full
https://www.sciencedirect.com/science/article/abs/pii/S1049964421000797
https://apsjournals.apsnet.org/doi/10.1094/PHYTO.2001.91.7.687
https://bepls.com/sept_2021/26.pdf
https://www.scirp.org/journal/paperinformation?paperid=80419
https://www.nature.com/articles/s41598-023-44296-z

Trichoderma viride

https://pmc.ncbi.nlm.nih.gov/articles/PMC10189891/
https://ejaj.journals.ekb.eg/article_257669_659d39a836a22f1e2650c678a589d37a.pdf

https://www.nature.com/articles/s41598-023-33669-z
https://www.sciencedirect.com/science/article/abs/pii/S0022201121000938
https://www.sciencedirect.com/science/article/pii/S2773078625000184
https://www.ijcmas.com/7-12-2018/Mumpi
https://pmc.ncbi.nlm.nih.gov/articles/PMC10189891/
https://www.longdom.org/open-access/optimal-physical-parameters-for-growth-of-trichoderma-species-atvarying-ph-temperature-and-agitation-9057.html
https://www.researchgate.net/publication/215805591_Influence_of_Environmental_Parameters_on_Trichoderma_Strains_with_Biocontrol_Potential
https://www.vidarbhabiotechlab.com/vbl-trichoderma-viride-fungicide.htm

Verticillium chlamydosporium

https://www.sciencedirect.com/science/article/pii/S2773078623000067
https://katyayanikrishidirect.com/products/katyayani-nemotode-plus
https://repository.rothamsted.ac.uk/item/88773/a-nematicidal-metabolite-from-verticillium-chlamydosporium
https://era.dpi.qld.gov.au/id/eprint/12216/
https://utkarshagro.com/products/utkarsh-nematoz-p-paecilomyces-lilacinus-1-0-w-p-bio-nematicide-bio-pesticides
https://patents.google.com/patent/CN102657235B/en
https://ajaybiotech.com/product-verti-guard.html
https://research.wur.nl/en/publications/the-significance-of-ecology-in-the-development-of-verticillium-ch/
https://www.researchgate.net/publication/237165620_Ecological_considerations_for_the_use_of_the_nematophagous_fungus_Verticillium_chlamydosporium_to_control_plant_parasitic_nematodes
https://pmc.ncbi.nlm.nih.gov/articles/PMC3625126/

Paecilomyces lilacinus

https://www.katyayaniorganics.com/product/paecilomyces-lilacinus-bio-insecticide/
https://www.indogulfbioag.com/post/the-complete-guide-to-paecilomyces-lilacinus-nature-s-powerful-biological-nematicide
https://pmc.ncbi.nlm.nih.gov/articles/PMC7763231/
https://www.nature.com/articles/s41598-024-74268-w
https://www.sciencedirect.com/science/article/abs/pii/S0882401014001156
https://patents.google.com/patent/CN102422844A/en
https://journals.plos.org/plosone/article?id=10.1371/journal.pone.0283550

NPV of *Helicoverpa armigera*

https://pmc.ncbi.nlm.nih.gov/articles/PMC9025457/
https://farmextensionmanager.com/nuclear-polyhedrosis-virus-npv-of-helicoverpa-armigera
https://www.researchgate.net/publication/354743030_Management_of_Helicoverpa_armigera_Hubner_Lepidoptera_Noctuidae_using_its_nucleopolyhedrovirus_HearNPv_formulations_applied_by_different_methods_on_tomato
https://agritech.tnau.ac.in/crop_protection/crop_prot_bio_mass_virus_3.html
https://www.cabidigitallibrary.org/doi/pdf/10.5555/20220118615
https://www.sciencedirect.com/science/article/pii/S0085562619300809

NPV of *Spodoptera litura*

https://bepls.com/beplsmay2021/3.pdf
https://farmextensionmanager.com/nuclear-polyhedrosis-virus-npv-of-spodoptera-litura

https://agriculture.vikaspedia.in/viewcontent/agriculture/agri-inputs/bio-inputs/production-of-ipm-inputs/biopesticides/mass-production-of-nuclear-polyhedrosis-virus-npv?lgn=en
https://pubmed.ncbi.nlm.nih.gov/2254753/
https://www.sciencedirect.com/science/article/abs/pii/S1049964400908776

PB Rope

https://www.piindustries.com/pi-products/pb-knot/
https://wcrc.confex.com/wcrc/2007/techprogram/P1712.HTM
https://www.researchgate.net/publication/306038782_Management_of_pink_bollworm_Pectinophora_gossypiella_Saunders_in_cotton_through_mating_disruption
https://www.downtoearth.org.in/agriculture/cotton-curse-pheromone-based-mating-disruption-technology-may-help-control-pink-bollworm-92547
https://www.researchgate.net/publication/362279275_BIO-EFFICACY_OF_PB-ROPE_L_AS_MATING_DISRUPTANT_AGAINST_PINK_BOLLWORM_Pectinophora_gossypiella_Saunders_IN_COTTON
https://vigyanvarta.in/adminpanel/upload_doc/VV_0424_42.pdf
https://researchjournal.co.in/upload/assignments/1_68-69.pdf
https://researcherslinks.com/current-issues/Eco-Friendly-Management-of-Pink-Bollworm-Pectinophora-gossypiella/24/1/6549

Brodifacoum

https://www.qld.gov.au/health/condition/accidents-injuries-and-poisonings/poisonings-and-chemical-exposures/rat-poison-contamination#:
https://guide.rrac.info/rodenticide-molecules/2nd-generation-anticoagulants.html
https://npic.orst.edu/factsheets/rodenticides.html
https://pmc.ncbi.nlm.nih.gov/articles/PMC6892705/
https://www.sciencedirect.com/topics/neuroscience/brodifacoum
https://www.pomais.com/brodifacoum-what-it-is-how-to-kill-rats/
https://sitem.herts.ac.uk/aeru/ppdb/en/Reports/87.htm

Bromadiolone

https://npic.orst.edu/factsheets/bromadgen.html
https://ppqs.gov.in/sites/default/files/bromadiolone_0.005_rb_cake93fpest_control_india_pvt._ltd_1.pdf
https://pubchem.ncbi.nlm.nih.gov/compound/Bromadiolone
https://www.sciencedirect.com/topics/chemistry/bromadiolone
https://sitem.herts.ac.uk/aeru/ppdb/en/Reports/89.htm
https://www.hpc-standards.com/shop/ReferenceMaterials/Pesticides/Bromadiolone

Coumatetralyl

https://pubchem.ncbi.nlm.nih.gov/compound/Coumatetralyl
https://www.sciencedirect.com/topics/agricultural-and-biological-sciences/coumatetralyl
https://sitem.herts.ac.uk/aeru/ppdb/en/Reports/182.htm
https://www.globalcropcare.com/product/rodent-coumatetralyl-0-75-tp/

Zinc phosphide

https://npic.orst.edu/factsheets/archive/znptech.html
https://pubchem.ncbi.nlm.nih.gov/compound/Zinc-Phosphide
https://www.mdpi.com/2305-6304/11/7/555
https://www.sciencedirect.com/topics/agricultural-and-biological-sciences/zinc-phosphide
https://sitem.herts.ac.uk/aeru/ppdb/en/Reports/1730.htm

https://en.wikipedia.org/wiki/Zinc_phosphide
Metaldehyde
https://www.sciencedirect.com/topics/agricultural-and-biological-sciences/metaldehyde
https://www.simsonpharma.com/blog-details/what-is-metaldehyde-used-for
https://www.piindustries.com/pi-products/snailkill/
https://ccmedia.fdacs.gov/content/download/32845/file/metaldehyde-for-controlling-snails-and-slugs-faq-2022.pdf
https://agritech.tnau.ac.in/crop_protection/pdf/Management%20of%20Giant%20African%20Snail.pdf
https://pubs.rsc.org/en/content/articlehtml/2017/ew/c7ew00039a
https://pmc.ncbi.nlm.nih.gov/articles/PMC7411606/
https://www.msdvetmanual.com/toxicology/metaldehyde-poisoning/metaldehyde-poisoning-in-animals
https://www.pomais.com/metaldehyde-insecticide-effective-slug-and-snail-control/

ANNEXURE

Crop Index

Pests of Cotton and Sunhemp

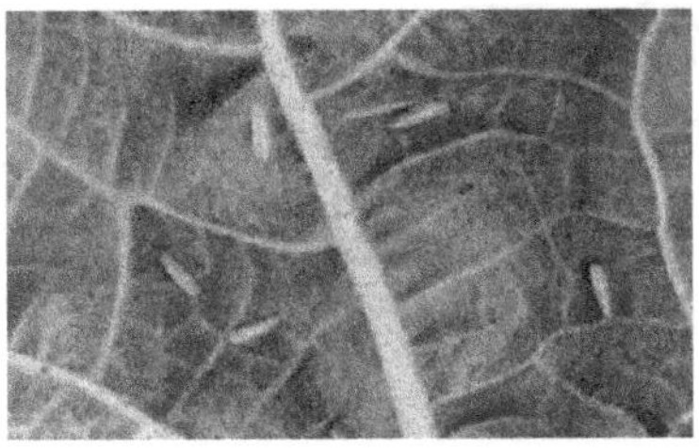
Cotton - Jassid

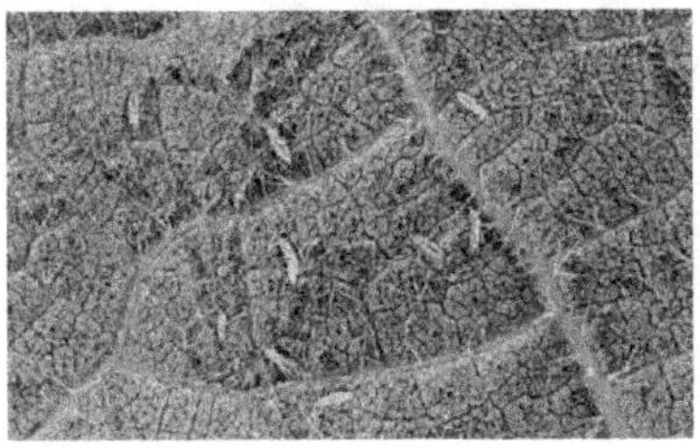
Cotton - Thrips

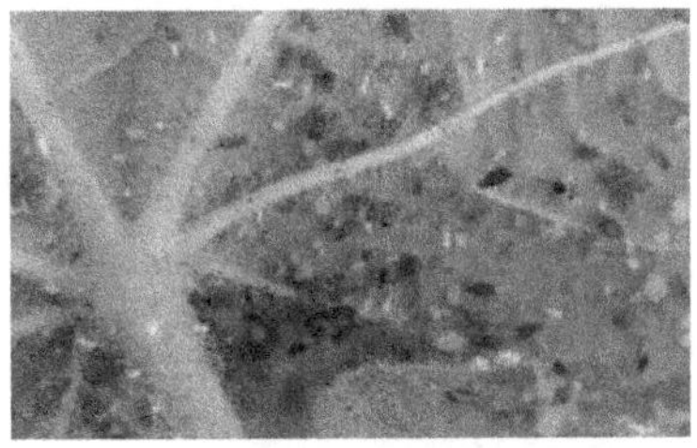
Cotton - Aphid

Cotton - Mealybug

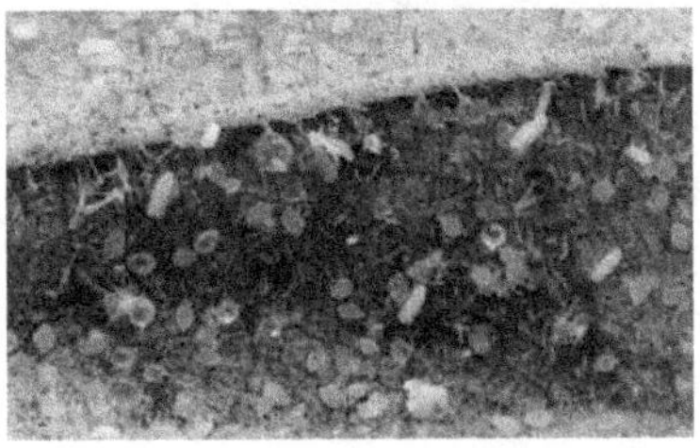
Cotton - Whitefly

Cotton - Spotted Bollworm

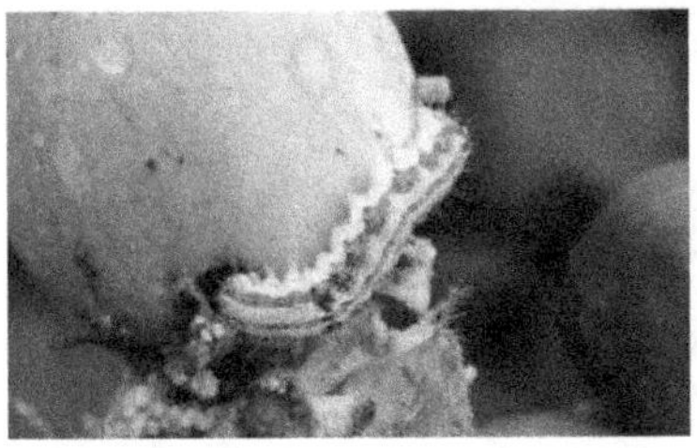
Cotton - American Bollworm

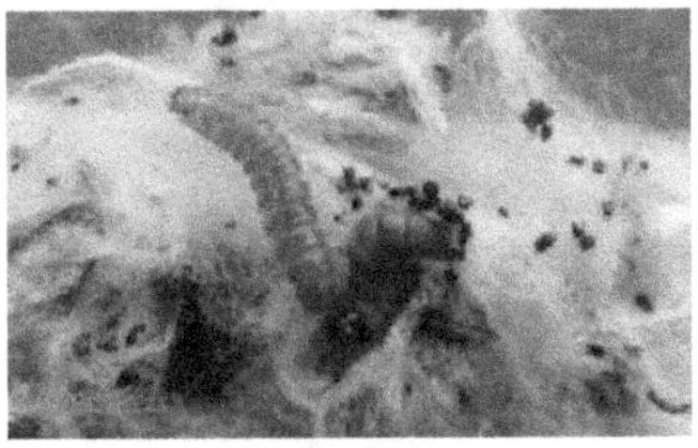
Cotton - Pink Bollworm

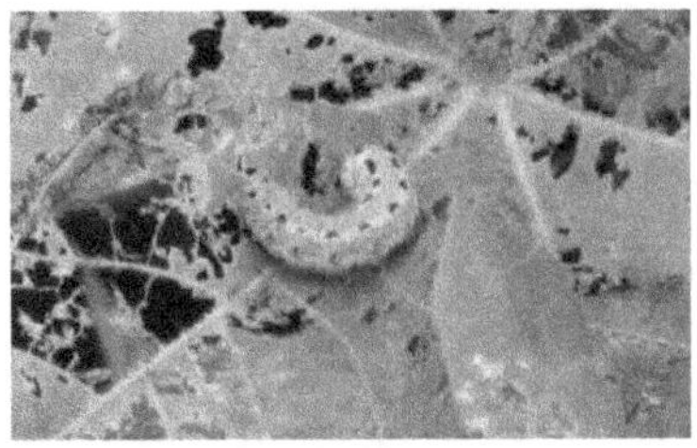
Cotton - Tobacco Caterpillar

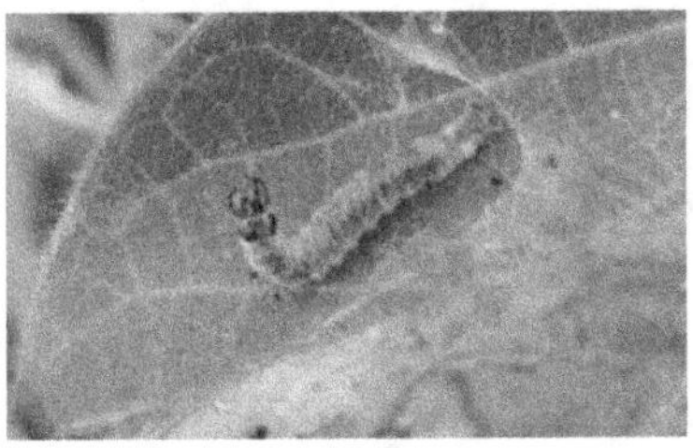
Cotton - Leaf Roller

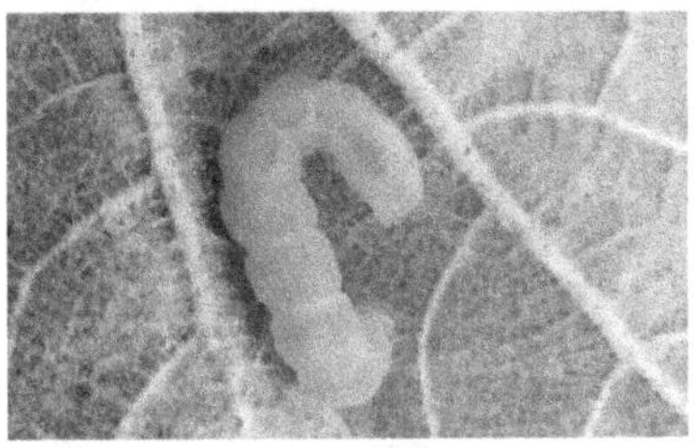
Cotton - Semilooper

Cotton - Grey Weevil

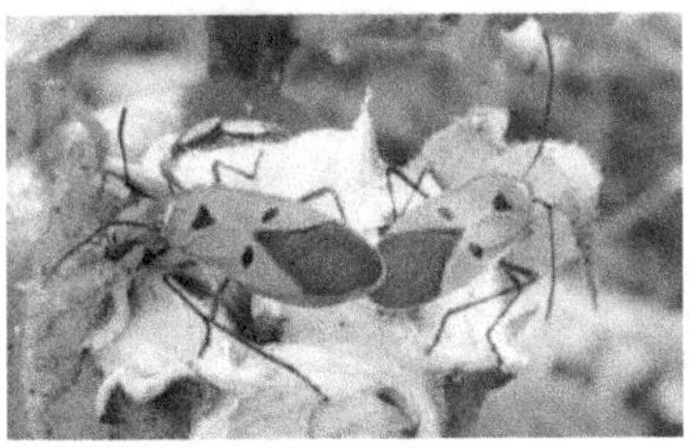
Cotton - Red Cotton Bug

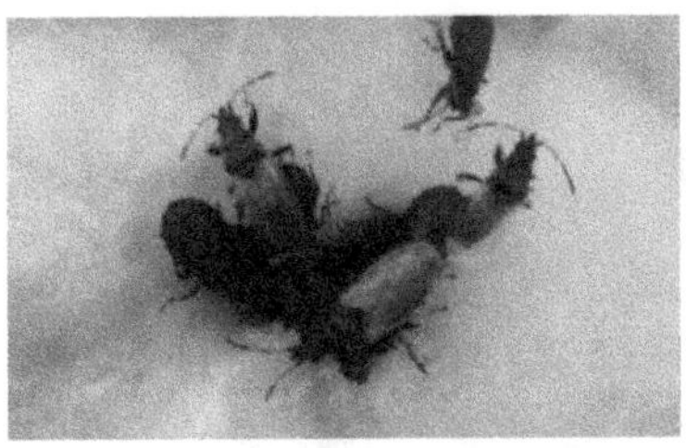
Cotton - Dusky Cotton Bug

Cotton - Green Stink Bug

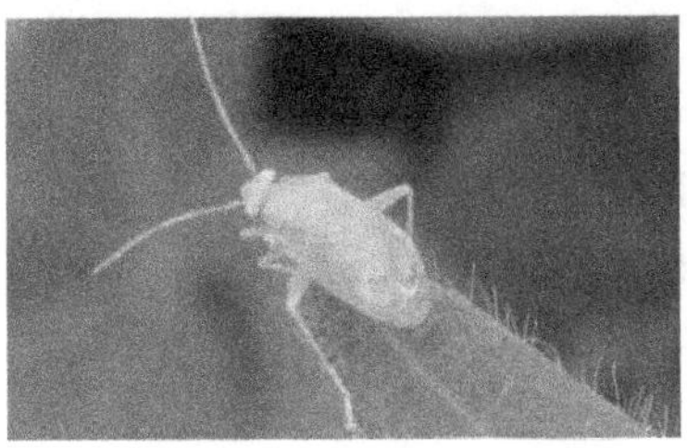
Cotton - Mirid Bug

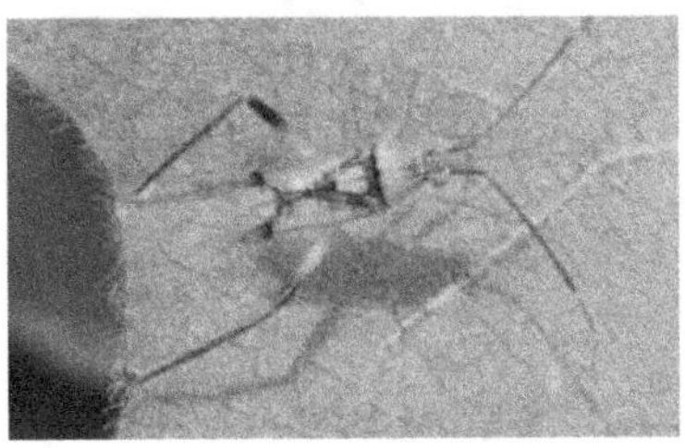
Cotton - Mirid Bug

Sunhemp - Hairy Caterpillar

Pests of Sugarcane and Rice

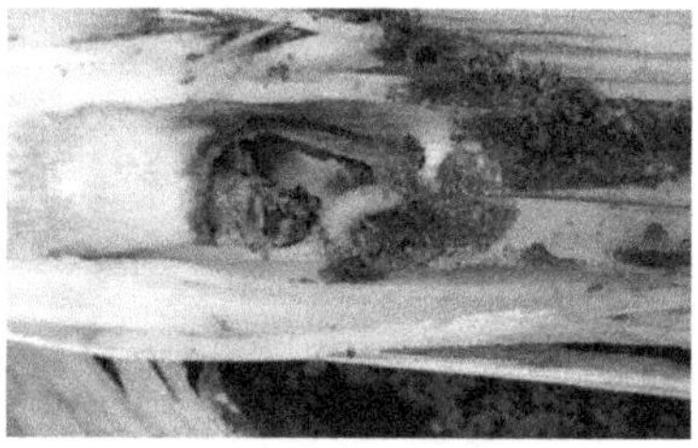
Sugarcane - Early Shoot Borer

Sugarcane - Internode Borer

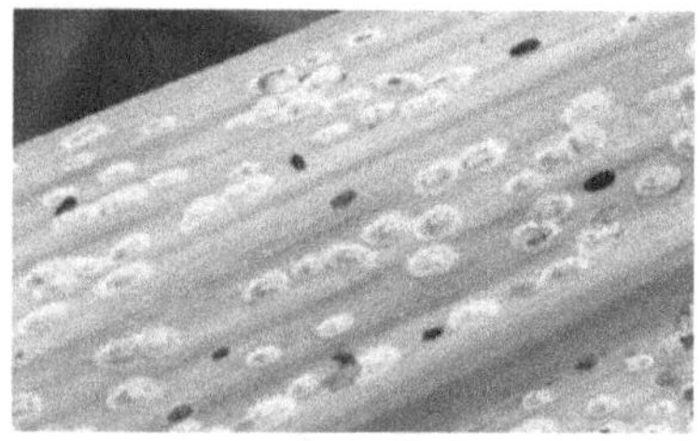
Sugarcane - Whitefly

Sugarcane - Pyrilla

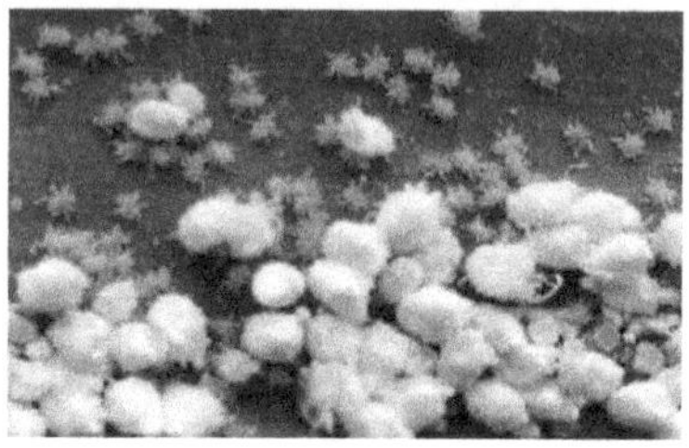
Sugarcane - Wooly Aphid

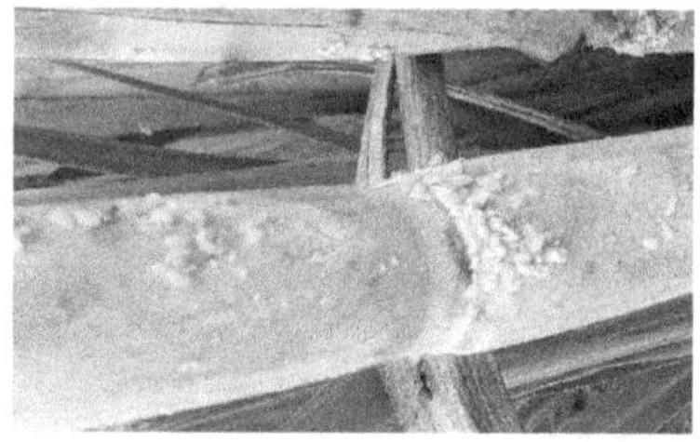
Sugarcane - Mealybug

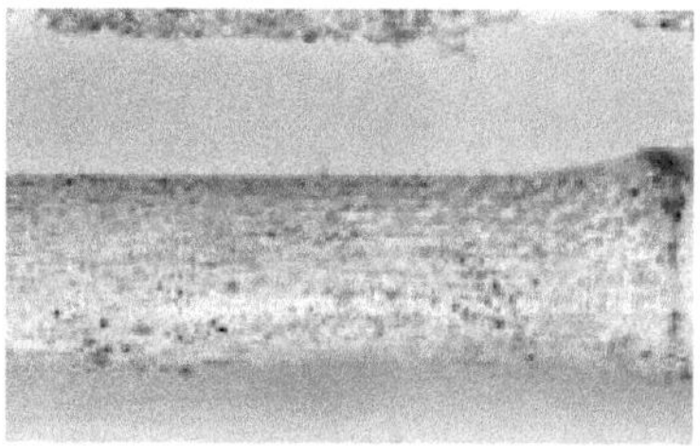
Sugarcane - Scale

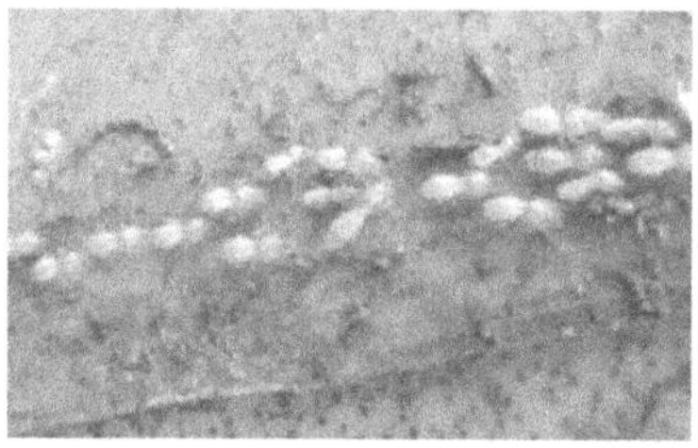
Sugarcane - Termite

Sugarcane - White Grub

Rice - Stem Borer

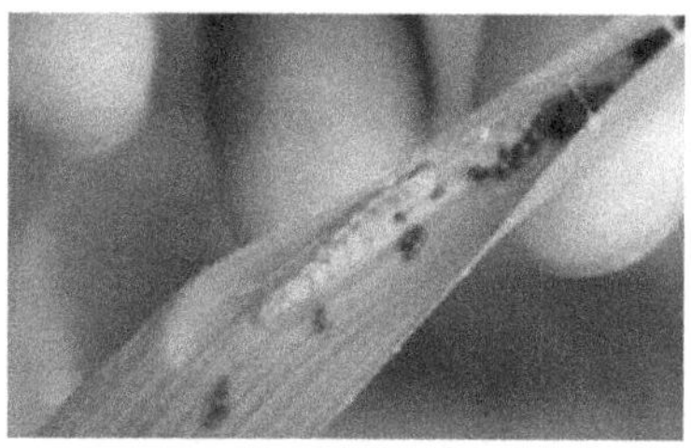
Rice - Leaf Roller

Rice - Green Leafhopper

Rice - Brown Planthopper

Rice - White Backed Planthopper

Rice - Gundhi Bug

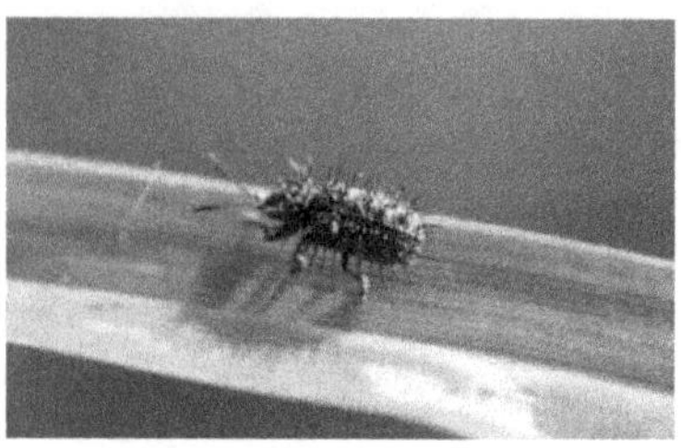
Rice - Hispa

Rice - Blue Beetle

Rice - Grasshopper

Pests of Sorghum, Maize, Pearl Millet, Wheat and Pigeonpea

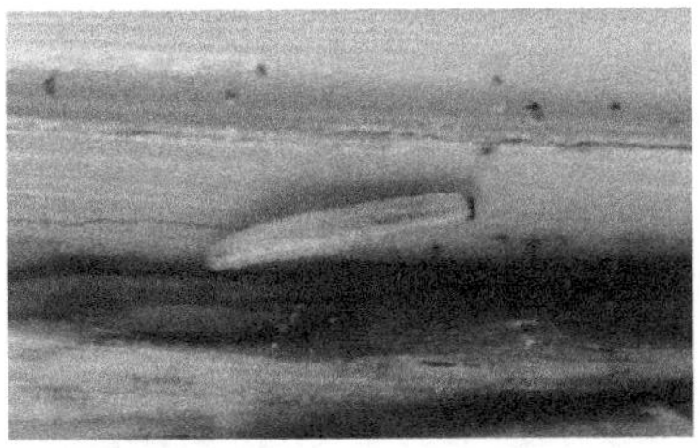

Sorghum - Stem Fly

Sorghum - Stem Borer

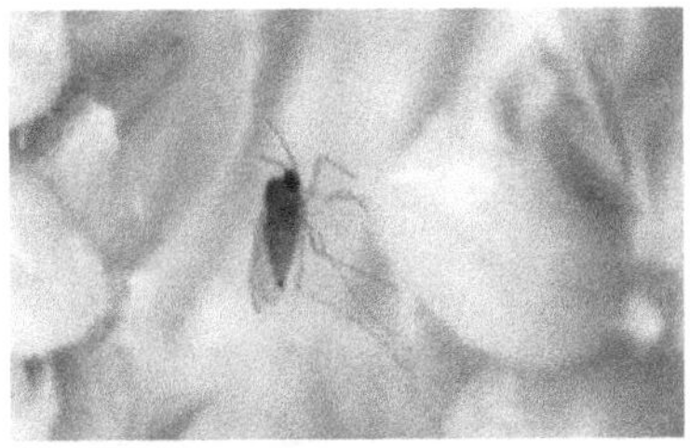

Sorghum – Midge Fly

Sorghum - Earhead Caterpillar

Sorghum - Aphid

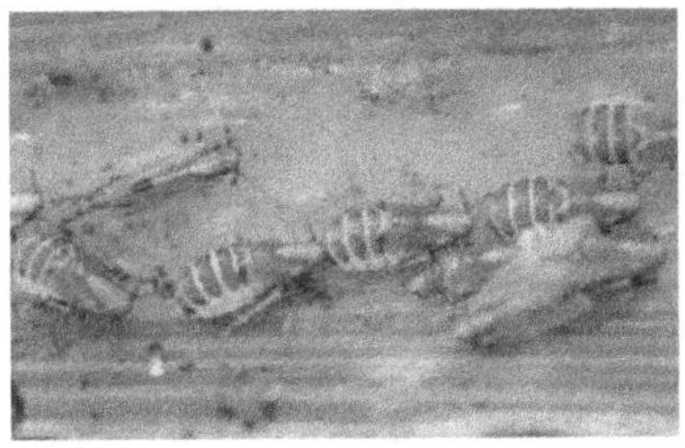

Sorghum - Delphacid

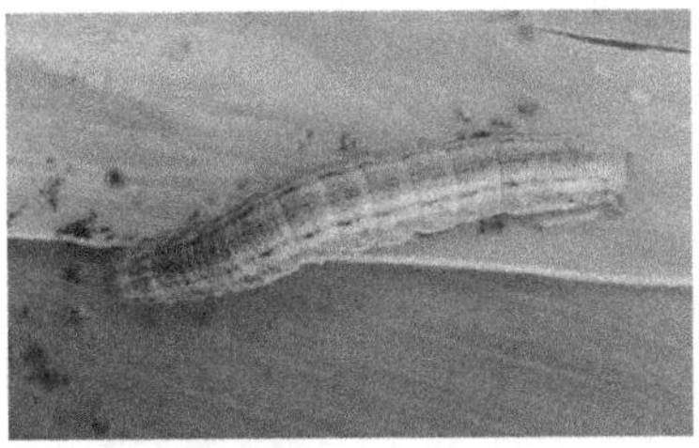

Maize - Armyworm

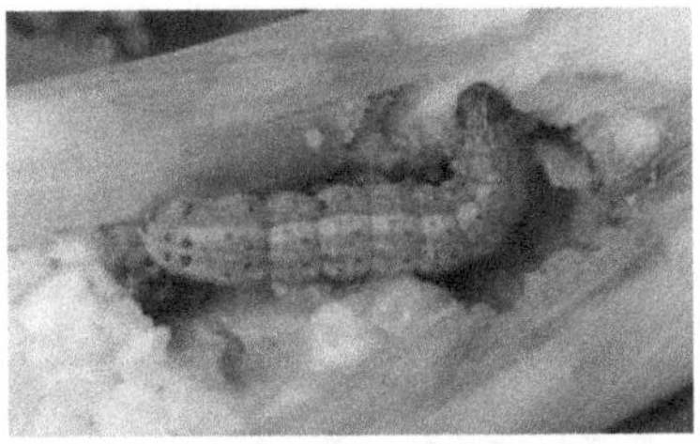

Maize - Fall Armyworm

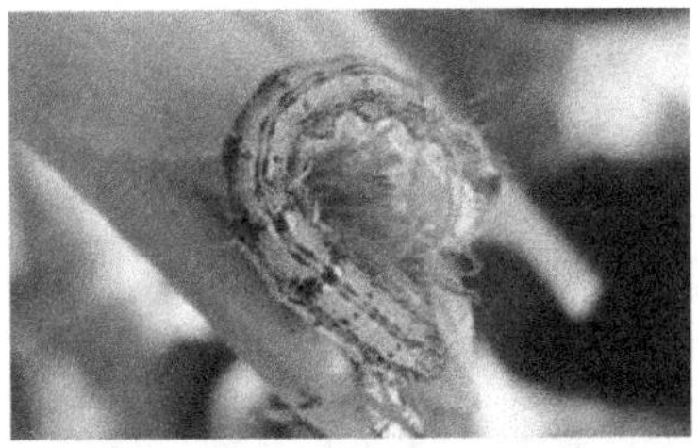

Maize - Cob Earworm

Pearl millet - Blister Beetle

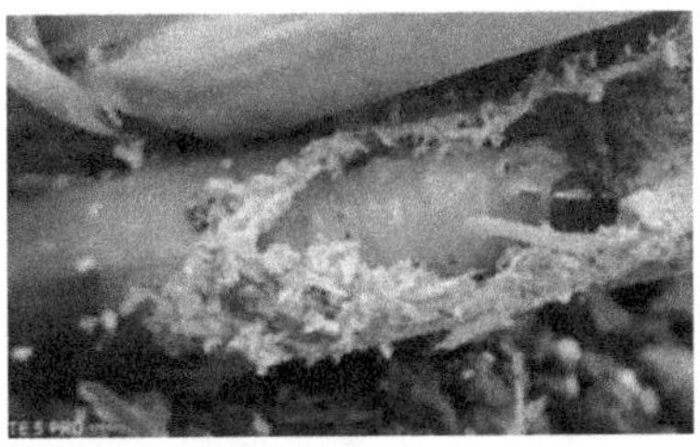

Wheat - Stem Borer

Wheat - Aphid

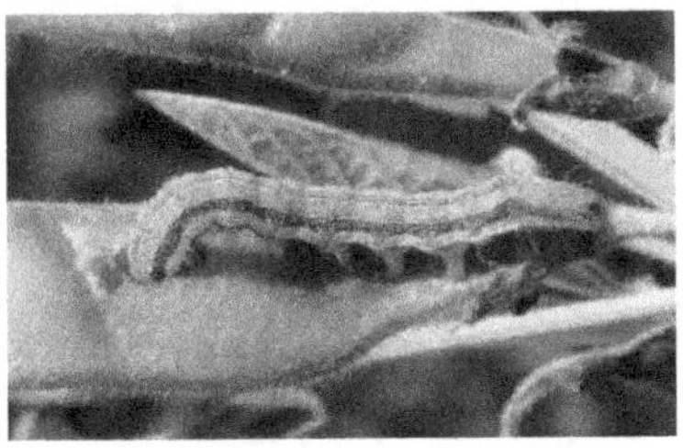

Pigeonpea - Pod Borer

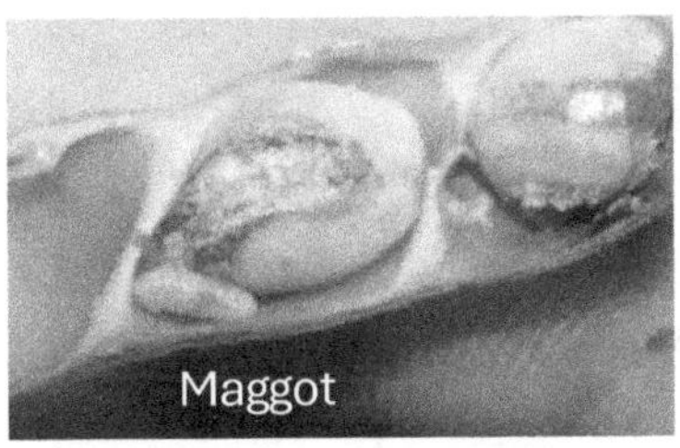

Pigeonpea - Pod Fly

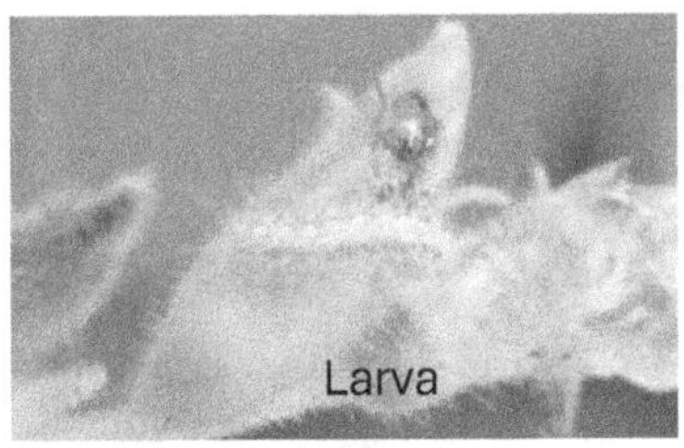

Pigeonpea - Plume Moth

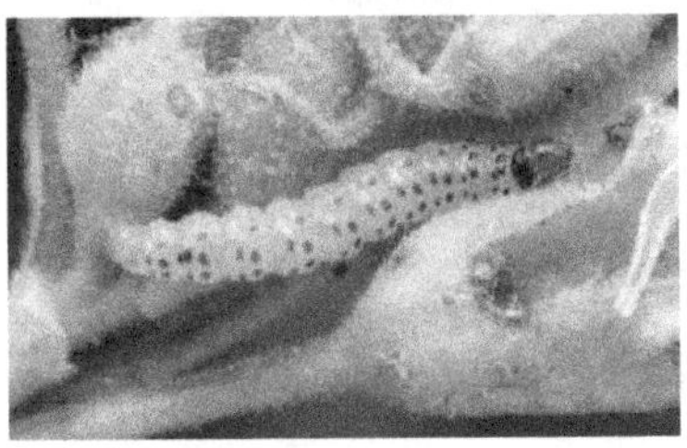

Pigeonpea - Spotted Pod Borer

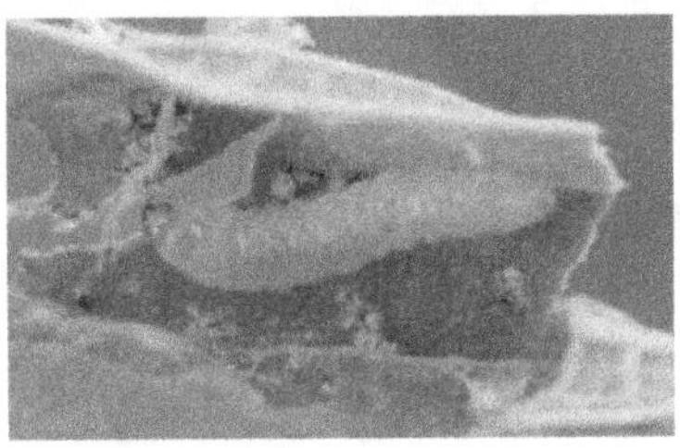

Pigeonpea - Leaf Webber

Pigeonpea - Pod Sucking Bug

Pests of Chickpea, Green Gram, Black Gram, Soybean and Groundnut

Chickpea - Pod Borer

Chickpea - Cutworm

Chickpea, Green & Black gram - Aphid

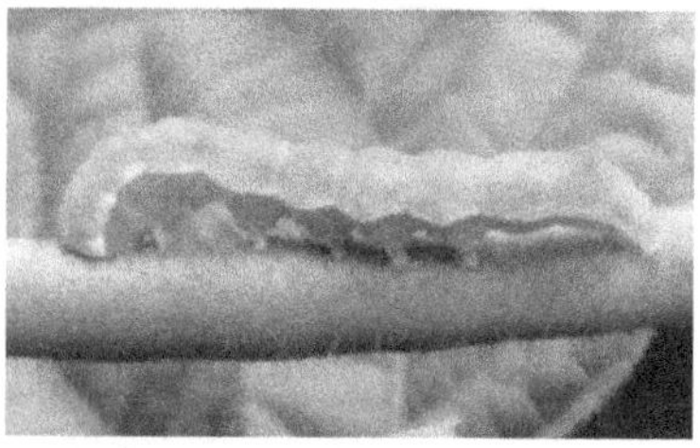
Green & Black gram – Pod Borer

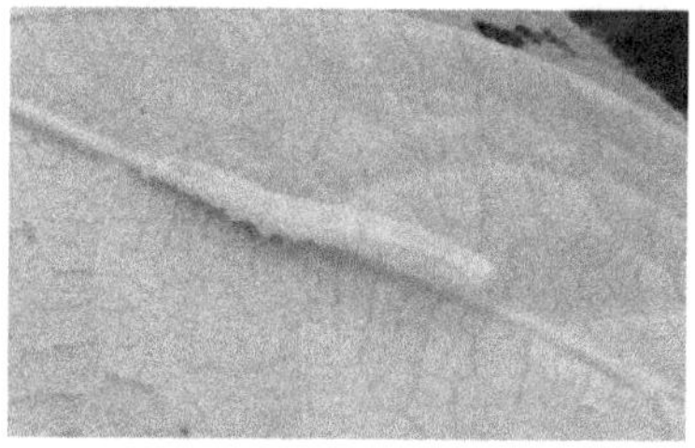
Green & Black gram - Semilooper

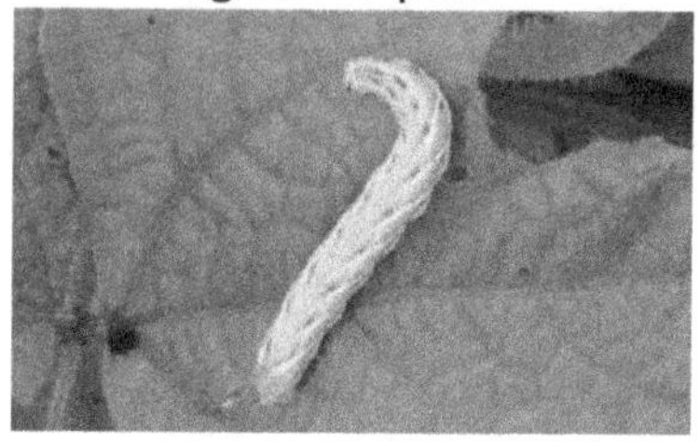
Green & Black gram – Leaf Eating Ctaerpillar

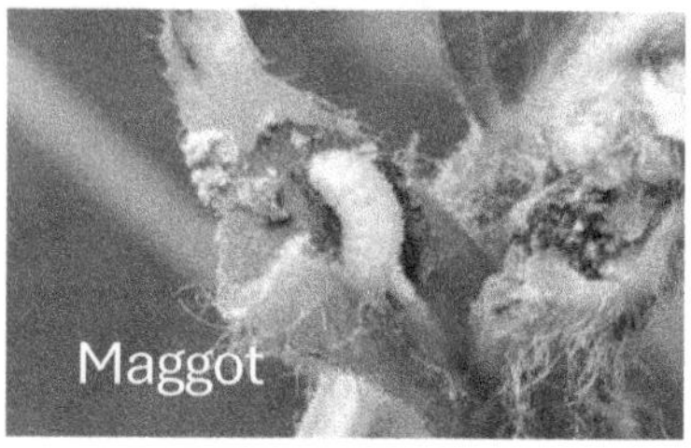

Soybean – Stem Fly

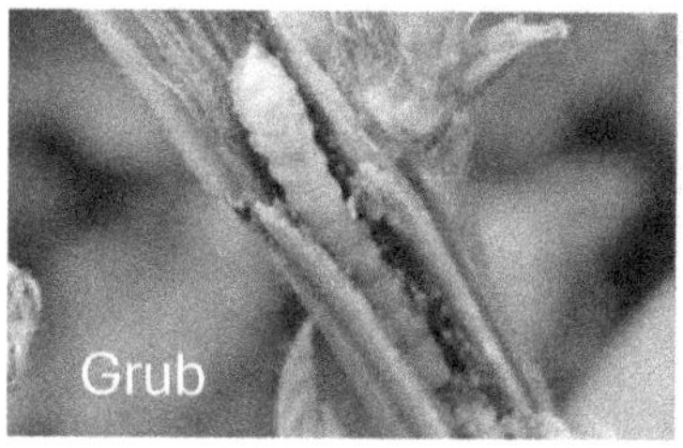

Soybean – Girdle Beetle

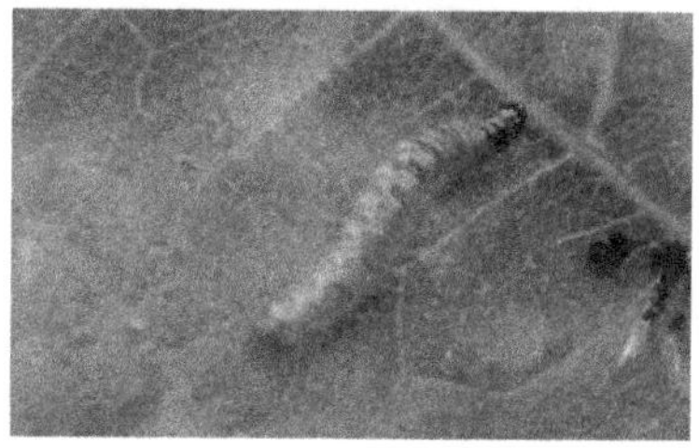
Soybean - Leaf Miner

Soybean – Tobacco Caterpillar

Soybean – Green Semilooper

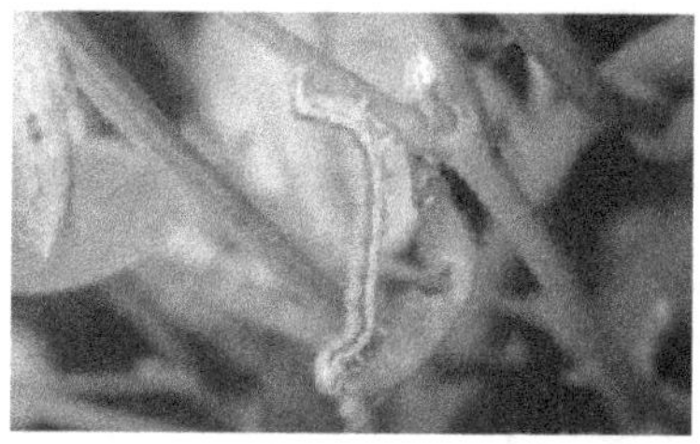
Soybean – Gram Pod Borer

Soybean - Whitefly

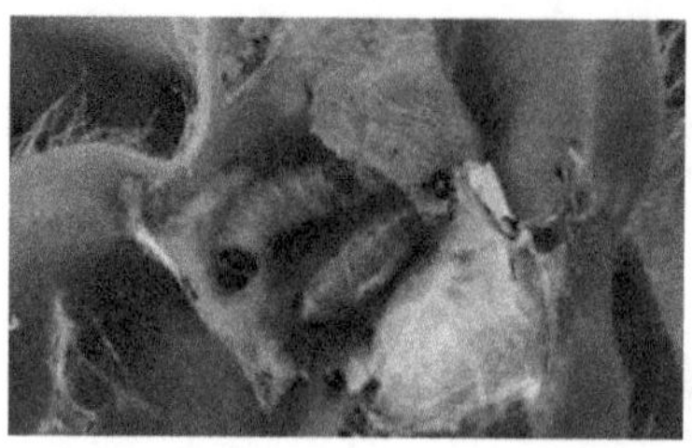
Groundnut – Leaf Miner

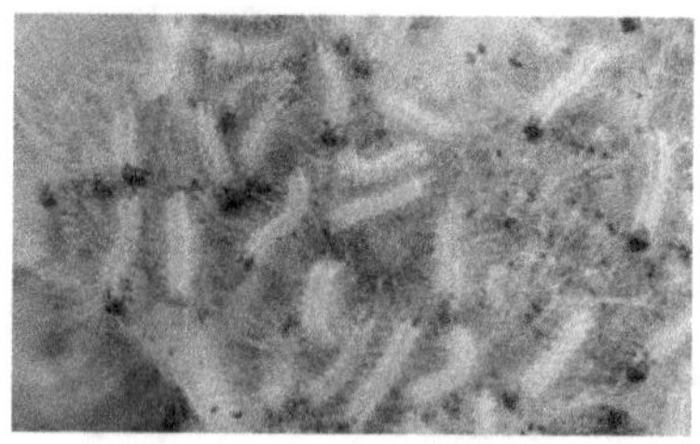
Groundnut – Bihar Hairy Ctaerpillar

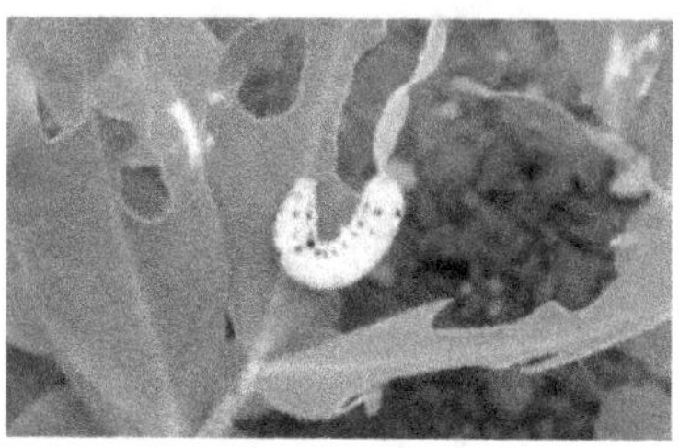
Groundnut – Tobacco Caterpillar

Groundnut - Aphid

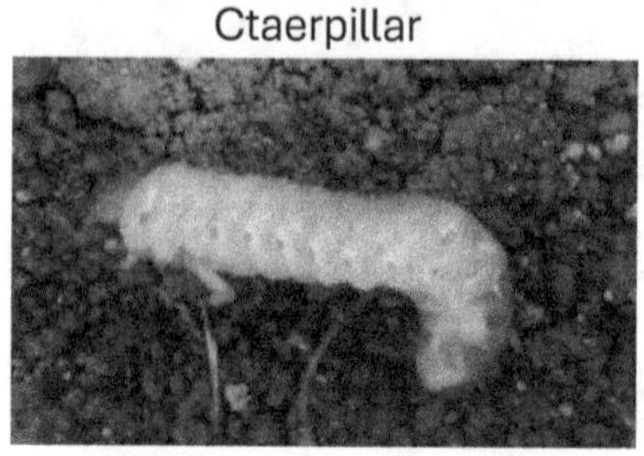
Groundnut – White Grub

Pests of Sunflower, Safflower, Mustard, Castor, Sesamum, Linseed and Niger

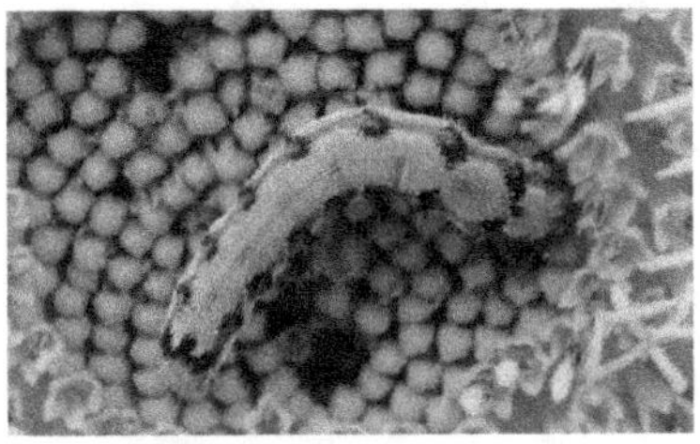

Sunflower - Capitulum Borer

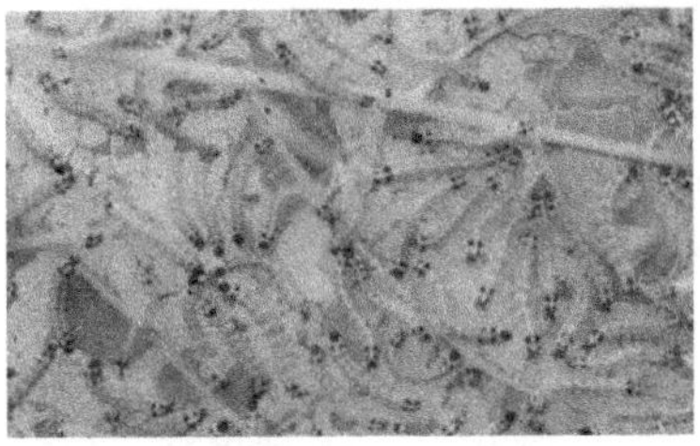

Sunflower – Hairy Ctaerpillar

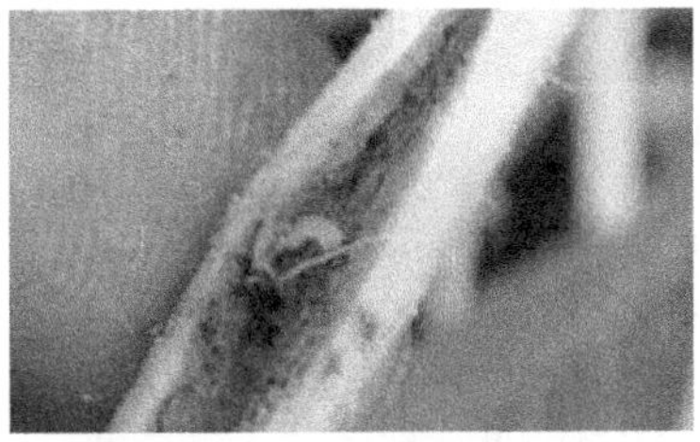

Sunflower – Stem Borer

Safflower - Aphid

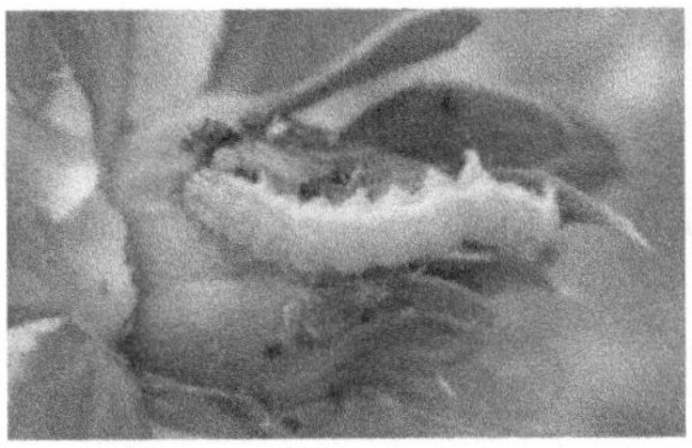

Safflower – Capitulum Borer

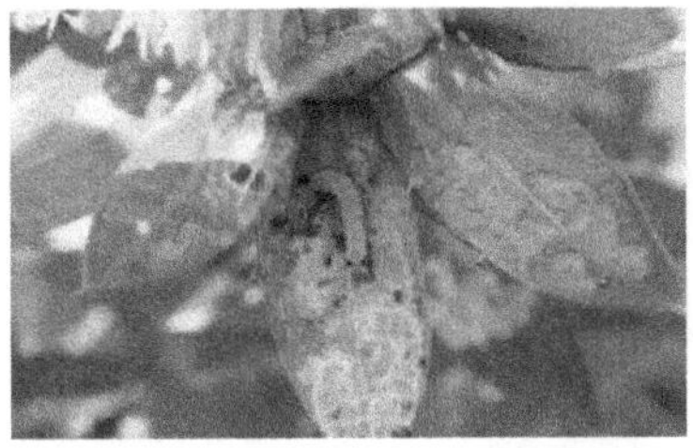

Safflower – Tobacco Caterpillar

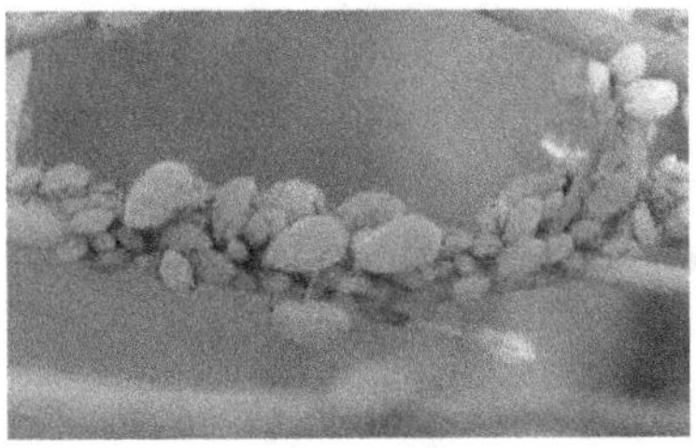

Mustard – Aphid

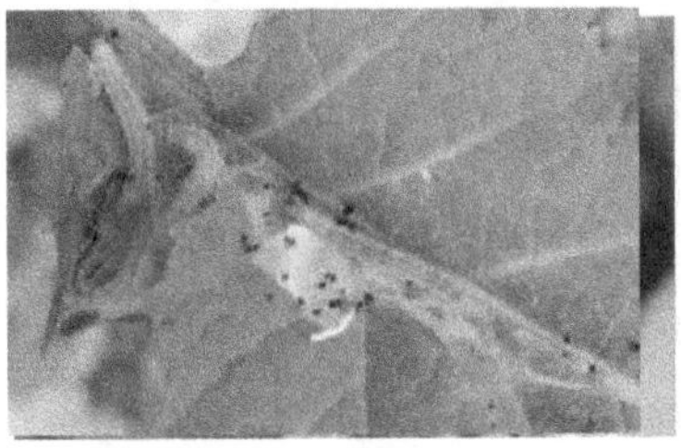

Mustard – Leaf Webber

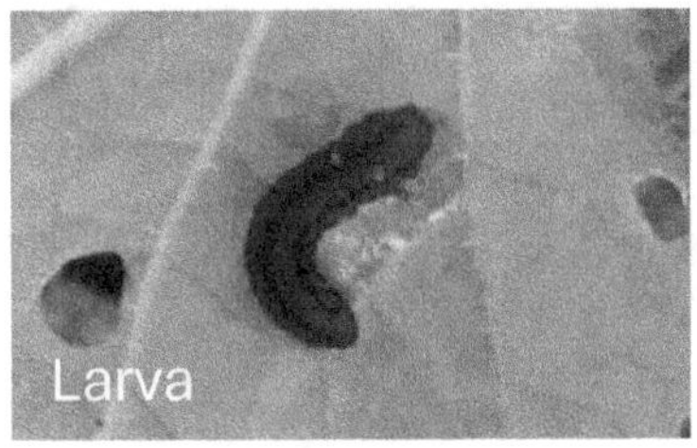

Mustard - Sawfly

Castor – Semilooper

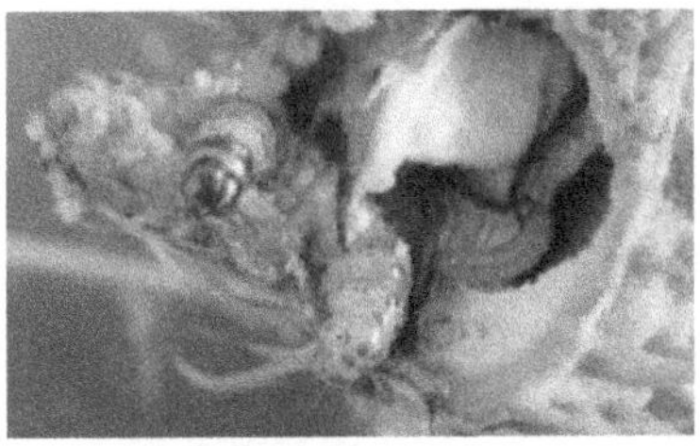

Castor – Capsule Borer

Castor – Jassid

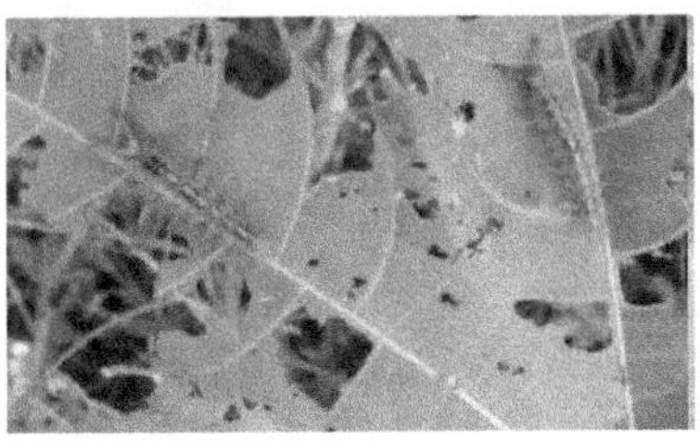

Castor – Slug Caterpillar

Sesamum – Til Hawk Moth

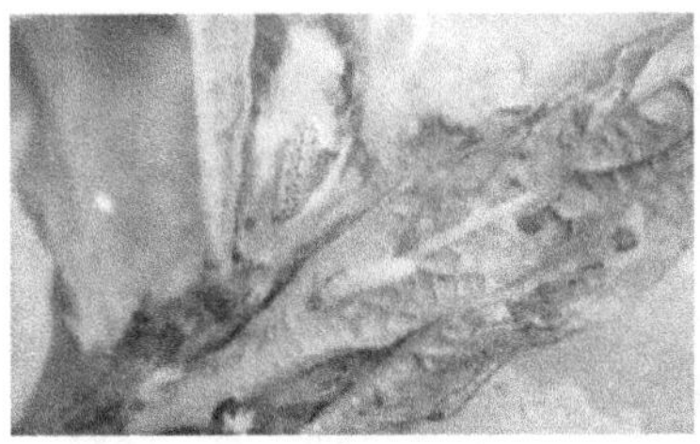

Sesamum – Leaf Webber

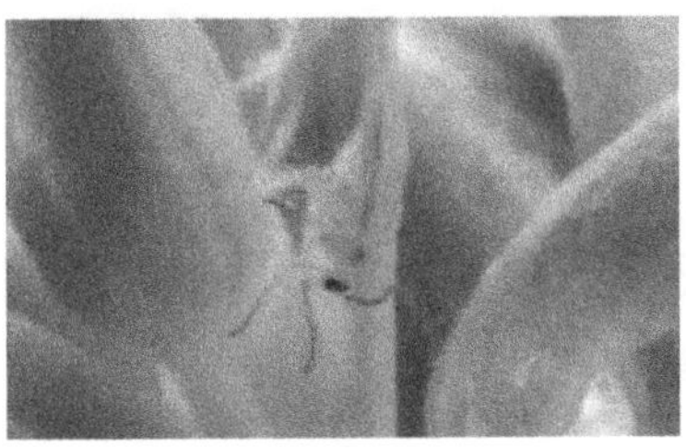

Linseed – Gall Gly

Niger – Pod Borer

Niger – Semilooper

Pests of Brinjal, Okra, Tomato and Chilli

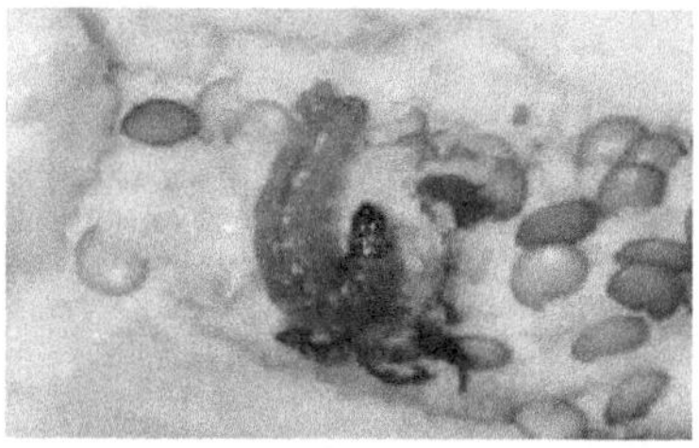
Brinjal – Shoot & Fruit Borer

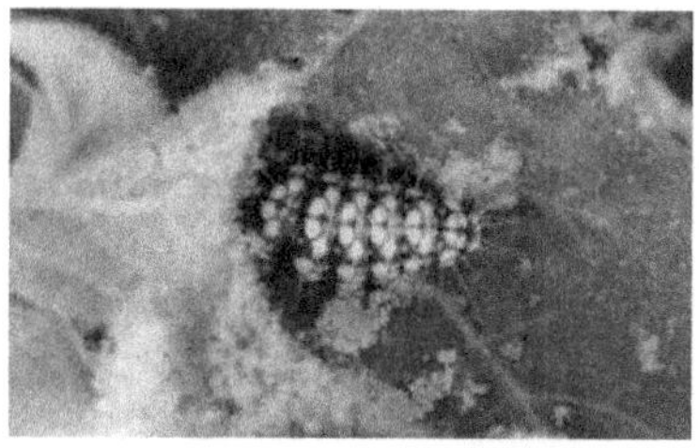
Brinjal – Leaf Webber

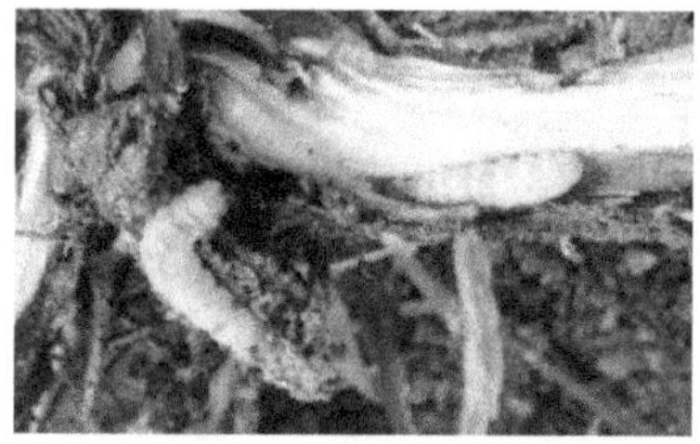
Brinjal – Stem Borer

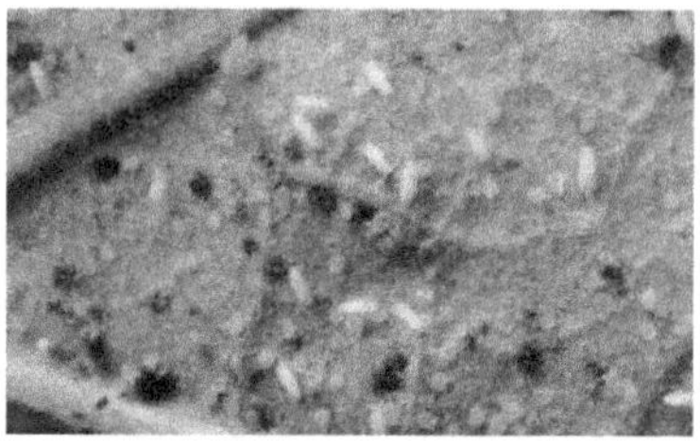
Brinjal - Whitefly

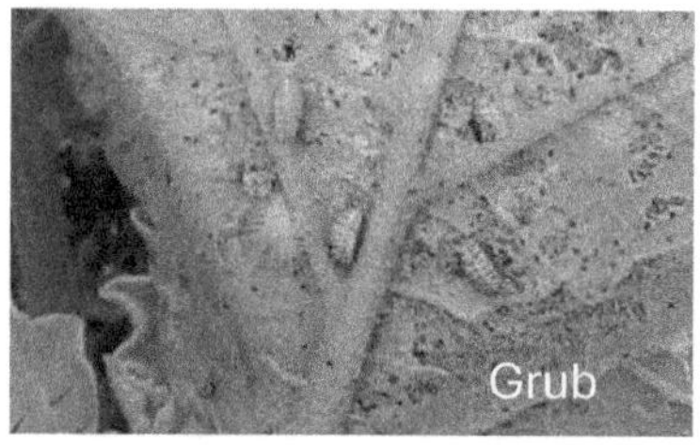

Brinjal – Hadda Beetle

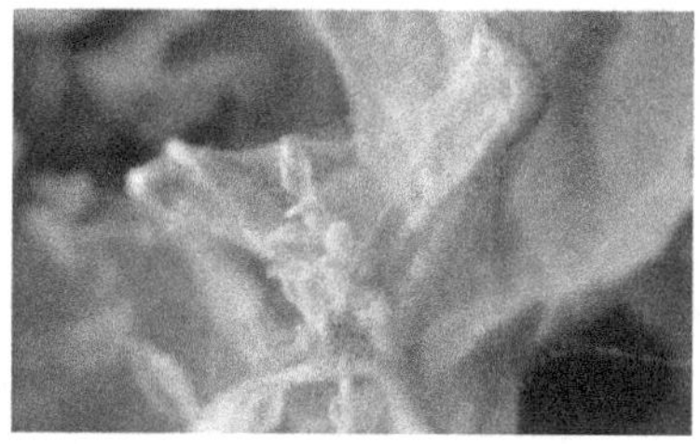
Brinjal – Red Spider Mite

Okra – Shoot & Fruit Borer

Okra – Fruit Borer

Okra - Jassid

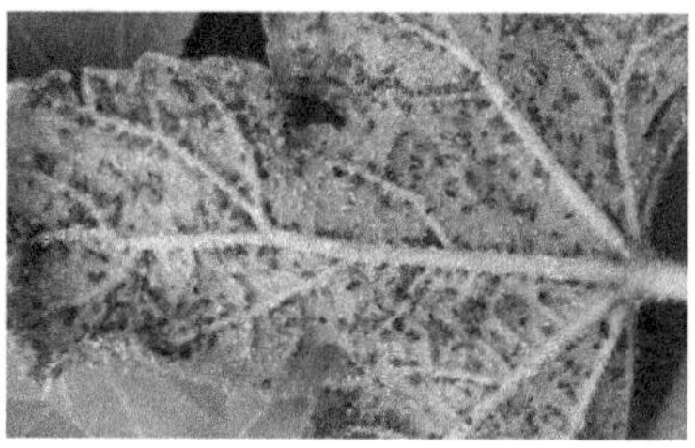
Okra - Aphid

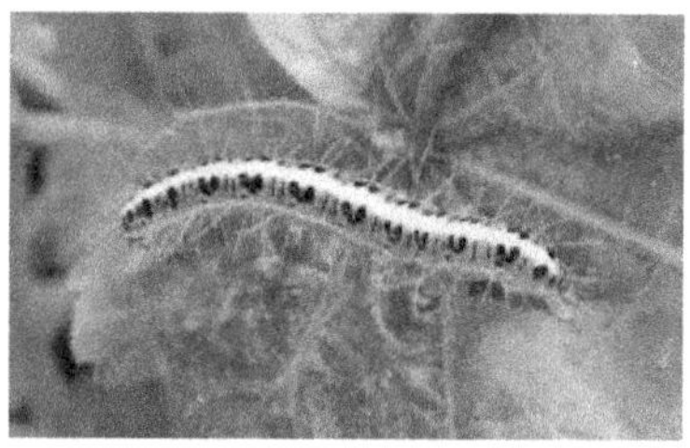
Okra - Semilooper

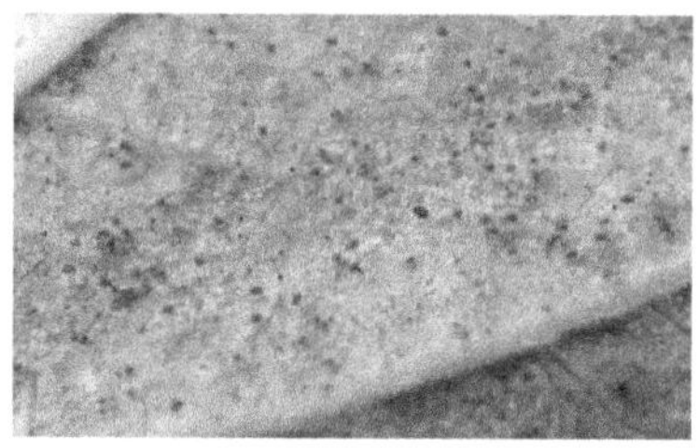
Okra – Red Spider Mite

Tomato – Fruit Borer

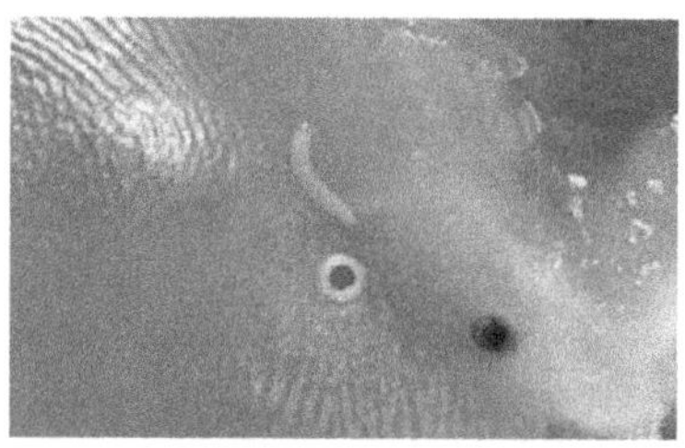
Tomato - Pinworm

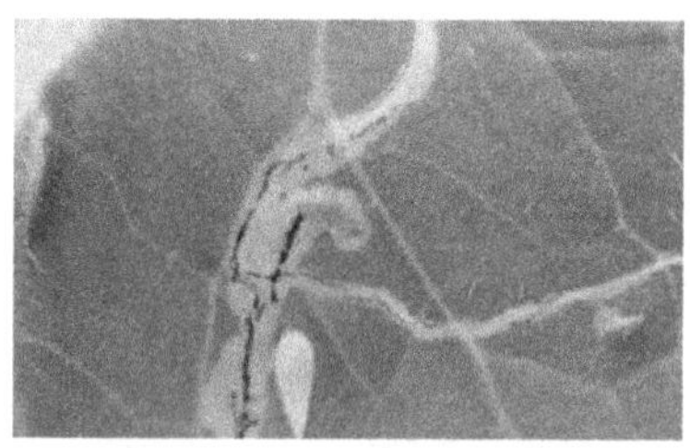
Tomato – Leaf Miner

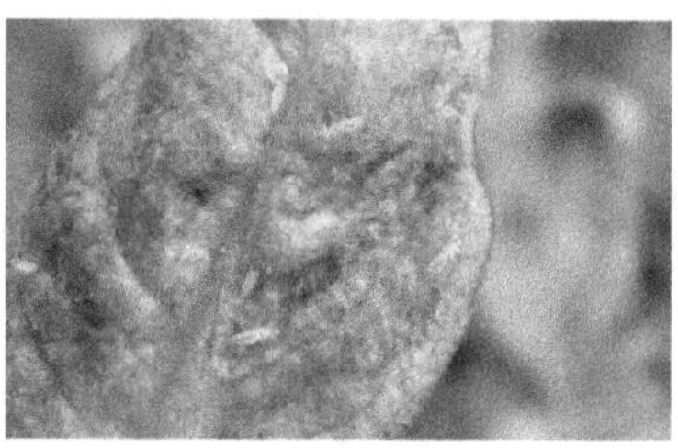
Chilli - Thrips

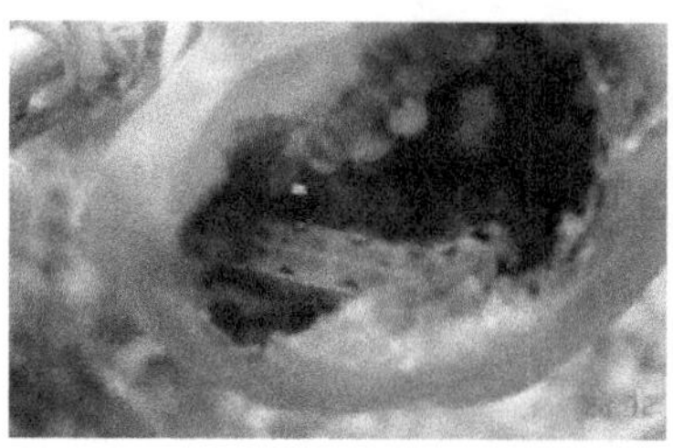
Capsicum – Tobacco Caterpillar

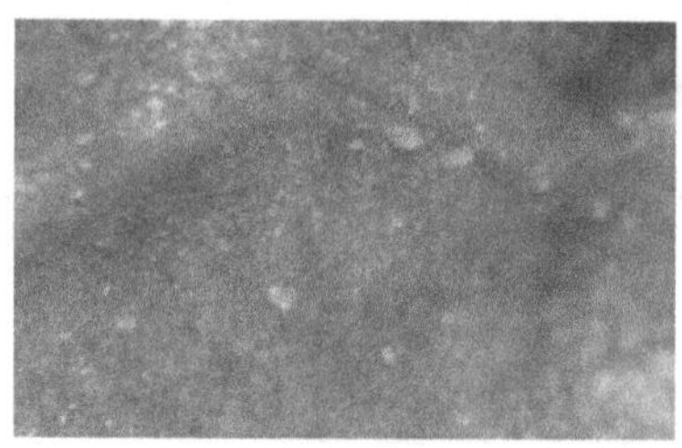
Chilli – Yellow Mite

Pests of Potato, Cruciferous, Cucurbitaceous and Bean Vegetable Crops

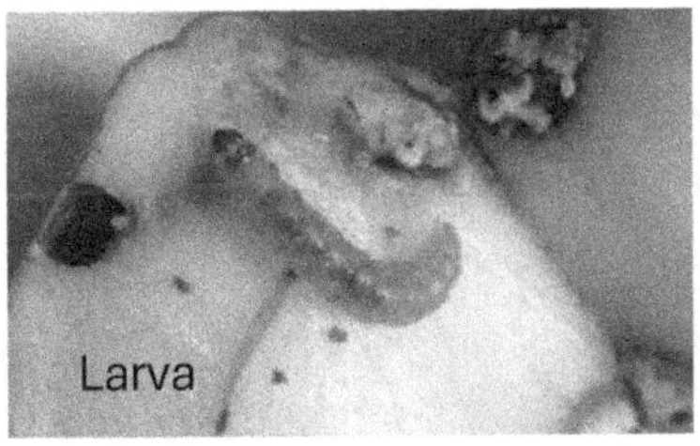

Potato – Tuber Moth

Potato - Cutworm

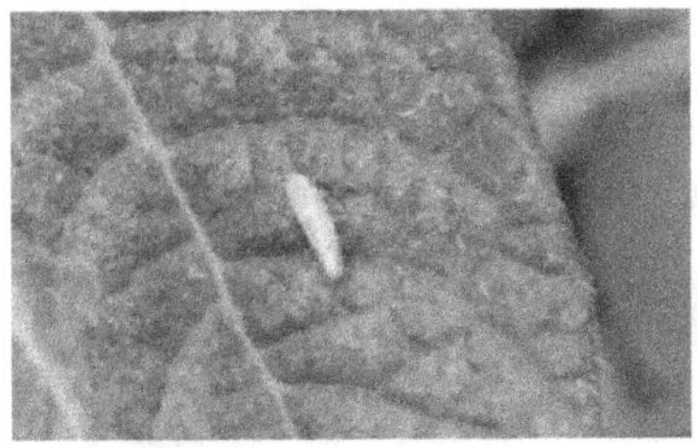

Potato - Jassid

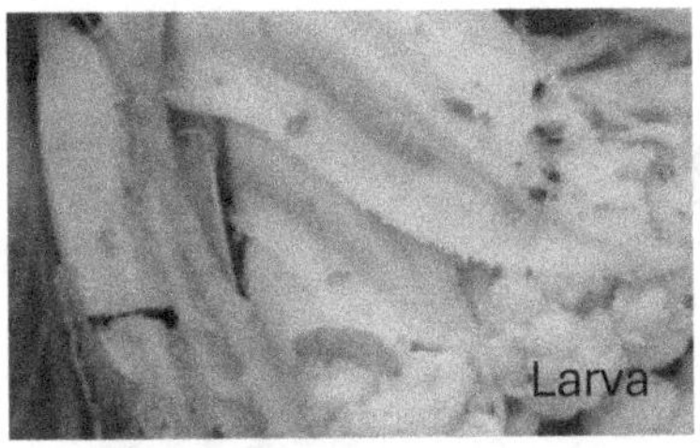

Cruciferous Crops – Diamond Back Moth

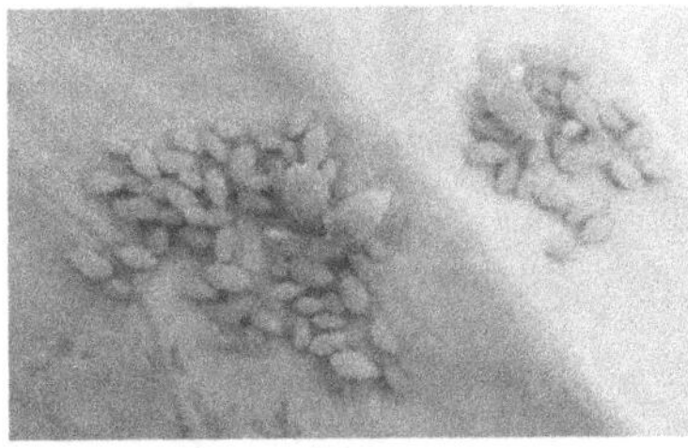

Cruciferous Crops - Aphid

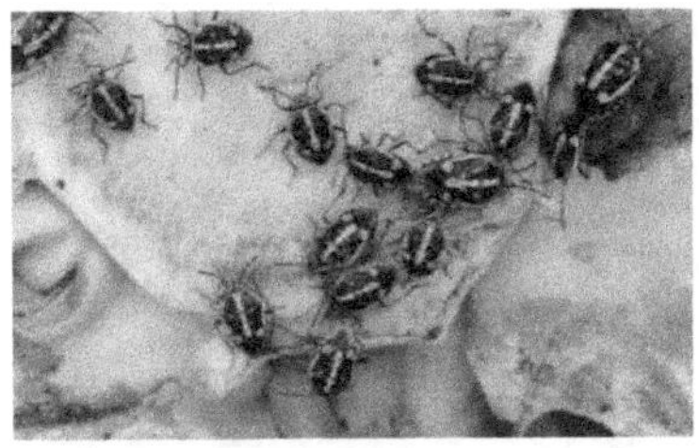

Cruciferous Crops - Painted Bug

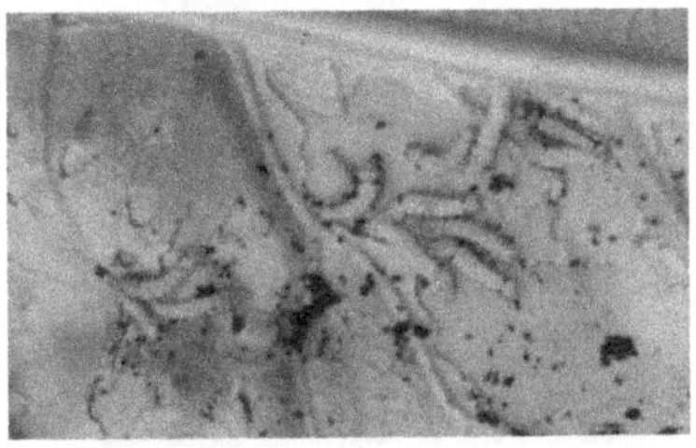

Cruciferous Crops - Tobacco Caterpillar

Cruciferous Crops – Head Borer

Cruciferous Crops - Semilooper

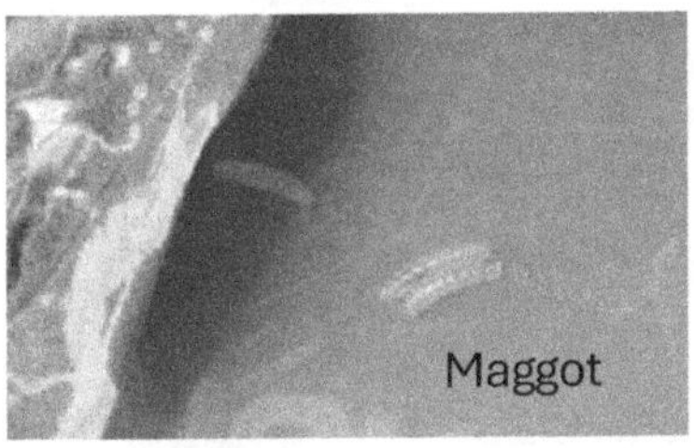

Cucurbitaceous Crops - Leaf Miner

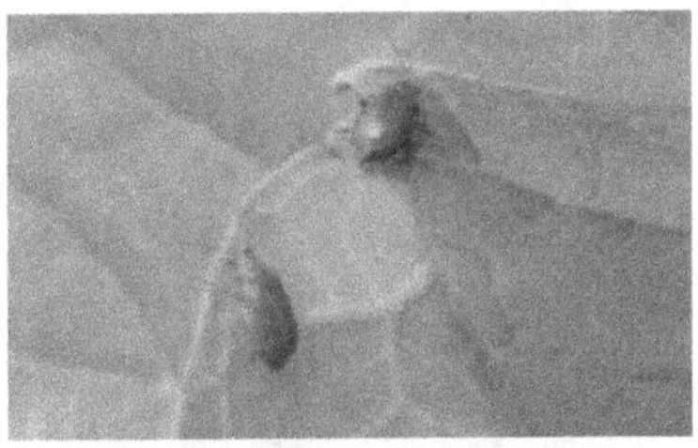

Cucurbitaceous Crops – Red Pumpkin Beetle

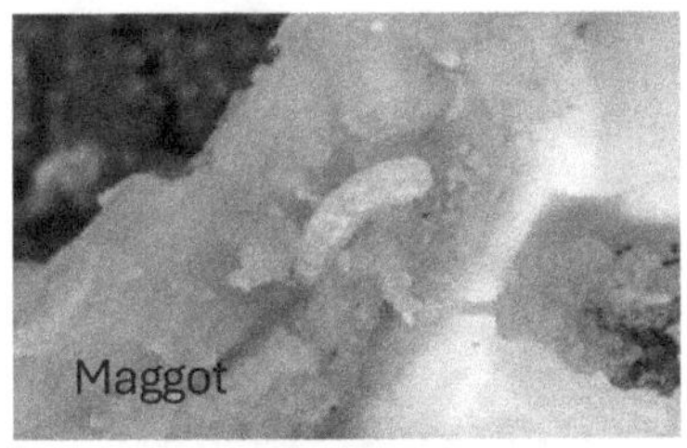

Cucurbitaceous Crops – Fruit Fly

Beans - Aphid

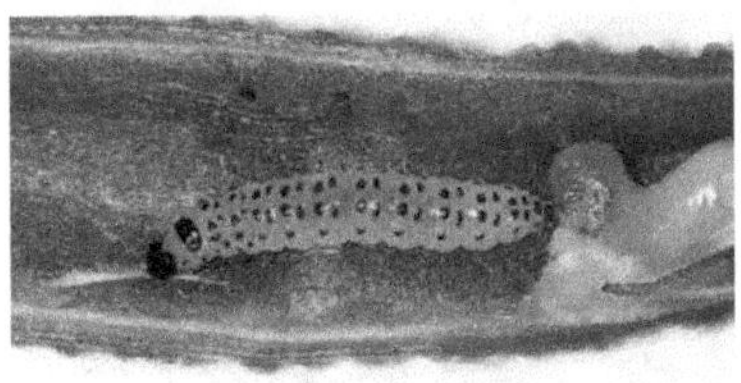

Beans – Spotted Pod Borer

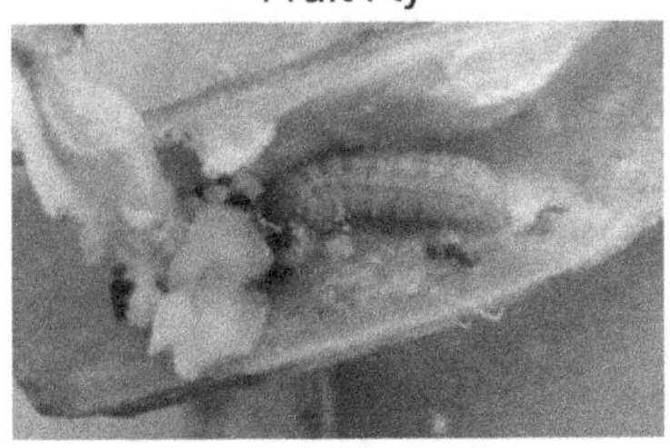

Beans – Pulse Pod Borer

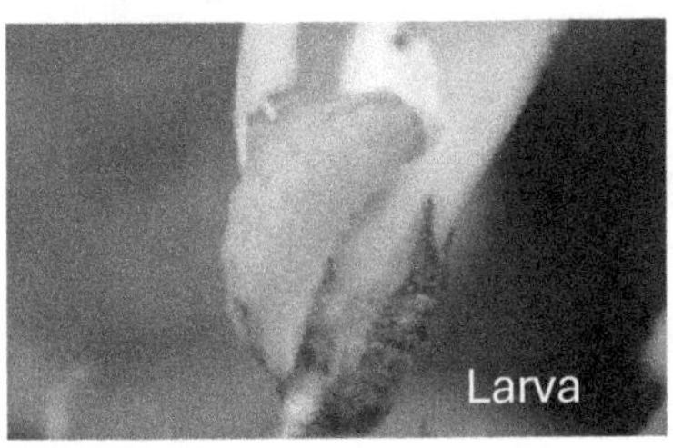

Beans - Blue Butterfly

Beans – Blister Beetle

Beans – Stem Fly

Pests of Citrus and Guava

Citrus – Lemon Butterfly

Citrus - Pyslla

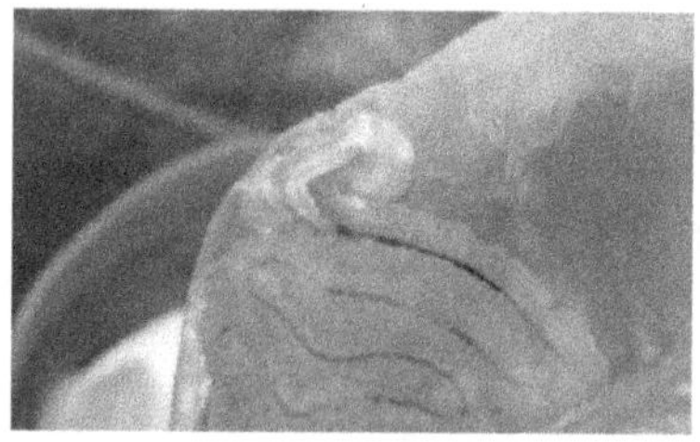

Citrus – Leaf Miner

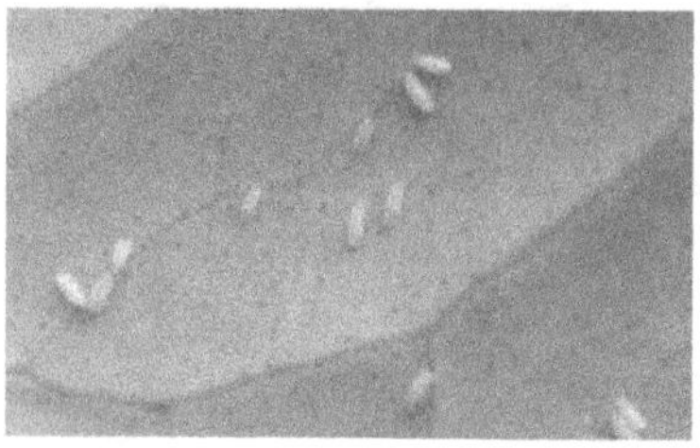

Citrus - Whitefly

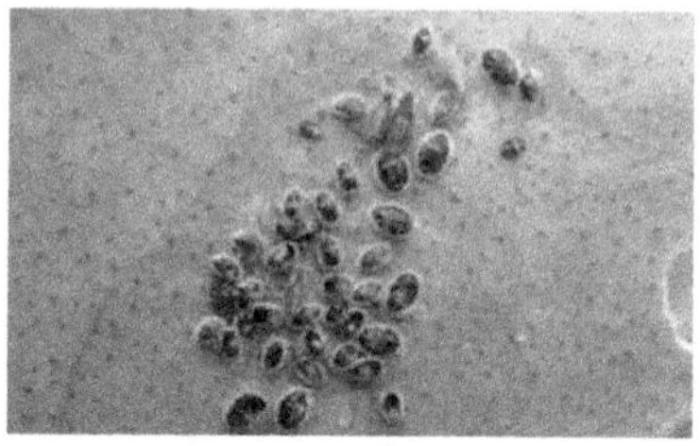

Citrus - Blackfly

Citrus - Aphid

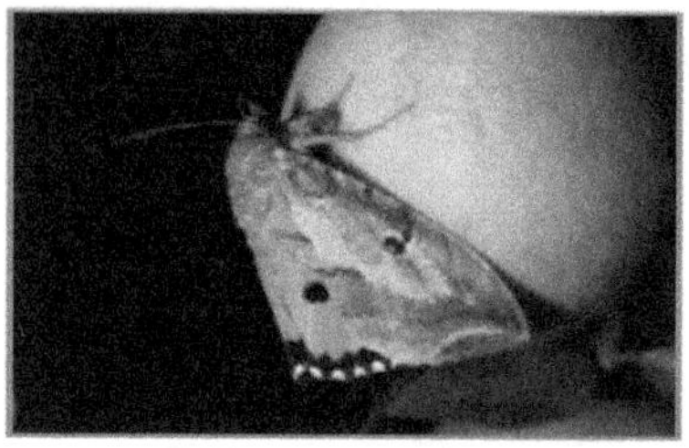

Citrus - Fruit Sucking Moth

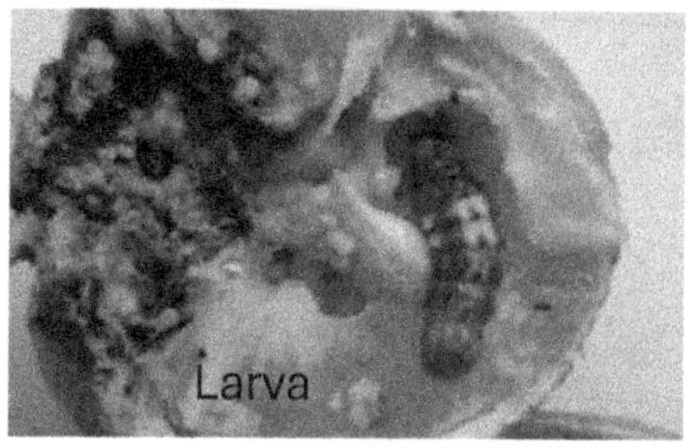

Citrus – Fruit Borer

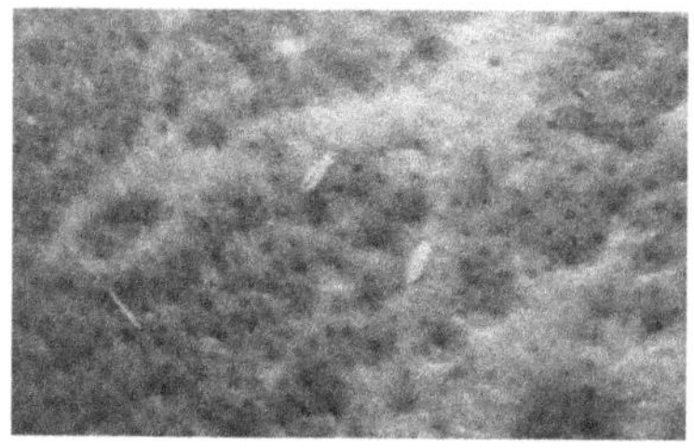

Citrus - Thrips

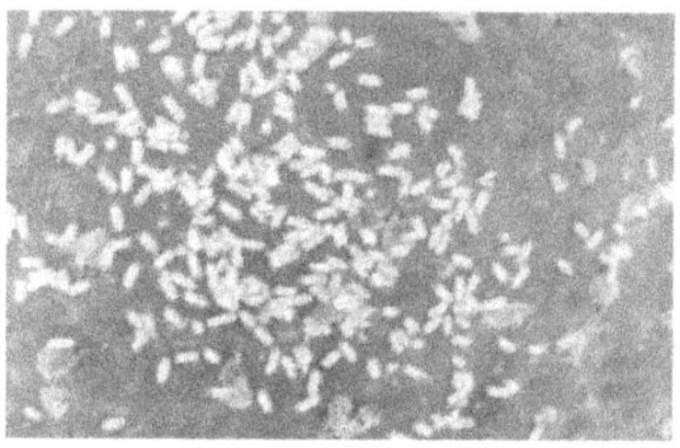

Citrus – Scale Insect

Citrus - Mealybug

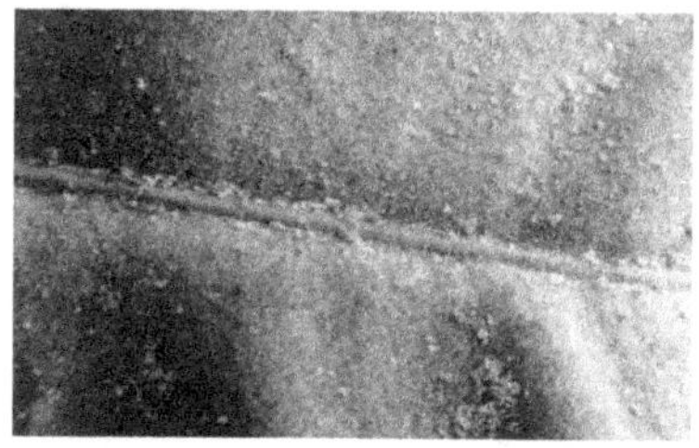

Citrus – Red Spider Mite

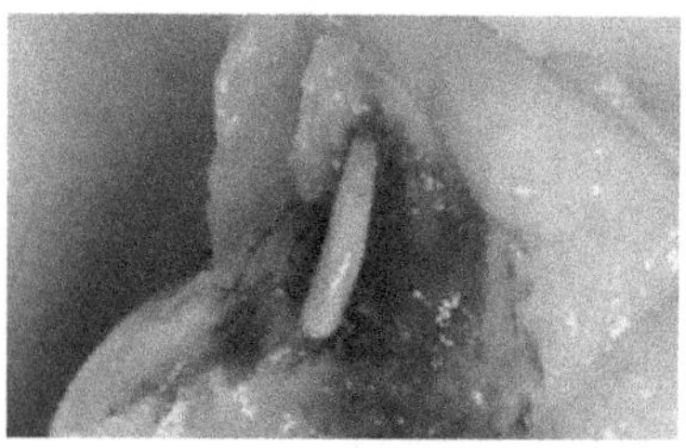

Guava – Fruit Fly

Guava – Fruit Borer

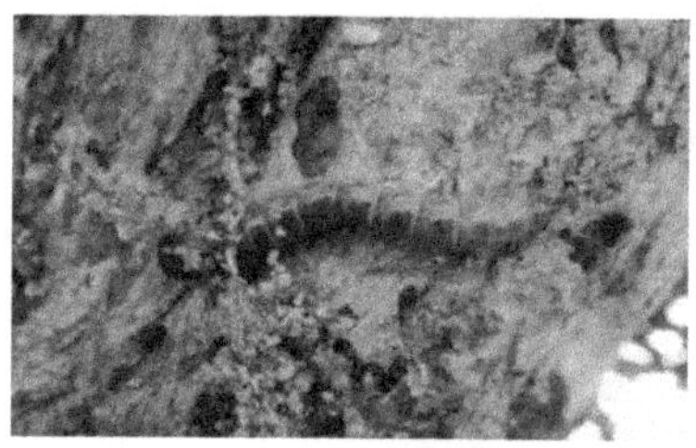

Guava – Bark Eating Caterpillar

Guava – Spiraling Whitefly

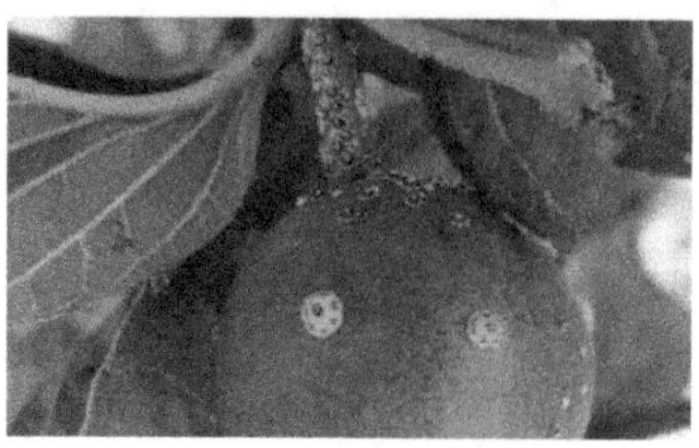

Guava - Tortoise Scale

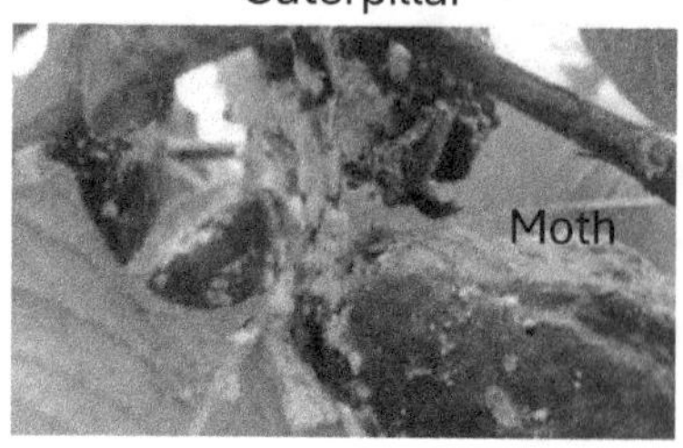

Guava - Mealybug

Pests of Mango and Banana

Mango - Jassid

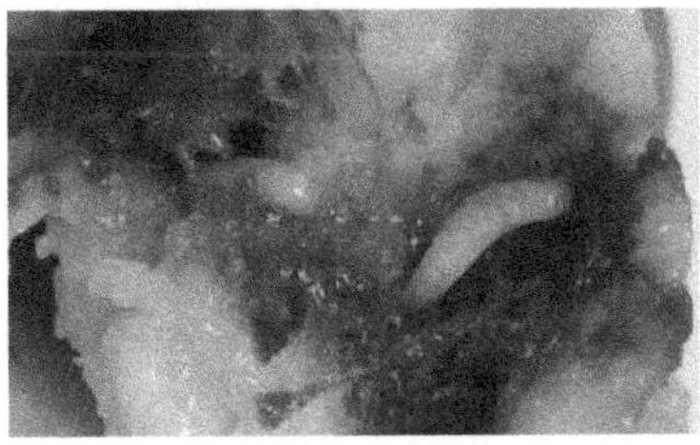
Mango – Fruit Fly

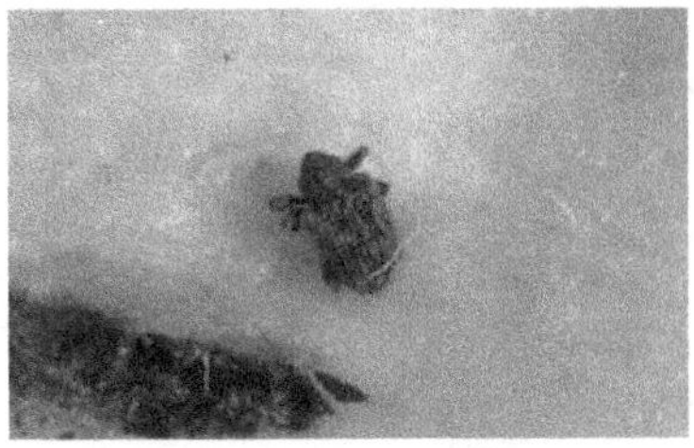
Mango – Stone Weevil

Mango - Stem Borer

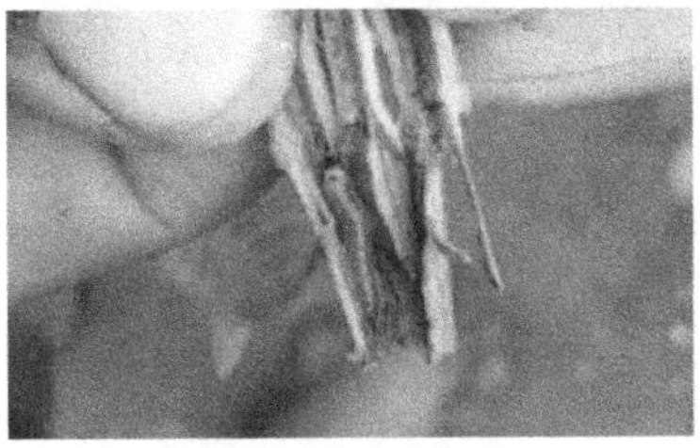
Mango – Shoot Borer

Mango – Leaf Webber

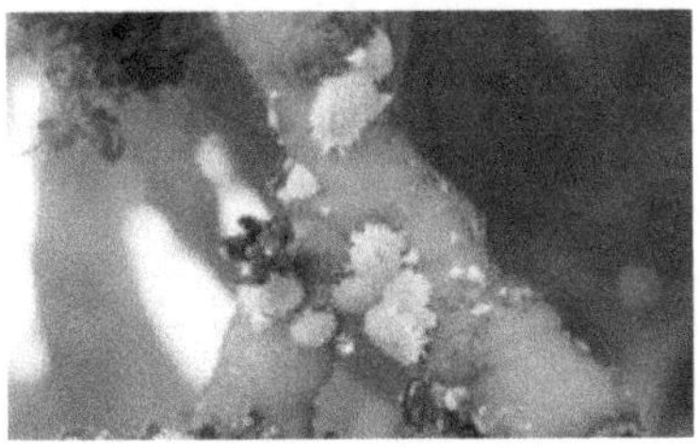
Mango - Mealybug

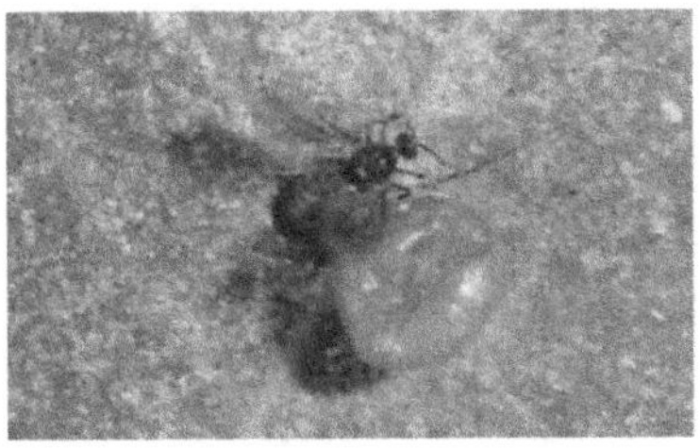
Mango – Leaf Gall Midge

Mango – Leaf Cutting Beetle

Mango - Slug Caterpillar

Mano - Scale Insect

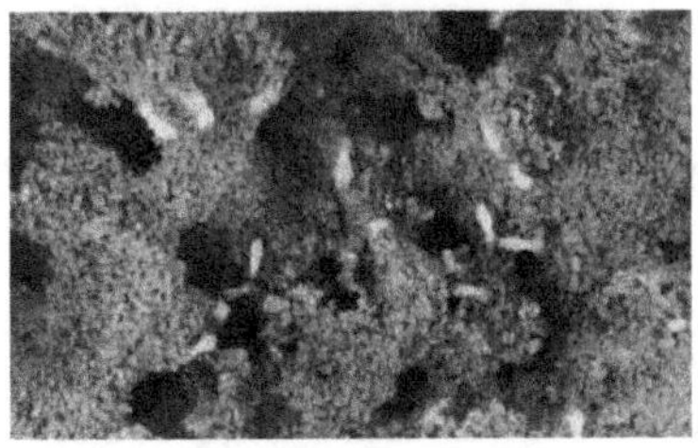
Mango - Termite

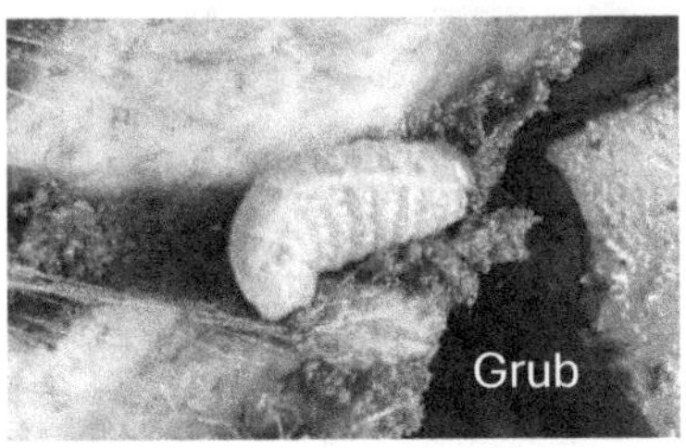

Banana – Rhizome Weevil

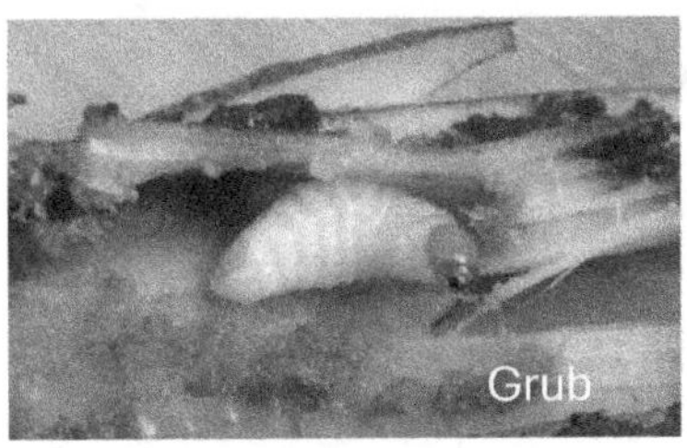

Banana – Pseudo Stem Borer

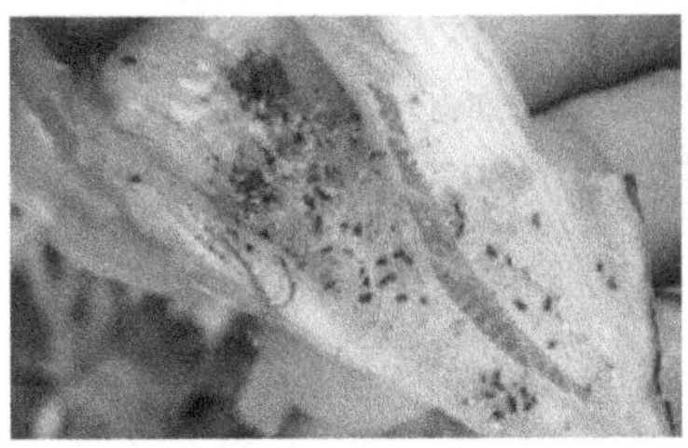
Banana - Aphid

Banana - Skipper

Banana – Leaf Eating Caterpillar

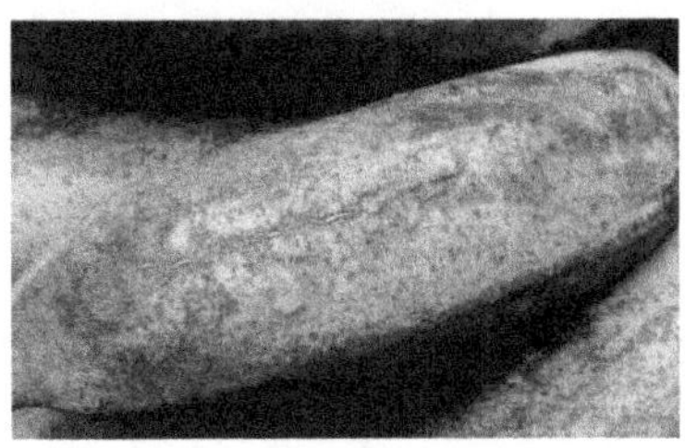
Banana – Rust Thrips Damage

Pests of Papaya, Sapota, Pomegranate, Ber, Aonla, Fig and Custard Apple

Papaya - Mealybug

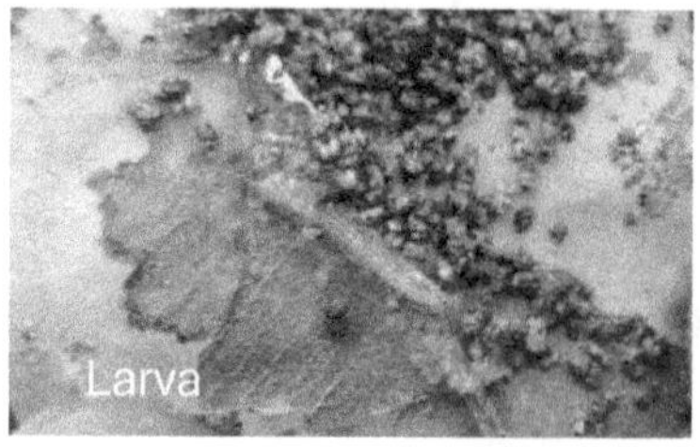

Sapota – Chiku Moth

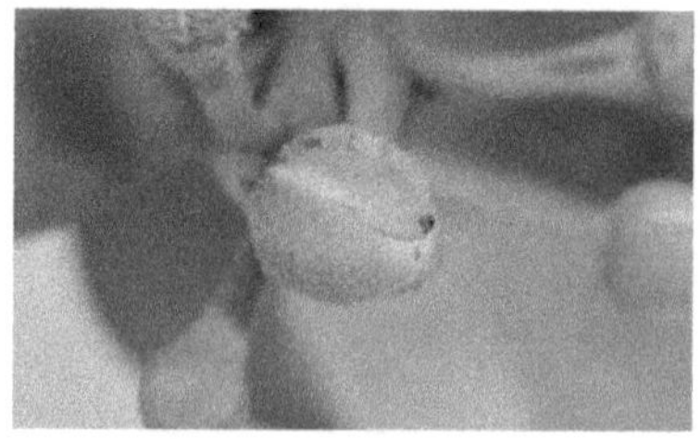

Sapota - Bud Borer

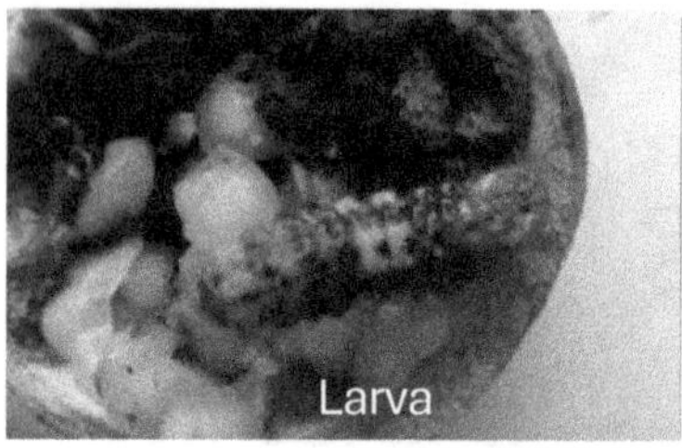

Pomegranate – Anar Butterfly

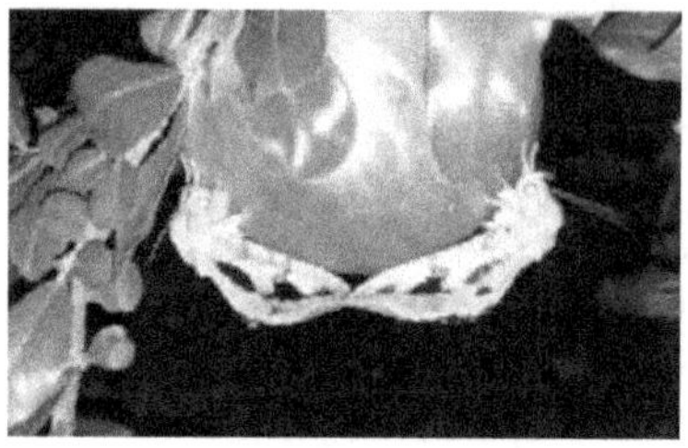

Pomegranate – Fruit Sucking Moth

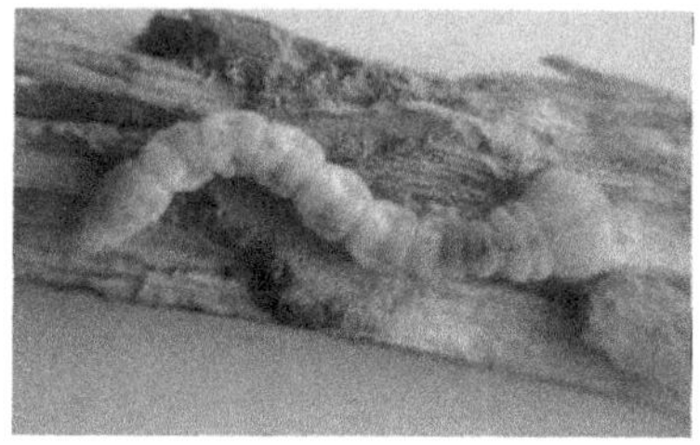

Pomegranate - Stem Borer

Pomegranate – Shot Hole Borer Damage

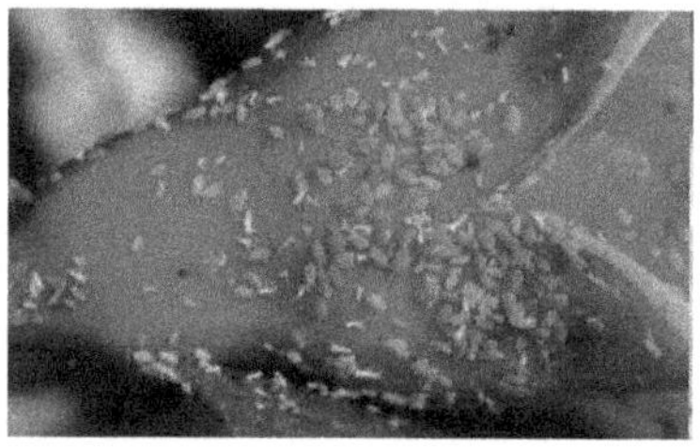

Pomegranate - Aphid

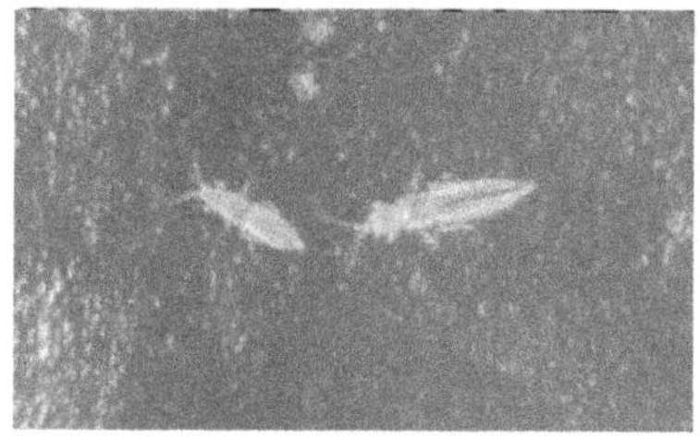

Pomegranate - Thrips

Ber – Fruit Fly

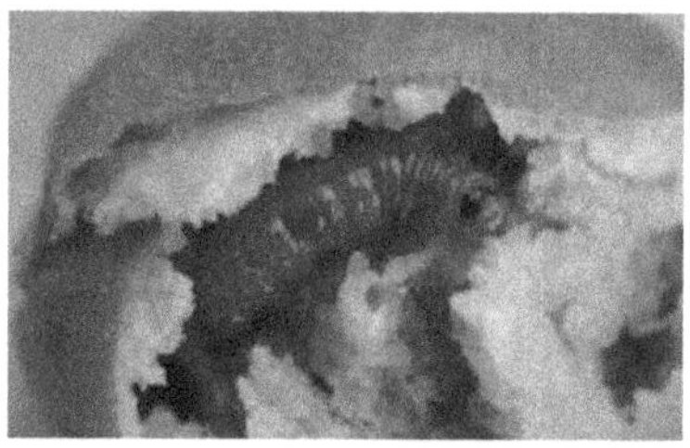

Ber – Fruit Borer

Ber - Hairy Caterpillar

Aonla – Apical Twig Gall Maker

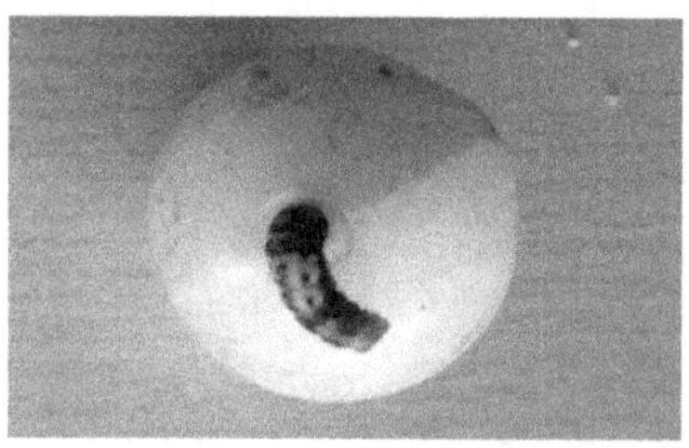

Aonla - Fruit Borer

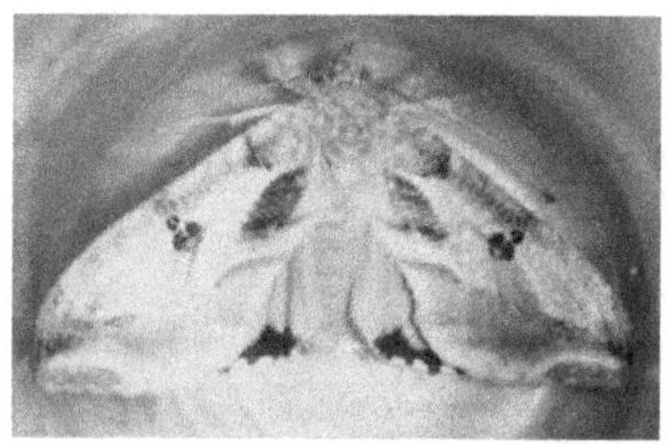

Aonla – Fruit Sucking Moth

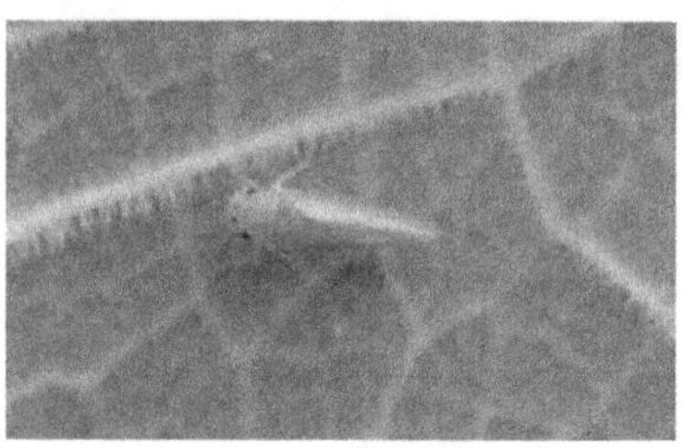

Fig - Jassid

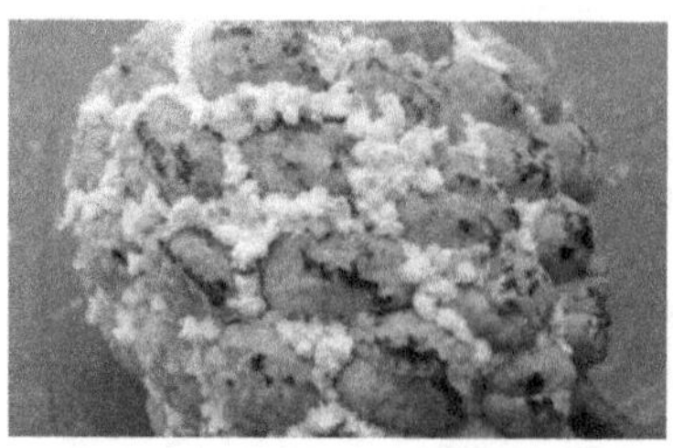

Custard Apple - Mealybug

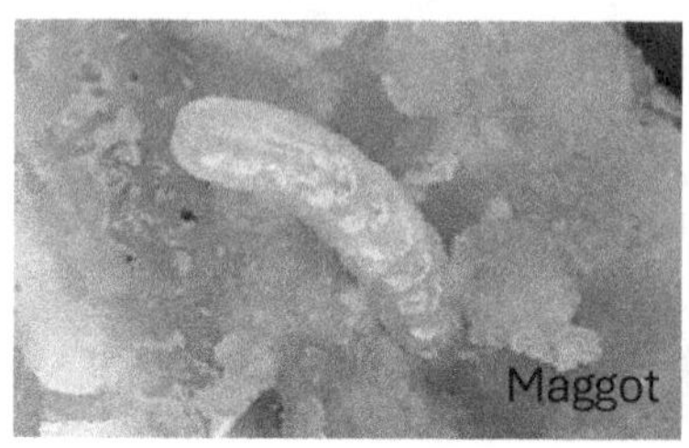

Custard Apple - Fruit Fly

Pests of Turmeric, Ginger, Onion, Garlic, Curry Leaf, Coriander and Flower Crops

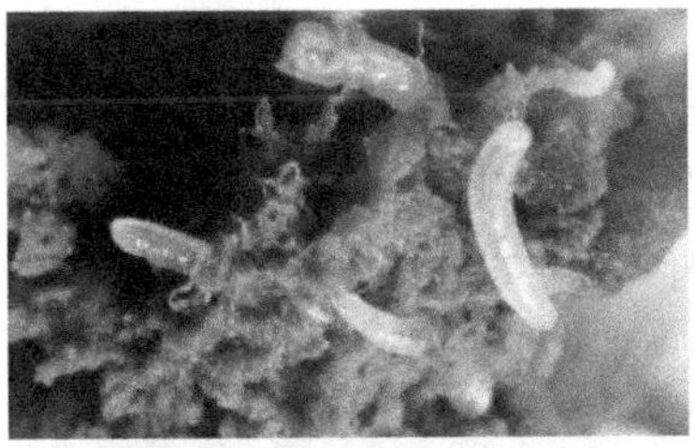

Turmeric & Ginger – Rhizome Fly

Turmeric & Ginger – Shoot Borer

Turmeric & Ginger - White Grub

Onion & Garlic - Thrips

Onion & Garlic - Cutworm

Onion & Garlic – Tobacco Caterpillar

Curry Leaf – Lemon Butterfly

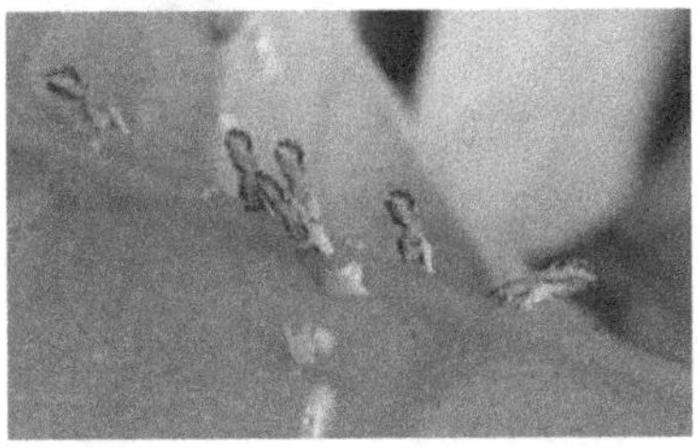

Curry Leaf - Psylla

Curry Leaf - Scale Insects

Coriander - Aphid

Coriander - Mite

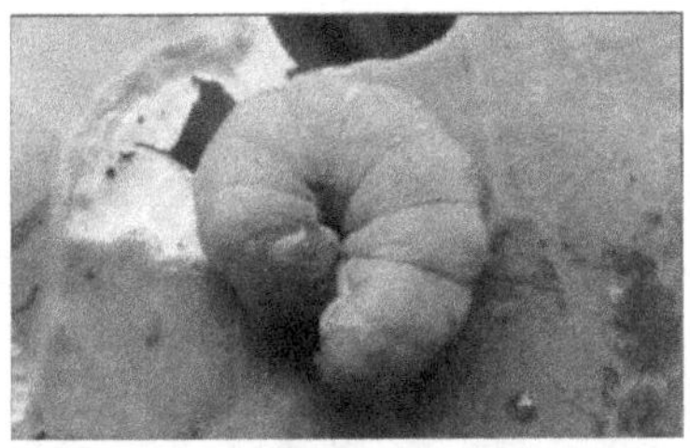

Coriander – Beet Armyworm

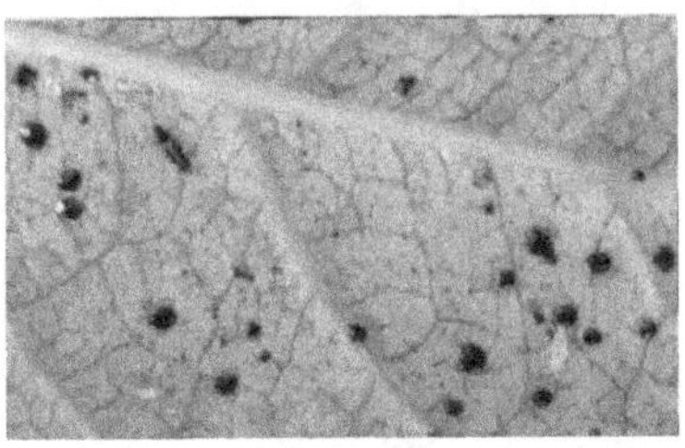

Rose - Thrips

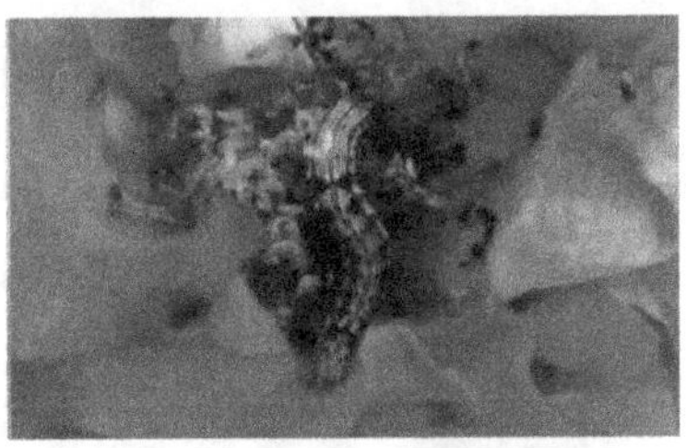

Rose – Flower / Bud Borer

Carnation - Whitefly

Carnation - Aphid

Rose - Scale Insects

Rose – Red Spider Mite

Pests of Stored Grains and Products

Rice Weevil

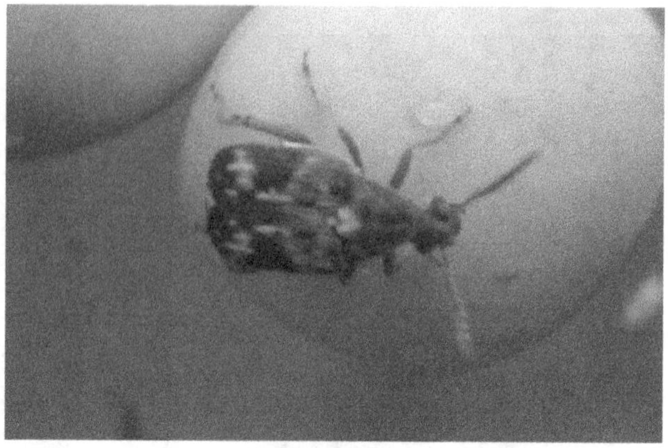

Pulse Beetle

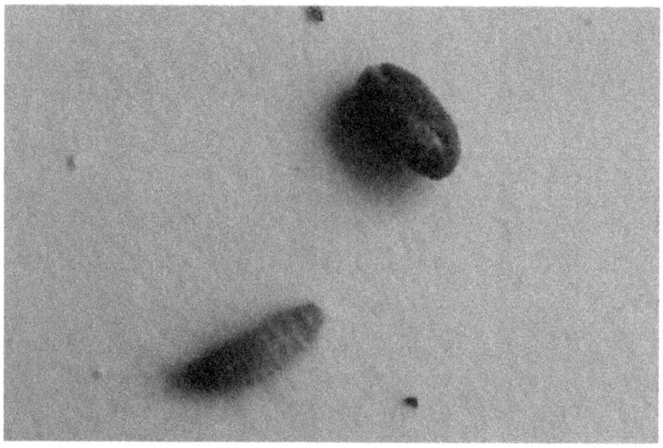

Khapra Beetle

Rust Red Flour Beetle

Rice Moth

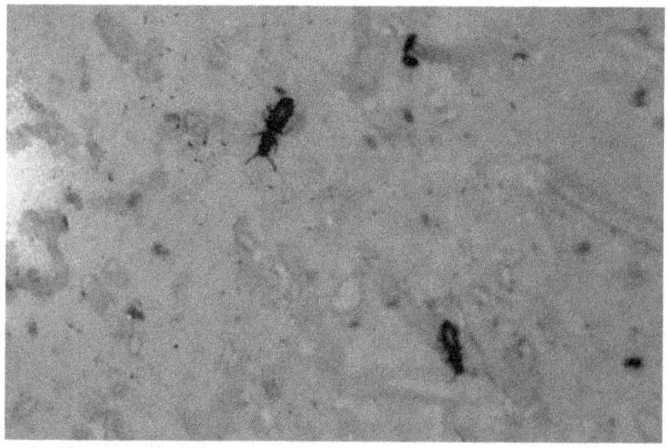

Saw Toothed Beetle

Almond Moth

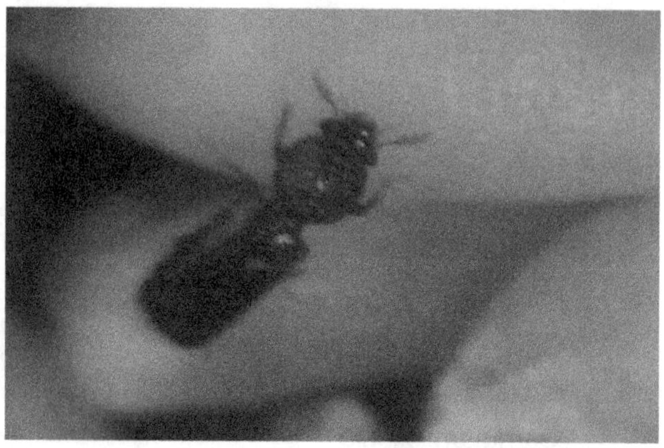

Lesser Grain Borer

www.ingramcontent.com/pod-product-compliance
Lightning Source LLC
LaVergne TN
LVHW081948200926
848484LV00009B/2526